AF788588

CONTINUUM THERMOMECHANICS

SOLID MECHANICS AND ITS APPLICATIONS
Volume 76

Series Editor: G.M.L. GLADWELL
Department of Civil Engineering
University of Waterloo
Waterloo, Ontario, Canada N2L 3GI

Aims and Scope of the Series

The fundamental questions arising in mechanics are: *Why?, How?,* and *How much?* The aim of this series is to provide lucid accounts written bij authoritative researchers giving vision and insight in answering these questions on the subject of mechanics as it relates to solids.

The scope of the series covers the entire spectrum of solid mechanics. Thus it includes the foundation of mechanics; variational formulations; computational mechanics; statics, kinematics and dynamics of rigid and elastic bodies: vibrations of solids and structures; dynamical systems and chaos; the theories of elasticity, plasticity and viscoelasticity; composite materials; rods, beams, shells and membranes; structural control and stability; soils, rocks and geomechanics; fracture; tribology; experimental mechanics; biomechanics and machine design.

The median level of presentation is the first year graduate student. Some texts are monographs defining the current state of the field; others are accessible to final year undergraduates; but essentially the emphasis is on readability and clarity.

Continuum Thermomechanics

The Art and Science of Modelling Material Behaviour

Paul Germain's Anniversary Volume

Edited by

GÉRARD A. MAUGIN (co-ordinator)
Laboratoire de Modélisation en Mécanique,
Université Pierre et Marie Curie,
Paris, France

RAYMONDE DROUOT
Centre de Mécanique de Milieux Continus,
Université de Versailles-St Quentin,
Versailles, France

and

FRANÇOIS SIDOROFF
Laboratoire de Tribologie et Dynamique des Systèmes,
Ecole Centrale de Lyon,
Ecully, France

SPRINGER-SCIENCE+BUSINESS MEDIA, B.V.

A C.I.P. Catalogue record for this book is available from the Library of Congress.

ISBN 978-90-481-5501-9   ISBN 978-0-306-46946-6 (eBook)
DOI 10.1007/978-0-306-46946-6

Printed on acid-free paper

All Rights Reserved
© 2000 Springer Science+Business Media Dordrecht
Originally published by Kluwer Academic Publishers in 2000
Softcover reprint of the hardcover 1st edition 2000
No part of the material protected by this copyright notice may be reproduced or
utilized in any form or by any means, electronic or mechanical,
including photocopying, recording or by any information storage and
retrieval system, without written permission from the copyright owner.

CONTENTS

FOREWORD

Professor Paul GERMAIN will be 80 years old on August 28, 2000. The day coïncides with the opening day of the 20th ICTAM in Chicago. This cannot be a mere coïncidence since Paul GERMAIN is a former President of IUTAM. But if it is a coïncidence at all, then the scientific children and grandchildren of Paul GERMAIN have decided to mark that memorable day with a special publication that reflects his direct influence on a successful scientific trend, the thermomechanics of materials.

Paul GERMAIN has had an extremely fruitful scientific carreer in which he contributed to so many various fields, including the study of peculiar partial differential equations, its application to various regimes of aerodynamics, transsonics, shock-wave structure, magnetohydrodynamics, the implementation of asymptotic methods in fluid mechanics, and the general formulation of continuum mechanics and thermodynamics. This he tells about beautifully in his own contribution to this volume, a rather easygoing scientific autobiography. What also emerges from that contribution is not only Paul GERMAIN's deep involvment in Aeronautical Research at a critical time (1960s-1970s), but also his passion for teaching, first mathematics, and then mechanics in its various guises. That a single person should so much influence the teaching of a discipline - continuum mechanics - in the whole of a medium-size country like France, and that, in turn, that spirit spread over to a large part of the profession, worldwide, is simply amazing. The thirty one contributions that make up the rest of this volume, reflect this influence to the utmost. This is a tribute paid by the contributors to the special talent and enthusiasm of a person they have tried to imitate, but often in vain. Professors Monique PIAU, Cristian TEODOSIU and André ZAOUI could not provide a contribution in time but they fully associate themselves to this *hommage.*

This volume has been prepared with the efficient help of Jean-Pierre GUIRAUD, Eleni MAUGIN, and Prof. Graham M.L. GLADWELL. Our heartfelt thanks go to them for their untiring efforts to improve all manuscripts.

The Editors

PAUL GERMAIN

My discovery of mechanics

Paul Germain
Emeritus Professor, Université Pierre et Marie Curie
Formerly, Professor of Mechanics at Ecole Polytechnique
Secrétaire Perpétuel Honoraire de l'Académie des Sciences de Paris

Some of my former students have been kind enough to dedicate to me for my eightieth anniversary a volume of their original contributions and they asked me to write a few lines as a preface. It was impossible to turn down this proposal. So I decided to try and tell how I discovered this discipline to which I have devoted my professional life, and how I became fascinated by some of its various aspects. The main reason is that these discoveries have been made mostly with them and for them. Therefore, it is quite natural to present this introduction to this volume as a testimony of my gratitude.

Ever since I was a teenager I have always intended to become a professor. I have been happy enough to realise my dream. I am ready to agree with the statement of a very interesting young boy who filled the tank of my car at the nearest gas station of Brown University and who told me one day, "Let me be straightforward, a professor is somebody who was put to school when he was five years old and who had not enough imagination to get out". It is true that there exists a great continuity between learning and teaching. I was lucky enough to have enjoyed sometimes the feeling of having discovered a new result or to be the first to solve a new problem. But I must confess that probably my greatest satisfaction came from teaching. I spent many hours trying to get a new and deeper understanding of a concept or of a method and to find the best way to make them easily understandable by the students. What can be actually a greater gratification than thc onc you fccl whcn you rcalisc that your teaching has been meaningful, that your students experience themselves the deep beauty of the discipline and, above all, when you see their joy when they obtain themselves new results, partly thanks to some ideas you have tried to give them.

1. STARTING SITUATION

1.1 Researcher in mathematics?

I was very happy to be admitted to the "Ecole Normale Supérieure" in 1939. This school has been created by the French revolution and its main purpose is to train young people who aspire to become professors, teaching normally in the upper grades of our "lycées", and in particular in those that prepare students for admission - always a matter of stiff competition - into one of the French "Grandes écoles": Ecole Polytechnique, Ecole Nationale des Ponts et Chaussées, Ecole Nationale d'Administration,.., and in particular, the Ecole Normale Supérieure (ENS). This last is a very famous school

G.A. Maugin et al. (eds.), Continuum Thermomechanics, 1-24.
© 2000 *Kluwer Academic Publishers.*

because the selection is very hard. At that time, for the whole country and for all scientific disciplines, only twenty students were admitted each year. I was delighted by the perspective of teaching mathematics to the candidates to these difficult competitions. The studies at ENS lasted three years. And then, war came. The times were distressing and, needless to say, not very favourable for absorbing the new mathematical and physical concepts, methods and knowledge. During the first year, we were very few; only those who were not drafted; and later on we suffered the hardships of the German occupation.

Consequently, in my class at ENS we missed many chapters of a good, modern mathematical training in comparison with our older or younger fellows. Roughly speaking, we covered the first two volumes of Goursat's "Cours d'analyse" published during WWI and also a few topics of differential geometry. In mechanics our knowledge was pretty poor: kinematics and dynamics of rigid bodies, and Lagrange's equations.

In November 1940, the Germans closed for a few weeks the Sorbonne, the sciences faculty where most of our courses were taught. The direction of the ENS advised us to get in touch with a professor to work on some research mini-project. I saw Georges Bouligand, a very enthousiastic professor, who suggested reading some chapters of the famous four-volume treatise of Gaston Darboux, "Théorie des surfaces". The presentation was old fashioned; but the contents fascinated me, since geometry was my pet topic. I succeeded to write a paper which improved and completed what was written in Darboux, on a family of surfaces presenting some curious properties. Georges Bouligand advised me not to accept, in October 1942, the position in a lycée which was offered me, and to try rather to do some research in view of becoming, later on, a university professor.

There were no regular seminars at the time. So, I spent many hours in one of the best libraries in mathematics and I studied some papers, looking for a domain of research opening more prospects than the classical theory of surfaces in three-dimensional space. In February, I was strongly recommended to replace immediately a professor, arrested by the Germans, who was teaching a class of students preparing the competition for admission to one of the most renowned engineering schools. Nine hours of mathematics and a lot of work! Nevertheless, I found the experience very interesting and stimulating. At the end of the academic year, I had to work in a factory in the eastern part of France, to fulfil my "Service du Travail Obligatoire"*. In September 1944, this area was liberated and I went back to Paris.

1.2 The choice of mechanics

During the last eighteen months, I was thinking hard on what I wanted to do. I became very dubious about my capacity to get the knowledge and the deep understanding of modern mathematics – in fact, the Bourbaki literature – which would be necessary in order to do fruitful research in this field. I was still tempted by teaching in a lycée, but also to follow the suggestion of my friend Raymond Siestrunck, a physicist of my class

* "Compulsory Work Duty", imposed by Vichy authorities in agreement with the Germans

at ENS, who was beginning to work in aerodynamics. He thought that theoretical fluid dynamics would be an attractive field for a student who was keen on geometry. He gave me the book of Joseph Pérès, a mathematician who wrote a few books with Voltera on functional analysis and who began to learn fluid mechanics when, at the age of forty, in 1930, he was appointed director of the newly founded "Fluid Mechanics Institute" at Marseille. The book deals with the theory of irrotational flows of an inviscid and incompressible fluid – two-dimensional and axisymmetric mostly – and also with Prandtl's lifting-line theory. Finally, I decided to try this last proposition. I met Joseph Pérès, Lucien Malavard and Raymond Siestrunck in their small laboratory in the basement of ENS, where they were doing interesting work using electrical analogy. Malavard proposed for me to study a numerical method described in a NACA Technical Note, in order to compute the pressure distribution along a given airfoil. It was based on a conformal mapping of the exterior domain of the airfoil onto the exterior domain of a circle. Everything was straightforward and easy, except for the computation of the imaginary part, on a circle, of an analytic function of a complex variable, defined outside this circle, when its real part on the circle was known. Analytically, the result is obtained through the Cauchy principal value of an integral. Numerically, it is not easy, and in applications, all the accuracy of the method is greatly affected by this operation. My contribution was to do this computation without using the representation through a Cauchy integral, but by appealing to the special and very simple behaviour of the operator acting on trigonometric functions. The gain of time and accuracy was very significant.

I was just obtaining this result, when Pérès received a letter from Jacques Valensi, who was a member of the French Scientific Mission, located at Carlton Gardens, in London. Valensi was writing to him that he was working at NPL, the National Physical Laboratory, and that he thought he might succeed to obtain for one of his research assistants or collaborators the possibility to work at NPL during a few months, as a temporary member of the French Scientific Mission. Malavard and Siestrunck declined the offer. "Do you speak English?" asked Pérès. "Not a single word", was my answer. "Are you ready to go?" said Pérès with a smile. "Yes", was my reply, quite surprisingly for Pérès, my friends and also for myself. Deciding to go to NPL during the war, knowing neither English nor fluid mechanics, was quite an adventure. In January 1945, after two weeks in London, Valensi succeeded to have an appointment with the director in charge of fluid mechanics, who sent me to the department dealing with theoretical aerodynamics, headed then by Sydney Goldstein. With the help of Valensi, I tried to describe the only thing I knew and I had done. "We also have a method to do this kind of numerical computations. Come Monday morning, we will give you an example to compute, and we will compare your result with ours". The test was satisfactory. Goldstein was not able to keep me in his laboratory that dealt with confidential problems. But he recommended to the direction of NPL that I be admitted as a visitor. This way, I was able to participate every day, for three months, at the NPL, in the team headed by Dr. Falkner and then, for two months, in the team of Dr. W.P. Jones. Fortunately, I had access to the library, where I spent much of my time. When discussing with people, my complete ignorance of the basic concepts and methods of fluid mechanics was hidden by my extreme difficulty to understand and to speak English. Back in Paris, in July, I finally wrote some reports on the questions I had studied and explained to Pérès and to my friends what I had learned, without being able to make the difference between what was new and what was known already for some years. Then, after a really incredible combination of happy circumstances, I knew that I could try to work in Mechanics.

2. MECHANICS, A GOOD FIELD FOR A YOUNG MATHEMATICIAN

During my first British experience, I had the happy occasion to read, on a mimeographed pre-print, what is now well known as "Supersonic flow and shock waves". This celebrated book by Courant and Friedrich was a wonderful way to convince me that a good classical mathematical training may lead to interesting contributions in theoretical fluid dynamics. When I began to work, I was inclined to attach more interest to the mathematical problems and their solution, than to their mechanical origins and significance. My thinking and my research were therefore rather mathematically oriented, but in a few years, progressively, my job and my relations with other scientists brought about a reversal in my attitude. First, as early as October 1946, I was appointed head of a small research team at the *Office National d'Etudes et de Recherches Aéronautiques* (ONERA.) Second, I attended in 1948 the International Congress of Mechanics in London, where, thanks to Goldstein I had the good fortune to meet a large number of talented colleagues. With some of them, particularly the British, it was the beginning of a long-lasting friendship. I must namely mention the two invitations to spend one or two weeks in the famous department of Applied Mathematics of the University of Manchester, headed in 1949 by Sydney Goldstein and in 1951 by my old, much admired, and now regretfully departed, friend, James Lighthill. With James I remained closely related until his dramatic death, sharing in common the status of professor, the professional experience of being directors of aeronautical research establishments in our respective country, and becoming both members of the IUTAM bureau. Let me finally mention my appointment as a senior lecturer at the University of Poitiers, to teach the subject of compressible and incompressible fluid flows in a school of mechanics and aeronautical engineering, a program which fitted very well with my work at the ONERA.

A paper devoted to my discovery of mechanics is not the place to spend too much time on my research activity during this period. The two main themes of my research were the mathematical linearised theory of supersonic aerodynamics and the theory of transsonic flows. But, in fact, for my teaching and for answering some questions at the ONERA. I was also led to work on other topics concerning subsonic aerodynamics and gas dynamics.

2.1 Linearised supersonic aerodynamics

In France, we had the quite old tradition of the "Doctorat ès Sciences", closer to the German Habilitationsschrift than to the American Ph.D. I worked hard to get my first professorship at Poitiers, writing a thesis of some 200 pages, published by the ONERA. and translated as a NACA Technical Memorandum, on the subject of conical supersonic flows. The field was opened by a pioneering work by Buseman who proved that mathematical solutions of the problem might be built through analytic functions of one complex variable. Of course I was lucky, because I had do work on a part of mathematics that I mastered. But I must recognise how much my training in fluid mechanics was poor, when I started to work in the fall of 1946. I spent a lot of time before understanding that, dealing with a linear problem, I had not to worry about the non-linearity of the boundary conditions on the wing, a delta one, for a typical application of special interest to engineers at O.N.E.R.A. I spent again a lot of time hesitating about the choice of the proper solution, because I was actually not aware of the role of the Kutta condition in supersonic flows and of the properties of the wake.

Anyway, when appointed in Poitiers, I was mature in fluid mechanics, but this is not the place for reporting on some other of my works on linearised supersonic aerodynamics. I will just mention the invitation to deliver a general lecture on that subject at the Brussels' 1956 International Congress. I chose to give a review of known results, mainly by others than myself, trying to improve and unify them, using Schwartz's theory of distributions, which was not familiar to most scientists of the mechanics' community. Needless to say, my purpose was to stress that distribution theory sheds light on many facets of supersonic wing theory, rather than to illustrate distribution theory with some problems, like the one of minimum drag. I take this opportunity to mention that most of my contributions were improvements of some results found in literature or new results soon to be improved by another scientist. Reading was very stimulating; it helped me to get a better understanding of the subject matter, which proved very useful in teaching and building closer links with foreign colleagues who became good friends.

2.2 Transsonic flow and equations of mixed type

During the early fifties I was involved in research about transsonic flows, a subject of interest for aeronautical engineers and one which fascinated me as a mathematician. I remember being enthusiastic while reading papers by the Soviet scientist Frankl, whom I never had a chance to meet. I worked mainly on mathematical aspects of linear equations of mixed type, which apply to steady, two-dimensional potential flow of an inviscid fluid, in the hodograph plane. I paid also attention to some special equations for which approximated solutions may be found, providing either special or approximated solutions to, or shedding light on, problems of technical interest: nozzle or jet flow, upstream flow around an airfoil; flow at Mach number one around a wedge; behaviour near special points, like the one at infinity, or special lines like the sonic one or the transsonic boundary. But my main interest was in mathematical problems proper, investigated with Roger Bader. They dealt with the Tricomi equation and its Euler-Poisson-Darboux solutions, their evolution when the singular point crosses the parabolic, i.e., the sonic, line, the singularity changing from logarithmic to Riemann's type; the theorem of maximum; a new proof of the existence theorem of the Tricomi problem and its Green's function.

All these topics gave me the great satisfaction of bringing me back to mathematics and the occasion to have close and friendly relations with mathematicians, especially at the Courant Institute. Mathematics, the beloved discipline of my youth has never ceased to fascinate me and excite my admiration. But I felt the necessity to be consistent with my choice and my decision and not to continue to use mechanics as an excuse to do mathematics. It was time to get a good knowledge of what really was this discipline.

3. GETTING A DEEPER INSIGHT INTO THE REALM OF MECHANICS

During the period 1952-1955, while pursuing my routine work, I was much concerned with thinking about the concepts, the trends, the understanding, and the reasons that sustained my vocation for mechanics. It was not only important for me, but also for the students and scientists working with me, now, and even more, in the future. One might feel that the example of my British colleagues could deliver an answer; but they would not be able to understand my uneasiness. They grew up in the best tradition since Newton's times. They were doing mechanics as a natural thing, just like breathing. I had to cope with the special situation of mechanics in France, at that time. During many decades, and especially after WWII, most talented young scientists were attracted either by pure mathematics or by hard physics, while mechanics was considered as some reminiscence of the 19th century. As a consequence, I felt myself rather isolated and I had to get, all by myself, an overview of what should be the discipline to which I wanted to devote my efforts in research and teaching. Of course, I was not fully conscious of this evolution in my mind, but at least three major events helped me in elaborating meaningful answers and I want to report on them.

3.1 First contact with Paco Lagerstrom.

The first of these was my meeting with Paco Lagerstrom during the Istanbul 1952 International Congress. He was, like myself, the author of an important report on conical supersonic flows. He started by studying Roman languages – I believe – in Sweden, and then attended courses in pure mathematics at Princeton. At the time I met him, he was professor at Caltech, in the department of aeronautical engineering, so that both our cultural backgrounds were similar. He spoke to me about some questions which he thought to be of the utmost importance for the understanding of fluid mechanics and which might be ripe for solution at that time. One of them was the mathematical basis of the boundary layer concept, discovered by Prandtl, nearly fifty years before. Another one concerned the steady flow of an inviscid fluid as the limit of a class of corresponding flows of the same fluid, involving a vanishingly small viscosity, so that, in some sense, solutions of the Euler equations might be related to a class of solutions of the Navier-Stokes equations, through some limiting process. He brought my attention to another of Prandtl's discoveries, namely the constancy of vorticity in a closed, streamlined, two-dimensional steady inviscid flow, as a result of vanishing viscosity, acting during an infinitely long time. I began to foresee a new link between mathematics and fluid mechanics, provided by asymptotic techniques; as a matter of fact, not simply a link, but a way of thinking at an enormous variety of problems. Needless to say that all this was opening avenues without a clear vision of getting the right way in.

3.2 Visiting professor at Brown University

During the spring 1952, being invited to present a paper on transsonic flows at a small colloquium held in Belgium, I had the chance to meet William Prager. A few days later he mailed me an invitation to deliver a course in gas dynamics at Brown University. I was grateful to him for this attractive and unforeseen proposal, but I asked to have it postponed for a year, to give myself, in the meantime, the opportunity to improve my English. I was very fortunate to spend the full academic year 1953-54 as a member of the graduate division of Applied mathematics, which was very close to the division of Engineering. For a mechanician, this was one of the best places in the world, providing many good courses, covering most of the main fields of mechanics and running a famous seminar attended by outstanding mechanicians from the United States and abroad. I took the measure of some fantastic gaps in my knowledge while attending these courses, mainly in solid mechanics, discovering plasticity, linear and non linear elasticity and studying Truesdell's papers, in which he gave the first systematic treatment of continuous mechanics, starting from fundamental concepts. I also learned a lot in fluid mechanics, boundary layer and viscous fluids, and even in hydrodynamics and incompressible inviscid fluid flows. I also had the opportunity to discuss many things with some of the best mechanicians in the world. Coming back to France, with so much new material, I felt worthy to be named professor of mechanics.

3.3 Professor of rational mechanics

This is what actually happened in October 1954. I was appointed professor in the chair of "rational mechanics" in the University of Lille – this is the third event that I have announced. At that time, in France at least, rational mechanics was considered as a branch of mathematics, dealing mostly with the application of the Newtonian theory to rigid body motions and Lagrange's analytical mechanics. Usually, a chair of rational mechanics was a position for a mathematician who worked there in anticipation of being appointed, as soon as possible, to a chair of calculus or of advanced geometry, or still, advanced analysis. With the choice I had done, this could not be my prospect. Actually, I had to understand more deeply the foundations of Newtonian mechanics as a mathematical model to the physics of the equilibrium and motions of bodies in the neighbourhood of the Earth, of the solar system, and of the Universe. The system of reference was a fundamental notion which was to be adapted to each situation. For the first time I realised how wonderful was the mathematical schematisation of the reciprocal action of one body on another. That the fact had required nearly twenty centuries to be recognised was not anymore a surprise to me! Lagrange equations for systems of rigid bodies with perfect constraints were also very attractive for a man who was very keen on geometry. I was able to cover nearly the whole classical programme in less than the full academic year and then, to have about six to eight weeks to present a topic, among the following ones: advanced analytical mechanics (Hamilton-Jacobi-Maupertuis); basic notions of relativity theory; basic concepts of continuous media with simple examples on incompressible inviscid fluid and classical linear elasticity; vibration theory. I mentioned these three experiences because they set up the starting point and the corner stone of what I have tried to build afterwards. I have discovered, above all, new facets of mathematics. It is not enough to solve problems and to study their solutions. One must also build the concepts and the models, which will be able to offer a deeper view of the phenomena than the one given by mere observation, and even by experiments. Mathematics has to build what I have called, in a talk at the Académie des Sciences, "la mathématique du monde". Moreover, these experiences have contributed to enrich my own cultural personality. Mechanics was not anymore a thing external to myself. It was becoming not only a part of my intellectual life but also a constituent of my spirit which, since that time, shaped my deepest convictions.

4. MODELLING AND ASYMPTOTICS IN FLUID MECHANICS

Let me repeat what I intend to describe in this introduction. It is not the discoveries I was able to make in the field of Mechanics – they are very limited. It is the progress of my understanding during all my life of what is this discipline, more precisely by discovering new problems, new ideas, new methods, new fields of applications. In fluid mechanics this progress of understanding was due mostly to the close relations with some colleagues and to the good knowledge of their works. Let me first mention some favourable occasions that offered these possibilities. These were, my stay at Caltech during four months near Paco Lagerstrom, Saul Kaplun, Julian Cole; later on, the academic year spent by Paco in my laboratory; the two visits in Paris, for one full year each, of Milton Van Dyke. I also benefited a lot from the long stay of W. Eckhaus in my laboratory. All of them theorised on asymptotic singular expansions which I came to use frequently. But, my best contributions to fluids were due to discovering, at the end of 1955, Jean-Pierre Guiraud. I succeeded to convince him that fluid mechanics was a very interesting topic and I asked the direction of ONERA. to recruit him to work in my small research team. During six years we met at least once a week, often for a full day. Jean-Pierre began to work for his doctoral thesis, on the small perturbation theory applied to hypersonic flows. But we did not limit our discussions to this topic. We exchanged our ideas about our recent readings, our small discoveries; we have learnt together, we have built our views together; our vision on mechanics was very similar. That was extremely stimulating and fruitful. In the beginning, I was the leader of the discussions; but very soon the positions were reversed.

4.1 Shock waves in gas dynamics and in MFD

Many of my personal studies dealt with flows of an inviscid fluid in which the variables defining its kinematics and its physics, suffer discontinuities on some manifolds in space-time. These are the shock waves. In order to explain my contributions, I must give some reminders. Burgers' equation, as it is well known, is the very simple mathematical model that gives the best physical meaning of this phenomenon. It involves one unknown u – which may be interpreted as a velocity – one space variable x, and one "viscosity" coefficient ν. As shown by Hopf, one may write explicitly the solution of this second order non-linear, partial differential equation which, for the initial time t=0, takes a given value $u_0(x)$. It is a continuous and differentiable function $u(x, t)$. Now, let ν tend toward zero; the limit is a "weak" solution of the inviscid Burgers' equation (ν=0) which, in general, presents discontinuities – that are shock waves (in a point of discontinuity the x-derivative of u has to be computed by an appropriate device, for instance, distribution theory). Locally, at a point of the shock, the jump of u must check not only an equation J, as any weak solution, but also an inequality which characterises a "shock solution" among all the "weak solutions". Let us recall also that, clearly, in this simple model, a shock wave appears as the result of two conflicting influences, first the non-linearities of the propagation for ν=0, the inviscid Burgers' equation, which tends to steepen the profile of the variations u along converging characteristics and, second, the weakening effect on this profile, due to viscosity.

One may now formulate the general situation, which is to be faced. It concerns the motions of a fluid in which are present some physical mechanisms of dissipation – let these motions be called P. These mechanisms may be mathematically schematised by some terms and then the motions we are looking for may be described as solutions of a general system of differential equations – let us call N this schematisation. But N also is too complicated. If the dissipations are very small, a further modelling may be considered by neglecting all the dissipations – call E the system of equations so obtained. Some solutions of E may involve surfaces of discontinuity. One wants to study the solutions of E which may be considered as limits of solutions of N for any vanishing dissipations. The shock S appears, then, as the limit of a small layer – called shock layer – of a solution of N which is the result of two mechanisms with opposite effects, the steepening due to the converging progressive waves of the limiting inviscid fluid and the weakening due to the dissipations. In order for it to be a shock solution of E, on each point of the shock, the jumps have to check some equations J and some inequality j.

In classical gas dynamics, the situation is clear and simple. The equations J are the Rankine-Hugoniot relations and the inequality says that the specific entropy cannot decrease when a particle crosses the shock. That is a necessary condition; it is also a sufficient one, as it may be seen when one studies "the shock structure". This can be done by the same method that was applied to study the ordinary boundary layer. At a fixed point of S and at a fixed time, a stretching variable ξ is introduced along the normal to S, becoming infinite when the dissipation coefficient tends towards zero, in order to obtain a "significant degeneracy" – see below. The values of the variables on the sides of the shock become the values for $\xi=+\infty$ and $\xi=-\infty$ according to the matching conditions between the distal (outer) expansion (the original solution) and the proximal (inner) expansion (for the stretched geometry). It is easy to show the existence of the structure which is the solution of the N equations in the stretched geometry, when ν tends towards zero.

Everything we recalled above is very classical. But I felt necessary to do it in order to understand the situation that is met when one wants to study shock waves within the frame of the classical magneto-fluid-dynamics theory (MFD).

In the classical MFD, when all the dissipative effects are neglected, the jump relations convey the conservation laws (mass, momentum, energy) and the Maxwell equations. They can be written with four variables $q_1.....q_4$ - say $\mathbf{q}$ – which are respectively the specific volume, the temperature, and the tangential components of the relative velocity and of the magnetic induction. The four constants $c_1....c_4$ - say $\mathbf{C}$ - represent the quantities conserved across the shock. The jump relations J may be written $L_k[\mathbf{q}, c_i] = 0$ (k=1, 2, 3, etc.) on both sides of the shock. One may check that for given values of the shock constants $\mathbf{C}$, it is possible to define a function P which takes in the $\mathbf{q}$ space stationary values at the two points, images of the values of the $\mathbf{q}$ on both sides of the shock. One may show also that for given values of the constants $\mathbf{C}$, there exist at most four points in the $\mathbf{q}$ space, S_1, S_2, S_3, S_4 in which P is stationary, the index of these points being chosen by non-decreasing values of the specific entropy.

One may be tempted to write the inequality condition j by imposing that the specific entropy cannot decrease when crossing the shock. That was the proposal of the first authors working on this question. In order to see if the statement is correct, one must investigate the structure of the shock, by taking account of the dissipation in the proximal representation of the shock layer with the ξ stretched distance to the inviscid shock. In the most simple representation, one has to take into account the four coefficients of dissipation – two for viscosity, one for Joule's dissipation, one for heat conduction – and assume the existence of a dissipative function as a linear combination of $\dot{q}_i^2 = \left(\frac{dq_i}{d\xi}\right)^2$, say $D(\dot{\mathbf{q}})$. Consequently the shock structure equations may be written

$$\frac{\partial D}{\partial \dot{q}_i} = \frac{\partial P}{\partial q_i} \qquad i = 1,2,3,4.$$

The structure of the shock, defined by the two points S_a and S_b $(a < b)$ is the integral of the system connecting S_a and S_b along which ξ increases from $-\infty$ to $+\infty$. One has to discuss the existence of such an integral and its limit when all the coefficients of dissipation tend towards zero, independently. The MFD shock $S_a \to S_b$ is admissible only if this limit is S_a for $\xi < 0$, and S_b for $\xi > 0$. What is found is that for given constants of the shock, only two are physically admissible: the fast shock $S_1 \rightarrow S_2$ and the slow shock $S_3 \rightarrow S_4$. Of course, if some of the dissipation coefficients vanish, the structure of the shock may present a discontinuity which is called a subshock.

I have presented this question in some detail, because its result led me to some conclusions of a more general nature which may be useful not only in fluid mechanics but also in other modellings of macroscopic theories of physics. The first concerns the significance of entropy. Roughly speaking, I would say: *The non-decreasing property of entropy is a necessary condition for a process to be physically admissible, but it is not always sufficient.*

Another comment concerns the validity of a simplified theory which gives rise to a mathematical system S of equations. Such a system admits *classical solutions* which must check the well-known Hadamard conditions: existence, uniqueness, continuous dependence on the data. Inviscid fluid dynamics shows that the set of classical solutions is too restrictive for the description of physical situations. One has also to consider *weak solutions*; but now, the set of weak solutions may be too large in order to have admissible solutions. Many papers have given conditions which may be imposed to weak solutions in order to achieve a satisfactory requirement. What was proposed above is to prescribe a "continuity" between the set of more or less refined theories which may describe the behaviour of a physical situation.

A solution of a mathematical schematisation S of a physical situation is not acceptable if it cannot be obtained by the limit of a solution of a more refined schematisation S' when S' tends to S.

One may try to apply this statement in order to see how the previous results are the limit of a more refined description. One possibility that has been investigated is to introduce a model with two fluids – ions and electrons – and, for simplicity, to neglect viscosity and heat-conduction dissipations. The differential system which rules the structure involves three parameters: one that rules the Joule's effect and the other two, β and χ, which are respectively proportional to the product and the differences of the densities of ions and electrons. When β and χ are zero, one recovers the differential system which governs the structure of a fast shock in the MFD model. Some interesting situations have been noted: first, of course, subshocks may be present. Then, oscillations may be found either in the front side or in the back side of the shock. Finally, if Joule's effect is neglected, one finds a structure which is a model of what is called a "collisionless shock", in plasma theory.

4.2 General theory of jump conditions and structures in gas dynamics

Kinetic heating during re-entry gave aeronautical engineers a strong impulse to improve the boundary layer theory. But if viscosity and heat conduction have to be taken into account outside the boundary layer, that means that the Rankine-Hugoniot equations which rule the jump across the bow-shock wave have to be rewritten. A first result given in the literature was roughly criticised by noting that one must take into account the thickness of the shock. But the proposed evaluation was not completely satisfactory. Both, Jean-Pierre Guiraud and I, we were convinced that the only way to get correct results was to apply matched asymptotic expansions. This is a quite complicated problem, because the powers of the inverse Reynolds number (R_e^{-1}) which arises, may be fractional and even more complex than the powers of the square root. I do not intend to discuss the whole matter here; I prefer to restrict myself to the special topic which has been worked out, namely the structure of the expansion when the sole presence of the bow shock wave is taken care of. Here one has to deal with simple R_e^{-n}, n integer, powers. Two expansions are needed, the so-called outer one, the leading term of which gives the inviscid solution, and the inner one, which provides the well known internal shock structure, to leading order. It is possible to build the whole expansions, at least formally, including matching. What is remarkable is that one may write out the jump conditions to be applied to the whole of the outer expansion. This comes from the conservative form of the Navier-Stokes equations. As a consequence, one may write out jump conditions by a very simple process, like the one that leads from inviscid conservation equations to jump conditions. Then viscous and heat conducting terms appear to have been taken care of. But the result is illusory because one has to add a contribution from the inner expansion. At least formally, this contribution may be written straight to any order R_e^{-n}. Of course, this is correct only for the terms of the expansion which are forced out by the shock. To order R_e^{-1}, the jump conditions are very easily written out when one knows the internal shock structure to leading order only. This is very well documented. Of course, the boundary layer brings in half powers. The result to order R_e^{-1} has found applications in kinetic re-entry heating.

4.3 Other topics involving singular asymptotics

I did not theorise on singular asymptotics but got a deep knowledge of this methodology through teaching and research work. Let me mention a few examples.

The first one arose with a few lectures on progressive waves I had to deliver in 1970 at Stanford and Berkeley. I read a number of outstanding papers. I do not want to choose among them here, but rather, to simply report on what was my view after that reading. A progressive wave occurs generally when a physical phenomenon is thought to be represented by the occurrence of steep gradients in one variable only, across three-dimensional manifolds, in four-dimensional space-time, with much smoother gradients in other directions. The mathematical structure of the representation looks like one of a phenomenon in five-dimensional space-time. We need some notation in order to avoid confusion. Let the phenomenon be quantified by an n-dimensional vector U and let t be the time, and $\boldsymbol{x}$ be the position vector in three-dimensional space. Assume that the manifold across which the gradients are steep is $F(t,\mathbf{x}) = \text{Const}$. Then, the mathematical progressive wave structure is $U(t,\mathbf{x},F/\varepsilon,\varepsilon)$ so that $\xi = F/\varepsilon$, is considered as a fifth variable. There is apparently nothing in the equations, which allows us to single out the dependency of U on ξ. But all is changed when we add the ansatz that the proper physical solution may be obtained as an expansion with respect to ξ, t and $\mathbf{x}$ being fixed when proceeding to the limit of vanishing ε. As a matter of fact, the multiple scale technique, through the requirement of vanishing of secular terms, provides the way, by means of which that dependency can be figured out. A key to the existence of progressive waves is that U, supposed to be dependent on ξ only, at leading order, exists as a planar wave solution ruled by a linear system with $\partial F/\partial t$, $\partial F/\partial \mathbf{x}$ obeying a dispersion relation. That relation defines a wave speed and may be considered as an eikonal or Hamilton-Jacobi equation, the solution of which is built by means of rays. The planar wave being a solution to a linear system is determined only up to a scalar amplitude factor. This amplitude obeys an equation which is got when going to higher order in the expansion and eliminating secular terms. The details depend on the particular phenomenon considered, and there exist quite a variety of situations that may be described mathematically by such a procedure. It is not my purpose to enter into the details, but let me frame a few remarks. The small parameter ε characterises the steepness of the transversal gradient. If the physical process is non-linear, and non-linearity is measured by the order of magnitude of the amplitude and, if, furthermore, the initial equations are first-order quasi-linear, as is the case with inviscid gas dynamics, then the amplitude obeys a partial differential equation which is generally an inviscid Burgers'equation along each ray. If there are second order derivatives present in the equations, with a small coefficient, then the amplitude obeys a partial-differential equation which is generally Burgers'. The role of time is played by the distance along each ray, while the role of space is played by ξ. One may deal with third order derivatives, and another small parameter, yielding then the Korteweg-de Vries equation. Both phenomena may occur simultaneously. The equation for the amplitude is called the transport equation. One may even treat cuspidal rays, corresponding to caustics of the wave, and get a kind of Tricomi equation for the amplitude but I have to stop here.

I have been too long and shall go faster with the other two examples. The second one was an invitation to give a course in theoretical fluid mechanics at the famous summer school in “Les Houches”, during the summer of 1973. I chose the topic “Asymptotic methods in Fluid Mechanics”. I lectured to brilliant young physicists who began to be somewhat attracted by mechanics and not simply by hard physics. As physicists, they knew the usefulness of approximations and of non-dimensional scaling. But they did not know that a systematic technique was available for building approximate mathematical models and trying to measure quantitatively their validity. I showed that the approximation is very often tied to the existence of a small parameter, coming out from the non-dimensional form of the equations, and I intended to show that the process is sustained by asymptotic singular expansions. I gave an account of the various methods of building the approximations, as asymptotic expansions, and insisted on the methodology, in particular the matching conditions and the concept of significant degeneracy, recently created by Eckhaus. I liked very much this last one, because it gives a systematic way to find out what should be the various stretchings. I thought that this might be attractive to physicists, because it is a quasi-systematic way of comparing the respective weights of various terms in the equations, which measure the physical importance of phenomena they are likely to describe.

The third and last example, which is quite recent, is issued from an invitation to write a paper on the extraordinary heritage of Prandtl. I was not long to focus on two topics, out of a large number initiated by this giant, namely, boundary layer and lifting line theories. It was wonderful to read anew Prandtl's original 1905 paper and to find in it, not only the essentials of the boundary-layer theory, but yet more wonderful, the query of Prandtl about separation, for which he run quite beautiful experiments. Of course, the history of the boundary-layer theory provides a lot of crucial events, but the most impressive one is the construction of the "triple deck" in 1969, independently, by Stewartson and Neiland, through matched asymptotic expansions. A more systematic alternative way to establish this beautiful "triple deck" would be to use the concept of significant degeneracy. This construction which would, I think, have been impossible without the matched asymptotic expansions gave, fifty-five years after the query by Prandtl, a satisfactory explanation of separation, at least for steady laminar flows. The lifting-line concept built by Prandtl in 1917-18 waited till 1964 to find, with Van Dyke using matched asymptotic expansions, not only a justification but also directly one approximate solution to the famous Prandtl's singular integral equation, which is consistent with the order of approximation at which Prandtl's equation itself is consistent. And it was yet more wonderful for me to discover that in 1991, eighty five ears after Prandtl's construction, two young French scientists, Guermond and Sellier, gave a fascinating asymptotic approach to full lifting surface theory of high aspect ratio, allowing, at least in principle, to build the expansion up to any order.

4.4 Final remarks on Fluid Mechanics

Looking back to the main ideas, methods and results I have related on above, I must recognise that I have benefited a lot from them. My research field covered a very limited part of the very large domain of Fluid Mechanics. There is nothing on very important topics: turbulence, rarefied gas dynamics, hydrodynamic stability, to cite a few. I have read many papers, particularly when I was the principal editor of the "Journal de Mécanique" during seventeen years; and I have listened to many talks, in particular at the regular seminar of my laboratory. But, after ten, fifteen, twenty years, what remains is just a feeling of how fascinating it could be to understand deeply the fundamental questions raised by many of these papers. Perhaps, I might say, what remains is a very pleasant "cultural" feeling, something close to the statement of a member of the Académie Française who said that "culture is what remains when everything else has been forgotten".

5. FORCES AND STRESSES VIA VIRTUAL POWER

5.1 General formulation

It is worth to relate the origin of this discovery. In a meeting with mathematicians and physicists which was organised in order to discuss who will assume the task of teaching mechanics to the students in their first two years of university, I had to explain the programme proposed by the mechanicians. For rigid body mechanics, our proposal was to introduce the torsor ("torseur" in French) concept, which gives an adequate mathematical representation of the action, exerted by exterior bodies on a rigid body. This concept is very seldom introduced in the English speaking universities, which prefer to start with the classical Newton laws. It is nevertheless an interesting concept, because it may also be used to describe the kinematics of a rigid body. A torsor is a field of moments and a resultant when it describes the forces. It is a velocity field and a rotation-rate vector, if it describes the kinematics. At that meeting somebody asked me: What is a "torsor" in a space of *n*-dimension? I was ashamed not to be able to answer and, above all, not to have thought, myself, to ponder on this question. A few weeks after this unhappy event, I found what I consider the best way to define the mathematical representation of the action. Let us consider the "forces" exerted on a given mechanical system by an outside system. It is based on the concept of virtual motion introduced nearly two, or more, centuries ago. Roughly speaking, the "mobility" of a system *B* at a fixed time *t*, in a given reference frame *R*, is the vector space *V* of all the possible velocity field of the virtual motions, which one has decided to consider. A system of actions *F* exerted on *B* is defined by a linear form on *V* – to each motion there corresponds a scalar *P* – which is its virtual power. In other words and briefly, *F* is an element of the dual of *V* defined by the linear form. The "force" is the dual of the "mobility"!

This definition may, at first sight, look a little abstract. In fact, it is not and, moreover, it presents many advantages. First, it is very natural: if you want to see if a suitcase is heavy, you try to raise it a little bit. Second, it gives immediately the known result for a system reduced to a material point [the virtual velocity is a vector; the force is a dual vector], or for a rigid body [the kinematics of a virtual motion which keeps the system rigid is a "distributor"defined by a field of the velocities vectors and its associated rotation rate anti-symmetric second-order tensor and the forces are then described by a "torsor", a field of antisymmetric second order tensors field of moments and its associated resultant force]. These two concepts, distributor and tensor, may be identified only in the 3-dimensional space.

The concept of virtual motion was used in analytical mechanics since Lagrange. So, it is not a new one. What is new, is, in addition to using it in order to write equations of motion, its use from the very beginning, that is, in order to define the mathematical representation of "forces". One must notice also that what is proposed is similar to what is done in distribution theory when a function (and its generalisation to a distribution) is defined by a linear and continuous functional in a space on "test functions". A virtual motion is a "test function".

The concept is also very flexible. Given one system, you may choose the definition of the mobility – if you refine the representation of the mobility, you will automatically refine the representation of "forces". You may also choose the linear functional. This remark will find its best application in continuum mechanics.

5.2 "Stresses " in continuum mechanics

When on deals with deformable bodies, it is convenient to introduce separately the (virtual) power of exterior action P^e – exerted on the given body *B* by the system exterior to B – and the power of interior actions P^i mutually exerted by the elements of B. In classical continuum mechanics the following axiom concerning interior actions plays a fundamental role: For *any rigid virtual motion of B, P^i is zero*.

It is equivalent to state: *The power P^i of internal forces is independent of the frame in which the virtual motion is defined.*

This axiom is another formulation of what is often called the principle of material indifference.

The best way to build a continuum theory is to start by writing P^i. The most simple choice is to consider P^i as the integral over B of a local p^i and to assume p^i as a linear function of the local values of the velocity and of its space derivatives. The coefficients of this function define the local representation of the interior forces, which may be called the "stresses" inside the body B.

If B is a 3-dimensional mechanical system, p^i is a linear function of the symmetric part of the gradient of the (virtual) velocity field – The velocity field itself and the antisymmetric part of the gradient cannot be present, on account of the axiom of internal forces. Then the "stresses" here are simply defined by the field of the stress tensors, i.e., symmetric second-order tensors. It is, of course, the classical result, but here it is defined directly and in a rather simple way.

If B is a compressible fluid, a medium sensitive only to the rate of the specific volume, the "stresses" are simply defined by a field of scalars – the pressure p.

If B is incompressible, for the virtual motions which satisfy this constraint, this rate of the specific volume must be zero. The "stresses" are reactions to this constraint and (if the constraint is ideal) are defined by a field of "Lagrange multipliers", i.e., pressures – but the latter are not of the same physical nature as the pressure in a compressible fluid.

The advantages are all the more important, when the situation is complex. For instance, they are very appreciable in plate and shell theories. In the natural theory of plates of small thickness, the mean plane being $x_3 = 0$, one may consider the virtual velocity fields whose components are

$$v_\alpha + x_3 \ell_\alpha, w \; (\alpha = 1,2)$$

which give for the components of the tensor of deformation rate

$$D_{\alpha\beta} = d_{\alpha\beta} + x_3 K_{\alpha\beta},\; D_{\alpha 3} = b_\alpha = \ell_\alpha + w_{,\alpha},\; D_{33} = 0 .$$

One may write

$$p^i = N_{\alpha\beta} d_{\alpha\beta} + M_{\alpha\beta} K_{\alpha\beta} + Q_\alpha b_\alpha$$

The "stresses" are defined by

$N_{\alpha\beta}$ membrane stresses, $M_{\alpha\beta}$ flexure stresses, Q_α shear forces

The Love-Kirchhoff theory considers virtual motions such that the small segments perpendicular to the plate remain perpendicular to the mean surface in the (virtual) deformation in such a way that $b_\alpha = 0$. Then, $p^i = N_{\alpha\beta} d_{\alpha\beta} - M_{\alpha\beta} w_{,\alpha\beta}$. This theory does not take shear forces into account.

In the previous examples, one starts with the definition of the mobility. As "mobility" and "forces" appear as a dual concept, one may also start by the definitions of "forces". In such a case, one must first derive what must be the rate of deformation of the virtual motions and after that, what is the deformation of the medium.

Let us emphasise what is, maybe, the greatest advantage of the method of virtual power. In classical presentations, one usually defines independently the "deformation" and the "stress" and then one is not sure that the assumptions made to define them are in agreement: one may be too refined compared to the other. The definition of "stress" via the virtual power avoids this difficulty.

In the above considered examples, p^i was a linear function of the first derivatives of the virtual velocity field – one says that these theories are of the first grade – except the theory of Love-Kirchhoff for plates, which is a theory of the second grade. It shows that one may refine a theory usually by assuming p^i to be a linear function of first and second derivatives of the velocity field.

One may also build the "stresses" of a micropolar medium, the local particle being not a point of matter, but an infinitely small rigid body whose kinematics are defined by a distributor (a velocity field and a skewsymmetric tensor, which represents the rotation rate of the particle). A liquid crystal is an example of such a medium. One may also build, by a similar process, a theory of micromorphic media by assuming that the particle is no more rigid but deformable. A polymeric solution is an example of such a medium.

Finally, let us note that Gérard Maugin and his co-workers have extended this theoretical scheme to study continuous media in which, physical interactions other than mechanical, are present, in particular electromagnetic interactions, as shown in many papers of this author.

6. CONTINUUM MECHANICS AND CONTINUUM THERMODYNAMICS

This is a new way to look at mechanics, a way by which one discovers its deep unity: mechanics of rigid bodies, fluid mechanics, solid mechanics, appear then as branches of a large tree. To have participated in this adventure is for me a great satisfaction. Within continuum mechanics, new physical phenomena receive a scientific treatment. Continuum mechanics appears today as the foundation of macroscopic physics. I owe Clifford Truesdell and Ronald Rivlin the first discovery of this field; in particular, I am grateful to them for pointing my attention to the works of Pierre Duhem who, at the beginning of the 20th century, had a clear and prophetic view of what it would be like, but unhappily, without being heard in France at that time.

As I try to tell how I have discovered this new important field, I must stress again that, above all for this point, everything that I have to report now was very closely connected with my teaching. I think I may distinguish three steps, which will be called: Mechanical interactions – Continuum thermodynamics – Mechanics of materials.

6. 1 Mechanical interactions

One very important decision for the development of continuum mechanics in France was the creation, in 1958, of a new curriculum called "licence de mathématiques appliquées" in which a semester optional course on mechanics of continuous media would be offered to the students in every university which would want to open such a possibility. I had just been appointed professor in Paris, and I was asked to teach this course. This was for me the occasion to develop a little what I had began to teach in Lille. The course comprised three parts: (i) a general introduction to the Euler and Lagrange representations of motion, deformation in a small-perturbation framework, (ii) stress tensor and the conservation laws of mass and momentum, and then, (iii) some examples of steady flows of an incompressible inviscid fluid and some examples of the classical linearised elasticity for homogeneous and isotropic bodies. The principal goal of these examples was to give the students an idea of the interest and usefulness of fluid mechanics and solid mechanics.

Another decision, a few years later, was to give to universities the possibility to organise more advanced courses in the curriculum towards a DEA (Diplôme d'études approfondies). In Paris, it was not difficult to create a DEA in fluid mechanics because competent professors were present in the department of mechanics. But solid mechanics was not a very well developed discipline. A DEA in solid mechanics was created as soon as the appointment of a new professor gave the department this possibility. I was then in charge of the creation of a more advanced course in continuum mechanics. To do that I had to learn new topics. Roughly speaking my teaching dealt with the general concept of the discipline, much inspired by Truesdell, Toupin, Noll, Coleman. I was using their notations, their reasoning and some of their examples, in particular the marvellous theory of simple fluids and materials with fading memory. But I included also, in the framework of small perturbations, applications of theories of linear viscoelasticity with the use of the Laplace transform, and elastoplasticity with a fixed yield surface in particular, in order to give notions of the beautiful limit analysis.

6.2 Continuum thermodynamics

It was clear that thermomechanical interactions are involved in most of the evolutions of the bodies that have to be considered in a general study of continuous media. It means then that something like thermodynamics was needed. But how to build such a satisfactory theory?

The principal basic question is to introduce the entropy and the absolute temperature. In classical thermodynamics, dealing with systems in equilibrium, many answers have been proposed, some of them, like that given by Caratheodory, very satisfactorily. But mechanics considers systems in motion. There a true *thermodynamics* is needed, and not the classical one, which, in fact, is thermostatics.

Two classes of answers were produced. In the first one, it is proposed to assume that entropy and absolute temperature are primitive concepts – or at least, entropy. This standpoint implies a drastic change in the concept of thermodynamics. In the second one, one wants to maintain what is known in thermostatics and to adapt the necessary requirements needed by the new situation.

The starting point of both approaches is to write the three conservation laws: mass and momentum, as above, and the conservation of energy, which implies the introduction of the specific internal energy, and the heat flux vector. Among the quantities involved in these five scalar density equations, some are the "principal unknowns", density, velocities and an (empirical) temperature; the others are "the complementary unknowns" – for instance, in the classical theory, internal energy, stress tensor and heat flux vector **q**. The main question is to write for a given medium the *constitutive equations* which, with the conservation laws, will allow one to find all the equations that need to be solved. In all the theories presented, the only basic statement appears to be the *Clausius-Duhem inequality*.

After a long hesitation, during two or three years, I decided to adopt for my future research and teaching the second standpoint. I have tried to explain this choice at an international seminar in Portugal in 1973 and also in a review paper I was invited to write with Quoc Son Nguyen and Pierre Suquet for the 50th Anniversary issue of the Journal of applied mechanics, in 1983.

It is, maybe, worth recalling how I reached that conclusion. First, it was the discovery, a few years after its publication, of a paper by H Ziegler in a volume of the *Progress in Solid Mechanics*. Writing, as Lord Rayleigh did in fluid mechanics, the function of dissipation, it was noted that it is a homogeneous function of order two for viscoelastic materials, like for a fluid, but of order one only, for a perfect elastic plastic material. For the professor I was, it was very interesting: it opened the possibility to use the same procedure in order to derive basic constitutive laws for two different types of materials. The second ingredient came from a wonderful note of Jean-Jacques Moreau in the *Comptes rendus de l'Académie des Sciences*. Once again I must confess that it took me at least some months in order to see that it contained what, in fact, I was, more or less unconsciously, looking for: the possibility, at least for a large class of materials, to introduce a pseudo-potential of dissipation. The third ingredient came from my colleague and friend, Joseph Kestin. In my opinion, he is one of the few scientists who had a deep understanding of what may be a correct extension of thermostatics to thermodynamics, in order to deal with complex situations in physics. Most of what I will report below finds its sources in the reading of his papers and in my fruitful discussions with him.

I do not want to enter into the system of equations and into the conclusions which permit to write the constitutive equations. But it may be worth answering the fundamental question about entropy and absolute temperature in this theory. And for that, I must comment a little on the significance of internal variables, a concept that appears in any theory of continuum thermodynamics. We state the possibility to introduce some variables describing some physical properties of the matter in the neighbourhood of any particle of the given system – they may be scalars, vectors or tensors – such that, with the variables which describe the deformation of the medium near this particle, it may be considered as the set of normal variables of a local thermostatic system associated to the particle, the l.a.s. (local accompanying state), when density and specific internal energy are the same as those of this particle. All the physical properties of this l.a.s. will be considered as physical properties of the neighbourhood of the particle of the given system. In particular, the entropy and the absolute temperature at each point of this system depend on the modelling chosen for describing the l.a.s.. Let us note the great flexibility offered for adapting this theory to a particular situation. One can choose

1. the internal variables $\alpha_1, \ldots\ldots \alpha_n$ and their geometrical nature;
2. one thermodynamical potential ψ, for instance, the free energy to describe the thermostatic properties of the l.a.s.. The derivatives of ψ with respect to the α are the forces A associated to the α. [*equations of state*] ;
3. one writes down the Clausius-Duhem inequality, which gives the dissipation - composed of heat-conduction dissipation and internal dissipation. In the simplest case [normal dissipation, standard material] one may choose a pseudo-potential of internal dissipation, for instance a function χ of the $\dot{\alpha}$, time rate of the α. The derivatives give A as function of $\dot{\alpha}$ - they are the *complementary constitutive equations*, which, together with the equations of state, give the complete constitutive laws of the material.

The dissipation mechanisms which have to be kept in mind for a good description of the system are those whose time rate derivatives have an evolution with time comparable to the rate of deformation. Namely, a dissipation mechanism described by internal variables α such that $\dot{\alpha}$ is very small, may be, approximately, neglected, because α will keep its initial value during the deformation of the particle. Now, a dissipative mechanism such that $\dot{\alpha}$ is very large, may also be neglected, because α reaches very quickly its asymptotic value and consequently, it is nearly constant during the deformation of the particle. The entropy and the absolute temperature of the l.a.s. depend on the number of mechanisms which are retained. Then, the entropy and the absolute temperature are not physical properties of the particle itself; but they are those of the l.a.s. which depends on the choice of the mechanisms of dissipation one wants to take into account.

6.3 Mechanics of materials

The continuum thermodynamics, which have been defined in the foregoing section, provide a frame that must be filled out by observations and experiments. The thermodynamic potential – the free energy – describes principally the reversible and elastic part of the behaviour. The pseudo-potential of dissipation describes principally the main physical properties of the material.

In the simplest cases of standard materials, ψ and χ are convex functions. Many unusual materials may be considered as standard. All their important physical properties must appear in the expression of ψ and χ. For instance, in damage mechanics, in the small-perturbation framework the ψ may be a quadratic function of the deformation, as in elasticity, but its coefficients will be affected by a damage variable, which is an internal variable. In plasticity, very often, the dual convex function of χ is the characteristic function of a closed convex set of the space of the A's, which, in many examples, is just the stress tensor. In such a case, one internal variable must be a tensor which represents a plastic deformation. Most of the physical properties of the material can be read off the two convex functions and their dual variables which describe completely the constitutive equations of the material.

7. TOWARD AN ANALYTICAL MECHANICS OF MATERIALS

I arrive now at a new facet of mechanics as a scientific discipline that I am presently discovering or I have just discovered recently. I am not able to organise them in a convincing way. Nevertheless, I will mention some of them briefly, without long comments, because I think they may become important and also because it is now for me a great satisfaction to learn something of these new developments.

I mention first what may be called the balance of material momentum, which may be considered as a primary notion and is a most adequate concept to exhibit nicely the material properties of a system, as shown by many recent publications of Gérard Maugin. For me, one of the best ways to get this new look at the mechanics of any system is due to Pierre Casal, in a not very well known paper of 1978, that I have, once more, really understood many years later: you define at a fixed time, forces and stresses, not by a virtual motion of the system, keeping fixed the reference configuration, but by the virtual motion of this reference configuration, keeping the position of the system itself, fixed. So, you obtain directly the Eshelby stress tensor and the suitable forces to describe singularities and inhomogeneities inside the material. Pierre Casal obtained directly a very elegant formulation and extension of the Rice integral to compute the stress intensity factor at the tip of a crack.

I indicate now what concerns the global formulations of statics and dynamics of structures, starting with the elegant presentation of energy theorems, variational equations, Castigliano theorems for Lagrangian and Hamiltonian integrals and equations in elasticity, including finite elasticity, and extended to a large class of materials, in particular to standard materials. It is impossible to mention all the questions which are treated: stability, buckling, rupture and all the industrial operations on materials, stamping, forging... On account of my personal interest I will just note the question of phase transition and shock waves and the possibility to extend significantly the concept of a shock generating function. One general result worth mentioning here is a kind of generalisation to systems of relations which have been introduced locally by the continuum thermodynamics.

It explains, at least partially, one of the reasons of the success of what is often called micro-macro description of the properties of the materials. The flawless case is the homogenisation of periodic structures, introduced by Sanchez-Palancia. The most evident is the study of a polycrystal as a collection of monocrystals. More generally, by a self-consistent scheme of localisation-homogenisation, one may relate the variables which appear in the constitutive equations at each point of the macrostructure, to average values (or global values) of the physical properties of the local microstructure – a representative volume element – in which one takes into account its own mechanical properties described as above with the convenient material variables. But, a last remark is in order. The scale of the microstructure is very large in comparison with the scale used by the people in solid-state physics, working with grains, dislocations and disclinations!

I would be tempted to evoke many other questions. But it would not be reasonable, because I cannot pretend to have for them a thorough understanding.

8. BEYOND THE SCIENTIFIC DISCIPLINE OF MECHANICS

Beyond the discipline or rather before the discipline, one has the community of men and women who are doing mechanics, professors and researchers. We had in France many important scientific societies: Société Mathématique de France, Société Française de Physique, but no similar association in Mechanics. Joseph Pérès was the founder of the "Association Universitaire de Mécanique des Fluides" a few years before his death in 1962. An "Association Universitaire de Mécanique des Solides" was created a few years after. The unity of the disciplines of mechanics was only recognised in 1973, with the foundation of AUM - "Association Universitaire de Mécanique", fifteen years after the introduction of continuum mechanics in the new curriculum of applied mathematics and the creation of a laboratory of theoretical mechanics, (now Laboratoire de Modélisation Mécanique) in the Sorbonne, the Paris faculty of science. That was, of course, favourable to my "discovery" of mechanics, which was the main project of my scientific activity.

But it was also in 1962 that a certain event has compromised greatly this project. Despite my firm resolution to devote all my professional activity to my job as a professor, I finally had to yield to friendly pressures and accept to become general director of ONERA. I was not prepared to assume such a function. During five years, this has taken up, approximately, two thirds of my time. I was able to continue with my teaching, to take care of some students and to give time to my duties as the principal editor of the Journal de Mécanique", a newly created journal, in order to give a tribune to the researchers in mechanics, particularly to the younger ones. Needless to say that during this period I have not been very active in research.

I tried to recover a little bit after this experience which gave me the opportunity to have a direct contact with the aeronautical industry and with the new activity of the country in space, by launching a new activity for increasing my knowledge and new research projects in mechanics. One sabbatical year as Visiting Professor in Stanford, thanks to an invitation of Nicholas Hoff, was very helpful. I had practically all the time to learn and to have fruitful discussions with my colleagues and friends, especially Lee and Van Dyke. I had, at that time, a big project: to publish a four-volume treatise for graduate students and researchers on the "Mécanique des milieux continus". The first of them appeared in 1973.

Something that I had never anticipated led me to a new serious desertion of mechanics. Our "Académie des Sciences" needed an important reform (statutes had received only slight modifications since 1816), A new *Secrétaire Perpétuel* had to be elected, after the resignation of Louis de Broglie, in 1975. I had accepted to be the candidate of the fellows who wished to move forward this reform. I was elected. I fulfilled the job during twenty years. I was, and still remain, convinced that a strong Academy must be able to deliver independent advices with the highest vision of what science must represent in the life and culture of a modern society. This decision meant for me the need to give up definitely some personal research activity. I was led to leave my laboratory and to take a professorship at the Ecole Polytechnique, without research obligations. I had an office near the Laboratoire de Mécanique des Solides de l'Ecole Polytechnique, one of the best research units in mechanics in the country and I used this wonderful possibility to talk with my young colleagues and to discuss with them, just as if I were a young student once again. This time I had to teach very well prepared students, who had passed the arduous entrance examinations to this school. Just a few hours were sufficient to give them the basic notions of mechanics, (rigid bodies, fluids, solids) at the level of the first year graduate studies. My colleagues of the Laboratory and the students helped me not to consume too quickly my capital of knowledge.

If I talk about this last period of my activity, it is because I had the opportunity to discover something I feel important about mechanics. Two weeks after the signature of the new statute of our *Académie*, in 1979, President Valéry Giscard d'Estaing asked our Fellowship to write a report about the strengths and the weaknesses of the mechanical sciences and industries in France, and to make the appropriate proposals. The last report of the *Académie* had been written in 1916! By this demand, the Président de la République wanted in particular to test whether the capacities of the *Académie* were at the same level as its claims. With a small team, we worked very hard, in order to give a satisfactory answer. But what is worth to be mentioned here is that I have discovered that mechanics was at the same time a science, a technology and an industry. Evident, of course! But an evidence I had never before realised. It is not the place to discuss the conclusions and the consequences of this report – nearly six hundred pages in length. It is certain that it had a big influence on the orientation of many people working in mechanics, on scientists in universities and research establishments, on engineers and directors of companies and on the orientation of long-term programmes. A committee *Haut Comité de Mécanique* of twenty people (one President, one secretary, six scientists, six engineers and six directors of companies) was created in order to study together the many problems of the activity of mechanics in France and to make suggestions in order to improve the mutual relations between these different groups represented. After a lapse of some fifteen years, in 1997, twenty small scientific and technical associations decided to join together in a single society, the *Association Française de Mécanique*. I have been very lucky to be a participant to this significant evolution of mechanics in France. Fifty years ago, you could not find mechanics among the basic Curricula in Universities; you found some courses, but no laboratories in most of the engineering schools, even in the most famous ones. The mechanical and aeronautical industries were very weak. Now mechanics is a scientific discipline which plays an important role in the present development of sciences and which is directly connected with the industries that have to build goods and equipment with the resources of modern technologies and new materials.

As the famous motto we have adopted in France goes: "*Mechanics? In the heart of a moving world!*" And a professor of mechanics? One of the best spots to look at and to participate in this moving world. That would be my answer today to the remark of the nice guy who filled my tank in the gas station near Brown University.

A selection of scientific works and publications by Paul Germain

(established by G.A.Maugin)

" La théorie générale des mouvements coniques et ses applications à l'aérodynamique supersonique " (Doctoral Thesis in Mathematics) **Publication ONERA** No.34, 197 pages , ONERA, Paris (1949) ; translated as " The general theory of conical motions with applications to supersonic aerodynamics ", **NACA Tech.Memo**. No. 1554 (1955).

" La théorie des mouvements homogènes et son application au calcul de certaines ailes delta en régime supersonique ", **La Recherche Aéronautique (Paris), 7**, 3-16 (1949).

" Solutions élémentaires de certaines équations aux dérivées partielles de type mixte ", **Bulletin de la Société Mathématique de France, 81**, 145-174 (1953).

" Sur le problème de Tricomi " (co-authored with R.Bader), **Rend.Circ.Mat.di Palermo, 11,** 53-72 (1953).

" Remarks on partial differential equations of mixed type and applications to transonic flow theory ", **Commun.Pure and Applied. Math., 7,** 117-143 (1954).

" Quelques progrès récents en aérodynamique théorique de grandes vitesses ", General lecture at the **IX.th International Congress of Mechanics** (Brussels, 1956), Vol.I, pp.5-44, Université de Bruxelles (1957).

" An expression for the Green's function for a particular Tricomi problem ", **Quart.Appl.Math., 14**, 113-123 (1956).

" Shock waves and shock-wave structure in magneto-fluid dynamics ", **Rev. Modern Physics, 32**, 951-958 (1960)

" Problèmes mathématiques posés par l'application de la méthode de l'hodographe à l'étude des écoulements transsoniques ", in : **Symposium Transsonicum**, pp.24-50, Springer-Verlag, Berlin (1962).

" Conditions de choc et structures des ondes de choc dans un écoulement stationnaire de fluide dissipatif " (co-authored with J.-P. Guiraud), **Publication ONERA, No.105,** 44 pages , ONERA, Paris (1962)

Mécanique des milieux continus (textbook affectuously nicknamed " the red Popaul's book " by the students at the University of Paris ; These were days of Chairman Mao's fame), Masson Editeurs, Paris (1962)(translated into several foreign languages).

G.A. Maugin et al. (eds.), Continuum Thermomechanics, 25-27.
© 2000 *Kluwer Academic Publishers.*

" Ecoulements transsoniques homogènes ", in : **Progress in Aeronautical Sciences**, ed. D.Küchmann, Vol.5, pp.143-273, Pergamon Press, Oxford (1964).

" Conditions de choc et structure des ondes de choc dans un écoulement non stationnaire de fluid dissipatif " (co-authored with J.-P. Guiraud), **Journ.Math.Pures et Appliquées, 45,** 313-358 (1966).

" A model of some plasma shock structure ", in : **Proceeding of Symposia in Applied mathematics,** Vol.18, (Symp.on Magneto-fluid and Plasma Dynamics, New York, 1965), ed.H.Grad, pp.17-45 , American Mathematical Society, Providence, RI (1967).

" Shock waves - Jump relations and structure ", in : **Advances in Applied Mechanics,** ed. C.-S.Yih, Vol.11, pp.132-194, Academic Press, New York (1972) .

" Progressive waves " (14th Ludwig Prandtl Memorial lecture) in : **Jahrbuch der Deut.Gesel. für Luft und Runfahrt**, pp.11-30, Köln (1971).

" La méthode des puissances virtuelles en mécanique des milieux continus, 1ère partie : la théorie du second gradient ", **Journal de Mécanique (Paris), 12,** 235-274 (1973).

" The method of virtual power in continuum mechanics. Part II. Application to continuous media with microstructure ", **SIAM J.Appl.Math., 25,** 556-575 (1973).

"The role of thermodynamics in continuum mechanics", in: **Foundations of Continuum Thermodynamics** (Proc.Intern.Symp., Bussaco, Portugal, July 1973), eds. J.J.D.Domingos, M.N.R.Nina, J.H.Whitelaw, pp.317-333, J.Wiley, New York (1973).

Cours de Mécanique des milieux continus- Théorie générale, Masson Editeurs, Paris (1973) (advanced textbook affectuously nicknamed " the green Popaul's book " by graduate students at the University of Paris in order to distinguish it from the more elementary " red one ").

" Méthodes asymptotiques en mécanique des fluides ", in : **Fluid Dynamics** (Les Houches Lectures), eds.R.Balian and J.L. Peube, pp. 3-147, Gordon and Breach, New York and Paris (1977).

" Duality et convection in continuum mechanics " in : **Trends in Applications of Pure Mathematics to Mechanics** (Proc.STAMM2, Lecce, 1975), ed.G.Fichera, pp.107-128, Pitman, London (1978).

" Sur certaines définitions liées à l'énergie ", **Int.J.Engng.Sci., 20,** 245-259 (A.C.Eringen Anniversary issue, ed. G.A.Maugin, 1982).

" Continuum Thermodynamics " (co-authored with Q.S.Nguyen and P.Suquet), **Trans.ASME. J.Appl .Mech.** (50th anniversary issue), **105**, 1010-1020 (1983).

Mécanique - Cours de l'Ecole Polytechnique, Two volumes, Editions Ellipses, Paris (1986).

" Toward an analytical mechanics of materials ", in : **Nonlinear Thermodynamical Processes in Continua**, eds. W.Muschik and G.A.Maugin, Heft 61, pp.198-216, TU Berlin-Dokumentation, Kongresse und Tagungen (1992).

" Functional concepts in continuum mechanics ", **Meccanica, 33**, 433-444 (1998).

Double diffusive aspects of the convection in moist-saturated air

Pierre-Antoine Bois and Agnès Kubicki
Laboratoire de Mécanique de Lille, URA CNRS 1441.
U.F.R de Mathématiques, Bât. M3, U.S.T.L., F-59655 Villeneuve d'Ascq
pbois@pop.univ-lille1.fr, kubicki@univ-lille1.fr

Abstract. - Equations of fluid mixtures are used in order to derive a set of equations governing the motion of a moist, viscous and heat conducting, saturated air in presence of gravity. From these equations, a model is constructed to exhibit the double diffusive behaviour of this system. As application, we consider free convection in a shallow layer, by assuming small water concentration : the medium is, in general, linearly stable, but two instability cases are exhibited : stationary instability exists for negative Rayleigh number, and oscillatory instability can exist for negative moist Rayleigh number. In realistic cases, where the moist Rayleigh number is always small, the stationary instability is the only occurring instability. Following the description of the problem of thermohaline convection, this instability may be interpreted as moisture fingers in the medium.
Keywords : atmospheric moisture, fluid mixtures, stratified fluids, shallow convection, double diffusive convection.

1. INTRODUCTION

For a long time, it was thought that, since the diffusivities of dry air and of water vapor are almost equal, double diffusion has no important effect on moist atmospheric flows [*see*, for instance, the review by Huppert and Turner, 1981]. In fact, this conclu-sion is true in unsaturated air. In saturated air, because of the changes of phase superim-posed to the diffusion, the role of the diffusivity of water vapor is considerably modified.

The role of saturated moisture in the atmospheric gravity waves has been studied from the 1960s : Kuo [1961, 1965], Ogura [1963] considered models of cloudy convection. Dudis [1972], Einaudi and Lalas [1973], Durran and Klemp [1982] considered the influence of moisture on the Brunt-Väisälä frequency and trapped mountain lee waves, more particularly taking into account the very important influence of the stratification and the inhomogeneity of a background state. Bougeault [1981] studied its effects on the atmospheric turbulence. In those works, however, the double diffusive properties of the medium are either absent (in nondissipative cases), or not really considered. Starting with the 1970s, there appeared papers exhibiting consequences of diffusivity in dissipative atmosphere : Deardorff [1976] set up a rheological model based on the microphysics of the medium, and exhibited the usefulness of the so-called "liquid-water potential temperature" as well as the role of the total water concentration in the equation of state. More recently, C.S. Bretherton and P.K. Smolarkiewicz [1987, 1988, 1989] considered motions in clouds, with view to study the appearance or disappearance of clouds due to saturation. This work, however, neglects both properties of smallness of

G.A. Maugin et al. (eds.), Continuum Thermomechanics, 29-42.
© 2000 *Kluwer Academic Publishers.*

the water mixing ratio in the medium, and the stratification of the equilibrium state.

Because of the heterogeneity of a cloud (the cloud is a *ternary* mixture), the combined influences of heat conduction and vapor diffusion make the thermomechanical phenomena in this medium multi-diffusive. Simplifying assumptions lead to a reduced double diffusive problem. The first simplification, classical in cloud theories, deals with the modeling of the liquid phase (water) as an *aerosol*. Practically, we assume that the velocity of the liquid water is the same as that of the whole mixture. By assuming, moreover, that the medium is always saturated, the problem of ternary convection *without* cross effects reduces to a convection in a binary mixture *with* cross effects. However, other properties of the ternary mixture remain present in the model, as we will see later.

The aim of this paper is to use the rheological model described above, to exhibit some cases where the double diffusive convection occurring in such a medium, combined with the stratification of the background state, causes original effects. The paper is organized as follows : in Section 2, the equations of motion are set up using the mixture theory developed by Bowen [1976] : this theory allows one to express the diffusion in a simple form, and it gives a systematic form of Fick's law. Moreover, the basic assumption that the moist atmosphere is a continuous medium is applied. Neglecting the density of the aerosol, we are led to describing it as a polytropic gas and the diffusion as double diffusion.

In the Section 3, a linearized wave equation governing the motions is derived. The study is made using the asymptotic framework of the Boussinesq approximation already used in previous papers [Bois, 1991, 1994]. In shallow convection, this assumption implies the existence of a (known) static state of the medium. This static solution is depicted using a single variable, namely the altitude referred to the atmospheric scale. The characteristic vertical scale of the perturbation motion is small, compared to the atmospheric scale, and the characteristic time of the motion is also scaled with the help of static data (namely $t_0 = c_0/g$) : these assumptions follow the analysis of Spiegel and Veronis [1960].

In Section 4, we consider free convection in the medium. A moist Rayleigh-Bénard problem is solved, using a Rayleigh number Ra and a moist Rayleigh number Rh. For small water concentration, Rh is always small. Two properties are derived : first, contrary to the case of pure fluids, stationary instabilities occur for negative values of the Rayleigh number ; second, an oscillatory instability may theoretically occur, but, since the moist Rayleigh number of this instability is large, it does not occur in realistic cases. The marginal regime exists only in conditions where the medium is statically stable. Moreover, the unstable motions are mainly motions of liquid water and dry air, so that they characterize the existence of "moisture fingers" along the medium

2. RHEOLOGICAL MODEL

2.1. Elements of thermodynamics of fluid mixtures

We consider a mixture of three constituents : dry air, water vapor and condensed water. The constituents are identified by the subscripts g (dry air), v (water vapor), L (liquid water). We also denote with the subscript w the quantities for *total water*. The densities are denoted ρ_α, $\alpha = g, v, L$. The density of the medium is $\rho = \rho_g + \rho_v + \rho_L$. The corresponding concentrations are denoted by $q_\alpha = \rho_\alpha / \rho$, and satisfy the relation $q_g + q_v + q_L = q_g + q_w = 1$. The velocity of each constituent is denoted $\mathbf{u}_\alpha$; the barycentric velocity of the fluid is $\mathbf{u}$. Denoting by $\mathbf{v}_\alpha$ the diffusion velocities $\mathbf{u}_\alpha - \mathbf{u}$, we have

$$(1) \qquad \rho\mathbf{u} = \rho_g\mathbf{u}_g + \rho_v\mathbf{u}_v + \rho_L\mathbf{u}_L, \qquad \rho_g\mathbf{v}_g + \rho_v\mathbf{v}_v + \rho_L\mathbf{v}_L = 0.$$

The variables defining the system satisfy the thermostatic Gibbs-Duhem equation

$$(2) \qquad de = Tds - p\, d(1/\rho) + g_L dq_L + g_v dq_v + g_g dq_g ,$$

where the g_α's denote the free enthalpies of the constituents. For a reversible transformation of the medium, dq_w is zero, and, because of the saturation hypothesis, the two free enthalpies g_L and g_v are equal. Since we have $dq_g + dq_w \equiv 0$, (2) may be written

$$(3) \qquad Tds - de = p\, d(1/\rho) + (g_v - g_g)dq_w .$$

For $dq_w = 0$, the left hand side of (3) is just the elementary reversible work in the medium. In the course of a reversible evolution of a parcel, the dissipation reduces, as in a pure fluid, to the term $-p\, d(1/\rho)$. Hence, the elementary reversible work in the medium is described using one equation of state only, say $p = p(\rho, T, q_w)$. Note that, even if the equations of state of the different constituents of the mixture are known, this equation of state is not explicitly given here.

Following a classical approximation, we now neglect the pressure p_L of the liquid phase : this assumption allows us to consider it as a polytropic gas whose adiabatic constant R_L is zero. Moreover, we assume that the dry air and the water vapor are polytropic gases, so that the whole medium is a mixture of polytropic gases. Under these assump-tions, the relation (3) may be rewritten after some rearrangement, using the enthalpy instead of internal energy, as

$$(4) \qquad c_p dT - dp/\rho + L_v\, dq_v = 0,$$

where c_p denotes the heat capacity at constant pressure and concentrations, and $L_v = h_v - h_L$ is the latent heat of vaporization.

2.2. The Clausius-Clapeyron equation

The saturation equation for a polytropic gas (Clausius-Clapeyron equation), is

$$(5) \qquad dq_v/q_v + d\rho/\rho = \left[L_v / R_v T - 1\right] dT/T,$$

[*see* Zemansky, 1968]. Equation (5) is valid for any state of the saturated system. It is of interest to examine particular forms of this formula, the partial pressure of the liquid phase being neglected. The dry air, the water vapor and the whole mixture follow the equations of state

$$(6) \qquad p_g = R_g \rho_g T, \qquad p_v = R_v \rho_v T, \qquad p = R_m \rho T, \qquad R_m = R_g q_g + R_v q_v ,$$

where R_g and R_v are two constants. The third relation (6) is the Dalton's law.

Let us now differentiate the third equation (6). Using the Clausius-Clapeyron equation to eliminate $d\rho$, we get the relation, valid everywhere

$$(7) \qquad dq_w = -(q_g/q_v)dq_v - (R_m/R_g)\ dp/p + (R_m/R_g)\ (L_v/R_v T^2)\ dT.$$

Along a reversible infinitesimal displacement of a parcel from an initial position, $dq_w \equiv 0$ in (3) and (7), and both equations (5) and (7) may be solved in order to provide dq_v and dT (for instance) in terms of dp and $d\rho$ only.

Now, let us denote the heat capacities of the water vapor and the liquid water by c_{p_v} and c_{p_L}, respectively. The latent heat of vaporization can be written as

$$(8) \qquad L_v = L_0 + (c_{p_v} - c_{p_L})(T - \Theta_0),$$

L_0 and Θ_0 being two constants. L_0 is the latent heat at the temperature Θ_0. Introducing (8) in (5), we get a differential relation which may be integrated

$$(9)\ q_v = q_{v_0} \frac{\rho_0 \Theta_0^{\alpha}}{\rho T^{\alpha}} \exp\left[\{\Lambda_0 + \alpha - 1\}\left[1 - \frac{\Theta_0}{T}\right]\right],\ \Lambda_0 = \frac{L_0}{(R_v \Theta_0)},\quad \alpha = 1 - \frac{c_{p_v} - c_{p_L}}{R_v}.$$

Another relation between these parameters can be obtained in order to describe the system. This relation proves the existence of a liquid-water potential temperature. A difficulty is that it cannot easily be written using the classical variables (*see* Bougeault [1981]). On the contrary, it is possible to deduce from Dalton's law and the equation (9) a relation defining the existence of a so-called *saturation concentration* of the water vapor in the gaseous phase, namely a function $Q_s(p, T)$, such that the property

$$(10) \qquad q_v / q_g = Q_s(p, T),$$

is identically satisfied. The role of a saturation concentration will be further investigated.

2.3. The diffusion equations

For a moving system, we depict the barycentric motion of the system using the material derivative $d/dt = \partial/\partial t + \mathbf{u}.\nabla$. We assume the validity of the local state hypothesis in the barycentric motion, so that the relations (2), (5), (7) remain valid for material derivatives. First, consider the balance of mass for the α-th constituent (*molecular diffusion equation*)

$$(11) \qquad \rho dq_\alpha/dt + \nabla.(\rho_\alpha \mathbf{v}_\alpha) = \hat{\rho}_\alpha ,$$

where $\hat{\rho}_\alpha$ denotes the supply of density gained by unit of time by the α-th constituent from the other constituents. Because of the necessary condition $\sum \hat{\rho}_\alpha = 0$, we deduce from the equations (11) for $\alpha = L, v, g$, the balance of mass in the barycentric motion

$$(12) \qquad \partial\rho/\partial t + \nabla.(\rho\mathbf{u}) = 0.$$

Since the supply $\hat{\rho}_g$ vanishes, the equation (11) for dry air becomes

$$(13) \qquad \rho dq_g/dt + \nabla.(\rho_g \mathbf{v}_g) = 0.$$

The diffusion velocities are themselves related to the gradients of concentrations by Fick's law. For the mixture considered here, this law can be written in the form

$$(14) \qquad \rho_g \mathbf{v}_g = \rho(D_{gv}\nabla q_v + D_{gL}\nabla q_L).$$

As it is classically done, we now assume that the liquid droplets travel in the medium with the barycentric velocity, and that their motion does not affect the motion of the mixture. Hence Fick's law reduces to

$$(15) \qquad \rho_g \mathbf{v}_g = \rho D \nabla q_v ,$$

where $D = D_{gv}$ is the only nonzero diffusion coefficient. Thus, eliminating the diffusion velocities, the relation (13) provides the equation

$$(16) \qquad \rho dq_g/dt + D\nabla.(\rho\nabla q_v) = 0.$$

Consider now the first law of thermodynamics : for a moving mixture of dissipative fluids, this law, written in terms of the enthalpy, provides the equation

$$(17) \qquad c_p\, dT/dt - (1/\rho)\, dp/dt + h_L\, dq_L/dt + h_v\, dq_v/dt + h_g\, dq_g/dt$$

$$= \Phi + (k/\rho)\, \Delta T - (1/\rho)\, \nabla.(\rho q_L h_L \mathbf{v}_L + \rho q_v h_v \mathbf{v}_v + \rho q_g h_g \mathbf{v}_g).$$

The left-hand side of (17) is just $dh + dp/\rho$. In the right hand side, the dissipation is the sum of *viscous dissipation* Φ, *thermal dissipation* $k\Delta T$ (k is the thermal conductivity), and *diffusive dissipation* (the last term). Finally, using (3), (15) and (16), the equation (17) becomes after some rearrangement

$$(18) \qquad c_p\, dT/dt \; - (1/\rho)dp/dt - L_v\, dq_L/dt \;=\; \Phi + (k/\rho)\,\Delta T - q_g \mathbf{v}_g . \nabla(h_g - h_v).$$

The simplifying assumptions made above and the equations (16) and (18) are currently adopted in the practice [Bougeault, 1981] from physical arguments. The equations (16) and (18) are the only diffusion equations involving these variables. It is convenient to write these equations with the same variables : hence, replacing dq_g by $-(dq_v + dq_L)$ in (16) and using (5), then (16) becomes

$$(19) \qquad -\frac{q_g}{q_v}\frac{dq_v}{dt} + \frac{R_m}{R_g}\frac{L_v}{R_v T^2}\frac{dT}{dt} - \frac{R_m}{R_g}\frac{1}{p}\frac{dp}{dt} = \frac{D}{\rho}\,\nabla.(\rho\nabla q_v),$$

In the same manner, eliminating dq_L in (18) with the help of (5) and (7) yields

$$(20) \quad c_p\frac{dT}{dt} - \frac{1}{\rho}\frac{dp}{dt} + L_v\frac{dq_v}{dt} = \frac{k}{\rho}\Delta T - q_g\,\mathbf{v}_g.\nabla(h_g - h_v) + L_v\frac{D}{\rho}\,\nabla.(\rho\nabla q_v) + \Phi.$$

The forms (19) and (20) are symmetrical forms of the diffusion equations : assuming that dp/dt is known (it is the approximate consequence of the Boussinesq equation, *see* Section 3) the real unknowns, in (19)-(20), are dT/dt and dq_v/dt. These equations can be solved in two independent combinations of these quantities. Molecular diffusion and heat conduction are present in the right hand sides, and they cannot be separately considered. Finally, the equations of motion for the saturated mixture are the third equation (6), and the equations (9), (12), (19), (20), to which we must add the balance of momentum

$$(21) \qquad \rho d\mathbf{u}/dt + \nabla p \;= \rho\mathbf{g} + \mu[\Delta\mathbf{u} + (1/3)\,\nabla(\nabla.\mathbf{u})]\,,$$

i.e. 8 scalar equations for $\mathbf{u}, p, \rho, T, q_v, q_g$. Since we have left the equation (11) outside the system, its coherence with the solutions must also be investigated.

3. BOUSSINESQ MOTIONS

3.1. Validity conditions and nondimensional equations

Because we are concerned with convective motions in shallow media, it is necessary to rewrite the equations using the Boussinesq approximation. This approximation refers the motion to a static state of the medium, in which the variables depend only on the altitude scaled by the so-called *atmospheric height.* The equilibrium equation $\nabla p = \rho \mathbf{g}$ shows that this scale is $H = p_0/(\rho_0 g)$, where p_0 and ρ_0 denote characteristic pressure and density, respectively. The validity conditions of the Boussinesq approximation [Bois, 1991] first imply that the characteristic length scale L of the motion is small compared to $H : L/H = \varepsilon << 1$, and, second, that the characteristic time of the motion is $t_0 = H\sqrt{\rho_0/p_0}$. We do not discuss these conditions, which are assumed to be satisfied. We note, that, rewritten with the relevant nondimensional parameters [*see* relations (29)], these conditions imply

$$\varepsilon << 1, \qquad F^2 = \varepsilon. \tag{22}$$

The variables are now scaled by characteristic values : $U, L, \rho_0, \Theta_0, c_{p_g}$. The scaling pressure is $p_0 = \rho_0 c_{p_g} \Theta_0$, and we have $R_g = (\gamma - 1)c_{p_g}/\gamma$, γ being the adiabatic constant of the dry air. Rewritten with nondimensional variables the equations become

$$\partial\rho/\partial t + \nabla.(\rho \mathbf{u}) = 0, \tag{23}$$

$$\rho\, d\mathbf{u}/dt + \varepsilon^{-2}\nabla p + \varepsilon^{-1}\rho\mathbf{k} = Re^{-1}[\Delta\mathbf{u} + (1/3)\nabla(\nabla.\mathbf{u})], \tag{24}$$

$$-(q_g/q_v)dq_v/dt + (q_g + r_v q_v)\left[(\Lambda_v/T^2)dT/dt - (1/p)dp/dt\right] = (ReS^*)^{-1}(1/\rho)\nabla.(\rho\nabla q_v), \tag{25}$$

$$\bar{c}_p\, dT/dt - (1/\rho)\, dp/dt + [(\gamma - 1)/\gamma] r_v\, \Lambda_v\, dq_v/dt = (PrRe)^{-1}(1/\rho)\Delta T \tag{26}$$

$$+ (\chi_v - 1)(ReS^*)^{-1}\nabla q_v.\nabla T + (\gamma - 1)/\gamma\, r_v\, \Lambda_v (ReS^*)^{-1}(1/\rho)\nabla.(\rho\nabla q_v) + \varepsilon^{-2}Re^{-1}\Phi,$$

$$q_v = q_{v_0}\rho^{-1}T^{-\alpha}\exp[\{\Lambda_0 + \alpha - 1\}(1 - 1/T)], \tag{27}$$

$$p = (\gamma - 1)(q_g + r_v q_v)\rho T/\gamma. \tag{28}$$

In the equations (23)-(27) we use the following abbreviations :

$$Re = LU\rho_0/\mu, \qquad Pr = \mu\, c_{p_0}/k, \qquad S^* = \mu/(\rho_0 D), \qquad F = U/\sqrt{gL},$$

$$\varepsilon = U/\sqrt{c_{p_0}\Theta_0}, \qquad \chi_v = c_{p_v}/c_{p_g}, \qquad \chi_L = c_{p_L}/c_{p_g}, \qquad r_v = R_v/R_g, \tag{29}$$

$$\Lambda_v = L_v/(R_v\Theta_0) = \Lambda_0 + \Lambda'_0(T - 1), \qquad \Lambda'_0 = 1 - \alpha = [\gamma(\chi_v - \chi_L)]/[(\gamma - 1)r_v],$$

$$c_p = q_g\, c_{p_g} + q_v\, c_{p_v} + q_L\, c_{p_L}, \qquad \bar{c}_p = q_g + \chi_v\, q_v + \chi_L\, q_L.$$

The diffusion parameters are the Prandtl number *Pr* and the Schmidt number S^*. Λ_0 is defined in (9). The last two relations (29) are valid for polytropic gases.

In absence of water, the system possesses equilibrium solutions. If we take into account the water condensation or vaporization, there is, in general, no exact equilibrium solution. However, if we neglect terms of order ε^2, for given $T_0(\zeta)$ and $Q_{v0}(\zeta)$, the equations (24) and (27) are satisfied by the pressure, the density and the dry air concentrations $p = P_0(\zeta)$, $\rho = R_0(\zeta)$, $q_g = Q_{g0}(\zeta)$, which satisfy the equations

$$(30)\quad P'_0(\zeta) + R_0(\zeta) = 0, \qquad P_0(\zeta) = [(\gamma - 1)/\gamma]\,(Q_{g0}(\zeta) + r_v Q_{v0}(\zeta))R_0(\zeta)T_0(\zeta)),$$

while the equations (9), (23), (25) and (26) are (at this order) identically satisfied. After some rearrangement, we obtain the static version of the equation (7) as

$$(31)\quad Q'_{w0}(\zeta) = -Q_{g0}(\zeta)Q'_{v0}(\zeta)/Q_{v0}(\zeta) + \gamma/[(\gamma - 1)T_0(\zeta))$$
$$+ [Q_{g0}(\zeta) + r_v Q_{v0}(\zeta)][(\Lambda_0 + \alpha - 1)/T_0(\zeta) + 1 - \alpha]T'_0(\zeta)/T_0(\zeta)\,.$$

3.2. Approximate equations

Following the Boussinesq procedure, we now expand the variables with respect to ε [Bois, 1991], in the following form :

$$(32)\quad \begin{aligned} p &= P_0(\zeta) + \varepsilon^2\bar{p}, & \rho &= R_0(\zeta) + \varepsilon\bar{\rho}, & T &= T_0(\zeta) + \varepsilon\bar{T}, \\ q_v &= Q_{v0}(\zeta) + \varepsilon\bar{q}_v, & q_g &= Q_{g0}(\zeta) + \varepsilon\bar{q}_g, & \mathbf{u} &= \bar{\mathbf{u}}, \end{aligned}$$

If the static terms in (23)-(28) are omitted, the perturbation equations form a quasi-linear system with slowly varying coefficients. Its solutions can be sought in two ways : (*i*) keeping the slowly varying coefficients in the system [*see* Kubicki 1999], we depict *deep convection behaviours* in the medium ; (*ii*) approximating the coefficients by their values at a given altitude, the system describes *shallow convection* at this altitude : this point of view is the one considered here. In what follows, we assume that the order of magnitude of the water concentration, say q_0, is small. We assume, moreo-ver, that the Schmidt number is also small of the same order q_0 ; we finally set

$$(33)\quad \bar{q}_v = q_0\,\bar{\bar{q}}_v, \qquad \bar{q}_L = q_0\,\bar{\bar{q}}_L, \qquad \bar{q}_g = q_0\,\bar{\bar{q}}_g, \qquad S^* = q_0 S.$$

4. SOME ASPECTS OF THE MOIST BENARD PROBLEM

4.1. The linearized model

We are now interested in free convection in a shallow layer. The thickness of the layer is chosen as length L in the equations. Since we look only for terms of order 0 with respect to ε, the coefficients of the equations are approximated by their values at $z = 0$:

$$- P'_0(0) = R_0(0) = T_0(0) = 1, \qquad P_0(0) = (\gamma - 1)/\gamma,$$

$$Q_{v0}(0) = q_0\,, \qquad Q_{L0}(0) = q_0(r_v - 1), \qquad Q_{g0}(0) = 1 - q_0\, r_v, \tag{34}$$

$$Q'_{v0}(0) = q_0 Q'_0, \quad \Lambda_{0v}(\zeta) = \Lambda_0, \qquad \bar{c}_{p_0}(0) = 1 + q_0(\chi_v - \chi_L + r_v(\chi_L - 1)).$$

The linearized equations associated with (23)-(27) are

$$\nabla.\bar{\mathbf{u}} = 0, \qquad \partial\bar{\mathbf{u}}/\partial t + \nabla\bar{p} + \bar{\rho}\mathbf{k} = Re^{-1}\,\Delta\bar{\mathbf{u}}, \tag{35}$$

$$\partial\bar{\bar{q}}_v/\partial t + \Gamma\bar{w} = (\Lambda_0\mu - 1)/(ReS)\,\Delta\bar{\bar{q}}_v + \Lambda_0/(Pr\,Re)\,\Delta\bar{T}, \tag{36}$$

$$\partial\bar{T}/\partial t + N^2\bar{w} = (Pr\,Re)^{-1}\Delta\bar{T} + (\mu/ReS)^{-1}\,\Delta\bar{\bar{q}}_v, \tag{37}$$

$$\bar{\bar{q}}_v + \bar{p} + (1 - \Lambda_0)\,\bar{T} = 0. \tag{38}$$

The values of the parameters Λ_0 and μ have been given previously. The stability parameters Γ and N^2 (square of the Brunt-Väisälä frequency of dry air) are defined by

$$\Gamma = Q'_0 - \gamma/(\gamma - 1) + \Lambda_0, \qquad N^2 = T'_0 + 1. \tag{39}$$

In a linear stability problem, we use a normal mode analysis to looking for solutions periodic with respect to time ; we set

$$(u, v, w, p, \rho, T, \bar{\bar{q}}_v) = (U, V, W, P, R, T, Q)\, e^{i(\omega t - kx - hy)}. \tag{40}$$

After some calculation, the equations reduce to an eighth-order differential equation for W. This equation reduces itself to a simpler form by setting

$$Ra = -Pr\,Re^2N^2, \qquad Rh = PrRe^2(\Gamma - \Lambda_0 N^2), \qquad \Omega = \omega Re\,,$$

$$A = (\Lambda_0\mu - 1)Pr + \tau Pr - 1, \qquad B = Pr\,(\Lambda_0\mu - 1 + \tau + \tau Pr), \tag{41}$$

$$D = Ra(\Lambda_0 - 1) - Rh[\mu\,(\Lambda_0 - 1) + \tau], \qquad \tau = S/Pr.$$

Ra is the Rayleigh number, Rh is the moist Rayleigh number, τ is the ratio of the diffusivities. Rewritten with these notations, the equation satisfied by W is

$$(42)\qquad D^8W + i\Omega AD^6W + \Omega^2BD^4W + (-i\Omega^3Pr^2\tau - K^2D)\,D^2W$$

$$= -\,i\Omega K^2\tau Pr\,(Ra - Rh)\,W.$$

Finally, we note that, because of the assumption of shallow convection, the static equation (31) becomes

$$(43)\qquad Q'_{w0} = -Q'_0 + \Lambda_0 T'_0 + \gamma/(\gamma - 1) = -(\Gamma - \Lambda_0 N^2) = -Rh.$$

Hence, the moist Rayleigh number is itself of order q_0.

4.2. Bénard convection with two free surfaces

Despite the academic character of free surface boundary conditions for the Bénard problem, their interest is to allow one to exhibit simply analytical solutions. If we require continuity of the temperature and the concentration at the boundaries, the other boundary conditions are standard [*see*, for instance, Drazin and Reid, 1981] so that we assume :

$$(44)\qquad W = D^2W = 0,\qquad T = 0,\qquad \approx{q}_v = 0,\qquad \text{in } z = 0 \text{ et } z = 1.$$

Setting $W = \mathcal{W}\sin n\pi z$, $n \geq 1$, $\mathcal{W}$ = const, the classical procedure for the Bénard problem may be applied (see Drazin and Reid). We get the following conclusions :
(*i*) *stationary solutions* : stationary solutions of (42), exist if the following inequality holds true

$$(45)\qquad [\mu(\Lambda_0 - 1) + \tau]Rh - (\Lambda_0 - 1)Ra \geq 27\pi^4/4,$$

For realistic values of the parameters governing the problem, this inequality defines a region of the (Ra, Rh) plane bounded by a straight line XY (see figure 1), the unstable region corresponding mainly to negative values of Ra and positive values of Rh.
(*ii*) *oscillatory solutions* : two conditions must be satisfied : first, Ω^2 being given, solutions exist only if the inequality

$$(46)\qquad \frac{\tau Ra[\Lambda_0(\mu - 1) + \tau(1 + Pr)] - \tau Rh(\tau Pr + \mu - 1)}{(\tau + \Lambda_0\mu - 1)\left(\tau + \Lambda_0\mu - 1 + \tau Pr - \frac{1}{Pr}\right)} \geq \frac{27\pi^4}{4},$$

is satisfied. Second, Ω^2 is itself positive if the inequality :

$$(47)\qquad Ra - \frac{[(\Lambda_0 - 1)\mu + \tau][(\Lambda_0\mu - 1)Pr + \tau Pr - 1] + \tau Pr}{(\Lambda_0 - 1)[(\Lambda_0\mu - 1)Pr + \tau Pr - 1] + \tau Pr}\,Rh \leq 0,$$

is satisfied. The domains of validity of the two instabilities are drawn on Figure 1 : the inequalities (46) and (47) delineate a region (I) (the angular region *UAW*) of the (*Ra*, *Rh*) plane. This region is the oscillatory instability region. The inequality (45) delineates a region (II) (the half plane above the straight line *XY*), which is the region of stationary instability. The three straight lines *XY*, *UV*, *TW* intersect at a single point A, which is the polycritical point of the problem. For numerical values corresponding to data in the terrestrial atmosphere, the coordinates of A are $Ra = 790$, $Rh = 13.7$: since the order of magnitude q_0 is currently 10^{-3}, the validity condition $Rh = O(q_0)$ is not satisfied in the region of oscillatory solutions, which are, hence, irrealistic. In the same manner, the only realistic part of the region I is a neighbourhood of the *Ra*-axis, of thickness $Rh = O(q_0)$.

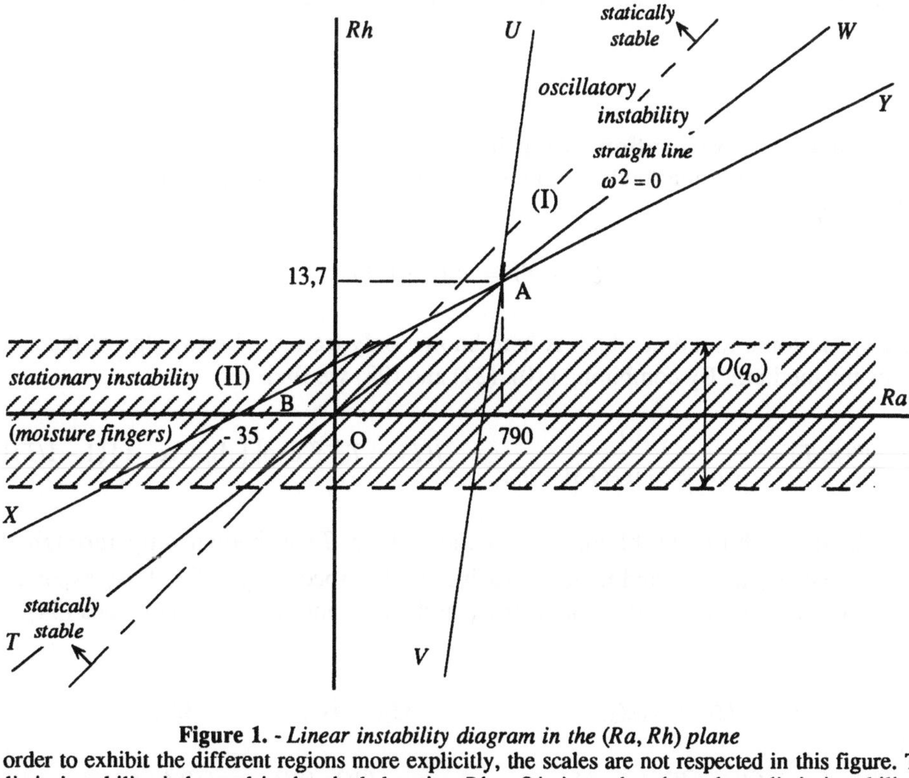

Figure 1. - *Linear instability diagram in the* (*Ra*, *Rh*) *plane*
In order to exhibit the different regions more explicitly, the scales are not respected in this figure. The realistic instability is located in the shaded region $Rh = O(q_0)$, so that the only realistic instability is stationary instability (moisture fingers). The straight line $Rh = Ra$ delineates the statically stable region, where the moisture fingers occur.

4.3. The equation of state

Let us now return to the equation (28) : with the same notations as those already used in the equations (35)-(38), the linearized Boussinesq form of this equation becomes

(48) $$\bar{p} + \bar{T} = -q_0(\tilde{\bar{q}}_g + r_v\tilde{\bar{q}}_v).$$

For vanishing q_0, the equation (48) becomes degenerate, because it becomes an overabundant equation satisfied by the solution of the system (35)-(38). In order to avoid this singu-larity, taking into account the homogeneous boundary conditions associated with the system (35)-(38), we look for solutions of this system as

$$(49) \qquad \bar{\rho} = q_0\rho^*, \qquad \bar{\bar{q}}_v = q_0 q_v^*, \qquad \bar{T} = q_0 T^*, \qquad \bar{p} = q_0 p^*, \qquad \bar{\mathbf{u}} = q_0 \mathbf{u}^*,$$

so that the perturbation of water vapor is assumed small, compared to the perturbation of dry air (and, subsequently the perturbation of liquid water).

The starred variables are solutions of the same system (35)-(38) as the former, and the degeneracy of (48) for vanishing q_0 now takes the nondegenerate form

$$(50) \qquad \rho^* + T^* = - \bar{\bar{q}}_g \, .$$

4.4. The saturation condition

It is now of interest to interpret the condition (10) : the function $Q_s(p, T)$, in this relation, is of order q_0, so that we set

$$(51) \qquad Q_s(p, T) = q_0 Q^*(p, T),$$

where Q^* is some function which is here known only from the knowledge of the static state of the medium. By combining (10), (27) and (28) we find

$$(52) \quad Q^*(p, T) = \frac{1}{\gamma p/[(\gamma - 1)G(T)] - q_0 r_v}, \qquad G(T) = T^{1-\alpha} \exp[\{\Lambda_0 + \alpha - 1\}(1 - 1/T)].$$

In the stability problem, the function $Q^*(p, T)$ is known only through the knowledge of q_v, q_g, etc. and their derivarives with respect to ζ in $\zeta = 0$. By expanding (52) with respect to the small parameter q_0, and taking into account the values (34), we get

$$(53) \qquad \begin{aligned} &Q^* = Q_{s0} + q_0 Q_{s1} + \dots, \qquad && Q_{s0} = 1, \qquad && Q_{s1} = r_v, \\ &(\partial Q_{s0}/\partial T)_{\substack{p=P_0\\ T=T_0}} = \Lambda_0, && (\partial Q_{s1}/\partial T)_{\substack{p=P_0\\ T=T_0}} = 2\Lambda_0\, r_v \, . \end{aligned}$$

On the other hand, by expanding the equilibrium equation (31) with respect to q_0, and taking into account the data (34), we get

$$(54) \quad (\partial Q_{s0}/\partial p)_{\substack{p=P_0\\ T=T_0}} = -\gamma/(\gamma - 1), \qquad (\partial Q_{s1}/\partial p)_{\substack{p=P_0\\ T=T_0}} = \Lambda_0\, r_v\, T'_0 - r_v\, \gamma/(\gamma - 1) - \tilde{Q}'_{w0}\, ,$$

where $Q'_{w0} = q_0\tilde{Q}'_{w0}$. Finally, the equation (10) allows one to relate the values of the given parameters to the values of Q^* and its derivatives at the level of altitude where the problem is set : the knowledge of Ra, Q'_0, and Rh is equivalent to that of the temperature gradient, the derivatives $(\partial Q_{s0}/\partial p)_{P_0,T_0}$ and $(\partial Q_{s1}/\partial p)_{P_0,T_0}$ respectively, at this altitude.

4.5. The moisture fingers configuration

As noticed above, the only instability of the system is a stationary instability occurring in the region $Ra \leq 0$ of the parameter plane : in fact, it is wellknown [Durran and Klemp 1982, Bois 1994] that the moist Brunt-Väisälä frequency of the medium, say N_m, is given by the formula

(55) $$N_m^2 = N^2(1 + O(q_0)) - Q'_{w0} \approx - Ra\,(1 + O(q_0)) + Rh,$$

so that the line of neutral static stability, in the (Ra, Rh) plane, is the straight line $Rh = Ra$. Hence, the instability occurs only in the statically stable region.

Under those conditions, the unstable regime which sets in, is mainly constituted by a perturbation of dry air and liquid water (the order of magnitude of the pertubations $\tilde{\tilde{q}}_g$ and $\tilde{\tilde{q}}_L$ is 1, while that of $\tilde{\tilde{q}}_v$ is q_0). Hence, this motion is mainly a mixing of dry air and liquid water, analogous to the wellknown "salt fingers regime" which occurs in the thermohaline instability problem : therefore, we can call it a "moisture fingers regime". Physically the cloud system organizes in cells or rolls as in the Bénard convection.

5. CONCLUDING REMARKS

To end with, note that some assumptions made in the present analysis are merely of mathematical nature : for instance, the assumption of small Schmidt number is a necessary condition for the existence of the double diffusive character of the problem, when the water concentration is weak. In practice, this assumption should be discus-sed. On the contrary, the assumption of small q_0 is very realistic. It is, sometimes, joined to the assumption of large Λ_0, in such manner that $\Lambda_0^2 q_0$ remains of order unity [Einaudi and Lalas, 1973] : such an assumption would not change the basis of our analysis, but the conclusions would be slightly modified.

Nonlinear aspects of this problems can also eventually be considered, in particular in the neighbourhood of the stationary instability : this problem must be studied using a weakly nonlinear analysis [Knobloch and Proctor, 1981].

REFERENCES

Bois P.A. ; Asymptotic aspects of the Boussinesq approximation for gases and liquids. - *Geoph. Astr. Fl. Dyn.*, **58**, 1991, 45-55.

Bois P.A. ; Boussinesq wave theory in fluid mixtures with application to cloudy atmosphere. - *Int. J. Eng. Sc.*, **32**, 1994, 281-290.

Bougeault P. ; Modeling the trade-wind cumulus boundary layer, Parts I and II. - *J. Atm. Sc.*, **38**, 1981, 2414-2428, and *J. Atm. Sc.*, **38**, 1981, 2429-2439.

Bowen R.M. ; Theory of mixtures. - in *Continuum Physics*, vol. 3, A.C. Eringen ed. Academic Press, New York, 1976, 1-127.

Bretherton C.S. ; A theory of nonprecipitating convection between two parallel plates. Parts I and II. - *J. Atm. Sc.*, **44**, 1987, 1809-1827, and **45**, 1988, 2391-2409.

Bretherton C.S. ; Smolarkiewicz P.K. ; Gravity waves, compensating subsidences and detrainment around cumulus clouds. - *J. Atm. Sc.*, **46**, 1989, 740-759.

Deardorff J.W. ; Usefulness of liquid-water potential temperature in a shallow cloud model. - *J. Appl. Meteor.*, **15**, 1976, 98-102.

Drazin P.G. ; Reid W.H.; "*Hydrodynamic Stability* ; Cambridge Univ. Press, Cambridge, 1981.

Dudis J.J. ; The stability of a saturated stably-stratified shear layer. - *J. Atm. Sc.*, **29**, 1972, 774-778.

Durran D.R. ; Klemp J.B. ; On the effects of moisture on the Brunt-Väisälä frequency. - *J. Atm. Sc.*, **39**, 1982, 2152-2158.

Einaudi F. ; Lalas D.P.; The propagation of acoustic-gravity waves in a moist atmosphere. - *J. Atm. Sc.*, **30**, 1973, 365-376.

Huppert H.E., Turner J.S. ; Double diffusive convection. - *J. Fl. Mech.*, **106**, 1981, 299-329.

Knobloch E. ; Proctor M.R.E.; Nonlinear periodic convection in double diffusive systems. - *J. Fl. Mech.*, **108**, 1981, 291-316.

Kubicki A. ; *Modèle rhéologique pour la diffusion d'ondes de gravité en air humide saturé et application à la convection nuageuse.* - Doctoral thesis, University of Lille, 1999.

Kuo H.L. ; Convection in a conditionally unstable atmosphere. - *Tellus*, **13**, 1961, 441-459

Kuo H.L. ; Further studies of the properties of the cellular convection in a conditionally unstable atmosphere. - *Tellus*, **17**, 1965, 413-433.

Ogura Y. ; The evolution of a moist convective element in a shallow conditionally unstable atmosphere : a numerical calculation. - *J. Atm. Sc.*, **20**, 1963, 407-424.

Spiegel E.A. ; Veronis G. ; On the Boussinesq approximation for a compressible fluid. - *Astr. J.*, **131**, 1960, 442-447.

Zemansky M.W. ; *Heat and Thermodynamics.* 5th ed. ; Mc Graw Hill, New York, 1968.

Application of the theory of Cosserat media to the elasto-plastic behaviour of polycrystals.

Maurizio Brocato*, Alain Ehrlacher†, Philippe Tamagny†

Abstract

Professor Germain has proposed in [Germain, 1973] a theory of Cosserat continua obtained through the principle of virtual power. We propose here an application of that theory to the elastoplastic behaviour of polycrystals. We put forward a model of continuum with microstructure representing a material body, the elements of which are single crystals. Crystals may deform with elasto-plastic behavior, with the plastic rules given through a multiple slip method.

The field of orientation of the crystal lattice is a microscopic kinematic descriptor of the system; correspondingly the lattice spin is a kinematic unknown. As customary in the theory of continua with microstructure, a balance condition for the micro-momentum is associated with this unknown.

First we recall the main equations of Cosserat continua, then we present a kinematic description of polycrystals and, finally, we establish the constitutive equations for the case at issue.

1 Introduction

The study of plastic deformations and texture evolutions in metals usually calls upon continuum models of polycrystals. They can be built through a Cauchy theory of continua, with hidden constitutive variables needed to represent the crystal structure (*cfr* *e.g.* [Rice, 1971], [Hill, 1972], [Mandel, 1972]), or through a theory of continua with microstructure (*cfr* [Germain, 1973], [Capriz, 1989], [Brocato, 1994], [Brocato et al., 1995]).

In this paper we investigate the second possibility; material elements are endowed with a crystal microstructure. In addition to the three coordinates usually giving the location of material elements in the Euclidean space, one needs a minimal set of Lagrangian coordinates to describe the microstructure. Correspondingly a field of kinematic

*IEI-CNR, Via S. Maria, 46 — 56126 Pisa, Italy.

†CERMMO-ENPC, 6 et 8 av. Blaise Pascal, Cité Descartes, Champs-sur-Marne, 77455 Marne-la-Vallée CEDEX 2, France.

G.A. Maugin et al. (eds.), Continuum Thermomechanics, 43-54.
© 2000 *Kluwer Academic Publishers.*

unknowns, the microstructural velocity, enriches the usual description given by the gross velocity field.

More precisely, we will adopt the point of view of multiple slip plasticity (*e.g. cfr* [Mandel, 1965], [Hill, 1966], [Brocato et al., 1999c]) and model polycrystals as continua of Cosserat type; the corresponding equations will be briefly presented in the next paragraph.

According to the multiple slip theory of plasticity, the plastic flow occurs, in crystals, along privileged crystallographic directions. Though plastic deformations most usually modify the defectiveness of the crystal lattice, it is customary to assume, within the multiple slip theory, that defects smooth out at the relevant scale of macroscopic processes, so that planes and directions of slip are not influenced by the plastic flow. Given a perfect, non defective crystal lattice, each material element of the polycrystal will be supposed to have, at any time and through some reversible unloading process, a similar set of slip systems. More precisely, following a classical approach, let us recall Mandel's *director* frame, *isoclinic* setting and plastic spin (*cfr* [Mandel, 1972], [Mandel, 1982]).

In perfect crystals (at least in the ideal multiple slip view) a privileged frame (or a class of equivalent frames due to symmetries of the crystal structure) for the constitutive description exists, which Mandel called the director frame: the crystal lattice—and thus planes and directions of possible slip within the crystal—is represented through a given list of parameters with respect to this frame.

For the constitutive description of polycrystals it is useful to refer to such a frame. We will introduce a suitable decomposition of the gradient of transplacement: material properties in any real setting result from a lattice structure which is stretched and rotated from the ideal perfect setting; material elements are equally stretched and rotated but also plastically deformed.

In the final part of this paper we discuss the consequences of this decomposition through the conditions given by the Clausius-Duhem inequality. For this purpose we assume the free energy to be an objective function of the elastic stretch and of the Lagrangian curvature of the lattice. Entering into further details, we will distinguish a reversible, though —in some sense— small, part of this curvature from a non reversible one. Finally we will give an expression for the dissipated energy, depending upon the usual list of plastic slips and the non reversible part of the curvature. An appropriate number of constitutive relations will arise from these steps.

In the following sections we will use two mathematical notations, intrinsic and tensorial, to avoid ambiguities when dealing with third order tensors. In tensorial notation greek indexes operate on variables defined in a reference microstructure (a concept that will be explained later), capital indexes operate on the initial configuration of the body, lowercase indexes on the actual configuration.

2 Cosserat continua

We briefly recall the theory of Cosserat (or micropolar) continua. In particular we refer to results of [Germain, 1973], where the balance equations are obtained through the principle of virtual power, and in [Eringen, 1999], where a general theory of continua with microstructure is given. The equivalence of the two presentations can be explained through arguments presented in [Capriz, 1989] and, under more general assumptions, in [Tamagny, 1996].

For clarity we recall the kinematic description which is considered in the model. Material points are placed in Euclidean space and are endowed with a lattice microstructure, which is given through a lattice orientation field $N \in \mathcal{M} \subseteq Orth^+$, measured from a reference microstructure; correspondingly their kinematics is described through a velocity of displacement u and a rate of lattice rotation—or lattice spin—$W := \dot{N}N^T \in Skw$.

Symmetries of the crystal lattice may reduce $\mathcal{M}$ to a proper subgroup of $Orth^+$ (see *e.g.* [Dluzewski, 1987], [Dluzewski, 1991] for fcc crystals).

Let grad denote differentiation in the present setting; the power of internal actions per unit volume is given by

$$\chi := -T \cdot \text{grad}u + Z \cdot \dot{N} - \text{s} \cdot \text{grad}\dot{N},$$
$$(= -T_{ij}u_{i,j} + Z_{i\alpha}\dot{N}_{i\alpha} - s_{i\alpha k}\dot{N}_{i\alpha,k}); \tag{1}$$

thus introducing the Cauchy stress tensor $T \in Lin^+$ (not necessarily symmetric), the equilibrated microforce $Z \in \mathcal{T}^*(N)$ (giving a torque per unit volume in the co-tangent space to the orthogonal tensor N, *i.e.* the dual of the set obtained right-multiplying Skw by N) and the microstress s $\in Lin^+(\mathcal{V}, \mathcal{T}^*(N))$ (a torque per unit surface).

Objectivity of χ implies and is implied by the following balance of moment of momentum

$$\text{skw}[T - ZN^T + {}^t[(\text{grad}N)^T]\text{s}] = 0,$$
$$(T_{ij} - Z_{i\alpha}N_{j\alpha} + N_{i\alpha,k}s_{j\alpha k} = T_{ji} - Z_{j\alpha}N_{i\alpha} + N_{j\alpha,k}s_{i\alpha k}) \tag{2}$$

(where the left—or right—lowercase t exponent denotes the minor right—left—transpose of a tensor and the capital T exponent denotes the complete transpose).

The external actions are modeled through the force fields b, acting on the unit mass, and t, acting on the unit surface, and through the couple field acting on the microstructure through the unit surface $C \in \mathcal{T}^*(N)$ (we do not introduce here the corresponding field per unit mass, null in most realistic cases when studying polycrystals).

To simplify matters we will not introduce inertial effects in the balance of momentum and micro-momentum, thus reducing our analysis to quasi-static circumstances. Notice that physical arguments on the purely geometric nature of the microstructure should anyway induce us to take its kinetic energy as null: the crystal lattice represents a network of sites, not a set of mass points, so that the kinetic energy related to its rotation should be, if any, negligible. Nevertheless the possible relevance of virtual inertial terms (as in the theory of liquid with bubbles) should be noted.

The balances of momentum and micromomentum, written in terms of bulk and boundary conditions, are at regular points:

$$\begin{cases} \mathrm{div}T + \rho b = 0 \\ \mathrm{div}\mathsf{s} + Z = 0 \quad (s_{i\alpha j,j} + Z_{i\alpha} = 0) \end{cases} \quad \text{in } \mathcal{B},$$
$$\begin{cases} Tn = f \\ \mathsf{s}n = C \quad (s_{i\alpha j}n_j = C_{i\alpha}) \end{cases} \quad \text{on } \partial\mathcal{B}. \tag{3}$$

Let ρ be the mass per unit volume in the present setting, ψ denote the free energy per unit mass and η the dissipation per unit present volume. The Clausius-Duhem inequality in isothermal conditions, for any possible evolution of the system, is

$$-\rho\dot{\psi} - \chi = \eta \geq 0. \tag{4}$$

Expressing ψ and η in terms of, respectively, internal and flow variables and investigating all possible consequences of inequality (4), one obtains the constitutive relations needed to complete the Cosserat model. In the following part of the paper we will apply this model to the study of polycrystals.

3 Kinematics of polycrystals

3.1 Decomposition of the gradient of the transplacement

Crystals display a periodic arrangement of matter, with an actual set of lattice directions which, in general under stretch, is obtained from that given in a perfect, natural state through an affine transformation.

In multiple slip plasticity the planes and directions of possible slip are related, through some average procedure, to the periodic structure of crystals. Supposedly the evolution of these planes and directions is given through an affine transformation which, in general, differs from the transformation of material elements. Therefore it is natural to introduce the assumption that a field of affine deformations of the lattice exists in any configuration of the polycrystal.

Let us call G_o the field of tensor giving the affine deformation of lattice in a reference configuration (which may be initial with respect to the studied process); let G be the same field in the present configuration.

Let F denote the usual gradient of the material transplacement from the initial to the present setting.

In general F has to fulfill the compatibility conditions $\mathrm{Grad}F = \mathrm{Grad}F^t$ (Grad denoting differentiation in the reference setting), but G has not.

If no plastic flow has occurred, then $F = GG_o^{-1}$. Otherwise we can define the plastic transformation of the reference crystal as

$$P := G^{-1} F G_o. \tag{5}$$

In multiple slip plasticity, the plastic flow presumedly does not alter the lattice directions, the changes of which are thus either reversible or due to rotation. To be precise, it is useful to adopt a polar decomposition of G into a stretch and a rotation of lattice

$$G = NE \;:\; E \in Sym,\; N \in Orth^{+}; \tag{6}$$

(same relations for $G_o = N_o E_o$) the stretch E will be assumed to be reversible (a statement which will be precised later) and thus

$$F = NEPE_o^{-1} N_o^{-1}. \tag{7}$$

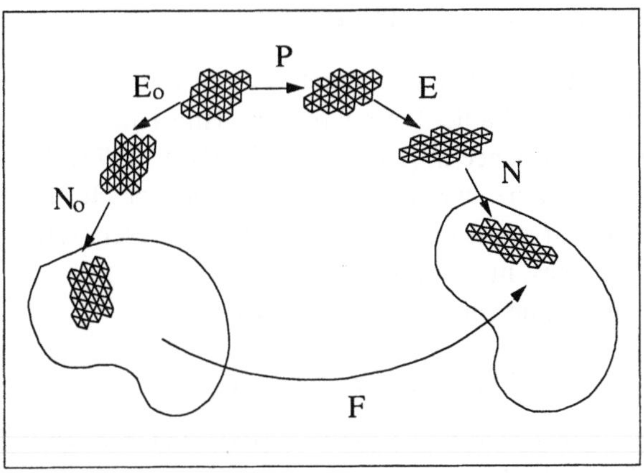

Figure 1: Sketch of the decomposition of the gradient of transplacement in terms of plastic and elastic parts and lattice rotation.

According to this decomposition, the velocity gradient is

$$\text{grad}u = \dot{F}F^{-1} = \dot{N}N^T + N\dot{E}E^{-1}N^T + NE\dot{P}P^{-1}E^{-1}N^T; \tag{8}$$

due to definition (5), $\dot{P}P^{-1}$ is the rate of plastic transformation in the reference crystals.

3.2 Plastic flow in multiple slip

In a multiple slip theory it is customary to assume

$$\dot{P}P^{-1} := \sum_{\kappa} \nu_\kappa m^*_{(\kappa)} \otimes n^*_{(\kappa)}, \tag{9}$$

where

- $\kappa \in \{1, \ldots \bar{\kappa}\}$, $\bar{\kappa}$ being the total number of possible slips within the crystal lattice,
- $m^*_{(\kappa)}$ and $n^*_{(\kappa)}$ are orthogonal unit vectors identifying the κ-th slip system in the reference microstructure ($n^*_{(\kappa)}$ is the normal to a slip plane and $m^*_{(\kappa)}$ a direction of slip within this plane),
- ν_κ is the rate of slip along this system.

To refer to previous works on polycrystals, note that, if the effects of E_o and E are negligible, NN_o^T is the rotation of the lattice from the initial to the present setting. In general, the unit vectors parallel to $m_{(\kappa)} := Gm^*_{(\kappa)}$ and $n_{(\kappa)} := G^{-T}n^*_{(\kappa)}$ give the set of slip systems in the present setting.

3.3 Lagrangian curvature of lattice

In continuum theories of dislocations and defects it is customary to introduce objective measures of the defectiveness of a field of micromorphisms (such as the tensor G of lattice deformation here). For instance in [Kröner, 1960], [Noll, 1972], [Capriz, 1979], one finds the definitions of curvature, wryness and Burgers tensor. In this paper we introduce a measure of the defectiveness of crystal lattice in the present setting which, though different from the quoted ones, fits with our viewpoint.

Let us define the Lagrangian curvature of lattice as

$$\mathrm{k} := N^T \,\mathrm{Grad} N \quad (k_{\alpha\beta K} := N_{h\alpha} N_{h\beta,K}); \tag{10}$$

the third order tensor $\mathrm{k} \in Lin^+(\mathcal{V}, Skw)$ can be represented through Ricci's tensor as a second order tensor: $K := \mathrm{ek}$ ($K_{\alpha K} := e_{\alpha\beta\gamma} k_{\beta\gamma K}$).

3.4 Elastic energy

We have briefly noted the reversible nature of transformation E; we make this statement precise by stating that the free energy depends on E locally and instantaneously.

Furthermore we assume that, if the Lagrangian curvature is almost null, the process is reversible, while if this curvature becomes large enough, there are irreversible consequences. To model this behaviour we take the following additive partition of curvature into (small) elastic and plastic parts

$$\mathrm{k} = \mathrm{k}^E + \mathrm{k}^P, \tag{11}$$

and presume the free energy to depend, again locally and instantaneously, on $\mathrm{k} - \mathrm{k}^P$.

In conclusion we have

$$\psi = \psi(E, \mathrm{k} - \mathrm{k}^P). \tag{12}$$

3.5 Dissipation

We have defined the plastic flow as a function of the slip rates along given slip systems (*cfr* (9)). We assume that the power dissipated by the process of slip is a positive homogeneous function of degree one of the slip rates. This assumption is consistent with a Schmid law of slip, with τ_c the critical shear stress;to simplify matters we take τ_c to be equal on all systems and constant during the process.

We assume the dissipation also depends on the rate of plastic curvature $\dot{\mathrm{k}}^P$. In analogy with classical theories of plasticity in Cauchy continua, we consider a convex set $\mathcal{C} \subset Lin^+(Skw, \mathcal{V})$, in the dual of the vectorial space to which $\dot{\mathrm{k}}^P$ belongs.

Let us denote $\mathrm{m} \in \mathcal{C}$ the running point into this set, $\partial\mathcal{C}$ its boundary and n the outward normal to $\partial\mathcal{C}$. Let $\bar{\mathrm{m}}(\dot{\mathrm{k}}^P) \in \partial\mathcal{C}$ be an element of the boundary of the convex set, such that $\dot{\mathrm{k}}^P$ is parallel to the outward normal n at this point. Uniqueness of $\bar{\mathrm{m}}(\dot{\mathrm{k}}^P)$ is implied by strict convexity of $\mathcal{C}$.

By assumption a convex set $\mathcal{C}$ exists such that the dissipative working of $\dot{\mathrm{k}}^P$ is

$$\pi(\dot{\mathrm{k}}^P) := \bar{\mathrm{m}}(\dot{\mathrm{k}}^P) \cdot \dot{\mathrm{k}}^P, \tag{13}$$

i.e. π is a positively homogeneous function of degree one of $\dot{\mathrm{k}}^P$.

Therefore the total dissipation is

$$\eta = \tau_c \sum_{\kappa} |\nu_\kappa| + \pi(\dot{\mathrm{k}}^P). \tag{14}$$

4 Constitutive equations

In this section we establish the constitutive equations through the Clausius-Duhem inequality (4). Through assumptions (12) and (14), this inequality can be written as

$$-\rho\frac{\partial\psi}{\partial E} \cdot \dot{E} - \rho\frac{\partial\psi}{\partial \mathrm{k}} \cdot (\dot{\mathrm{k}} - \dot{\mathrm{k}}^P) - \chi = \tau_c \sum_{\kappa} |\nu_\kappa| + \pi(\dot{\mathrm{k}}^P) \geq 0\,. \tag{15}$$

For the purpose of establishing the constitutive consequences of (15) it is useful to group terms as factors of the flow variables $\dot{E}$, $\dot{N}$, $\dot{\mathrm{k}}$, ν_κ and $\dot{\mathrm{k}}^P$; therefore, in particular

$$\begin{gathered} \operatorname{grad}\dot{N} = N\dot{\mathrm{k}}F^{-1} - N\dot{N}^T\,\operatorname{Grad}N F^{-1} \\ (\dot{N}_{h\beta,j} = N_{h\alpha}\dot{k}_{\alpha\beta K}F^{-1}_{Kj} - N_{h\alpha}\dot{N}_{i\alpha}N_{i\beta,K}F^{-1}_{Kj}). \end{gathered} \tag{16}$$

Inequality (15) being true for all possible process, each term of the factorisation must necessarily vanish.

Take the following decomposition of the working of Cauchy stress according to the kinematic decomposition (8) and to assumption (9):

$$T \cdot \mathrm{grad} u = \\ T \cdot (\dot{N} N^T) + (N^T T N) \cdot (\dot{E} E^{-1}) + \textstyle\sum_\kappa \left[(N E m^*_{(\kappa)}) \cdot T (N E^{-T} n^*_{(\kappa)}) \right] \nu_\kappa \; ; \tag{17}$$

similarly, we take for the working of microstresses the decomposition

$$\mathrm{s} \cdot \mathrm{grad} \dot{N} = (N^T \mathrm{s} F^{-T}) \cdot \dot{\mathrm{k}} - [(\,\mathrm{Grad} N)^t (F^{-1} \mathrm{s}^T N)] \cdot \dot{N} \\ (s_{h\beta k} \dot{N}_{h\beta,k} = N_{h\alpha} s_{h\beta k} F^{-1}_{Kk} \dot{k}_{\alpha\beta K} - N_{h\alpha} s_{h\beta k} F^{-1}_{Kk} N_{j\beta,K} \dot{N}_{j\alpha}), \tag{18}$$

and express the power of internal actions χ, given through definition (1), accordingly.

Equation (4) becomes

$$\begin{aligned} &\left(T - Z N^T + {}^t[(\mathrm{grad} N)^T] \mathrm{s}\right) \cdot (\dot{N} N^T) \\ &+ \left(N^T T N E^{-T} - \rho \tfrac{\partial \psi}{\partial E}\right) \cdot \dot{E} \\ &+ \left(N^T \mathrm{s} F^{-T} - \rho \tfrac{\partial \psi}{\partial \mathbf{k}}\right) \cdot \dot{\mathrm{k}} \\ &+ \textstyle\sum_\kappa \left[(N E m^*_{(\kappa)}) \cdot T (N E^{-T} n^*_{(\kappa)}) \nu_\kappa - \tau_c |\nu_\kappa| \right] \\ &+ \rho \tfrac{\partial \psi}{\partial \mathbf{k}} \cdot \dot{\mathrm{k}}^P - \pi(\dot{\mathrm{k}}^P) = 0, \end{aligned} \tag{19}$$

which, being true for all processes, implies that each term of the sum vanishes:

- vanishing of the first term, due to arbitrariness of $\dot{N} N^T \in Skw$, implies and is implied by the objectivity condition (2),
- vanishing of the second term, due to arbitrariness of $\dot{E} \in Sym$, implies the first law of elasticity:

$$\mathrm{sym}(N^T T N E^{-T}) = \rho \frac{\partial \psi}{\partial E}, \tag{20}$$

- vanishing of the third, due to arbitrariness of $\dot{\mathrm{k}} \in Lin^+(\mathcal{V}, Skw)$, implies the second law of elasticity (notice that $N^T \mathrm{s} F^{-T} n^o \in Skw \; \forall n^o$):

$$N^T \mathrm{s} F^{-T} = \rho \frac{\partial \psi}{\partial \mathbf{k}} \quad (N_{h\alpha} s_{h\beta k} F^{-1}_{Kk} = \rho \frac{\partial \psi}{\partial k_{\alpha\beta K}}). \tag{21}$$

The remaining terms of (19) give the laws of plasticity as it will be explained in the remaining part of this section.

To express these laws in brief, let us first introduce the definition of the resolved shear stress on system κ:

$$\tau_\kappa := (NEm^*_{(\kappa)}) \cdot T(NE^{-T}n^*_{(\kappa)}), \tag{22}$$

and the stress of curvature (a material force which allows capturing the effects of lattice defects, *cfr* [Maugin, 1995]):

$$\mathrm{m} := \rho\frac{\partial\psi}{\partial \mathrm{k}} \quad (m_{\alpha\beta I} = \rho\frac{\partial\psi}{\partial k_{\alpha\beta I}}); \tag{23}$$

this definition with (21) implies

$$\mathrm{m} = N^T \mathbf{s} F^{-T}. \tag{24}$$

Vanishing of the fourth term in (19) leads to the criterion of plasticity:

$$\nu_\kappa \tau_\kappa - |\nu_\kappa|\tau_c = 0, \tag{25}$$

and thus to the flow rule:

$$\nu_\kappa \begin{cases} < 0 & \text{if} \quad \tau_\kappa = -\tau_c \\ = 0 & \text{if} \quad |\tau_\kappa| < \tau_c \\ > 0 & \text{if} \quad \tau_\kappa = \tau_c. \end{cases} \tag{26}$$

Vanishing of the fifth term of (19) gives the criterion of plasticity:

$$\mathrm{m} \cdot \dot{\mathrm{k}}^P - \pi(\dot{\mathrm{k}}^P) = 0, \tag{27}$$

and thus the flow rule:

$$\dot{\mathrm{k}}^P \begin{cases} = 0 & \text{if} \quad \mathrm{m} \in \overset{\circ}{\mathcal{C}} \\ = \lambda \mathrm{n} & \text{if} \quad \mathrm{m} \in \partial\mathcal{C}, \end{cases} \tag{28}$$

where $\overset{\circ}{\mathcal{C}}$ denotes the interior of $\mathcal{C}$, $\lambda \in \mathbb{R}^+$ and n is the outward normal to $\mathcal{C}$ at m.

5 Relative rotations across a grain boundary

A grain boudary is a discontinuity surface of the orientation field. Let us call n^o the normal vector to the surface in the initial setting and $[\![a]\!]$ the jump of a field a across the surface (supposedly a continuous function on the surface).

Let us assume from now on that the grain boundary is a material surface: the discontinuity is supported by a fixed surface in the initial setting and n^o does not change during the process.

Refering to the plastic lattice curvature behavior we have presented in the previous paragraph, the function

$$\mathrm{h} := A \otimes n^o, \quad A := [\![N^T \dot{N}]\!]$$
$$(h_{\alpha\beta K} := N_{i\alpha} \dot{N}_{i\beta} n^o_K), \tag{29}$$

is a Dirac density of the time rate of plastic curvature $\overset{\bullet}{\mathrm{k}}^P$ on the discontinuity surface (the elastic part of the curvature cannot induce lattice orientation discontinuities), and the plastic dissipative working per unit surface of the grain boundary is given through the same function π (and thus $\bar{\mathrm{m}}$) as in (13):

$$\pi(\mathrm{h}) := \bar{\mathrm{m}}(\mathrm{h}) \cdot \mathrm{h} \quad (= \bar{m}_{\alpha\beta K}(\mathrm{h}) h_{\alpha\beta K}), \tag{30}$$

or, introducing (29) in (30), through the function π_s:

$$\pi(\mathrm{h}) = \pi_s(A) := [\bar{\mathrm{m}}(\mathrm{h}) n^o] \cdot A \quad (= \bar{m}_{\alpha\beta K}(\mathrm{h}) n^o_K A_{\alpha\beta}), \tag{31}$$

where $[\bar{\mathrm{m}}(\mathrm{h}) n^o] \in Skw$ and π_s defines a convex set $\mathcal{C}_s \in Skw$.

The balance of micromomentum and the Clausius-Duhem inequality written on the grain boundary in the particular case we are studying (*i.e.* if the boundary is a material surface and F is continuous across it) give (*cfr* [Brocato et al., 1999b])

$$[\![\mathrm{m}]\!] n^o = 0 \quad ([\![m_{\alpha\beta K}]\!] n^o_K = 0),$$
$$0 \leq \eta_s = [\![N^T \dot{N}]\!] \cdot (\mathrm{m} n^o) \quad (= [\![N_{h\alpha} \dot{N}_{h\beta}]\!] m_{\alpha\beta K} n^o_K). \tag{32}$$

Assumption $\eta_s = \pi_s$ gives the criterion of plastic evolution of the relative lattice rotation across the grain boundary:

$$[\![N^T \dot{N}]\!] \cdot [\mathrm{m} - \bar{\mathrm{m}}(\mathrm{h})] n^o = 0$$
$$\left([\![N_{h\alpha} \dot{N}_{h\beta}]\!] [m_{\alpha\beta J} - \bar{m}_{\alpha\beta J}(\mathrm{h})] n^o_J\right), \tag{33}$$

and the associated flow rule:

$$[\![N^T \dot{N}]\!] \begin{cases} = 0 & \text{if } \ \mathrm{m} n^o \in \overset{\circ}{\mathcal{C}}_s \\ = \lambda W_{\partial \mathcal{C}_s} & \text{if } \ \mathrm{m} n^o \in \partial \mathcal{C}_s, \end{cases} \tag{34}$$

where $W_{\partial \mathcal{C}_s} \in Skw$ is the normal to $\mathcal{C}_s$ at point $\mathrm{m} n^o$.

6 Conclusion

We have given a Cosserat continuum model of a polycrystal, with constitutive equations which take into account classical elasticity and elastic curvatures of lattice, plasticity—through a multiple slip theory, and lattice curvature plasticity.

The plastic curvature term represents a non reversible, non uniform evolution of crystal directions which leads, in processes, to the formation of grain patterning. Then the model needs the analysis of discontinuous solutions, and, particularly, of the evolution of grain boundaries in the polycrystal. Steps have been taken in [Brocato et al., 1998], [Brocato et al., 1999a], [Brocato et al., 1999c].

Acknowledgement

This research was carried out within the project *"Modelli Matematici per la Scienza dei Materiali"* of the Italian *Ministero dell'Università e della Ricerca Scientifica e Tecnologica* and with funding of the French *Ecole Nationale des Ponts et Chaussées.*

References

[Aifantis, 1987] E. C. Aifantis, The physics of plastic deformation, *Int. J. Plasticity*, Vol. 3, pp. 211-247.

[Brocato, 1994] M. Brocato, *Un modèle thermo-mécanique de polycristal pour l'étude des bandes de cisaillement* Ph. D. Dissertation, ENPC, Paris.

[Brocato et al., 1995] M. Brocato, A. Ehrlacher, P. Tamagny, A two dimensional polycrystal model in a 'deep' space representation, *Meccanica*, Vol. 30, pp. 641-663.

[Brocato et al., 1998] M. Brocato, A. Ehrlacher, Ph. Tamagny, Description eulerienne de la mécanique d'un front de discontnuité avec apparition d'une localisation de la déformation, *C. R. Acad. Sci. Paris*, t. 326, Sé rie II *b*, p. 693-698.

[Brocato et al., 1999a] M. Brocato, A. Ehrlacher, Ph. Tamagny, Détermination de la dissipation caractéristique dans la propagation d'un front de recristallisation, *C. R. Acad. Sci. Paris*, t. 327, Sé rie II *b*, p. 179-184.

[Brocato et al., 1999b] M. Brocato, A. Ehrlacher, P. Tamagny, Stability of discontinuities in polycrystals, *Proc. of the Xth International Conference on Waves and Stability in Continuous Media—WASCOM '99*, Vulcano (Italy) 7/12-6-1999, in print.

[Brocato et al., 1999c] M. Brocato, A. Ehrlacher, P. Tamagny, *Texture evolution and rotational hardening in multiple slip plasticity — a two dimensional study* , submitted.

[Capriz, 1979] Capriz, G., *Bodies with microstructure*, Riv. Mat. Univ. Parma, (4) Vol. 5, 673-691.

[Capriz, 1989] G. Capriz *Continua with Microstructure*, Springer-Verlag, New York–Berlin–Heidelberg–London–Paris–Tokyo.

[Dluzewski, 1987] P. H. Dluzewski, Some remarks on the integration domain of the slip orientations, *Arch. Mech.*, Vol. 39, No. 4, pp. 419-422.

[Dluzewski, 1991] P. H. Dluzewski, Crystal orientation spaces and remarks on the modelling of polycrystal anisotropy, *J. Mech. Phys. Solids*, Vol. 39, No. 5, pp. 651-661.

[Eringen, 1999] A. C. Eringen, *Microcontinuum Field Theories*, Springer-Verlag, New York–Berlin–Heidelberg–London–Paris–Tokyo.

[Germain, 1973] P. Germain, The method of virtual power in continuum mechanics. Part 2: Microstructure. *SIAM J. Appl. Math.*, Vol. 25, No. 3, pp. 556-575.

[Hill, 1966] R. Hill, Generalized constitutive relations for incremental deformation of metal crystals by multislip, *J. Mech. Phys. Solids*, Vol. 14, pp. 95-102.

[Hill, 1972] R. Hill, J. R. Rice, Constitutive analysis of elastic-plastic crystals at arbitrary strain, *J. Mech. Phys. Solids*, No. 20, pp. 401-413. .

[Kröner, 1960] Kröner, E., Allgemeine Kontinuumstheorie der Versetzungen und Eigenspannungen, *Arch. Rational Mech. Anal.*, Vol. 4, 273-334.

[Mandel, 1965] J. Mandel, Généralisation de la théorie de la plasticité de W. T. Koiter, *Int. J. Solids Structures*, Vol. 1, pp. 273-295.

[Mandel, 1972] J. Mandel, *Plasticité Classique et Viscoplasticité*, International Centre for Mechanical Sciences, Courses and Lectures – No. 97, Springer-Verlag, Wien-New York.

[Mandel, 1982] J. Mandel, Définition d'un repère privilégié pour l'étude des transformations anélastiques du polycristal. *J. Méc. Th. Appl.*, Vol. 1, No. 1, pp. 7-23.

[Maugin, 1995] G. A. Maugin,Material forces: Concepts and applications, *Appl. Mech. Rev.*, vol. 48, No. 5, pp. 213-245.

[Noll, 1972] Noll, W., A new mathematical theory of simple materials, *Arch. Rational Mech. Anal.*, Vol. 48, 1-50.

[Rice, 1971] J. R. Rice, Inelastic constitutive relations for solids: an internal-variable theory and its applications to metal plasticity, *J. Mech. Phys. Solids*, No. 19, pp. 433-455. .

[Tamagny, 1996] P. Tamagny, *Les milieux à microstructure; approches par la méthode des puissances virtuelles*, Ph. D. Dissertation, ENPC, Paris.

From Clausius to finite anelasticity, *via* Bridgman, Eckart and Ziegler

Louis C.J.Brun
LMS, Ecole polytechnique, F-91128 Palaiseau
CEA/DIF/PMC, F-91680 Bruyères-le-Châtel
louis.brun@bruyeres.cea.fr

Abstract: A general model of thermomechanical behaviour is introduced, based on Bridgman's perception of the role of thermodynamic forces, Eckart's finite anelastic deformation concept, Ziegler's dissipation potential and a greatly simplifying assumption of 'small elastic deformation'. An apparently new non-linear extension to the 'thermorheologically simple' viscoelastic model is obtained and a fairly comprehensive elastoviscoplastic model presented.
Keywords: released state , anelastic potential, quasi-incompressibility, standard model.

1. INTRODUCTION

A great variety of physical mechanisms may contribute to the irreversible thermomechanical behaviour of condensed matter, which continuum modelling broadly divides into transport (or transfer) processes between neighbouring elements, such as heat conduction, viscosity, mass diffusion and a wealth of more or less complex processes confined to the material element viewed as a *microsystem* [Germain, 1986], such as chemical reactions, phase changes, relaxation, solid friction, damage.

Whenever some of these mechanisms occur simultaneously, thermodynamics offers, even by uniform temperature, the only rational approach to the study of their combined effects. Up to now, all such fruitful thermodynamic models rest upon the local form of the *entropy inequality* of the thermodynamics of irreversible processes between the entropy density rate $\dot{s}$, the entropy flux and the entropy production density. In the absence of mass transfer, the entropy flux is quite generally taken to be $\underline{q}/T$, where T is the absolute temperature and $\underline{q}$ the heat flux. Both relationships combine into the so-called *Clausius - Duhem inequality*, after Clausius (1865), who coined the (global) entropy and stated its famous growth law for an isolated system, and Duhem (1903), who arrived at the inequality from phenomenological arguments [Woods, 1982].

Since the fifties, the introduction of *internal variables*, first with and then without reference to springs and dashpots ([Staverman and Schwarzl, 1952; Biot, 1954; Meixner, a. 1954, b. 1972]) laid the thermodynamic foundation of isothermal linear

G.A. Maugin et al. (eds.), Continuum Thermomechanics, 55-62.
© 2000 *Kluwer Academic Publishers.*

viscoelasticity according to Onsager relations. Ziegler proposed a non-linear extension of these, while Valanis emphasised the importance of the *frozen elastic response* concept for solids and brought to light the corresponding non-linear constitutive laws more specially applicable to viscoelastic polymers [Ziegler, 1962; Valanis, 1968]. Along these lines, Sidoroff systematically studied the consequences of material isotropy on the mechanical behaviour of viscoelastic materials with an 'intermediate' configuration [Sidoroff, 1974, 1975]. Independently, the 'relaxed state' concept, of importance for plasticity, had been given a firm geometrical status within the frame of finite deformation, although not such a convincing physical treatment in [Eckart, 1948]. Later on, new considerations on the elastoviscoplastic behaviour of metal and the metallic single-crystal began to fill the gap [Kratochvil *et al.*, 1970; Rice, 1971]. Since then, many other constitutive properties and even damage as related to internal variables have entered the realm of thermodynamics. Distributed heterogeneity, a difficult matter under permanent investigation, still lags behind.

Oddly enough, Eckart applies on the basis of Cl-D inequality a formalism explicitly intended for ductile matter to "one particular set of laws" which turn out to be, if any, of viscoelastic type. Besides, cancelling for a check the mechanical dissipation in his results fails to give back the thermoelastic equations! No matter, on the working assumption of *small (thermo)elastic deformation* we shall in the following demonstrate the usefulness of the formalism on the ground of both viscoelastic and elastoviscoplastic behaviours. The material description will be preferred to the spatial one used by Eckart.

2. GENERAL FRAME

2.1 Frozen thermoelastic response and accompanying state

A widely adopted frame for constitutive laws for solids presupposes mechanical and thermal state laws derived from a free energy $\psi(\underline{\underline{e}}, \underline{\alpha}, T; \underline{X})$:

(1) $$\underline{\underline{\pi}} = \rho_0 \frac{\partial \psi}{\partial \underline{\underline{e}}} \qquad s = -\rho_0 \frac{\partial \psi}{\partial T}.$$

Here, $\underline{\underline{\pi}}$ is the symmetric Piola-Kirchhoff stress tensor, $\underline{X}$ the initial location of the particle, $\underline{\underline{e}} = (1/2)\ (\underline{\underline{C}} - \underline{\underline{1}})$ the deformation (tensor), $\underline{\underline{C}} = \underline{\underline{F}}^T \cdot \underline{\underline{F}}$ the Cauchy tensor and $\underline{\underline{F}}$ the transformation gradient from the reference configuration. The set of internal variables $\underline{\alpha} = (\alpha_1, \alpha_2, .\ \alpha_n)$ obeys *evolution laws*, which according to [Bridgman, 1943] should only depend upon the mechanical control variables through the *thermodynamic forces*:

(2) $$A_i = -\rho_0 \frac{\partial \psi}{\partial \alpha_i}.$$

Thanks to the independent satisfaction of the heat conduction inequality $\underline{q} \cdot \underline{\nabla T} \leq 0$ and to the conservation equations, the Clausius-Duhem inequality is *automatically* fulfilled when the evolution laws are of the form:

(3) $$\underline{\dot{\alpha}} = \underline{r}(\underline{A}, \underline{\alpha}, T; \underline{X}), \quad \text{with} \quad \underline{r}(\underline{Y}, \underline{\alpha}, T; \underline{X}).\underline{Y} \geq 0,$$

for any $\underline{Y}$. The argument of $\underline{r}$ conforms to Bridgman's views.

While keeping $\underline{\alpha}$ frozen, we reassign by thought to the local stress and temperature the values $\underline{\underline{\pi}}_0(\underline{X})$, $T_0(\underline{X})$ that they had in some previous "0" configuration of the system (possibly different from the reference one). The deformation $\underline{\underline{e}}^{ac}$ in the *accompanying* state: $\underline{\underline{\pi}}^{ac} = \underline{\underline{\pi}}_0$, $T^{ac} = T_0$, $\underline{\alpha}^{ac} = \underline{\alpha}(t)$ is such that:

(4) $$\rho_0 \frac{\partial \psi}{\partial \underline{\underline{e}}} (\underline{\underline{e}}^{ac}, \underline{\alpha}, T_0(\underline{X}); \underline{X}) = \underline{\underline{\pi}}_0(\underline{X}),$$

while $\psi^{ac}(\underline{\alpha}, T_0(\underline{X}); \underline{X})$ and the 'accompanying' value of any other quantity is obtained on substituting $\underline{\underline{e}}^{ac}(\alpha, T_0(\underline{X}); \underline{X})$ and $T_0(\underline{X})$ for $\underline{\underline{e}}$ and T in the arguments of ψ and the other quantity. The present definition of the accompanying state differs from that in [Meixner, 1971].

2.2 Small thermoelastic deformation

We now drop T_0 and $\underline{X}$, since we are concentrating on a given element. Due to the high frozen stiffness of the (supposedly undamaged) material compared to the stress and the small temperature variation in most engineering applications, the frozen thermoelastic deformation

$$\underline{\underline{\delta e}}(\underline{\underline{e}}, \underline{\alpha}) = \underline{\underline{e}} - \underline{\underline{e}}^{ac}(\underline{\alpha})$$

often happens to remain small. Letting $\underline{\underline{\underline{A}}}^{ac} = (\partial^2 \psi / \underline{\underline{\delta e}}\,\underline{\underline{\delta e}})^{ac}$ and similar expressions for the coupling constant and specific heat, the expansion

$$\rho_0 \psi^s(\underline{\underline{e}}, \underline{\alpha}) = \rho_0 \psi^{ac}(\underline{\alpha}) + \underline{\underline{\pi}}_0 : \underline{\underline{\delta e}} + \frac{1}{2} \underline{\underline{\delta e}} : \underline{\underline{\underline{A}}}^{ac}(\underline{\alpha}) : \underline{\underline{\delta e}} - \tau \underline{\underline{k}}^{ac}(\underline{\alpha}) : \underline{\underline{\delta e}} - \frac{1}{2} \rho_0 \frac{C^{ac}(\underline{\alpha})}{T_0} \tau^2$$

up to second order with respect to $\underline{\underline{\delta e}}$ and $\tau = T - T_0$ then becomes a 'simplified' approximation to $\rho_0 \psi$. The simplified thermomechanical and 'internal' state laws follow:

(5) $$\underline{\underline{\delta \pi}} \equiv \underline{\underline{\pi}}^s - \underline{\underline{\pi}}_0 = \underline{\underline{\underline{A}}}^{ac} : \underline{\underline{\delta e}} - \tau \underline{\underline{k}}^{ac} \qquad s = s^{ac} + \frac{1}{\rho_0} \underline{\underline{k}}^{ac} : \underline{\underline{\delta e}} + C^{ac} \frac{\tau}{T_0}$$
$$A_i^s = \frac{\partial}{\partial \alpha_i} \underline{\underline{e}}^{ac} : \underline{\underline{\delta \pi}} - \rho_0 \frac{\partial \psi^{ac}}{\partial \alpha_i} + \rho_0 \tau \frac{\partial s^{ac}}{\partial \alpha_i}.$$

In case of a stress free "0"state, relations (5) are found in [Brun, 1992]. The accompanying quantities f will then be termed *released*, instead of the ambiguous 'relaxed' of Eckart, and be indexed accordingly. The released deformation becomes *anelastic*:

$$f^{ac}\left(\underline{\underline{\pi}}_0=0\right) \equiv f^r \qquad \underline{\underline{e}}^{ac}\left(\underline{\underline{\pi}}_0=0\right) \equiv \underline{\underline{e}}^{an}.$$

In the sequel, we shall confine to isothermal processes from a stress free state.

3. THE VISCOELASTIC SOLID

Experimental qualitative viscoelastic solid behaviours of immediate significance are the existence of a specific unstressed equilibrium state at any fixed temperature, which deserves to be called *natural*, as in elasticity, and the inclination of the element to spontaneously tend towards the natural state $S_{Nat}(T)$, without oscillation, after the stress has been released.

The modelling will achieve the existence and uniqueness property of the natural state, if it requires that the free energy be *bounded from below* and *strictly convex with respect to* $\underline{\underline{e}}$-$\underline{\alpha}$ in some neighbourhood of S_{Nat}. The thermodynamic forces will consequently vanish at the natural state as well:

(6) $$\underline{\underline{\pi}}_{Nat}(T)=0 \qquad \underline{A}_{Nat}(T)=0.$$

Satisfaction of (3) by the evolution laws does not imply relaxation towards S_{Nat} if solid friction ($\underline{r}=0$ for $\underline{A}\neq 0$) is present. Likewise, it does not ensure the non-oscillatory behaviour if only internal viscous mechanisms are present, as simple counterexamples show. A well-known sufficient condition to agree with the observations consists in having $\underline{r}$ derive from a *dissipation potential*:

(7) $$\underline{r}=\frac{\partial}{\partial \underline{A}}D\left(\underline{A},\underline{\alpha},T\right),$$

with D being a *strictly convex* function of $\underline{A}$, *minimal* at $\underline{A}=0$.

To be specific, we adopt the symmetric positive definite form :

(8) $$D=(1/2)\,d_{ij}(\underline{\alpha},T)A_iA_j.$$

On the small elastic deformation assumption, the combination of (8) and $(5)_2$ yields:

(9) $$\dot{\alpha}_i=d_{ij}(T_0)\left(\frac{\partial}{\partial \alpha_j}\underline{\underline{e}}^{an}(\underline{\alpha},T_0):\underline{\underline{\pi}}-\rho_0\frac{\partial}{\partial \alpha_j}\psi^r(\underline{\alpha},T_0)\right).$$

If $\underline{\underline{\pi}}$ is the control variable, *i.e.* a given function of *t*, this is a set of differential equations for the α_i's which allow their determination from the start. It still is, on account of $(5)_1$, if $\underline{\underline{e}}$ is the control variable. The determination of the complete state follows.

In both applications to come, the system is referred to the "0" configuration and the latter is supposed to be natural.

3.1 Linear isothermal viscoelasticity

The evolution of the system is further confined to the vicinity of S_{Nat}, where $\underline{\alpha}_0 = 0$ (say), reducing $\rho_0\psi^s$ to the quadratic form $\rho_0\psi^r + (1{:}/2)\underline{\underline{\delta e}}{:}\underline{\underline{\underline{A}}}{:}\underline{\underline{\delta e}}$ in $\underline{e}$ - $\underline{\alpha}$, with:

$$(10) \qquad \rho_0\psi^r = (1/2)\,a_{ij}\alpha_i\alpha_j \; > 0 \text{ for } \underline{\alpha} \neq 0, \quad \underline{\underline{\delta e}} = \underline{\underline{e}} - \underline{\underline{e}}^{an}, \quad \underline{\underline{e}}^{an} = \alpha_i \underline{\underline{e}}_i \,.$$

The coefficients $A_{ij,hk}$, a_{ij}, $\underline{\underline{e}}_i$ are *independent* constants. Some, not all of course, of the $\underline{\underline{e}}_i$ may be zero. Substituting from (10) into (9) results in equations which agree with those obtained by [Biot, 1954] and, as a bonus, connects the terms of the latter to released quantities.

3.2 Finite viscoelasticity: the anelastic deformation, internal variable

In the case of finite viscoelasticity, the deformation may not be small and relations $(10)_1$ and $(10)_3$ no longer necessarily hold. Instead, the assumption is now made that the released deformation is an internal variable and, for the sake of simplicity, that there are no others: $\underline{\alpha} = \underline{\underline{e}}^{an}$. The kinematic equations accordingly become:

$$(11) \qquad \dot{e}^{an}_{ij} = d_{ij,hk}(\underline{\underline{e}}^{an}, T_0)[\pi_{hk} - \pi^{back}_{hk}] \qquad \underline{\underline{\pi}}^{back} = \rho_0 \frac{\partial}{\partial \underline{\underline{e}}^r}\psi^r(\underline{\underline{e}}^{an}, T_0),$$

where $\underline{\underline{\pi}}^{back}$ is the *back stress*. Since ψ^r is a strictly convex function of $\underline{\underline{e}}^{an}$, in a retardation test ($\underline{\underline{\pi}}(t)$ held constant), $\underline{\underline{e}}^{an}$ tends toward the unique solution of $\underline{\underline{\pi}} - \underline{\underline{\pi}}^{back} = 0$ and the deformation toward its corresponding equilibrium value.

Whenever the natural "0" configuration is one of material isotropy, ψ is an isotropic function of $\underline{\underline{e}}$ and $\underline{\underline{e}}^{an}$ (or $\underline{\underline{\delta e}}$). The frozen moduli accordingly take on the form $A_{ij,hk} = \lambda\,\delta_{ij}\delta_{hk} + \mu\,(\delta_{ih}\delta_{jk} + \delta_{ik}\delta_{jh}) + \nu\, e^{an}_{ij}\, e^{an}_{hk} + \lambda'(\delta_{ij} + \alpha' e^{an}_{ij})(\delta_{hk} + \alpha' e^{an}_{hk}) + \mu'(\delta_{ij} + \beta'$ $e^{an}_{ij} + \gamma' e^{an}_{il} e^{an}_{lj})(\delta_{hk} + \beta' e^{an}_{hk} + \gamma' e^{an}_{hl} e^{an}_{lj})$, where $\lambda, \mu, \nu, \lambda', \alpha', \mu', \beta', \gamma'$ are functions of the invariants $I^r_N = \operatorname{tr} \underline{\underline{C}}^{an\,N}$ of $\underline{\underline{C}}^{an}$ and of T_0 (as is ψ^r). The $d_{ij,hk}$'s take on a similar form.

Let the anelastic Cauchy deformation be constrained by the incompressibility condition: $J^{an} = 1$ ($J = \det \underline{\underline{C}}$). The deformation will accordingly be *quasi-isochoric*. There follows from the kinematic counterpart of the identity $6J = I_1^3 + 2I_3 - 3I_1I_2$, *viz.*: $\dot{J} = 2\,\mathrm{tr}(\underline{\underline{I}}.\dot{\underline{\underline{C}}})$, with $\underline{\underline{I}} = (I_1^2 - I_2)\,\underline{\underline{1}} + 2(\underline{\underline{C}}^2 - I_1\underline{\underline{C}})$, that only such evolution laws are to bring about incompressibility as satisfy : $I^{an}_{ij} d_{ij,hk} = 0$ identically for any h,k. In case of isotropy, the restricted form in terms of the functions a,b,d of I^{an}_1, I^{an}_2, T results:

$$(12) \qquad d_{ij,hk} = d(\delta_{ij} + aC^{an}_{ij} + bC^{an}_{il}C^{an}_{lj})(\delta_{hk} + aC^{an}_{hk} + bC^{an}_{hl}C^{an}_{lj}),$$

with $d > 0$ and a,b bound to satisfy identically, whatever I^{an}_1, I^{an}_2, the relationship:

$$(13) \qquad a(I^{an}_1 I^{an}_{-1} - 3) + b(I^{an}_2 I^{an}_{-1} + 1) + 2I^{an}_{-1} = 0 \qquad (2I^{an}_{-1} = I^{an\,2}_1 - I^{an}_2).$$

Choosing $d = d(T)$ and λ, μ constant as only non zero moduli results in a simple non-linear extension of the linear model of quasi-incompressible thermorheologically simple isotropic solid. Besides, the constancy of the $A_{ij,hk}$'s makes the model *thermodynamically exact*, instead of approximate.

4. THE STANDARD ANELASTIC SOLID

Viscoelasticity is the most elementary type of anelastic behaviour encountered in solid polymers. Remarkably enough, loosening on specific points the assumptions that make up the model described by Eqs (6) to (9) will suffice to set up an elementary 'standard' model, meant to reflect typical aspects of the, primarily plastic, behaviour of metallic substances. Such aspects need be succinctly reviewed first.

4.1 About elastoviscoplasticity

The absence of a definite stress-free shape, hence of a natural state, differentiates the plastic from the viscoelastic isothermal behaviour. This feature will be accounted for by relaxing the strict convexity property of the relaxed free energy.

In viscoelasticity, no natural distinction can be operated among the internal variables on the sole basis of the evolution laws. By contrast, nearly *zero plastic volume change* is common place in engineering conditions, a property which solid state physics of the metal traces back to glide mechanisms [Hill, 1950]. To these last the modelling associates a set $\underline{a}$ of internal *kinematic* variables [Brun, 1992] governed by a specific set of evolution laws.

The existence of a yield surface in the stress space is another distinctive feature of metal plasticity, which suggests the occurring of internal *solid friction* along the anelastic deformation process. Demanding that $\dot{\underline{a}}$ alone derive from a potential $\widetilde{D}$ with less strict properties than the viscoelastic one will allow for this feature. Despite the difficulty of defining the reference frame suited to follow the evolution of the yield surface [Mandel, 1966], the shape of the surface is observed to change. A set E of *intrinsic hardening* variables is introduced as possible shape parameters next to the a_i's responsible for the *kinematic hardening*.

A third set Z of variables, coupled with $\underline{a}$ and E, may be necessary to specify the evolution laws. Non kinematic variables will form the *complementary* set $\underline{b}$. In brief:

$$\underline{\alpha} = (\underline{a}, \underline{b}), \text{ with } \underline{\underline{e}}^{an}(\underline{a}) \text{ and } \underline{b} = (\mathsf{E}, \mathsf{Z}). \tag{14}$$

Parallel to (14), the set of conjugate thermodynamic forces splits up into kinematic and complementary forces: $\underline{A} = (A_{\underline{a}}, A_{\underline{b}})$. In the s.e.d. approximation, the internal state laws $(5)_2$ apply and render the $A_{\underline{b}}$'s independent of the stress.

4.2 The standard non-linear anelastic model

The foresaid facts combine into the following constitutive elements:

- The state laws: $\underline{\underline{\pi}} = \underline{\underline{\underline{\underline{A}}}}^r(\underline{\alpha}):(\underline{\underline{e}} - \underline{\underline{e}}^{an}(\underline{\alpha}))$, $\quad A_i = \dfrac{\partial}{\partial \alpha_i}\underline{\underline{e}}^{an}:\underline{\underline{\pi}} - \rho_0 \dfrac{\partial \psi^r}{\partial \alpha_i}$;
- The relaxed free energy $\psi^r(\underline{\alpha})$, a *convex* function *in the broad sense* (ductility);
- The function $\underline{\underline{e}}^{an}(\underline{a})$, which I tentatively call 'kinematic scheme';
- The evolution laws governing the a_i's. They derive from a *kinematic dissipation potential*:

(15) $$\dot{\underline{a}} = \frac{\partial}{\partial A_{\underline{a}}}\widetilde{D}(A_{\underline{a}}; A_{\underline{b},}\, \underline{\alpha}, T),$$

where $\widetilde{D}$ is a *differentiable, convex* function of the A_i's, *minimal in a broad sense* when these last vanish. Dissipativity of the kinematic mechanisms: $\dot{\underline{a}} \cdot A_{\underline{a}} \geq 0$ is implied, viscoplastic mechanisms allowed by the loose minimum. Anelastic volume preservation requires from $\widetilde{D}$ the further condition: $\partial J^{an}(\underline{a}) / \partial \underline{a} \cdot \partial \widetilde{D} / \partial A_{\underline{a}} = 0$, for any $\underline{A}, \underline{\alpha}$;

- The evolution laws governing the b_i's, of the form: $\dot{\underline{b}} = \underline{r}_b(A_{\underline{a}}; A_{\underline{b},}\, \underline{\alpha}, T)$, involved in the dissipation inequality: $A_{\underline{a}} \cdot \partial \widetilde{D} / \partial A_{\underline{a}} + A_{\underline{b}} \cdot \underline{r}_b \geq 0$.

4.3 Flow rule

The evolution law for the anelastic deformation, or flow rule, is obtained by substituting for the evolution laws into $\partial \underline{\underline{e}}^{an} / \partial \underline{a} \cdot \dot{\underline{a}}$. When performing this operation, an important property of the model easily follows from the thermodynamic forces being *linear* on the stress, namely:

(16) $$\dot{\underline{\underline{e}}}^{an} = \frac{\partial}{\partial \underline{\underline{\pi}}}\Omega(\underline{\underline{\pi}}; \underline{\alpha}, T).$$

Here, the scalar Ω results from substituting for the kinematic forces from the internal state laws into $\widetilde{D}$. The *flow potential* Ω, or *anelastic potential*, is a *convex* function of $\underline{\underline{\pi}}$. It takes on the same sequence of values as the kinematic dissipation potential along the motion. The property applies to viscoelastic and elastoviscoplastic constitutive laws as well. For the latter, it was first established in [Rice, 1971], on more restrictive assumptions.

In case they are of viscoplastic type and provided the evolution laws satisfy conditions which make them degenerate into plasticity when the control variable $\underline{\underline{\pi}}$ accompanies the yield surface motion, the anelastic potential becomes the *plastic potential* and the property the *principle of maximum work* [Hill, 1950].

5. FINAL REMARKS

The frame adopted has proved useful for modelling viscoelastic behaviour at little cost – witness the non-linear extension to the thermorheologically simple model. Such immediate outcomes seem improbable in elastoviscoplasticity, due to the complex interweaving of geometry and physics. Existing applications to the metal, with the plastic deformation being *the* kinematic variable, to the single-crystal (for a lagrangian approach, see *e.g.* [Teodosiu *et al.*, 1976; Brun, 1992]), with the non symmetric plastic transformation tensor playing the same role would be here out of place.

REFERENCES

Biot, M.A., 1954; Theory of stress-strain relations in anisotropic viscoelasticity and relaxation phenomena, *J. App. Phys.* **25**, pp.1385-1391

Bridgman, P.W., 1943; *The Nature of Thermodynamics*, Harvard University Press

Brun, L.C.J., 1992; *Energétique et Modélisation du Comportement Thermomécanique (une Introduction)*, Cours de l'Ecole polytechnique

Eckart, C., 1948; The Thermodynamics of irreversible processes. IV. The theory of elasticity and anelasticity, *Phys. Rev.*, **73**, pp. 373-382

Germain, P., 1986; *Mécanique*, Tome II ; Ellipses, Ed.; ISBN 2-7298-8654-0

Hill, R., 1950; *The Mathematical Theory of Plasticity*, Oxford

Kratochvil, J. and **Dillon, O.W.**, 1970; Thermodynamics of elastic viscoplastic materials, *J. App. Phys.*, **41**, pp.1470-1479

Mandel, J., 1966; *Cours de Mécanique des Milieux Continus*, Tome II, Gauthier-Villars

Meixner, J., a. 1954; Thermodynamische Theorie der elastischen Relaxation, Z. Naturforschg. **9a**, , pp. 654-665; b. 1972; The fundamental inequality in thermodynamics, *Physica*, **59**, pp.305-313

Rice, J.R., 1971; Inelastic constitutive relations for solids : an internal variable theory and its application to metal plasticity, *J. Mech. Phys. Solids*, **19**, 433-455

Sidoroff., F., a. 1974, Un modèle viscoélastique non linéaire avec configuration intermédiaire, *J. de Mec.*, **13,** pp.679-713; b. 1975, Variables internes en viscoélasticité, *J. de Mec.*, **14**, pp. 546-566 (part I); pp.572-595 (part II)

Staverman, A.J. and **Schwarzl, F.**, 1952; Thermodynamics of viscoelastic behaviour (model theory), *Proc. Roy. Ac. Sc. Amsterdam*, **55**, pp. 474-485

Teodosiu, C. and **Sidoroff., F.**, 1976, A theory of finite elastoviscoplasticity of single crystals, *Int. J. Engng. Sc.*, **14**, pp.165-176

Valanis, K.C., 1968; The viscoelastic potential and its thermodynamic foundations; *J. Math. and Phys.* **47**, pp. 262-275

Woods, L.C., 1981; Thermodynamic inequalities in continuum mechanics; *IMA J. App. Math.* **29**, pp. 221-246

On viscous fluid flow near a moving crack tip

Huy Duong Bui [1,2], *Corinne Guyon and Bernadette Thomas* [2]
(1) Laboratoire de Mécanique des Solides, CNRS UMR 7649, Ecole Polytechnique, 91128 Palaiseau, France
(2) Division Recherche & Développement, Electricité de France, Clamart

Abstract : We consider a crack partially filled with a fluid. We show that the presence of a lag avoids the appearance of pressure and velocity singularities. For the static equilibrium, we recall the previous result on the Capillary Stress Intensity Factor which provides a purely mechanical explanation of the Rehbinder effect, according to which the toughness of the material can be lowered by humidity. For the steady state propagation of a crack due to viscous fluid flow, we set the coupled system of non-linear equations.

Keywords:

1. INTRODUCTION

Most applications of hydraulic fracturing of rocks are found in the petroleum industry. To increase the out-flow of petroleum, one generates a multiple fracture in the rock-oil by injecting a pressurized fluid in the well. This problem involves the analysis of a fluid-crack interaction, for initiation as well as for crack propagation. Most works known in the literature are related to the Boussinesq model of 1D-fluid flow inside the crack [Abe et al., 1976], [Bui et al., 1982], [Balueva et al., 1985]. One-dimensional flow models are justified on almost all the crack length because of the smallness of the crack opening displacement compared to the crack size. However it leads to some difficulties in the vicinity of the crack tip. For example, the pressure in the 1D flow model is unbounded at the tip. Therefore, to avoid a pressure singularity, one cannot ignore the two-dimensional nature of flow near the crack tip, [Bui et al., 1982]. There is another possibility for having a finite pressure, by assuming that a lag exists between the fluid and the crack tip [Advani et al., 1997], [Garash et al., 1999].

The existence of a lag is a necessity for some physical reasons. As discussed in [Bui, 1996], there is some physical and/or geometrical incompatibility near the crack tip zone when the fluid completely fills the crack. This can be illustrated by an example. Let us consider the following constants of a rock : Toughness ($K_{IC}=10^6\ Nm^{-3/2}$), Poisson's

G.A. Maugin et al. (eds.), Continuum Thermomechanics, 63-74.
© 2000 *Kluwer Academic Publishers.*

ratio (ν=0.3), Young's modulus (E=5. 10^{10} Nm^{-2}). Typically, the radius of curvature of the crack opening displacement, corresponding to the critical toughness K_{Ic} is given by $R_c=4K_{Ic}^2(1-\nu^2)^2/\pi E^2$ or $R_c=5.10^{-10}$m = 0.5nm, i.e the size of two water molecules. Therefore, from the physical point of view, a continuous description of the fluid-solid may be questionable in models without a lag between the fluid and the crack tip.

The second reason is purely geometrical. Suppose that the incompressible viscous fluid completely fills the opening crack and adheres to the solid wall. The no-slip condition for fluid particles in contact with both crack surfaces implies singular velocity and singular resistant hydrodynamic forces on the crack tip region, precluding its propagation.

Some experimental results confirmed the lag assumption near the crack tip. This is schematically shown in Figure 1, adapted from [Van Dam et al., 1999], representing a hydraulic fracture in plaster, after splitting the sample over the fracture surface. One can observe that the fluid front and the crack front are clearly separated by a lag, which is about one tenth of the crack radius.

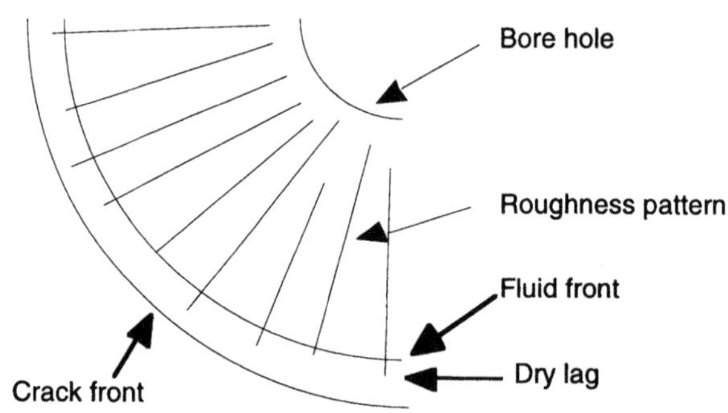

Figure 1. Surface roughness pattern and dry lag at the tip of a hydraulic fracture (after Van Dam et al, 1999).

In this paper we will discuss qualitatively different kinds of incompatibility, physical or geometrical, and inconsistencies which exist near the crack due to simplified assumptions introduced by models. We shall set up general equations for investigating the particular case of Newtonian incompressible non-wetting fluid flow near the moving crack tip, in the presence of surface forces.

2. WHAT MODELS OF FLUID-FILLED CRACKS?

Coupled problems of fluid-filled cracks can be found either in elasticity or in poro-elasticity. In a poro-elastic cracked medium, the system of equations are provided by Biot's consolidation theory for the porous solid. The crack is treated as a cavity, with a sink or a source located inside the crack. The fluid flow in the crack is then analysed by one-dimensional Boussinesq's approximation. The whole crack faces are permeable and the fluid can be assumed to occupy the whole crack and to be subjected to the difference between the pore pressure of the crack and the fluid pressure in the medium. No particular difficulties arise in the formulations and modelings of coupled hydro-poro-elasticity systems even if the equations are more complicate than those in hydro-elasticity, [Balueva et al., 1985].

When the crack surfaces are unpermeable, the viscous fluid flow is parallel to the crack. Some specific difficulties arise in the modelling of the coupled elastic-hydrodynamic problems because there exists a moving interface S between the fluid and the vacuum near the crack tip, where surface forces are present. If the flow is almost one-dimensional in the main part of the crack, it becomes two-dimensional near the crack tip zone. With a lag, in both cases of the Griffith crack and the Barenblatt crack, the fluid rolls on the crack surface and can adhere to the solid at points B, B' far away from the fluid stagnation point and the opening crack tip. Very high velocity gradient and high hydrodynamic stresses are then expected at the interface region S, of high curvature, where surface force can not be neglected. Another difficulty arises in the modeling of the common line between the fluid interface and the solid (the triple point in 2D).

2.1 **The equilibrium Griffith crack**

A fluid-filled Griffith crack in an infinite elastic medium, subjected to a remote tension σ_{22}^{∞} and a pore pressure p, has been investigated in [Bui, 1996], with a lag and a surface tension force. The case of a non-wetting fluid-solid system, defined by a surface tension coefficient α and the contact angle θ at the triple point FSV (fluid-solid-vapor or fluid-solid-vacuum) less than π has been worked out. The contact angle θ at the triple point, Figure 2, the slope of the crack opening displacement $-\partial u_2/\partial x_1$ and the angle ψ of the convex meniscus are related by

$$(1) \qquad \theta = -\partial u_2/\partial x_1 + \psi + \pi/2.$$

Neglecting the slope $-\partial u_2/\partial x_1$ of the crack opening displacement, we take

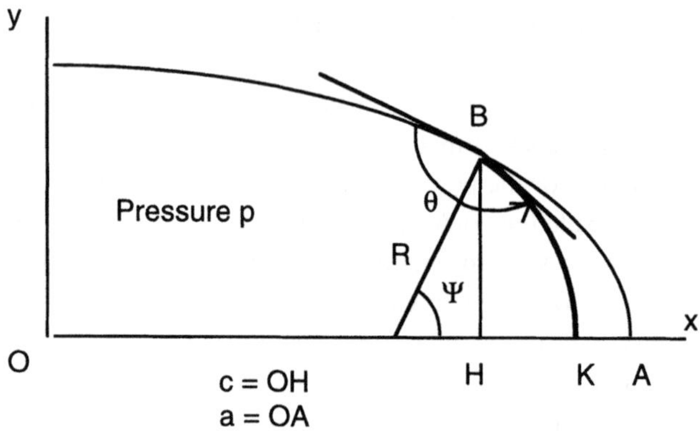

Fig. 2. The equilibrium of a fluid-filled crack of size 2a, with the meniscus BK

(2) $\theta \approx \psi + \pi/2.$

This approximate equation makes it possible to derive a closed form solution of the coupled systems of equations, given in [Bui, 1996]. The wet zone size 2c (< 2a the crack length) is found to be a function of p. It is found that for vanishing remote tension, $\sigma^{\infty}_{22}=0$, a wetting crack $c(p)\neq 0$ exists for a well-defined pore pressure p which gives rise to an initial stress-intensity factor,

$$K_I^0(c) = \frac{2p(c)}{\pi} \arcsin\left(\frac{c}{a}\right) (\pi a)^{1/2}.$$

This stress-intensity factor is due solely to the existence of a wet zone in the crack and is refered to as the Capillary Stress Intensity Factor (CSIF). The plot of CSIF versus the wet zone size c is shown in Figure 3. It increases as the wet zone size increases up to a maximum value $c_{max}<a$.

This simple model explains why humidy can lower the toughness of materials significantly, $K_{Ic}(wet) \leq K_{Ic}(dry)$. This is nothing but the well-known phenomenon called Rehbinder's effect. The explanation provided here by the CSIF, which lowers the apparent critical stress intensity factor of wet materials is purely mechanical. It differs from classical arguments based upon Van der Vaals's interaction forces and chemical in teractions between water molecules and atoms. In the classical interpretation of Rehbin-

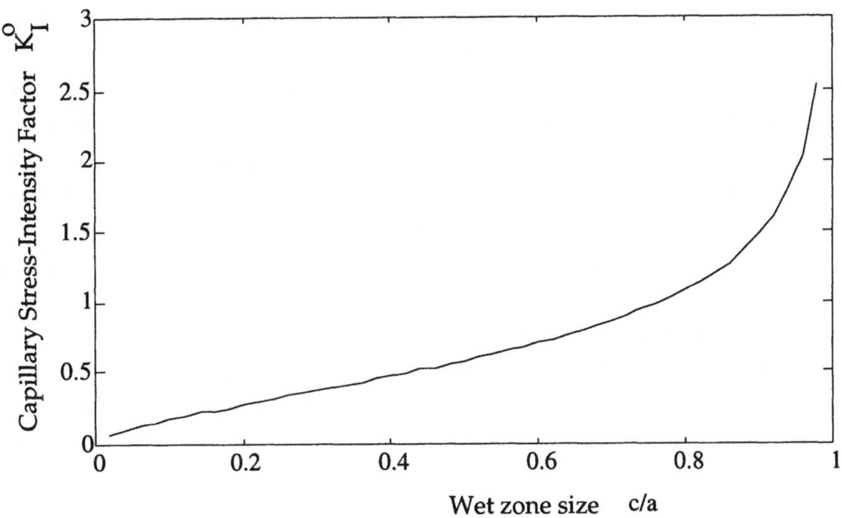

Figure 3. Capillary stress-intensity factor versus the wet zone size, after [Bui, 1996]

der's effect, a water molecule can penetrate the crack tip cavity in glass because of its size smaller than R_c. A link is broken between the atoms Si and O of ..Si-OH; another one is broken in H-OH of water molecule. Then the two hydroxyls OH make new chemical bonds with the silicium atoms giving two separate -Si-OH and resulting in a new cavity or an additional crack length, [Michalske et al., 1988]. This mechanism requires less energy than a direct separation of the bond of --Si-O-. The result is a separation of atoms in glass. Water molecule is small compared to methanol or aniline. As a matter of fact, the chemical influence of bigger molecules in the toughness of materials is negligible since there is no room for interactions at the atomic level. In this case, the lowering of toughess by the CSIF is one of the possibility.

There are approximations inherent to classical models of surface tension. Due to surface tension force on the meniscus S, the solid deforms as a dimple at the FSV point. In 2D elasticity, the displacement due to point force should be logarithmically singular. However the ratio of surface force acting on S to the pressure acting on the wet zone is of the order $O(v(c)/a) \ll 1$. Hence the effect on the stress intensity factor due to surface force acting on the dimples is *negligible*, compared to the effect due to pressure. Perhaps the model is inconsistent regarding the appearance of a *local* singularity at the triple points, but it gives a good idea of *global* physical quantities such as the stress-intensity factors etc. The deficiency of the theory about the behavior near the triple point is simply due to the approximations used in the model, such as linear elasticity, simple theory of surface force etc. A better description of the triple point zone may be possible with more elaborate models (non linear elasticity, second gradient theory of fluid/solid interaction). But it is difficult to know a priori what kind of model to use for a better modeling of specific quantities encountered in the coupled systems fluid-solid.

2.1 The moving triple point

A more elaborate model of the triple point has been provided in [Seppecher, 1989] where the second gradient theory is used. The theory introduces a third order stress tensor s, symmetric in its first two indices, generalizing the power of internal forces in a fluid domain Ω as follows

$$P_{int} = -\int_{\Omega} [\sigma : \nabla\mathbf{v} + s \therefore \nabla\nabla\mathbf{v}]\, d\mathbf{x} \tag{3}$$

By relaxing the incompressibility condition, he succeeded in analysing the structure of the density field $\rho(\mathbf{x})$ in the vicinity of the stationary triple point. The iso-density lines have a constant contact angle, outside the triple point zone, Figure 4.

Such a refined model, when extended to the moving interface, is expected to provide better informations about the structure of the triple point.

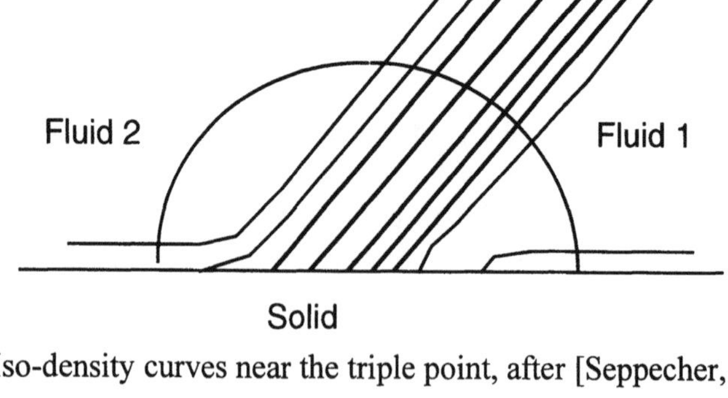

Figure 4. Iso-density curves near the triple point, after [Seppecher, 1989].

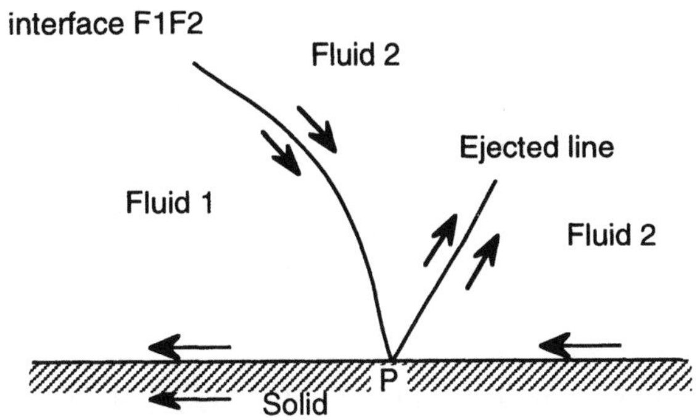

Figure 5 Structure of flow near the moving interface Fluid-Fluid-Solid. (after [Dussan et al., 1974].)

Analyses of the moving interface fluid 1-fluid 2 along a solid wall have been worked out theoretically and experimentally by many investigators, [Moffath, 1964], [Dussan et al., 1974], [Durbin, 1989]. For example fluid 1 is the water injected in the well and fluid 2 is the oil extracted from the crack tip. This system is very intricate even outside the common line, as is shown in Figure 5, adapted from [Dussan et al., 1974]. Fluid 1 is rolling *on* the rigid wall while fluid 2 is rolling *off* at the triple point. Since the fluid points of the moving interface F_1F_2 map to the triple point P at certain angle, an *ejected* line of fluid 2 is predicted (and observed experimentally) when rolling off the solid wall, going inside the domain F_2. There exists a multi-valued velocity singularity at the triple point due to model assumptions: incompressibility, Newtonian fluid and no-slip condition. It is observed experimentally that the no-slip condition does not mean that fluid particles can never leave the solid at some instant. The adherence condition is stronger than the no-slip condition because it states that fluid particles can never leave the solid wall.

In [Durbin, 1989] the no-slip condition is relaxed by introducing a slip zone to be determined, with a constant shear. Durbin's boundary condition avoids the appearance of singularity in the fluid, just as the Barenblatt assumption of constant yield stress eliminates the stress singularity in linear fracture mechanics.

For large surface tension or small capillary number $c_a=V\mu/\alpha$, the constant contact angle θ may be a good approximation. Some arguments are given in [Dussan et al., 1974] for a rapid change of the tangent to the interface F_1F_2 near the wall, suggesting that the fluid rolls on the solid wall smoothly and that the contact angle is constant just outside the contact line zone. Thus the structure of the fluid-fluid interface line near the contact line seems to be very complex, since it depends very much on the boundary condition used. In this discussion, we do not consider the deformation of the solid at the triple point which can modify the flow.

Can the structure of flow near the contact line be predicted by a refined theory ? Is such a theory a necessity or not, regarding the question of determining the macroscopic behaviour of the fluid-crack system ? These are still open problems.

3. A SIMPLIFIED MODEL OF FLUID-FILLED PROPAGATING CRACK

We consider a steady state propagation of a crack partially filled by an incompressible viscous fluid. The absolute crack velocity is V. The geometry of the coupled system in the moving frame is shown in Figure 6. Since the crack is propagating (inertial force neglected), the geometry of the crack opening displacement in the vicinity of the crack tip is well-defined by the toughness K_{Ic} of the solid.

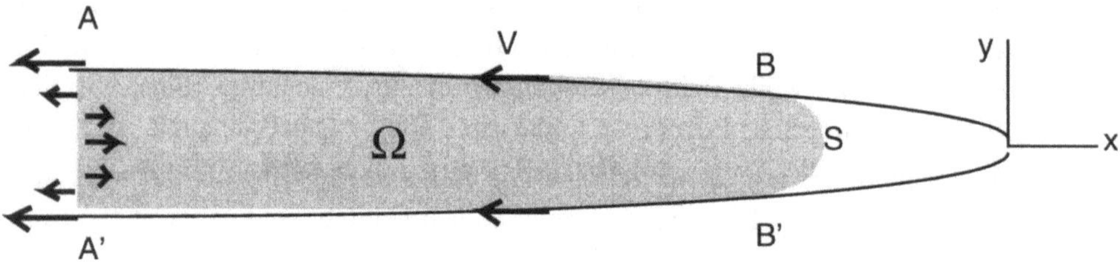

Figure 6. The fluid-filled moving crack with a convex meniscus

The fluid is separated from the crack tip by a lag (a vacuum), with the surface tension α, the contact angle θ. The boundary condition at the section AA' is given by a parabolic profile velocity, with a vanishing total flow rate and the velocity at the wall equal to $-V$. The equations of the fluid-solid system are summarized below:

- Navier-Stokes equations :

(4) $\quad \operatorname{div} \mathbf{v} = 0, \quad \text{in } \Omega,$

(5) $\quad (\mathbf{v}\,.\mathrm{grad})\,\mathbf{v} + \dfrac{1}{\rho}\ \mathrm{grad}\ p - \eta\,\Delta\,\mathbf{v} = 0, \quad \text{in } \Omega,$

(where η is the kinematic viscosity, p the pressure, ρ the constant density)

- Capillary force:

(6) $p = \alpha \operatorname{div} \mathbf{n}$ on S ($\mathbf{n}$ is the unit outward normal to S),

(7) $\mathbf{v} \cdot \mathbf{n} = 0$ on S,

- In-flow velocity (2h is the crack opening at A'A)

(8) $v_1 = \frac{3V}{2}\left(1 - \frac{x_2^2}{h^2}\right) - V, \qquad v_2 = 0$ on A'A,

- Boundary conditions at the wall:

(9) $v_1 = -V, \quad v_2 = 0$ on AB and A'B',

- Parabolic crack profile:

(10) $x_1 = -x_2^2/2R_o$, with $R_o = 4K_{Ic}^2(1-\nu^2)^2/\pi E^2$,

- Global equilibrium equation :

(11) $K_{Ic} = f(p, \sigma_{22}^{\infty})$, ($\sigma_{22}^{\infty}$: remote applied stress).

Equation (11) expresses the global relation between the stress-intensity factor in the symmetric mode I, the applied loads on the cracked body and the pressure p on the crack, without the point force at FSV. It corresponds to the matching between the outer expansion solution with the asymptotic inner solution defined by Eqs (4) - (10). Such a matching can only be made when the solution of (4)-(10) is obtained for any K_{Ic} and p.

For low velocity, small crack opening and high viscosity, the Reynolds number $\mathfrak{R}$ is small and the Stokes approximation may be justified. The above equations are still non-linear, even if the linearized fluid equations are used, because the contact points B, B' are yet unknown and the pressure is coupled to the curvature of the meniscus.

4. NUMERICAL RESULTS

In numerical computations, the value $\theta=2\pi$ of the contact angle has been used, in order to avoid either the appearance of velocity singularity at the triple point (no-slip condition) or the consideration of a slip zone between the fluid and the solid (Durbin's boundary condition). This restriction is not essential since the magnitude of the point force at FSV is small in comparison with the total force due to pressure acting on the wet zone. The point force and the velocity singularity should have a little effect on the overall behavior of the fluid-solid system. Instead of solving the complete nonlinear system of equations, we are searching an approximate solution of Equations (4), (5), (7), (8), (9), (10), for a fluid domain having an elliptical geometry and for given triple point positions. Since the pressure and the radius of the meniscus are practically constant near the stagnation point, the equation (6) is satisfied approximately, a posteriori, by adjusting some constant α. This inverse fitting applies also to (11) where the toughness is assumed to be related to the meniscus radius R_0 and to the loading.

The non-linear problem, without the Stokes approximation, is solved by an iterative procedure. The computations are carried out, using the finite element method provided by the non-linear computer code N3S of Electricité de France (3D Navier-Stokes).

Numerical results are shown in Figure 7 for the stream lines and the pressure, which decreases while approaching the crack tip.

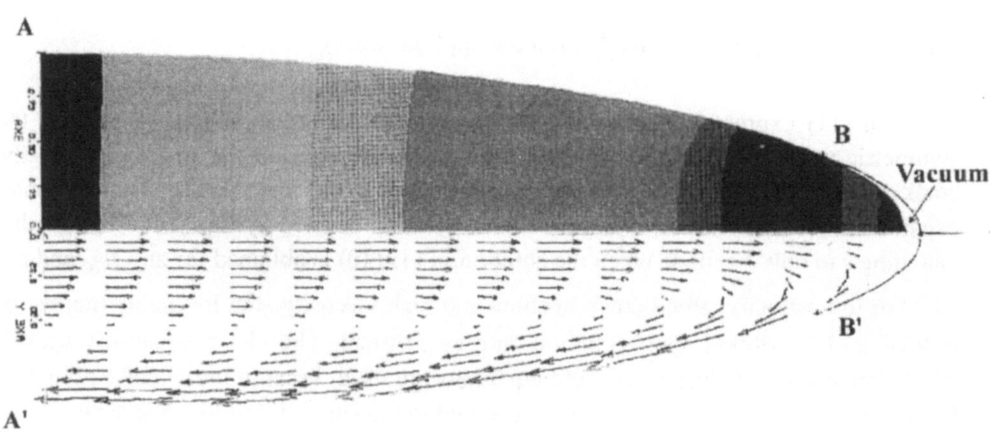

Figure 7. Pressure and stream lines near the crack tip

The stream lines patterns show clearly an interfacial motion similar to the moving tractor tread described in [de Gennes, 1985] .

5. CONCLUSIONS

1. In studying the coupling between a crack and a fluid, it seems reasonable to consider only *microscopic* surface force phenomena having a significant effect on the *macroscopic* response of the cracked solid. One can neglect microscopic details with short range effects such as the dimple, the singular velocity at the triple point.
2. The pressure which is related to the curvature of the meniscus has long range effects because it applies to the crack surface.
3. Simplified theory of surface force is used for investigating the main features of the solution.

6. REFERENCES

[Abe et al., 1976] Abe, T.; Mura, T.; Keer, L.M.; Growth rate of a renny-shaped crack in hydraulic fracturing of rock; In: *J. Geophys. Res*., p. 5325; **81**, n° 29.

[Advani et al., 1997] Advani, S.; Lee, T.; Dean, R.; Pak, C.; Avasthi, J.; Similarity solution of a semi-infinite fluid-driven fracture in a linear elastic solid ; In: *Int. J. Numer. Anal. Meth. Geomech*, p. 229-240; **21**.

[Balueva et al., 1985] A.V. Baluaeva, A.V.; Zazovskii, A.F.; Elastic-hydrodynamic problem of inflow of fluid in a crack in a porous medium ; In: *Izv. AN SSSR. Mekhanika Tverdogo Tela*, pp. 157-166; **20**, n° 5.

[Bui et al., 1982] Bui, H.D.; Parnès, R.; A reexamination of the pressure at the tip of a fluid-filled crack; In: *Int. J Engng. Sci*., p. 1215-1220; **20**, n° 11.

[Bui , 1996] Bui, H.D.; Interaction between the Griffith crack and a fluid: theory of Rehbinder's effect; In: FRACTURE*: A Topical Encyclopedia of Current Knowledge*; Ed. Cherepanov, G.P.; p. 99-105; Chap. 6.

[Conca et al., 1994] Conca, C.; Planchard, J.; Thomas, B.; Vaninathan, M.; Problèmes mathématiques en couplage fluide-structure; *Collection des Etudes & Recherches*; **85**, Eyrolles, Paris, 1994.

[De Gennes, 1985] De Gennes, P.G.; Wetting: statics and dynamics. In: *Reviews of Modern Physics*, p. 827; **57**, 3, I.

[Durbin, 1989] Durbin, P.A. ; Stokes flow near a moving contact line with yield -stress boundary condition ; In : *Quart. J. Mech. & Math.*, p. 99-113; **42**.

[Dussan et al., 1974] Dussan, E.B.; Davis, S.H.; On the motion of a fluid-fluid interface along a solid surface; In: *J. Fluid. Mech.*, p. 71; **65**, I.

[Garash et al., 1999] Garash, D.; Detournay, E.; Similarity solution of a semi- infinite fluid-driven fracture in a linear elastic solid; Private Comm. (to appear).

[Michalske et al., 1988] Michalske, T.; Bunker, B.; La fracture du verre ; In: *Pour la Science*, p.52-59; Février 1988.

[Moffatt, 1964] Moffatt, H.K.; Viscous and resistive eddies near a sharp corner; In: *J. Fluid. Mech.*, p. 1-18; **18**.

[Monla, 1990] Monla, M. ; Sur la détermination de la forme d'une bulle en ascension dans un liquide visqueux; *Thèse de l'Ecole Centrale de Lyon*; n° 90-08.

[Rehbinder, et al. 1957] Rehbinder, P.; Lichtman, V.; Effect of surface active media on strain and rupture of solid ; In: *Proc. 2nd Int. Cong. Surf. Act.*, p. 563; **3**.

[Seppecher, 1989] Seppecher, P.; Etudes des conditions aux limites en théorie du second gradient: cas de la capillarité; In: *Comptes Rendus Acad. Sci. Paris*, p. 497-502; **II 309**.

[Van Dam et al., 1999] Van Dam, C.J.; De Pater, C.J.; Romijn, R.; Reopening of dynamic fractures in laboratory experiments; In: *Cong. Int. Mech. Rocks*, Paris August 1999, p. 792-794; **II**.

Elastoviscoplasticity with aging in aluminium alloys

*G. Cailletaud****, *C. Depoid**, *D. Massinon***, *E. Nicouleau-Bourles****
* *Renault, Rueil*, ** *Montupet, Nogent*
*** *Centre des Matériaux de l'Ecole des Mines de Paris / UMR CNRS 7633*
BP 87 - 91003 Evry France
Georges.Cailletaud@mat.ensmp.fr

Abstract: After two decades of intense activity on constitutive equations and integration algorithms in the research laboratories, news methods are now available for engineers who need numerical tools for the design of critical components working at high temperature. Due to the increase of computer power, this approach (computation of the actual stress-strain state, then creep–fatigue damage modeling) should progressively replace the classical design method, based on elastic computations followed by plastic corrections, and reduce the number of prototypes, thus the duration of the development. The present paper shows an application of the method to the cylinder head of car engines.
Keywords: Elastoviscoplasticity, thermomechanical fatigue, aging, aluminium alloy, F.E. structural calculation.

1. INTRODUCTION

The thermodynamic of irreversible processes [Ger73, GNS83] is a powerful tool which allowed both famous theoretical developments, and a pragmatic approach to the modeling of the materials. The courses taught by Paul Germain in the seventies [Ger75] can easily be seen as an invitation to develop new material models in the laboratories and export them to industrial applications. The purpose of the present paper relates to the second part of *Fifty Years of Continuum Thermomechanics*, illustrating the development of the material modeling in the eighties and of the numerical algorithms in the last decade. The achievement of this cycle is now a new generation of Finite Element (F.E.) codes available for industrial use.

The class of constitutive equations considered in the paper introduces internal variables to define the actual state of the material. It has been successfuly applied to the modeling of viscoplastic behaviour, including cyclic thermomechanical loadings. The present paper is devoted to the modeling of aging, with an application to an aluminium alloy for a cylinder head, AS7U3G (ASTM 319). This material is classically used, due to its low density, good thermal conductivity, and good casting properties. Nevertheless it is also very sensitive to high temperature exposure, and microstructural evolution drastically alters mechanical properties. The first section is then devoted to the description of the cyclic viscoplastic behaviour including aging effects. The second section recalls the principles of the numerical implementation in a finite element code. The third section shows the calibration of the material parameters and the validation of the model on simple anisothermal tests. F.E. calculations of a Diesel cylinder head are finally shown in the fourth section.

G.A. Maugin et al. (eds.), Continuum Thermomechanics, 75-86.
© 2000 *Kluwer Academic Publishers.*

2. CONSTITUTIVE EQUATIONS

Due to microstructural evolutions, many materials may exhibit strong variations of their mechanical properties. Phase change is the most dramatical event (for instance austenite-martensite transformation), and many theories are still in development to represent their effect under complex temperature or load histories. The present paper is concerned by precipitation and coalescence phenomena in a copper bearing aluminium alloy, affecting the hardening mechanisms, then the mechanical properties. Examples of such studies can be found in the literature. In Ni-Base superalloys, transient overheating periods produce very large hardening due to the dissolution of large precipitates followed by a fine precipitation [CC79]. The case of aluminium alloys has also been studied in the past, for instance aging at room temperature of a AU4G1 alloy [Mar89], or hardening-softening effect under temperature exposure of the same material [CEP95]. More recently, the thermomechanical fatigue of AS7U3G alloy has been studied from a mechanical and microstructural point of view [SMS+99].

2.1 Experimental facts

After the classical work by Guinier [Gui56], the metallurgy of Al-Cu alloys has been studied in details. This aspect is omitted here for the sake of brevity. The material of the study has been submitted to a T5 treatment (5 hours at 210°C). The subsequent evolution of the microstructure during operation corresponds to the sequence:

$$\theta'' \rightarrow \theta' \rightarrow \theta(Al_2Cu)$$

The evolution takes place for temperatures higher than 200°C , and the final product depends on maximum temperature [ECG98]. The successive transformations produce larger and larger precipitates, so that the matrix hardening becomes lower and lower.

As shown in fig.1, the effect of the transformations can be characterized either by mechanical testing (fig.1a) or by means of hardness measurements (fig.1b). The stress-strain loops of figure 1a are obtained at 250°C , in three strain controlled tests at $\Delta\varepsilon = 0.8\%$. The reference test is denoted T5, and corresponds to the initial state. After aging periods at 250°C or 320°C , the maximum stress during the test at 250°C is only 100MPa or 60MPa instead of 155MPa. On each loop, the center of the yield domain is marked by a star: a large softening can be seen on isotropic hardening, but kinematic hardening is affected too, the total amount of softening representing more than 150% of the initial stress for aging at 320°C .

The purpose of figure 1b is to illustrate the evolution of hardening versus time. Hardness measurements were found to be correlated with yield stress values, and give a good view of the transformation rate. They are performed at room temperature, during aging periods at various temperatures. The measurements confirm that the maximum softening is reached for aging at 320°C . On the other hand, a two-level test at 200°C then 320°C clearly demonstrates that the asymptotic value obtained during the preheating at 200°C is "forgotten" by the material at the second level, and that the asymptotic value at 320°C is reached. Other tests [NB99] show that, in a complex temperature history, the final asymptotic value depends first on the maximum temperature.

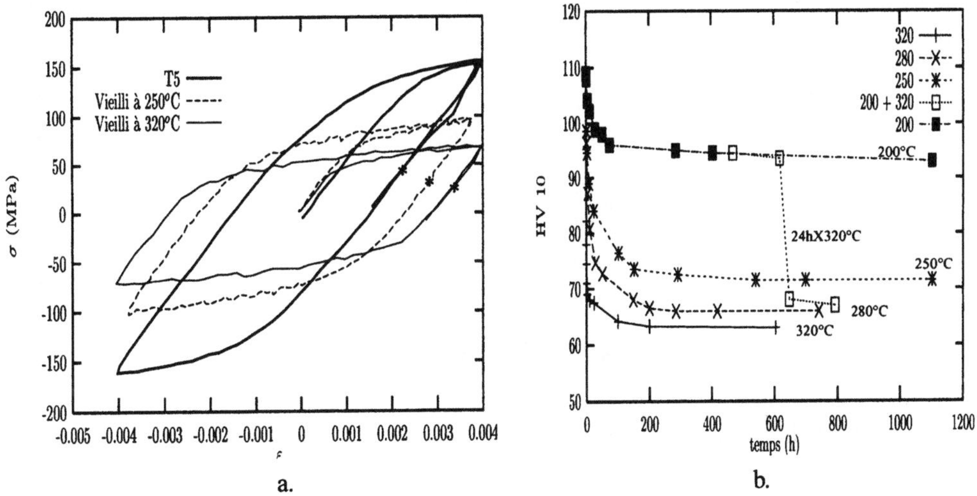

Figure 1: Effect of aging on mechanical properties: (a) Evolution of Vickers hardness for various temperature histories; (b) Stress–strain loops after several aging periods

2.2 Thermodynamical framework

The models previously proposed to represent the behavior of aluminium alloys were designed to represent hardening[Mar89], or hardening and softening[CEP95]. In the present study, the model has to represent softening only. A very simple formulation is then chosen, having in mind that the computational efficiency is a key point for industrial use. Aging is represented by a scalar variable a, starting from zero, and tending to an asymptotic value a_∞, depending on temperature. The model is written using the small perturbation formalism, and assuming the partition between elastic and viscoplastic strains. The thermoelastic part of the specific free energy takes its classical form [LC87], which is not reproduced here; it is assumed to be decoupled from the viscoplastic+aging part Ψ_p, which is expressed as

$$\rho\Psi_p = \frac{1}{3} C \underset{\sim}{\alpha} : \underset{\sim}{\alpha} + \frac{1}{2} bQr^2 + \frac{L}{2}\left(a_\infty - a\right)^2 \qquad (1)$$

where C, b, Q, L and a_∞ are material coefficients, $\underset{\sim}{\alpha}$ is the state variable for kinematic hardening, and r the state variable for isotropic hardening. At initial state, there is no hardening, and the asymptotic value a_∞ is also equal to zero, provided the temperature is below the threshold which produces a microstructural evolution. On the other hand, when temperature increases, the free energy increases also, and the microstructure of the material element must change to recover a more stable shape. The state equations define the kinematic variable $\underset{\sim}{X}$, the isotropic variable R and the energy Z associated with aging:

$$\underset{\sim}{X} = \frac{2}{3} C \underset{\sim}{\alpha} \quad ; \quad R = bQr \quad ; \quad Z = -L((a_\infty - a) \qquad (2)$$

The evolution of viscoplastic flow and of the hardening variables is defined in the framework of standard models, using the successive definitions and notations:

$$\text{yield function f}: f = J(\underset{\sim}{\sigma} - \underset{\sim}{\mathbf{X}}) - R - k \tag{3}$$

$$\text{with}: J(\underset{\sim}{\mathbf{A}}) = \left((3/2)\underset{\sim}{\mathbf{A}}^d : \underset{\sim}{\mathbf{A}}^d\right)^{0.5} \quad \text{and } \underset{\sim}{\mathbf{A}}^d \text{ deviatoric part of } \underset{\sim}{\mathbf{A}} \tag{4}$$

$$\text{viscoplastic flow: } \underset{\sim}{\dot{\varepsilon}}^p = \dot{p}\,\frac{\partial f}{\partial \underset{\sim}{\sigma}} = \dot{p}\,\underset{\sim}{\mathbf{n}} \tag{5}$$

$$\text{with}: \dot{p} = \left\langle \frac{f}{K} \right\rangle^n \tag{6}$$

$$\text{kinematic hardening: } \underset{\sim}{\dot{\alpha}} = \dot{p}\left(\underset{\sim}{\mathbf{n}} - \frac{3D}{2C}\underset{\sim}{\mathbf{X}}\right) \tag{7}$$

$$\text{isotropic hardening: } \dot{r} = \dot{p}\left(1 - \frac{R}{Q}\right) \tag{8}$$

The evolution of the aging variable linearly depends on Z, the aging phenomenon being active only when the actual value of Z is negative (that is aging is lower than the current asymptotic value):

$$\dot{a} = \left\langle \frac{-Z}{L_T} \right\rangle \tag{9}$$

Mechanical and aging terms are both present in the intrinsic dissipation Φ:

$$\Phi = \sigma : \underset{\sim}{\dot{\varepsilon}}^p - \underset{\sim}{\mathbf{X}} : \underset{\sim}{\dot{\alpha}} - R\dot{p} - Z\dot{a} \tag{10}$$

$$= (J(\underset{\sim}{\sigma} - \underset{\sim}{\mathbf{X}}) - R)\dot{p} + \frac{3D}{2C}\underset{\sim}{\mathbf{X}} : \underset{\sim}{\mathbf{X}}\,\dot{p} + \frac{R^2}{Q}\dot{p} + \frac{Z^2}{L_T}\dot{a} \tag{11}$$

Denoting by σ_v the viscous stress ($\sigma_v = J(\underset{\sim}{\sigma} - \underset{\sim}{\mathbf{X}}) - R - k$), we can decompose the first term in eqn.11 into a viscous part $\sigma_v\dot{p}$ and a friction part $k\dot{p}$. The second and the third terms represent the dissipation associated with the nonlinearity of kinematic and isotropic hardening terms. The last term represents the contribution of aging to the dissipation. Each of these terms is positive, so that the dissipation is always positive.

In the previous equations, the material coefficients defining aging depend on temperature, and the material coefficients in the purely mechanical part depend on temperature and aging. An explicit expression is chosen for $k(a, T)$:

$$k = R_0(T) + R_0^*(T)(1 - a) \tag{12}$$

Two kinematic variables are introduced, only one of them depending on aging:

$$\underset{\sim}{\mathbf{X}} = \underset{\sim}{\mathbf{X}}_1 + \underset{\sim}{\mathbf{X}}_2 \tag{13}$$

so that the relation between the kinematic hardening variable and the corresponding state variable in equation 2 uses $C \equiv C_1(T)$ and $D \equiv D_1(T)$ for the first variable, $C \equiv$

$(1-a)C_2(T)$ and $D \equiv D_2(T)$ for the second one:

$$\underset{\sim}{\mathbf{X}}_1 = \frac{2}{3} C_1(T) \underset{\sim}{\alpha}_1 \quad ; \quad \underset{\sim}{\mathbf{X}}_2 = \frac{2}{3} C_2(T)(1-a) \underset{\sim}{\alpha}_2 \tag{14}$$

Note that this new dependence introduces an additional term $-(1/3)C_2 \underset{\sim}{\alpha}_2 : \underset{\sim}{\alpha}_2$ in the expression of Z (equation 2), nevertheless the corresponding term in $\dot{a}$ (equation 9), $(C_2/3L\tau)\underset{\sim}{\alpha}_2 : \underset{\sim}{\alpha}_2$ can be made very small for large values of L, and will be neglected. The problem could be reconsidered if a coupling between hardening and aging was present in the experiments.

In a tension test, the tension curve is defined by σ_I for the initial material ($a = 0$), and σ_∞ for a fully aged material ($a = 1$):

$$\sigma_\infty(\varepsilon^p, \dot{\varepsilon}^p) = (C_1/D_1)(1 - \exp(-D_1\varepsilon^p)) + Q(1 - \exp(-b\varepsilon^p) + R_0 + K\,(\dot{\varepsilon}^p)^{1/n} \tag{15}$$

$$\sigma_I(\varepsilon^p, \dot{\varepsilon}^p) = \sigma_\infty(\varepsilon^p, \dot{\varepsilon}^p) + (C_2/D_2)(1 - \exp(-D_2\varepsilon^p)) + R_0^* \tag{16}$$

3. NUMERICAL IMPLEMENTATION

The key point for a correct numerical implementation valid for non isothermal loadings is to integrate the state variables, and not the hardening variables [CC96]. The model is implemented in the F.E. system Z-set[BF98], using the user interface for the development of constitutive equations Z-front[BLFC98]. In a finite element code, the section concerning constitutive equations should be carefully kept independent from the "global" system, and can be treated at the level of the integration point. The purpose of the integration box at this level is only to start with the actual value of the state variables, the increment of external prescribed parameters (like temperature in a mechanical computation) and the strain increment, and to provide the increment of state variables, the updated stress value, and the jacobian matrix $\partial\underset{\sim}{\sigma}/\partial\underset{\sim}{\varepsilon}$ to evaluate the consistent stiffness matrix. In the present case, the user interface is specially devoted to the implementation of models using internal variables, offering a number of "building bricks" to the developer. Several keywords are available:

- the name of the material coefficients;
- the name of the internal variables;
- the section concerning the strain partition, defining the actual stress state from the elastic strain;
- the section defining the explicit shape of the model, that is the value of the derivative of the state variables;
- the (optional) section defining the jacobian matrix for a possible solution of the global system.

In the present case, the variables are the elastic strain $\underset{\sim}{\varepsilon}^e$, the two kinematic state variables $\underset{\sim}{\alpha}_1$ and $\underset{\sim}{\alpha}_2$, the isotropic state variable r (or the acumulated viscoplastic strain p), and the aging variable a. Note that, the evolution of aging being independent of the mechanical variables, its history could also be obtained by a post-processing of the thermal computation, then aging would be considered as an "external parameter" for the mechanical computation. Due to the low amount of computation needed to compute this evolution, the direct computation of a during the mechanical calculation, which avoids an additional step in the numerical procedure, is the best solution. Two integration strategies can be chosen in the code: Runge–Kutta integration with automatic time step definition, and a θ–method. In the first case, only the explicit expression of the derivatives has to be introduced in the software: for the second one, a Newton method is used to solve the following implicit system in terms of $\Delta\underset{\sim}{\epsilon}^e$, $\Delta\alpha_1$, $\Delta\alpha_2$, Δp, Δa, for given values of $\Delta\underset{\sim}{\epsilon}$ (increment of mechanical strain) and Δt:

$$\underset{\sim}{\mathbf{R}}_e = \Delta\underset{\sim}{\epsilon}^e - \Delta\underset{\sim}{\epsilon} + \Delta p\, \underset{\sim}{\mathbf{n}} = 0 \tag{17}$$

$$\underset{\sim}{\mathbf{R}}_{a1} = \Delta\alpha_1 - (\underset{\sim}{\mathbf{n}} - D_1\underset{\sim}{\alpha}_1)\Delta p = 0 \tag{18}$$

$$\underset{\sim}{\mathbf{R}}_{a2} = \Delta\alpha_2 - (\underset{\sim}{\mathbf{n}} - D_2\underset{\sim}{\alpha}_2)\Delta p = 0 \tag{19}$$

$$R_v = K\left(\frac{\Delta p}{\Delta t}\right)^{1/n} - \langle f\rangle = 0 \tag{20}$$

$$R_a = \Delta a - \left\langle \frac{a_\infty - a}{\tau} \right\rangle \Delta t = 0 \tag{21}$$

In the previous equations, variables like α_i are expressed as $\alpha_i(t) + \theta\Delta\alpha_i$, and quantities like f and $\underset{\sim}{\mathbf{n}}$ are also expressed at the intermediate time $t + \theta\Delta t$, so that the final system may be quite complex. Note that time independent plasticity can be obtained as a limiting case provided $K = 0$ in equation 20.

4. IDENTIFICATION AND VALIDATION TESTS

4.1 Identification on isothermal tests

The data base needed to model the behavior of the real component covers a temperature range from room temperature to 320°C. LCF tests are made until failure at constant strain rate, and more complex histories, including several blocks of cycles at different temperatures are also performed [NBEMMC99]. The cyclic hardening curve is obtained either by incremental tests, with increasing strain range, or by a direct testing at a given level. Due to aging, the incremental curves present a temperature history effect, the larger strain ranges producing a lower maximum stress due to the longer temperature exposure. This effect can be included in the modeling. The identification procedure consists in two steps. In the first step, all the temperature levels are considered independently; the results obtained in this case usually produce a set of coefficients which does not present a consistent temperature dependence. A second step is then applied to finally get a smooth variation of each coefficient versus temperature.

A typical simulation result is reproduced in figure 2, which shows a good agreement with the experiment hysteresis loops at 320°C, in a test at $\Delta\varepsilon = 0.8\%$, for the initial state, an intermediate value of aging, and for fully aged material.

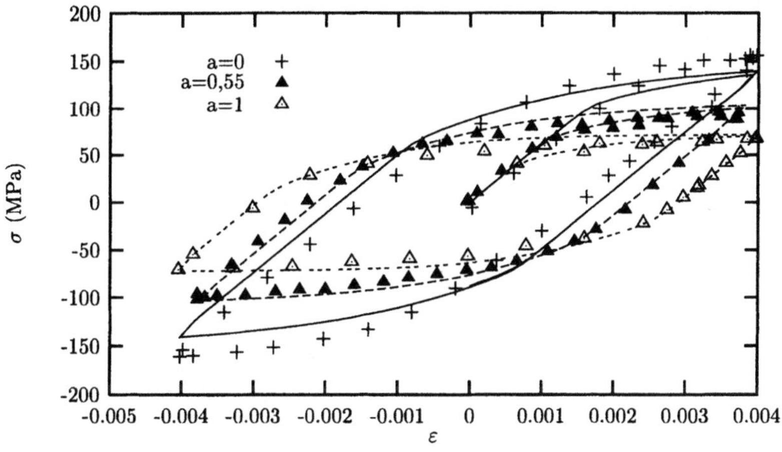

Figure 2:

4.2 Validation on anisothermal tests

Having in hand a good modeling of isothermal loading, the next step consists in checking that the non isothermal behavior is also correctly represented. This will be the case if there is no temperature history effect, so that the behavior of the material at temperature T during a temperature transient is equivalent to the behavior in an isothermal test at the same temperature, provided the proper value of the aging variable is introduced. The assumption is then that the scalar variable a takes into account all the possible temperature histories starting from T5 state.

Tests at variable temperature can classically be obtained on hydraulic machines, with special procedures in order to impose thermal cycling and mechanical cycling. In the present study, an alternative solution is proposed, adapting a thermomechanical fatigue (TMF) testing device initially developed at ONERA [CCK81]. As shown by figure 3, the specimen is a classical LCF specimen, heated by means of a radiation furnace. The sample is constrained by three springs, and the mechanical load is produced only by the constrained thermal expansion. In the hot part of the cycle, compressive stresses are present, producing a compressive viscoplastic flow, so that the sample is in tension during the subsequent cooling period. Three columns, with an adjustable cylinders, provide an unilateral contact during tensile load. The test is run without any servohydraulic machine: it can then be used for very long tests, but it is fully instrumented, with an extensometer on the specimen, and strain gages on the springs to get the stress value.

The simulation can be made on a material element, using either the experimental strain or the experimental stress in the center of the specimen to check the constitutive equation. A full modeling can also be performed, with a Finite Element computation of the specimen (axisymmetric conditions), taking into account the axial temperature gradient and the stiffnesses of the springs and of the columns, with the unilateral contact condition [NB99]. An example of such a simulation is given in figure 4. The prescribed temperature cycle is 70°C–320°C. The duration of the transient periods is 1 min, and the hold time is 3 minutes. It can be seen that the width of the hysteresis loop, corresponding to the viscoplastic strain range, increases during the test, due to aging effect.

Figure 3: View of the TMF device

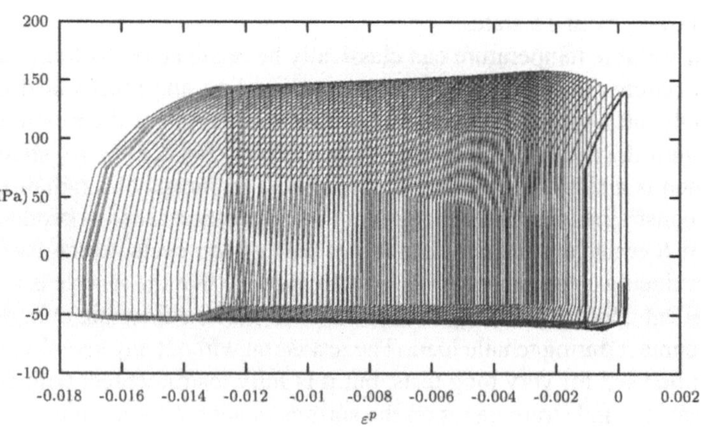

Figure 4: A typical FE simulation of the stress–strain loop on the TMF device

5. APPLICATION IN STRUCTURAL COMPUTATIONS

Having in mind an industrial use of the model, the target CPU time for one computation should typically be about one night, that is about 15 hours. Performing many cycles with a large number of degrees of freedom is nowadays out of reach with classical computers. This is why parallel computation is needed. The solver Zébulon of Z-set code has been parallelised [FCR98] using a subdomain decomposition technique, with the FETI method [FR94] to reconnect the different parts of the mesh. With such a method, the CPU time for the solution of the global problem if proportional to the number of processors. It has to be noted that the method remains efficient even if many processors are used.

5.1 Stress and strain fields

The model is now applied to the cylinder head of a Renault engine. The calculations have been made on the IBM-SP2 of Ecole des Mines de Paris, using 7 processors. The mesh, shown in figure 5a, with the subdomain decomposition, represents one half of the engine head, and has more than 70000 degrees of freedom. The boundary conditions were taken from a preliminary elastic computation on the whole engine, and correspond to prescribed displacements on specific node sets. The temperature fields are also prescribed. The installation of the seats is simulated first, in order to get the initial residual stresses, then the aging+mechanical computation is performed on 5 cycles. The duration of the computation is about 10 hours per cycle. Three calculations have been made, one with accelerated aging, in order to reach the saturated value after 5 cycles, named "asymptotic aging"; one with the properties of the initial material; one with an "overaged" material ($a = 1$).

Since the maximum temperature in this test is only 250°C on aluminium, the maximum aging value (fig.5b) is 0.55. The aging distribution clearly relates to maximum temperature; note that it would not be reasonnable to avoid aging calculation, as long as some areas keep the mechanical properties of the initial material, and other ones are submitted to aging. The von Mises stress contours confirm the location of the most loaded zone. It can be checked [NB99] that the larger principal stress is perpendicular to the axis between admission and exhaustion valves, and that the stress level is maximum in the chamber. The maximum stress value is 160MPa. It is quite different in the two other computations (200MPa for the initial material, 110MPa for "overaged" material), proving again that aging effect must be taken into account.

5.2 Life prediction

The life prediction is performed using a classical creep–fatigue interaction model [LC87]. The number of cycles obtained corresponds to the initiation of a short crack of 0.2–0.5mm. The value obtained (600 cycles, fig.6a) is consistent with tests made on the engine, with very agressive temperature and regime cycles. The equations used are recalled in fig.6b. This approach does not take into account the physical aspect of the crack initiation mechanism, which should be related with a microcrack growth from porosities and casting defects, nevertheless it can be considered as a good engineering method.

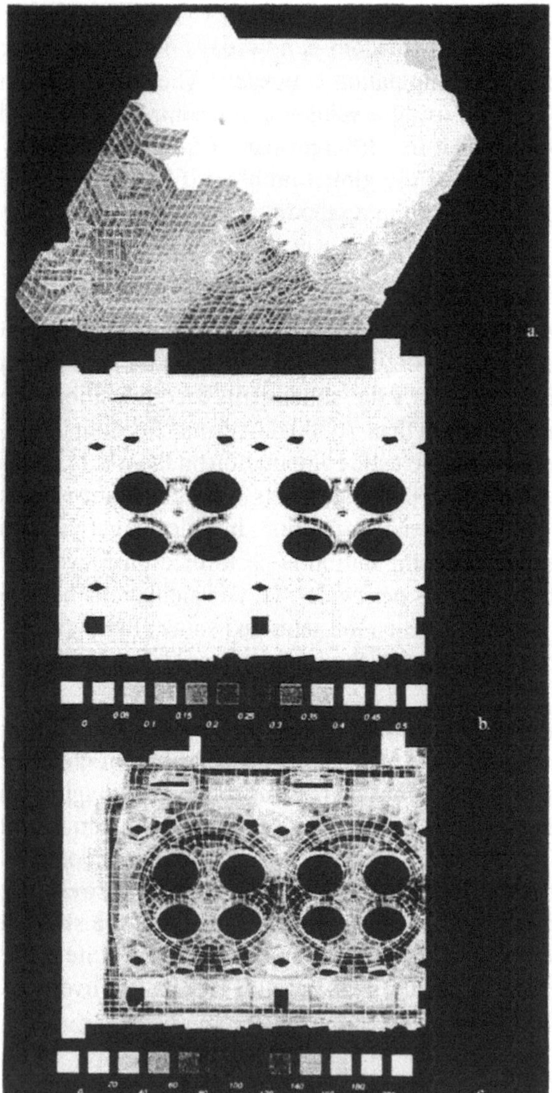

Figure 5: Computation of a cylinder head. (a) View of the mesh (b) Aging contour (c) von Mises stress at maximum temperature

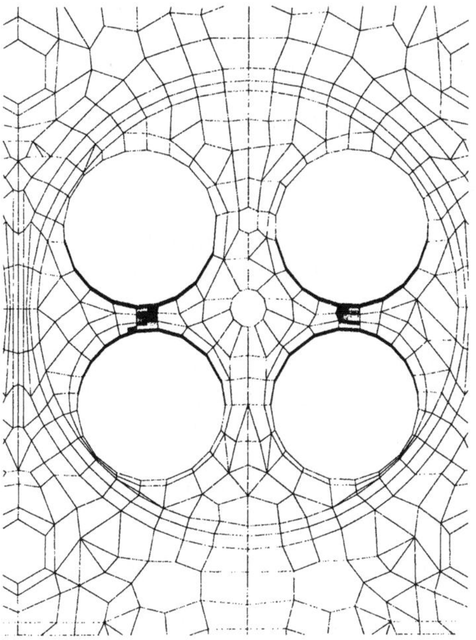

$N_R = 600$ cycles

a.

N_C computed on a period P :

$$C = \frac{1}{N_C} = \frac{1}{\kappa+1} \int_P \left(\frac{\chi(\underset{\sim}{\sigma})}{A} \right)^r dt$$

A, κ, r are material coefficients,

$\chi(\underset{\sim}{\sigma})$ = linear combination of von Mises, eigen stress and hydrostatic pressure

N_F computed for one cycle :

$$F = \frac{1}{N_F} = (\beta + 1)(1 - \alpha) \left(\frac{\Delta J^*/2}{M^*} \right)^\beta$$

$$\text{with } 1 - \alpha = \gamma \left\langle \frac{\Delta J^*/2 - \sigma_l^*}{1 - S_{eq}^*} \right\rangle$$

$\gamma, \beta, \sigma_l^*, M^*$ are material coefficients,

ΔJ^* and S_{eq}^* are stress amplitude and maximum equivalent stress normalized by the actual (aging dependent) ultimate stress

N_R obtained by damage cumulation:

$$(1 - D_0)^{\kappa+1} - (1 - D_1)^{\kappa+1} = C$$

$$\left(1 - (1 - D_1)^{\beta+1}\right)^{1-\alpha} -$$

$$\left(1 - (1 - D_2)^{\beta+1}\right)^{1-\alpha} = F$$

$$D_0 = D_2, \text{ etc} \ldots$$

b.

Figure 6: Prediction of the number of cycles to crack initiation. (a) Life contours, (b) Model used

References

[BF98] J. Besson and R. Foerch. Object-oriented programming applied to the finite element method, part i: General concepts. *Revue Européenne des Éléments Finis*, 7(5):535–566, 1998.

[BLFC98] J. Besson, R. Leriche, R. Foerch, and G. Cailletaud. Object-oriented programming applied to the finite element method, part ii: Application to material behaviors. *Revue Européenne des Éléments Finis*, 7(5):567–588, 1998.

[CC79] G. Cailletaud and J.-L. Chaboche. Macroscopic description of the microstructural changes induced by varying temperature : Example of in100 cyclic behaviour. In K.J. Miller and R.F. Smith, editors, *Int. Conf. on Mechanical behaviour of metals, ICM3*, pages 23–32. Cambridge, august 1979.

[CC96] G. Cailletaud and J.-L. Chaboche. Integration methods for complex constitutive equations. *Comput. Methods Appl. Mech. Engrg*, 133:125–155, 1996.

[CCK81] G. Cailletaud, J.-P. Culié, and H. Kaczmarek. Description et simulation numérique d'essais de fatigue thermique sur éprouvettes lisses. *La Recherche Aérospatiale*, 1981(2):85–97, 1981.

[CEP95] Chaboche, J.-L., El-Mayas, N., and Paulmier, P. Modélisation thermodynamique des phénomènes de viscoplasticité, restauration et vieillissement. *C. R. Acad. Sci. Paris*, 320(II):9–16, 1995.

[ECG98] C. Evrard, M. Clavel, and I. Guillot. Etude du comportement cyclique d'alliages d'aluminium de fonderie. Technical Report UTC-98/3, Univ. Tech. Compiègne, 1998.

[FCR98] F. Feyel, G. Cailletaud, and F.-X. Roux. Application of parallel computation to material models with a large number of internal variables. *Revue Européenne des Éléments Finis*, 7(1-2-3):55–72, 1998.

[FR94] C. Fahrat and F.X. Roux. Implicit parallel processing in structural mechanics. *Computational Mechanics Advances*, 2(1):1–101, 1994.

[Ger73] P. Germain. *Cours de Mécanique des Milieux Continus*. Masson, 1973.

[Ger75] P. Germain. Dea de mécanique appliquée à la construction. Technical report, Univ. Paris VI, 1975.

[GNS83] Germain, P., Nguyen, Q.S., and Suquet, P. Continuum thermodynamics. *ASME J. of Appl. Mech.*, 50:1010–1020, 1983.

[Gui56] A. Guinier. Précipitation et vieillissement. *Sc. et Tech. de l'ingénieur*, M1(240):1–9, 1956.

[LC87] J. Lemaitre and J.L. Chaboche. *Mécanique des Matériaux Solides*. Dunod, 1987.

[Mar89] D. Marquis. *Phénoménologie et thermodynamique : couplage entre thermoélasticité, plasticité, vieillissement et endommagement*. PhD thesis, Univ. Paris VI, 1989.

[NB99] E. Nicouleau-Bourles. *Etude expérimentale et numérique du vieillissement d'un alliage d'aluminium. Application aux culasses automobiles*. PhD thesis, École Nationale Supérieure des Mines de Paris, 1999.

[NBEMMC99] E. Nicouleau-Bourles, N. El-Mayas, D. Massinon, and G. Cailletaud. Thermomechanical fatigue of aluminium alloys: experimental study and numerical simulation. In J.J. Skrypek and R.B. Hetnarski, editors, *Thermal Stresses '99*, pages 241–244. Cracow, june 1999.

[SMS+99] Smith, T.J., Maier, H.J., Sehitoglu, H., Fleury, E., and Allison, J. Modeling high-temperature stress-strain behavior of cast aluminium alloys. *Metall. and Mat. Trans.*, 30A:133–146, 1999.

The application of the irreversible thermodynamics to the development of constitutive equations

J.L. Chaboche
ONERA/DMSE, 29, Av de la Division Leclerc
BP72, F-92322 Châtillon Cedex, France
Tel: 33 1 46 73 46 64, Fax: 33 1 46 73 48 91, e-mail: chaboche@onera.fr

16th February 2000

Abstract: The thermodynamics of irreversible processes, using the notion of local state and internal state variables, is used for developing consistent constitutive and damage equations. The Generalized Standard Models are recalled and some of the induced limitations are discussed, both for non linear kinematic hardening in viscoplasticity and for elasto-plastic damage couplings. In order to release some of these constraints, a pseudo-standard approach is formulated and discussed, that allows us to recover most of the classically used constitutive equations and damage models.
Keywords: Thermodynamics - Internal variables - Constitutive equations - Damage Mechanics - Viscoplasticity

1 INTRODUCTION

The use of the framework given by the thermodynamics of irreversible processes for developing constitutive and damage equations has been an important progress in the past 25 years. Since the pioneering theoretical works [Carathéodory, 1909], [Truesdell and Noll, 1965], [Coleman and Gurtin, 1967], the development of a continuum thermodynamic framework , within a French school initiated by Germain [Germain, 1973], [Germain, 1974], contributed to propose better and more consistent inelastic and damage models of solid materials and structures. In particular, the use of internal variables and the notion of Generalized Standard Models (GSM) [Halphen and Nguyen, 1975] offer several significant advantages : (i) - some limitations regarding the too wide space of possibilities for choosing material models, (ii) - a guideline for incorporating temperature effects in the constitutive equations, (iii) - the direct expression of energy dissipated as heat during irreversible processes, an important aspect when dealing with coupled thermomechanical situations, (iv) - stability properties of the corresponding boundary value problems.

However, there are some situations where the GSM approach introduces too stiff constraints, so that there is need for a slightly softer definition of the generalized normality rules. In the present paper, we discuss some of the corresponding

G.A. Maugin et al. (eds.), Continuum Thermomechanics, 87-100.
© 2000 *Kluwer Academic Publishers.*

limitations in the degrees of freedom, both for constitutive equations and for damage models. Then we push towards a slightly less restrictive approach, called here the *Pseudo Standard Models*, using several independent dissipation potentials and several associated independent scalar multipliers. It was used implicitly in several theories [Benallal, 1989], [Chow and Wei, 1991], [Hansen and Schreyer, 1992], [Zhu and Cescotto, 1995] and is more general in the sense of less restrictive conditions leading to larger modeling capabilities.

2 THE CLASSICAL FRAMEWORK

2.1 The state law

We assume the small perturbation quasistatic theory, with the additive decomposition of total strain :

$$\underset{\sim}{\varepsilon} = \underset{\sim}{\varepsilon}_{e\theta} + \underset{\sim}{\varepsilon}_{p} = \underset{\sim}{\varepsilon}_{e} + \underset{\sim}{\varepsilon}_{\theta} + \underset{\sim}{\varepsilon}_{p}, \tag{1}$$

where $\underset{\sim}{\varepsilon}_{e\theta}$ is the thermoelastic strain, $\underset{\sim}{\varepsilon}_{\theta}$ is the thermal expansion and $\underset{\sim}{\varepsilon}_{p}$ is the plastic or viscoplastic strain. We follow the classical framework of Continuum Thermodynamics of Irreversible Processes (CTIP) based on internal variables (that store the effect of the whole history of observable variables), [Germain, 1974], [Sidoroff, 1975]. We use also the axiom of *locality*, the state of a material point (surrounded by an infinitesimal volume element) being considered as independent of the state at neighboring points. In other words, we do not consider any of the *material lengths* that are introduced in Generalized Continua.

As irreversible processes we consider inelasticity (plasticity, viscoplasticity), associated hardening/recovery processes, aging effects corresponding to metallurgical changes, and damage effects. Classical viscoelasticity is not discussed in itself, even if it could be considered as a limit case of viscoplasticity [Chaboche, 1997].

The present state of the material is characterized by the knowledge of observable variables, total strain $\underset{\sim}{\varepsilon}$; temperature T; and by the internal state variables associated with inelasticity and hardening, denoted a_j; the damage variables, denoted d_k, and a set of variables that define the evolution of the non-damage domain, called b_l. We do not consider the thermal gradient $\underline{\nabla}\mathbf{T}$ in the state potential, consistently with the local state restriction.

The observable variable is the total strain, but for nonviscous elastic strain, provided we use the strain partition (1), valid for small strains, it is sufficient to consider the thermoelastic strain as the independent observable variable. We choose the Helmholtz free energy as the thermodynamic potential : $\psi = u - TS$, where u is the internal energy, S the entropy, and ψ is assumed to depend on all the independent state variables : $\psi = \psi\left(\underset{\sim}{\varepsilon}_{e\theta},\, a_j,\, d_k,\, b_l,\, T\right)$. Taking into account the conservation of energy, we reduce the second principle of thermodynamics to the Clausius-Duhem inequality :

$$\underset{\sim}{\sigma} : \dot{\underset{\sim}{\varepsilon}} - \dot{\psi} - \mathcal{S}\dot{T} - \frac{1}{T}\underline{\mathbf{q}}.\underline{\nabla}\mathbf{T} \geq 0 . \tag{2}$$

From a classical treatment [Coleman and Gurtin, 1967], we obtain the laws of state :

$$\underset{\sim}{\sigma} = \frac{\partial \psi}{\partial \underset{\sim}{\varepsilon}_{e\theta}} = \frac{\partial \psi}{\partial \underset{\sim}{\varepsilon}}, \qquad \mathcal{S} = -\frac{\partial \psi}{\partial T}, \tag{3}$$

and by analogy, we set the following expressions for the conjugate thermodynamic forces (or affinities) associated with the internal variables :

$$A_j = \frac{\partial \psi}{\partial a_j}, \qquad y_k = -\frac{\partial \psi}{\partial d_k}, \qquad B_l = \frac{\partial \psi}{\partial b_l}. \tag{4}$$

The choice of the signs in (4) is arbitrary. We set a minus sign for the affinities y_k because damage decreases strength. The remaining Clausius-Duhem inequality is assumed to be separated into the intrinsic and thermal dissipation inequalities [Truesdell and Noll, 1965] :

$$\underset{\sim}{\sigma} : \dot{\underset{\sim}{\varepsilon}}_p - A_j \dot{a}_j + y_k \dot{d}_k - B_l \dot{b}_l \geq 0 \qquad -\frac{1}{T}\underline{\mathbf{q}}.\underline{\nabla}\mathbf{T} \geq 0 \tag{5}$$

2.2 The Generalized Standard Model

In standard associated plasticity, Hill's postulate of maximum dissipation, written in the stress space, leads to the quasi-convexity of the yield surface, and defines the direction of plastic strain rate as the outer normal to the yield surface. The same applies for viscoplasticity, leading to the normality to the equipotential surface.

Considering thermoviscoplasticity, we can generalize such a postulate in order to check the second principle automatically. Constitutive theories obeying such a principle are called *Generalized Standard Models* [Halphen and Nguyen, 1975]. In the dissipation inequality (5) we observe the duality between rates of dissipative fluxes and the corresponding thermodynamic forces. Following Germain [Germain, 1973], [Germain et al., 1983], we assume the existence of a pseudo-potential of dissipations : $\phi = \phi\left(\dot{\underset{\sim}{\varepsilon}}_p, -\dot{a}_j, \dot{d}_k, -\dot{b}_l\right)$. Here we do not incorporate the heat flux, which has already been taken into account, leading to Fourier's Law. ϕ is assumed to be positive, convex in its variables, and to contain the origin. The thermodynamic forces are obtained as follows :

$$\underset{\sim}{\sigma} = \frac{\partial \phi}{\partial \dot{\underset{\sim}{\varepsilon}}_p}, \qquad A_j = -\frac{\partial \phi}{\partial \dot{a}_j}, \qquad y_k = \frac{\partial \phi}{\partial \dot{d}_k}, \qquad B_l = -\frac{\partial \phi}{\partial \dot{b}_l}. \tag{6}$$

Applying to ϕ the Legendre-Fenschel transform, $\phi^*(A) = \min_{\dot{a}}(A\,\dot{a} - \phi(\dot{a}))$, we obtain the complementary dissipative potential ϕ^* in the generalized space of thermodynamic forces. Equations (6) are then replaced by

$$\dot{\underset{\sim}{\varepsilon}}_p = \frac{\partial \phi^*}{\partial \underset{\sim}{\sigma}}, \qquad \dot{a}_j = -\frac{\partial \phi^*}{\partial A_j}, \qquad \dot{d}_k = \frac{\partial \phi^*}{\partial y_k}, \qquad \dot{b}_l = -\frac{\partial \phi^*}{\partial B_l}. \tag{7}$$

Such a generalized normality confers interesting properties on the constitutive equations. In particular, the second principle is automatically satisfied, provided ϕ^* is positive and convex. Using also $\phi^*(0,0,0,0) = 0$, we can rewrite the intrinsic dissipation (5-a) as

$$\underset{\sim}{\sigma} : \dot{\underset{\sim}{\varepsilon}}_p - A_j\dot{a}_j + y_k\dot{d}_k - B_l\dot{b}_l = \underset{\sim}{\sigma} : \frac{\partial \phi^*}{\partial \underset{\sim}{\sigma}} + A_j\frac{\partial \phi^*}{\partial A_j} + y_k\frac{\partial \phi^*}{\partial y_k} + B_l\frac{\partial \phi^*}{\partial B_l} \geq \phi^* \geq 0\,. \tag{8}$$

Remark : From the GSM, in the simple case where we neglect damage and recovery effects (both dynamic and static recovery), it can be easily shown that the state variables associated to the hardening (in a unified viscoplastic framework) should be respectively the following [Chaboche, 1996] :

- the plastic strain itself, associated with the back-stress (kinematic hardening),
- the accumulated plastic strain, $\dot{p} = ||\dot{\underset{\sim}{\varepsilon}}_p||$, associated with the isotropic hardening through the yield stress evolution,
- the square root of the accumulated viscoplastic work $\int \sigma_v dp$ for the isotropic hardening that uses the drag stress evolution (viscoplasticity).

For that, let us assume the isotropic case, a von Mises yield function and the simplest viscoplastic behavior, using a power function for the viscoplastic potential. It leads to

$$\phi^* = \frac{D}{n+1}\left\langle \frac{||\underset{\sim}{\sigma} - \underset{\sim}{\mathbf{X}}|| - R - k}{D} \right\rangle^{n+1} = \frac{D}{n+1}\left\langle \frac{\sigma_v}{D} \right\rangle^{n+1}, \tag{9}$$

where the back stress $\underset{\sim}{\mathbf{X}}$, the yield stress increase R, and the drag stress D are the thermodynamic forces associated with some internal variables denoted respectively $\underset{\sim}{\alpha}$, r, δ.; The parameter k is the initial yield stress. The generalized normality indicates that

$$\dot{\underset{\sim}{\varepsilon}}_p = \frac{\partial \phi^*}{\partial \underset{\sim}{\sigma}} = \frac{3}{2}\dot{p}\,\frac{\underset{\sim}{\sigma}' - \underset{\sim}{\mathbf{X}}'}{||\underset{\sim}{\sigma} - \underset{\sim}{\mathbf{X}}||}, \qquad \text{with} \quad \dot{p} = \left\langle \frac{\sigma_v}{D} \right\rangle^n, \tag{10}$$

$$\dot{\underset{\sim}{\alpha}} = -\frac{\partial \phi^*}{\partial \underset{\sim}{\mathbf{X}}} = \dot{\underset{\sim}{\varepsilon}}_p, \qquad \dot{r} = -\frac{\partial \phi^*}{\partial R} = \dot{p}, \qquad \dot{\delta} = -\frac{\partial \phi^*}{\partial D} = \frac{n}{n+1}\frac{\sigma_v \dot{p}}{D}, \tag{11}$$

leading to the announced results (if the drag stress D is assumed to be proportional to the variable δ).

3 SOME OVERRESTRICTIVE CONSEQUENCIES OF **GSM**

3.1 Non linear kinematic hardening

The GSM framework is perfectly consistent with the use of linear kinematic hardening. However, introducing non linear rules in the same framework is difficult. The simple Armstrong-Frederick format, with a *dynamic recovery* term, constitutes a typical example. In this form, widely used in many constitutive models (see [Chaboche, 1989]), as the Non Linear Kinematic (NLK) hardening model, we replace (11-a) by

$$\dot{\underset{\sim}{\alpha}} = \dot{\underset{\sim}{\varepsilon}}_p - \underset{\approx}{\mathbf{Q}} : \underset{\sim}{\mathbf{X}}\,\dot{p}, \tag{12}$$

with $\underset{\sim}{\mathbf{X}} = \underset{\approx}{\mathbf{N}} : \underset{\sim}{\alpha}$, where $\underset{\approx}{\mathbf{N}}$ and $\underset{\approx}{\mathbf{Q}}$ are two fourth order positive definite tensors (the classical formulations consider scalars).

In the rate independent formulation, we can assume the indicatrix function of the convex dissipative potential to be of the form :

$$F = f + \frac{1}{2}\underset{\sim}{\mathbf{X}} : \underset{\approx}{\mathbf{Q}} : \underset{\sim}{\mathbf{X}}. \tag{13}$$

The function F is not identical with the function that defines the elastic domain. If we assume $\underset{\sim}{\mathbf{X}}$ to be the back stress we can choose f of the simplest Hill form :

$$f = ||\underset{\sim}{\sigma} - \underset{\sim}{\mathbf{X}}||_H - k, \tag{14}$$

where the norm $||\,.\,||_H$ is defined by $||\underset{\sim}{\mathbf{x}}||_H^2 = \underset{\sim}{\mathbf{x}} : \underset{\approx}{\mathbf{H}} : \underset{\sim}{\mathbf{x}}$. Here we neglect isotropic hardening. The generalized normality rule is then :

$$\dot{\underset{\sim}{\varepsilon}}_p = \dot{\lambda}\frac{\partial F}{\partial \underset{\sim}{\sigma}}, \qquad \dot{\underset{\sim}{\alpha}} = -\dot{\lambda}\frac{\partial F}{\partial \underset{\sim}{\mathbf{X}}} = \dot{\underset{\sim}{\varepsilon}}_p - \underset{\approx}{\mathbf{Q}} : \underset{\sim}{\mathbf{X}}\,\dot{\lambda}, \tag{15}$$

and it can be demonstrated that $\dot{\lambda} = \dot{p}$, which leads to the expected form (12). However, in the viscoplastic case we usually have (9), with $f \equiv \sigma_v$, then (11-a). [Chaboche, 1983], [Germain et al., 1983] (see also [Lemaître and Chaboche, 1985]) used a viscoplastic potential of the form

$$\phi_p^* = \frac{D}{n+1}\left\langle \frac{F}{D} \right\rangle^{n+1}, \tag{16}$$

with F given by

$$F = f + \frac{1}{2}\underset{\sim}{\mathbf{X}} : \underset{\approx}{\mathbf{Q}} : \underset{\sim}{\mathbf{X}} - \frac{1}{2}\underset{\sim}{\alpha} : \underset{\approx}{\mathbf{N}} : \underset{\approx}{\mathbf{Q}} : \underset{\approx}{\mathbf{N}} : \underset{\sim}{\alpha}, \tag{17}$$

in which $\underset{\sim}{\alpha}$ is considered as a parameter. In that case we still obtain :

$$\dot{\underset{\sim}{\varepsilon}}_p = \frac{\partial \phi_p^*}{\partial \underset{\sim}{\sigma}} = \dot{p}\,\underset{\sim}{\mathbf{n}}, \qquad \text{with} \quad \underset{\sim}{\mathbf{n}} = \frac{\underset{\approx}{\mathbf{H}} : (\underset{\sim}{\sigma} - \underset{\sim}{\mathbf{X}})}{||\underset{\sim}{\sigma} - \underset{\sim}{\mathbf{X}}||_H}, \tag{18}$$

$$\dot{\underset{\sim}{\alpha}} = -\frac{\partial \phi_p^*}{\partial \underset{\sim}{\mathbf{X}}} = \dot{\underset{\sim}{\varepsilon}}_p - \underset{\approx}{\mathbf{Q}} : \underset{\sim}{\mathbf{X}}\,\dot{p}. \tag{19}$$

Provided the equality $\underset{\sim}{\mathbf{X}} = \underset{\approx}{\mathbf{N}} : \underset{\sim}{\alpha}$ can be re-used in (17), respecting at each instant $F = f$, we recover the normal viscoplastic rule with

$$\dot{p} = \left\langle \frac{F}{D} \right\rangle^n = \left\langle \frac{f}{D} \right\rangle^n. \tag{20}$$

However such a treatment appears as quite artificial. To some extent it obeys GSM, using state variables as parameters [Germain et al., 1983].

3.2 The pure GSM form of the viscoplastic NLK rule

The choice made by Ladevèze [Ladevèze, 1992], within a purely GSM framework, consists in retaining (13) in the viscoplastic case, with (16) as the potential. It produces the expected NLK hardening rule (19) but at the price of a non vanishing extra term in the expression for the true elastic domain :

$$F = f + \frac{1}{2}\underset{\sim}{\mathbf{X}} : \underset{\approx}{\mathbf{Q}} : \underset{\sim}{\mathbf{X}} = ||\underset{\sim}{\sigma} - \underset{\sim}{\mathbf{X}}||_H - k^* \leq 0, \tag{21}$$

in which the true elastic limit (when isotropic hardening is neglected) is : $k^* = k - \frac{1}{2}\underset{\sim}{\mathbf{X}} : \underset{\approx}{\mathbf{Q}} : \underset{\sim}{\mathbf{X}}$. Such a coupling term between the size of the elastic domain and the back stress has a significant shortcoming. k^* in (21) must be restricted to be always positive, for obvious reasons. There are only two ways to satisfy this constraint :

- take a very large value for k. This leads to a significant modification in the mechanical response of the model compared to the reference version. Moreover it greatly increases the initial value of the elastic limit (for $\underset{\sim}{\mathbf{X}}$ around 0), which is exactly the contrary to the line to follow if we want to respect experiments and to be able to describe a very smooth elastic-plastic transition at the very beginning of plastic flow.
- accept an experimental value for k ; in that case k^* may vanish, and must be artificially maintained at 0 for situations where $1/2\,\underset{\sim}{\mathbf{X}} : \underset{\approx}{\mathbf{Q}} : \underset{\sim}{\mathbf{X}} > k$. In these situations, even if the mechanical response could be acceptable, the model is

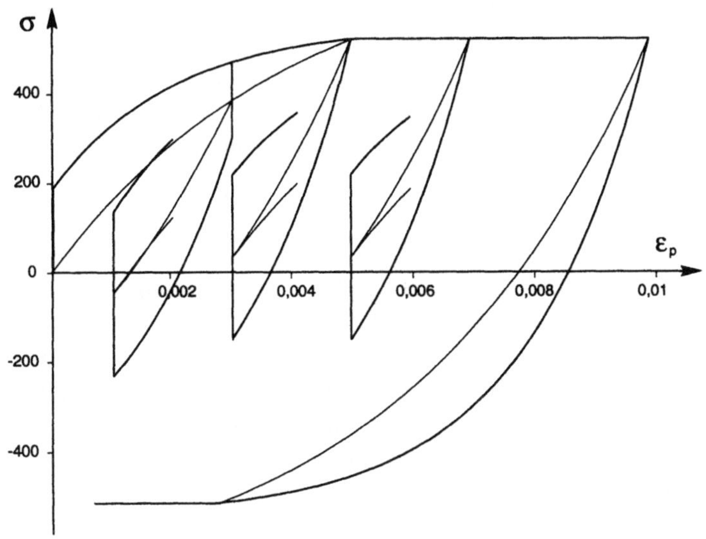

Figure 1: Application of the completely standard model of Ladevèze for various cyclic loadings (one back-stress : $C = 180000MPa$, $\gamma = 250$, $k = 185MPa$)

no longer working in the GSM framework. We need to meet an additional inequality like a flow condition :

$$g = \frac{1}{2}\underset{\sim}{\mathbf{X}} : \underset{\approx}{\mathbf{Q}} : \underset{\sim}{\mathbf{X}} - k \leq 0\,; \tag{22}$$

This limits the evolution of the back stress, which retains a constant level (given by $g = 0$). In that asymptotic situation, flow occurs without further hardening, the elastic domain being reduced to a single point (eqn. (21) with $k^* = 0$). Figure 1 illustrates the uniaxial response of the model with only one back stress. There is a discontinuity in the back-stress rate, as well as in the stress rate itself (at least when the condition $g = 0$ is attained before saturation of the back stress). Moreover, such a description, under these asymptotic conditions, does not respect GSM. Consistency imposes $\underset{\sim}{\mathbf{X}} : \underset{\approx}{\mathbf{Q}} : \dot{\underset{\sim}{\mathbf{X}}} = 0$, but GSM leads to $\dot{\underset{\sim}{\mathbf{X}}} = \underset{\approx}{\mathbf{N}} : \dot{\underset{\sim}{\varepsilon}}_p$

3.3 Damage modeling

Using the GSM approach for damage modeling introduces important restrictions on the set of possibilities. They affect both the shape of the multiaxial criterion and the capability to separate inelastic processes and damage processes.

If we consider a dissipation potential depending both on the conjugate forces associated with plasticity ($\underset{\sim}{\sigma}$, A_j) and on those associated with damage processes (y_k, B_l), we can assume the following additive separation : $\phi^* = \phi_p^*(\underset{\sim}{\sigma}, A_j\,;\, d_k) + \phi_d^*(y_k, B_l\,;\, d_k)$, where d_k plays role of parameters. In that case, the use of the generalized normality (GSM) in the rate independent situation will lead to

$$\dot{\underset{\sim}{\varepsilon}}_p = \dot{\lambda}\,\frac{\partial \phi_p^*}{\partial \underset{\sim}{\sigma}}, \qquad \dot{a}_j = -\dot{\lambda}\,\frac{\partial \phi_p^*}{\partial A_j}, \qquad \dot{d}_k = \dot{\lambda}\,\frac{\partial \phi_d^*}{\partial y_k}, \qquad \dot{b}_l = -\dot{\lambda}\,\frac{\partial \phi_d^*}{\partial B_l}, \tag{23}$$

in which we have the same plastic multiplier $\dot{\lambda}$, most often identical to the norm of the plastic strain rate. Such a writing automatically induces, as in the models developed by Lemaître [Lemaître, 1985] [Lemaître, 1992], a simultaneous evolution of damage and plastic strain, at least when the thresholds appearing in ϕ_p^* and ϕ_d^* are exceeded. Even if acceptable for ductile damage, this is certainly not true for other situations. For instance, brittle creep damage can take place without significant strain; damage in composites (or in concrete) very often develops (as microcracks) without any plastic strain.

In order to dissociate plasticity and damage evolutions, we could consider, but only for the rate dependent formulation, a dissipation potential being the sum of two non linear functions, F_p and F_d, constructed from f_p and f_d, for instance with power functions :

$$\phi^* = F_p(\underset{\sim}{\sigma}, A_j\,;\, d_k) + F_d(y_k, B_l\,;\, \underset{\sim}{\varepsilon}^e, d_k) \tag{24}$$

$$= \frac{K}{n+1}\left\langle \frac{f_p}{K} \right\rangle^{n+1} + \frac{A}{r+1}\left\langle \frac{f_d}{A} \right\rangle^{r+1}. \tag{25}$$

The rates of the state variables are then derived from this potential by the generalized normality rule :

$$\dot{\underset{\sim}{\varepsilon}}^p = \frac{\partial \phi^*}{\partial \underset{\sim}{\sigma}} = \left\langle \frac{f_p}{K} \right\rangle^n \frac{\partial f_p}{\partial \underset{\sim}{\sigma}}, \qquad \dot{a}_j = -\frac{\partial \phi^*}{\partial A_j} = -\left\langle \frac{f_p}{K} \right\rangle^n \frac{\partial f_p}{\partial A_j}, \tag{26}$$

$$\dot{d}_k = \frac{\partial \phi^*}{\partial y_k} = \left\langle \frac{f_d}{A} \right\rangle^r \frac{\partial f_d}{\partial y_k}, \qquad \dot{b}_l = -\frac{\partial \phi^*}{\partial B_l} = -\left\langle \frac{f_d}{A} \right\rangle^r \frac{\partial f_d}{\partial B_l}. \tag{27}$$

By proceeding in this way, we distinguish between the nonlinear effects contained in the power functions (with, for instance, different exponents r and n). However, this approach is restrictive as regards the form of the multiaxial damage criterion, as in the Hayhurst multiaxial criterion [Hayhurst, 1972]. In practice, it is relatively difficult to reconstruct a criterion in the space of forces $\underset{\sim}{\mathbf{y}}$ which gives the isochronous surface forms observed in creep. The only way of proceeding is then to consider that f_d also depends on the elastic strains, acting as parameters, and allowing the stress to be introduced indirectly by the elastic constitutive law. In this case, the choice of the expression of criterion f_d in the stress space is again free. This approach is purely standard, but is rather artificial.

4 A PSEUDO STANDARD MODEL

4.1 Multiple independent potentials

The use of a single global potential in the GSM leads to overrestrictive conditions. This is why we suggest the idea of a somewhat broader framework [Chaboche, 1997], in which we accept the possibility of several independent potentials, corresponding to different physical processes. In fact there is no reason why phenomena as different as thermal dissipation, plastic dissipation, damage or metallurgical evolutions should all derive from the same single potential.

To some extent the considered formulation represents a generalization of the classical formalism of plasticity with multiple criteria [Mandel, 1964] and was already implicitly used in a certain number of works [Benallal, 1989], [Hansen and Schreyer, 1992], [Zhu and Cescotto, 1995]. Leaving aside the thermal dissipation potential, we therefore define three independent potentials expressed in the thermodynamic forces space :

- Ω_p, associated with the plastic strain and related hardening processes, and depending on $(\underset{\sim}{\sigma}, A_j)$;
- Ω_s, the slow (static) microstructural evolution potential, such as the static recovery and aging (dissolution of precipitates or precipitations, for example), depending on hardening forces A_j ;
- Ω_d, the damage potential, in which damage mechanisms are considered as acting on a scale larger than those of the plastic strains. This potential Ω_d depends on (y_k, b_l).

Each potential may eventually depend also on state variables, considered as parameters. The main hypothesis of this extension is to express the generalized normality rule in the following form :

$$\dot{\underset{\sim}{\varepsilon}}_p = \dot{\lambda}_p \frac{\partial \Omega_p}{\partial \underset{\sim}{\sigma}} + \dot{\lambda}_s \frac{\partial \Omega_s}{\partial \underset{\sim}{\sigma}} + \dot{\lambda}_d \frac{\partial \Omega_d}{\partial \underset{\sim}{\sigma}}, \tag{28}$$

$$-\dot{\alpha}_j = \dot{\lambda}_p \frac{\partial \Omega_p}{\partial A_j} + \dot{\lambda}_s \frac{\partial \Omega_s}{\partial A_j} + \dot{\lambda}_d \frac{\partial \Omega_d}{\partial A_j}, \tag{29}$$

$$\dot{d}_k = \dot{\lambda}_p \frac{\partial \Omega_p}{\partial y_k} + \dot{\lambda}_s \frac{\partial \Omega_s}{\partial y_k} + \dot{\lambda}_d \frac{\partial \Omega_d}{\partial y_k}, \tag{30}$$

$$-\dot{b}_l = \dot{\lambda}_p \frac{\partial \Omega_p}{\partial B_l} + \dot{\lambda}_s \frac{\partial \Omega_s}{\partial B_l} + \dot{\lambda}_d \frac{\partial \Omega_d}{\partial B_l}. \tag{31}$$

Of course, with the dependencies considered in the potentials, certain derivatives are zero, and we get

$$\dot{\underset{\sim}{\varepsilon}}_p = \dot{\lambda}_p \frac{\partial \Omega_p}{\partial \underset{\sim}{\sigma}}, \qquad -\dot{a}_j = \dot{\lambda}_p \frac{\partial \Omega_p}{\partial A_j} + \dot{\lambda}_s \frac{\partial \Omega_s}{\partial A_j}, \tag{32}$$

$$\dot{d}_k = \dot{\lambda}_d \frac{\partial \Omega_d}{\partial y_k}, \qquad -\dot{b}_l = \dot{\lambda}_d \frac{\partial \Omega_d}{\partial B_l}. \tag{33}$$

As exemplified below, the multipliers $\dot{\lambda}_p$, $\dot{\lambda}_s$, $\dot{\lambda}_d$, can be defined independently as Lagrange multipliers [Hansen and Schreyer, 1992], [Chow and Wei, 1991]. They can depend on all variables, conjugate forces or state variables.

4.2 Example for the Unified Viscoplasticity

In order to recover the classical formulation of viscoplasticity, with Non Linear Kinematic hardening, we can choose the particular form :

$$\Omega_p = f_p + \frac{1}{2} \underset{\sim}{\mathbf{X}} : \underset{\approx}{\mathbf{Q}} : \underset{\sim}{\mathbf{X}}, \qquad \dot{\lambda}_p = \left\langle \frac{f_p}{D} \right\rangle^n, \tag{34}$$

where $f_p \leq 0$ defines the elastic domain by (14). For the static potential, we may assume a single thermal recovery format :

$$\Omega_s = \Omega_s(\underset{\sim}{\mathbf{X}}) \qquad \text{and} \qquad \dot{\lambda}_s = 1\,. \tag{35}$$

Using the Pseudo Standard Model, we then have

$$\dot{\underset{\sim}{\varepsilon}}_p = \dot{\lambda}_p \frac{\partial \Omega_p}{\partial \underset{\sim}{\sigma}} = \dot{p}\, \frac{\underset{\approx}{\mathbf{H}} : (\underset{\sim}{\sigma} - \underset{\sim}{\mathbf{X}})}{\|\underset{\sim}{\sigma} - \underset{\sim}{\mathbf{X}}\|_H}, \qquad \text{with} \quad \dot{p} = \left\langle \frac{f_p}{D} \right\rangle^n, \tag{36}$$

$$\dot{\underset{\sim}{\alpha}} = -\dot{\lambda}_p \frac{\partial \Omega_p}{\partial \underset{\sim}{\mathbf{X}}} - \dot{\lambda}_s \frac{\partial \Omega_s}{\partial \underset{\sim}{\mathbf{X}}} = \dot{\underset{\sim}{\varepsilon}}_p - \underset{\approx}{\mathbf{Q}} : \underset{\sim}{\mathbf{X}}\, \dot{p} - \frac{\partial \Omega_s}{\partial \underset{\sim}{\mathbf{X}}}. \tag{37}$$

We note the easy separation between dynamic recovery and static recovery effects and the fact that, contrary to the pure GSM, we have the non-modified elastic domain in the viscoplastic strain rate (cf. the discussion in section 3.2). The formulation easily degenerates to the rate independent plasticity format ($f_p \leq 0$, $D \to 0$) with $\dot{\lambda}_p$ obtained from the consistency condition $\dot{f}_p = 0$. The method is valid for more complicated hardening rules, including isotropic hardening or multi-kinematic hardening with thresholds, as shown in [Chaboche, 1996].

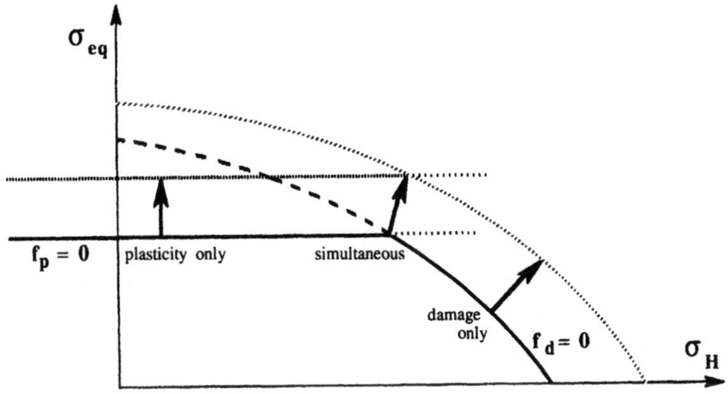

Figure 2: Coupling of a plasticity criterion and a damage criterion

4.3 Example for plasticity coupled with damage

Let us recall the discussion of section 3.3, where the rate independent formulation was unable to separate the damage growth from the plastic strain evolution. Now, with the quasi-standard approach, we have two independent potentials, plasticity and damage, associated respectively with the elastic domain (in the stress space) and with the nondamage domain (in the space of damage conjugate forces) :

$$f_p(\underset{\sim}{\sigma},\, A_j\,;\, d_k) \leq 0\,, \qquad\qquad f_d(y_k,\, B_l\,;\, d_k) \leq 0\,. \tag{38}$$

B_l is the set of variables that describes the actual position, size and shape of the non damage domain, in the space of damage conjugate forces y_k. Some recent theories [Voyiadjis and Kattan, 1999] introduce both the dilatation and the translation of that domain. The damage variable d_k can be used as a parameter in these functions. For instance d_k in f_p can play a role through the effective stress concept [Rabotnov, 1969], [Chaboche, 1977] (not discussed here). The indicatrix functions of the two potentials are chosen as $F_p = f_p + q(A_j;\, d_k)$ and $F_d \equiv f_d$. The normality rules then become

$$\dot{\underset{\sim}{\varepsilon}}_p = \dot{\lambda}_p \frac{\partial f_p}{\partial \underset{\sim}{\sigma}}, \qquad \dot{a}_j = -\dot{\lambda}_p \frac{\partial f_p}{\partial A_j} - \dot{\lambda}_p \frac{\partial q}{\partial A_j}, \qquad \dot{d}_k = \dot{\lambda}_d \frac{\partial f_d}{\partial y_k}, \qquad \dot{b}_l = -\dot{\lambda}_d \frac{\partial f_d}{\partial B_l}. \tag{39}$$

Multipliers $\dot{\lambda}_p$ and $\dot{\lambda}_d$ will have to be found by solving the consistency conditions $f_p = \dot{f}_p = 0$ and $f_d = \dot{f}_d = 0$. Referring to Figure 2, they are determined independently in the cases where plasticity is involved alone ($f_d < 0 \to \dot{\lambda}_d = 0$) or when purely brittle damage occurs without any plastic strain ($f_p < 0 \to \dot{\lambda}_p = 0$). When both processes are involved simultaneously, it is necessary to consider the possibility of coupling between the two.

In the rate dependent (or viscous) case there are also further possibilities. For instance, when modeling creep damage processes, we can now express the multiaxial damage rate directly in the stress space, using $\Omega_d = \Omega_d(y_k;\, d_k)$ and $f_d = f_d(\underset{\sim}{\sigma};\, d_k)$; this leads to

$$\dot{d}_k = \dot{\lambda}_d \frac{\partial \Omega_d}{\partial y_k}, \qquad \dot{\lambda}_d = \left\langle \frac{f_d}{A} \right\rangle^r, \tag{40}$$

where r and A are material and temperature dependent parameters. For example the expression for f_d can be chosen exactly to meet Hayhurst's multiaxial criterion [Hayhurst, 1972] (in the stress space). We can even take Ω_d as linear in the space of damage conjugate forces (let us assume only one tensorial damage variable) : $\Omega_d = \underset{\approx}{\Gamma}(\underset{\sim}{\mathbf{d}}) : \underset{\sim}{\mathbf{y}}$, so that the damage rate becomes independent of $\underset{\sim}{\mathbf{y}}$. Here the damage variable has been supposed to be a second rank tensor and $\underset{\approx}{\Gamma}(\underset{\sim}{\mathbf{d}})$ is a fourth rank tensor, giving the effect of damage rate anisotropy (see [Lemaître and Chaboche, 1985], [Chaboche, 1999]). In that case the anisotropic creep damage equation used by Murakami & Ohno [Murakami and Ohno, 1980] can easily be recovered.

5 CONCLUDING REMARKS

A thermodynamic framework for developing constitutive and damage equations has been revisited. Based on the notion of internal variables and the local state hypothesis, it limits the range of possibilities for these constitutive equations, in order to meet the thermodynamic requirements. Within this framework, the Generalized Standard Model is particularly discussed, considering some overrestrictive induced properties, both for kinematic hardening and damage modeling.

In order to release these restrictions, a slightly less constraining approach is formulated, using several independent potentials and several independent scalar multipliers, associated with the various considered physical proceses. Such a *Pseudo Standard Model* allows us to incorporate some models that were not entering easily into the GSM framework :

- the non linear kinematic hardening (dynamic recovery), both in the rate independent plasticity case and in the viscoplasticity case,
- the thermal recovery, as well as aging effects or other metallurgical changes,
- the separation between damage evolution and plastic strain rate, important for brittle damage description,
- the independence between directionality of damage rate and the shape of the isochronous creep rupture surfaces.

References

[Benallal, 1989] Benallal, A. (1989). *Thermoviscoplasticité et endommagement des structures*. Doctorat d'Etat, Université Pierre et Marie Curie, Paris 6.

[Carathéodory, 1909] Carathéodory, C. (1909). Untersuchungen über die Grundlagen der Thermodynamik. *Mathematische Annalen*, 67:355–386.

[Chaboche, 1977] Chaboche, J. L. (1977). Sur l'utilisation des variables d'état interne pour la description de la viscoplasticité cyclique avec endommagement. In *Problèmes Non Linéaires de Mécanique*, pages 137–159. Symposium Franco-Polonais de Rhéologie et Mécanique, Cracovie.

[Chaboche, 1983] Chaboche, J. L. (1983). On the constitutive equations of materials under monotonic or cyclic loadings. *La Recherche Aérospatiale*, (5):31–43.

[Chaboche, 1989] Chaboche, J. L. (1989). Constitutive equations for cyclic plasticity and cyclic viscoplasticity. *Int. J. of Plasticity*, 5:247–302.

[Chaboche, 1996] Chaboche, J. L. (1996). Unified cyclic viscoplastic constitutive equations : development, capabilities and thermodynamic framework. In Krauss, A. S. and Krauss, K., editors, *Unified Constitutive Laws of Plastic Deformation*, pages 1–68. Academic Press Inc.

[Chaboche, 1997] Chaboche, J. L. (1997). Thermodynamic formulation of constitutive equations and application to the viscoplasticity and viscoelasticity of metals and polymers. *Int. J. Solids Structures*, 34(18):2239–2254.

[Chaboche, 1999] Chaboche, J. L. (1999). Thermodynamically founded CDM models for creep and other conditions. In Skrzypek, J. and Altenbach, H., editors, *Modeling of Material Damage and Failure of Structures*, page in press. Springer.

[Chow and Wei, 1991] Chow, C. L. and Wei, Y. (1991). A model of Continuum Damage Mechanics for fatigue failure. *Int. J. Fracture*, 50:301–316.

[Coleman and Gurtin, 1967] Coleman, B. D. and Gurtin, M. E. (1967). Thermodynamics with internal state variables. *J. Chem. Phys.*, 47:597–613.

[Germain, 1973] Germain, P. (1973). *Cours de Mécanique des Milieux Continus*, volume I. Masson, Paris.

[Germain, 1974] Germain, P. (1974). Thermodynamique des milieux continus. *Entropie*, 55:7–14. Congrès Français de Mécanique.

[Germain et al., 1983] Germain, P., Nguyen, Q. S., and Suquet, P. (1983). Continuum thermodynamics. *J. of Applied Mechanics*, 50:1010–1020.

[Halphen and Nguyen, 1975] Halphen, B. and Nguyen, Q. S. (1975). Sur les matériaux standards généralisés. *J. de Mécanique*, 14(1):39–63.

[Hansen and Schreyer, 1992] Hansen, N. R. and Schreyer, H. L. (1992). Thermodynamically consistent theories for elastoplasticity coupled with damage. In *Damage Mechanics and Localization*, volume 142/AMD, pages 53–67. ASME.

[Hayhurst, 1972] Hayhurst, D. R. (1972). Creep rupture under multiaxial state of stress. *J. Mech. Phys. Solids*, 20(6):381–390.

[Ladevèze, 1992] Ladevèze, P. (1992). On the role of creep continuum damage in structural mechanics. In Ladevèze, P. and Zienkiewiecz, O., editors, *New Advances in Computational Structural Mechanics*, pages 3–22, Oxford. Elsevier.

[Lemaître, 1985] Lemaître, J. (1985). A Continuum Damage Mechanics model for ductile fracture. *J. of Engng. Mat. Technol.*, 107:83–89.

[Lemaître, 1992] Lemaître, J. (1992). *A course on Damage Mechanics.* Springer.

[Lemaître and Chaboche, 1985] Lemaître, J. and Chaboche, J. L. (1985). *Mécanique des Matériaux Solides.* Dunod, Paris.

[Mandel, 1964] Mandel, J. (1964). Contribution théorique à l'étude de l'écrouissage et des lois de l'écoulement plastique. Munich. 11th Congress ICTAM.

[Murakami and Ohno, 1980] Murakami, S. and Ohno, N. (1980). A continuum theory of creep and creep damage. In Ponter, A. and Hayhurst, D. R., editors, *Creep in Structures*, pages 422–443. 3rd IUTAM Symp., Springer-Verlag. Leicester.

[Rabotnov, 1969] Rabotnov, Y. N. (1969). *Creep problems in structural members.* North-Holland.

[Sidoroff, 1975] Sidoroff, F. (1975). On the formulation of plasticity and viscoplasticity with internal variables. *Arch. Mech., Poland*, 27(5-7):807–819.

[Truesdell and Noll, 1965] Truesdell, C. and Noll, W. (1965). The non-linear field theories of mechanics. In Flügge, S., editor, *Encyclopedia of Physics*, volume 3/3, Sec.79, New York. Springer.

[Voyiadjis and Kattan, 1999] Voyiadjis, G. Z. and Kattan, P. I. (1999). *Advances in Damage Mechanics : Metals and Metal Matrix Composites.* Elsevier.

[Zhu and Cescotto, 1995] Zhu, Y. Y. and Cescotto, S. (1995). Fully coupled elasto-visco-plastic damage theory for anisotropic materials. *Int. J. Solids Structures*, 32(11):1607–1641.

On micromechanics of martensitic transformation in SMA and TRIP steels

M. Cherkaoui, M. Berveiller and E. Patoor
Laboratoire de Physique et Mécanique des Matériaux,
CNRS-ISGMP, Ile de Saulcy F-57045 Metz Cedex 01 France
cherk@Lpmm.univ-metz.fr

Abstract: A micromechanical model is developed to derive the overall behavior of solids during martensitic transformation. The physical nature of martensitic transformation allows us to introduce the concept of moving boundaries in the framework of thermodynamics and micromechanics. The obtained results are specified both in SMA and TRIP steels.

1.INTRODUCTION

In most heterogeneous materials, the thermophysical properties may be assumed piecewise uniform at the mesoscale. These materials are constituted of different phases and/or grains separated by interfaces called grain or phase boundaries. The interfaces are in general stationary with respect to particles and often considered as perfect, at least at low temperatures. This hypothesis corresponds to the continuity of displacement ($[\boldsymbol{u}]=0$) and/or velocity ($[\boldsymbol{v}]=0$) fields as well as the stress vector ($[\boldsymbol{\sigma}].\boldsymbol{n}=0$), where $[x]=x^{+}-x^{-}$ denotes the jump of $\boldsymbol{x}$ across the interface. The continuity assumptions extended to equivalent forms for the volume lead to the usual localization and homogenization relations, from which the classical scale transition methods are developed. The overall behavior of the Representative Volume Element (RVE) is then deduced from the microstructure and the local behavior.

In various situations where the inelastic strains result from discrete physical mechanisms like twinning, martensitic transformation, or recrystallization, the previous hypotheses are partially not valid since one deals with the problem of evolutive microstructures or moving boundaries whose velocities are different from those of the particles. Consequently, the strain field and/or thermophysical properties undergo discontinuities across moving boundaries and therefore:

- Additional terms in localization and homogenization relations have to be introduced in the framework of local thermodynamics and micromechanics to account for the moving boundaries.
- Driving forces as well as the germination and growing laws describing the evolutive microstructure need to be determined.

G.A. Maugin et al. (eds.), Continuum Thermomechanics, 101-112.
© 2000 *Kluwer Academic Publishers.*

The purpose of this paper is to present the basic results concerning the modeling of martensitic transformation in Shape Memory Alloys (SMA) [Patoor et al., 1987] and TRIP steels (TRansformation Induced Plasticity) [Cherkaoui et al., 1998].

1.1 Kinematics and martensitic phase transformation kinetics

Inelastic strain called transformation strain ε^{T}, produced by martensitic transformation, results from two contributions: the Bain strain and the lattice invariant strain. The Bain strain describes the geometrical transformation from the parent lattice (austenite) to the product lattice (martensite). The lattice invariant strain is an accommodation step relaxing, at least partially, the internal stresses generated by the Bain strain through a shear along an invariant plane resulting in the formation of a habit plane. Wechsler, Liebermann and Read were the first to formulate a geometrical theory giving the possible crystallographic habit planes. This theory is based on the concept of inelastic compatibility of the transformation strain $\varepsilon^{\mathrm{tr}}(r)$ assumed to be uniform within an elementary transformed volume (plate or lath). Due to the high symmetry of the austenitic lattice, several transformation strains are possible so that different martensitic variants may be formed with the corresponding habit planes and transformation strains given by

$$\varepsilon_{ij}^{\mathrm{tr}_I} = \frac{1}{2} g\left(M_i^I N_j^I + M_j^I N_i^I\right) = g\, R_{ij}^I \tag{1}$$

where $\vec{N}$ and $\vec{M}$ are the habit plane normal and the direction of the transformation, respectively. g is the amplitude of the transformation considered as a material constant. The local transformation strain $\varepsilon^{\mathrm{tr}}(r)$ is then piecewise uniform and written as

$$\varepsilon_{ij}^{\mathrm{tr}}(r) = \sum_{I=1}^{N} \varepsilon_{ij}^{\mathrm{tr}I} \theta^I(r) \tag{2}$$

where $\theta^I(r)$ are the Heaviside step functions for the different transformed domains I and N their number.

For a single crystal with volume V subjected at its external boundary ∂V to a displacement $u_i = E_{ij} x_j$, and undergoing an inelastic field $\varepsilon^{\mathrm{tr}}(r) + \varepsilon^{\mathrm{p}}(r)$ within V, the total strain is

$$E = \frac{1}{V} \int_V \left(\varepsilon^{\mathrm{e}}(r) + \varepsilon^{\mathrm{tr}}(r) + \varepsilon^{\mathrm{p}}(r)\right) dV = \left\langle \varepsilon^{\mathrm{e}}(r) + \varepsilon^{\mathrm{tr}}(r) + \varepsilon^{\mathrm{p}}(r) \right\rangle \tag{3}$$

where the elastic field $\varepsilon^{\mathrm{e}}(r)$ arises both from E and the incompatibility of fields $\varepsilon^{\mathrm{tr}}(r)$ and $\varepsilon^{\mathrm{p}}(r)$.

For homogeneous elasticity with elastic constant C, one obtains directly the elastic part E^{e} of E as follows:

$$E = C^{-1} : \Sigma + E^{\mathrm{tp}} \tag{4}$$

where $\Sigma = \langle \sigma(r) \rangle$, and $E^{\text{tp}} = \langle \varepsilon^{\text{tr}}(r) + \varepsilon^{\text{p}}(r) \rangle$ corresponds to the macroscopic inelastic strain arising from the plastic flow and the phase transformation.

Case of SMA

In such a case, one assumes that $\varepsilon^{\text{p}}(r) = 0$ and the transformation strain becomes

$$E^{\text{tr}} = \frac{1}{V} \int_V \varepsilon^{\text{tr}}(r) dV \tag{5}$$

Due to typical properties of the field $\varepsilon^{\text{tr}}(r)$, one can distinguish two simplified forms for E^{tr}:

- The microstructure of variants as well as the fact that $\varepsilon^{\text{tr}}(r)$ is piecewise uniform lead to

$$E^{\text{tr}} = \frac{1}{V} \sum_I \int_{V^I} \varepsilon^{\text{tr}_I} dV = \sum_I \varepsilon^{\text{tr}_I} f^I \tag{6}$$

where $\varepsilon^{\text{tr}_I}$ are material constants, and f^I internal variables subjected to the following constraints:

$$0 \le \sum_I f^I \le 1, 0 \le f^I \le 1 \tag{7}$$

- Since the transformation strain vanishes within the austenitic phase, (5) can be written as

$$E^{\text{tr}} = \frac{1}{V} \int_{V^{\text{M}}} \varepsilon^{\text{tr}}(r) dV = \frac{V^{\text{M}}}{V} \left\{ \frac{1}{V^{\text{M}}} \int_{V^{\text{M}}} \varepsilon^{\text{tr}}(r) dV \right\} = f \bar{\varepsilon}^{\text{tr}} \tag{8}$$

where $f = \sum_I f^I$ is the total volume fraction of martensite, and $\bar{\varepsilon}^{\text{tr}}$ the average transformation strain over V^{M}, which is now an unknown of the problem.

The time derivative of (5) shows the different mechanisms associated with martensitic transformation in SMA which are not revealed through a simple "static" comparison between the actual configuration and the reference one (austenite). Taking into account the discontinuities $[\varepsilon^{\text{tr}}(r)] = \varepsilon^{\text{tr}_+} - \varepsilon^{\text{tr}_-}$ of the field $\varepsilon^{\text{tr}}(r)$ along the moving boundaries, one has

$$\dot{E}^{\text{tr}} = \frac{1}{V} \int_V \dot{\varepsilon}^{\text{tr}}(r) dV - \frac{1}{V} \int_S [\varepsilon^{\text{tr}}(r)] \omega_\alpha n_\alpha dS \tag{9}$$

and since $\dot{\varepsilon}^{\text{tr}}(r) = 0$, it follows that

$$\dot{E}^{\text{tr}} = -\frac{1}{V} \int_S [\varepsilon^{\text{tr}}(r)] \omega_\alpha n_\alpha dS \tag{10}$$

S represents all of the moving boundaries (austenite-martensite as well as boundaries between the martensitic variants), and $\vec{\omega}$ the velocity of the interface.
The "crystalline" description of the martensitic transformation (in analogy with crystal plasticity) comes directly from (10):

$$\dot{E}^{\text{tr}} = \sum_I \varepsilon^{\text{tr}_I} \dot{f}^I \tag{11}$$

where, actually, the distinction between transformation and exchange between variants is not explicitly shown.
The different mechanisms are more explicitly described when one expands the time derivative of (8)

$$\dot{E}^{\text{tr}} = \dot{f}\bar{\varepsilon}^{\text{tr}} + f\dot{\bar{\varepsilon}}^{\text{tr}} \tag{12}$$

In fact, the analysis of (12) allows us to distinguish the following phenomena:

- Transformation ($\dot{f} \neq 0$) without deformation ($\bar{\varepsilon}^{\text{tr}} = 0, \dot{\bar{\varepsilon}}^{\text{tr}} = 0$).
- Transformation ($\dot{f} \neq 0$) with deformation ($\bar{\varepsilon}^{\text{tr}} \neq 0$) but without reorientation ($\dot{\bar{\varepsilon}}^{\text{tr}} = 0$).
- Deformation by reorientation ($\dot{\bar{\varepsilon}}^{\text{tr}} \neq 0$) without transformation ($\dot{f} = 0$).
- Transformation, deformation and reorientation ($\dot{f} \neq 0$, $\bar{\varepsilon}^{\text{tr}} \neq 0$, $\dot{\bar{\varepsilon}}^{\text{tr}} \neq 0$).

Case of TRIP Steels

In such a case, the inelastic strain of the single crystal is

$$E^{\text{tp}} = \frac{1}{V}\int_V \left(\varepsilon^{\text{tr}}(r) + \varepsilon^{\text{p}}(r)\right) dV \tag{13}$$

where $\varepsilon^{\text{p}}(r)$ describes simultaneously the plastic strain of an elementary volume element at austenitic and martensitic states. For TRIP steels, the mechanisms of reorientation and inverse transformation being insignificant, the progression of the transformation can be described by the instantaneous growth of new plates or laths in the austenitic phase. Therefore, one obtains the expression [Cherkaoui et al., 1998]:

$$\dot{E}^{\text{tp}} = (1-f)\dot{\bar{\varepsilon}}^{\text{p}_A} + \sum_I f^I \dot{\bar{\varepsilon}}^{\text{M}_I} + \sum_I \varepsilon^{\text{tr}_I} \dot{f}^I \tag{14}$$

- The first term describes the average plastic flow in the residual austenitic phase.
- The second term corresponds to the plastic flow in the pre-existing martensitic phase.
- The last term expresses the formation of new plates or laths.

1.2 Thermodynamics and driving forces.

By limiting the study to quasistatic and isothermal processes, the change of free energy associated with a martensitic transformation involves both volume and surface terms. For the free energy, one has to account for the elastic energy (due to internal and applied stresses), the chemical or crystallographic energy (due to phase transition) and the energy of interfaces between austenite-martensite and martensite-martensite. The last contribution is usually neglected in martensitic transformations.

If we denote by φ and $w = 1/2\sigma : \varepsilon^{e}$ the chemical and elastic energy densities, the Helmholtz free energy Φ per unit volume V is given by

$$\Phi = \frac{1}{V}\int_{V} (\varphi + w) dV \tag{15}$$

Since φ and w are discontinuous across the moving boundaries S, one has for $\dot{\Phi} = d\Phi/dt$ the expression

$$\dot{\Phi} = \frac{1}{V}\int_{V} (\dot{\varphi} + \dot{w}) dV - \frac{1}{V}\int_{S} [\varphi + w] \omega_{\alpha} n_{\alpha}\, dS \tag{16}$$

The volume integral in (16) is due to the change of Helmholtz free energy during the evolution of the inelastic strain of austenite and pre-existing martensitic phase at the current configuration. During this mechanism, the chemical energy remains constant (no lattice change), i.e. $\dot{\varphi} = 0$. Furthermore, the jump $[\varphi]$ is calculated as a linear approximation with respect to temperature T in the vicinity of the equilibrium temperature T^{o}: $[\varphi] = B(T - T^{o})$. B is a material constant.

By taking into account the properties of the field w, we may express the jump $[w]$ for a homogeneous elastic behavior as follows:

$$[w] = \frac{1}{2}(\sigma^{+} + \sigma^{-}) : [\varepsilon - \varepsilon^{tp}] \tag{17}$$

and due to the fact that $\dot{\varepsilon}^{tr} = 0$, one has for dw/dt

$$\dot{w} = \sigma : (\dot{\varepsilon} - \dot{\varepsilon}^{p}) \tag{18}$$

With the above results, (16) becomes

$$\dot{\Phi} = \frac{1}{V}\int_{V} \sigma : (\dot{\varepsilon} - \dot{\varepsilon}^{p}) dV - \frac{1}{V}\int_{S} \left\{ \frac{1}{2}(\sigma^{+} + \sigma^{-}) : [\varepsilon - \varepsilon^{tp}] + [\varphi] \right\} \omega_{\alpha} n_{\alpha}\, dS \tag{19}$$

where $S = S_{AM} + S_{MM}$, S_{AM} corresponds to the moving austenite-martensite boundaries, and S_{MM} those between the martensitic variants.

The calculation of the power of external forces

$$P^{e} = \frac{1}{V}\int_{\partial V} \Sigma_{ij} n_{j} v_{i} dS \tag{20}$$

is transformed into a volume integral by taking into account the static conditions

$$[u_i] = 0, \; [\sigma_{ij} n_j] = 0 \tag{21}$$

as also the dynamics ones (Hadamard's relations)

$$[v_i] = -[u_{i,k}] n_k \omega_{\alpha} n_{\alpha} \tag{22}$$

on the moving boundaries.

It follows that

$$P^{\mathrm{e}} = \frac{1}{V}\int_V \sigma : \dot{\varepsilon}\, dV - \frac{1}{V}\int_S \frac{1}{2}\left(\sigma^+ + \sigma^-\right) : [\varepsilon]\omega_\alpha n_\alpha \, dS \tag{23}$$

and the intrinsic dissipation $D = P^{\mathrm{e}} - \dot{\Phi}$ becomes

$$D = \frac{1}{V}\int_V \sigma : \dot{\varepsilon}^{\mathrm{p}}\, dV - \frac{1}{V}\int_S \left\{\frac{1}{2}\left(\sigma^+ + \sigma^-\right) : \left[\varepsilon^{\mathrm{tp}}\right] + [\varphi]\right\}\omega_\alpha n_\alpha \, dS \tag{24}$$

This equation involves two quantities:

- the volumic driving forces σ associated with the plastic flow of both phases,
- the surface driving force $-\left\{\frac{1}{2}\left(\sigma^+ + \sigma^-\right) : \left[\varepsilon^{\mathrm{tp}}\right] + [\varphi]\right\}$ associated with the moving boundaries.

In (24), the surface forces are associated with the progression of a boundary where the inelastic discontinuity arises both from plasticity and transformation. It is also possible to deduce from (24) the driving force linked with a "flow" $\dot{f}$ if one simplifies the fields σ and $\varepsilon^{\mathrm{tp}}$ or (and) if further indications on nucleation and growth of a transformed domain are specified.

2. BEHAVIOR OF SOLIDS DURING MARTENSITIC TRANSFORMATION

2.1 Mechanisms and different cases

Martensitic transformation is characterized by the existence of a macroscopic shape change associated with modification of the crystalline structure. Growth of martensitic plates within the parent phase produces an important stress field, and plastic accommodation occurs if the yield stress of the parent phase (or the martensite one) is reached. When plastic yield stresses are large enough, or when the strain energy is weak, the transformation strain is accommodated in a fully elastic way and the plate growth proceeds by a succession of thermoelastic balances. Different mechanical behaviors associated with transformation plasticity find their origin in three basic mechanisms:

- Orientation of the martensite variants by the applied stress. This effect is related to the shear component of the phase transformation.
- Reorientation of the martensite variants by the applied stress. Interfaces between these variants have a great mobility, and reorientation process may occur under applied loading.
- Orientation of the plastic yielding around plates of martensite.

2.2 Pseudoelasticity of Shape Memory alloys

During martensitic transformation, the microstructure developed within the grain may be analyzed at two levels:

- The shape of the elementary transformed volume corresponds generally to that of a thin plate. The corresponding internal variables are the volume fraction of each possible variant and the aspect ratios of the plates.
- In the most common case, several domains, such as subgrains in which only one variant is activated, are developed. The boundaries between these domains have to be described by additional internal variables.

Using the compatibility conditions, the equilibrium equations and boundary conditions, we find that eqn (15) leads to

$$\Phi\left(E,T,\varepsilon^{\mathrm{tr}}\right)=\frac{1}{2}\left(E-E^{\mathrm{tr}}\right):C:\left(E-E^{\mathrm{tr}}\right)+B\left(T-T^{\mathrm{o}}\right)f-\frac{1}{2V}\int_V \tau(r):\varepsilon^{\mathrm{tr}}(r)dV \quad (25)$$

where $\tau=\sigma-\Sigma$ describes the internal stress field resulting from the incompatibilities of the field $\varepsilon^{\mathrm{tr}}(r)$.

Due to the properties of $\varepsilon^{\mathrm{tr}}(r)$, the interaction energy

$$W_{\mathrm{int}}=\frac{1}{2V}\int_V \tau(r):\varepsilon^{\mathrm{tr}}(r)dV \quad (26)$$

may be also related to the mean internal stress $\bar{\tau}^I$ for a variant I and the corresponding volume fraction f^I, so that

$$W_{\mathrm{int}}=\frac{1}{2}\sum_I \bar{\tau}^I:\varepsilon^{\mathrm{tr}_I} f^I \quad (27)$$

Finally, the free energy $\Phi\left(E,T,f^I\right)$ depends only on the control variables E and T, and the internal variables f^I, which are subjected to the conditions (7).

For a transformation microstructure in domains or subgrains, one has

$$W_{\mathrm{int}}=-\frac{1}{2}F^n F^m\left(\bar{\varepsilon}^n-\bar{\varepsilon}^m\right):Q^{nm}:\left(\bar{\varepsilon}^n-\bar{\varepsilon}^m\right) \quad (28)$$

where F^n and F^m are the volume fractions of the domains n and m; $\bar{\varepsilon}^n$ and $\bar{\varepsilon}^m$ are the mean transformation strains over the corresponding volumes V_n and V_m; Q^{nm} is the interface operator depending on the unit normal to the interface between n and m.

Due to the high mobility of interfaces between variants, W_{int} is minimized with respect to the volume fractions F^n and F^m and the unit normal $\vec{N}$.

The result now is an interaction energy $W_{\mathrm{int}}=-1/2\, f^n H^{nm} f^m$ depending only on the volume fractions f^n and f^m. The interaction matrix H^{nm} is related only to the different transformation strains $\varepsilon^{\mathrm{tr}_n}$. Finally, the Helmholtz free energy is given by

$$\Phi\left(E,T,f^n\right)=\frac{1}{2}\left(E-E^{\mathrm{tr}}\right):C:\left(E-E^{\mathrm{tr}}\right)+B\left(T-T^{\mathrm{o}}\right)f+1/2\sum_{n,m} f^n H^{nm} f^m \quad (29)$$

where $\varepsilon^{\mathrm{tr}_n}, C, B, T^{\mathrm{o}}$ are material parameters and f^n are a total of 24 scalar internal variables. The derivation of the thermomechanical behavior of a grain follows now from a classical analysis where the driving forces $\Im^n = \partial L / \partial f^n$ on the volume fractions are compared with critical values $\Im^{n_c}$ (assumed constant: no hardening). L is the Lagrangian built from the free energy and the N+1 Lagrange multipliers introduced from the constraints (7). Using a classical thermomechanical self-consistent model for the transition toward the polycrystal, we find that the theoretical results are in good agreement with the experimental ones on copper base alloys (Fig.1, Fig.2). Note that the results have been obtained with no adjusting parameters.

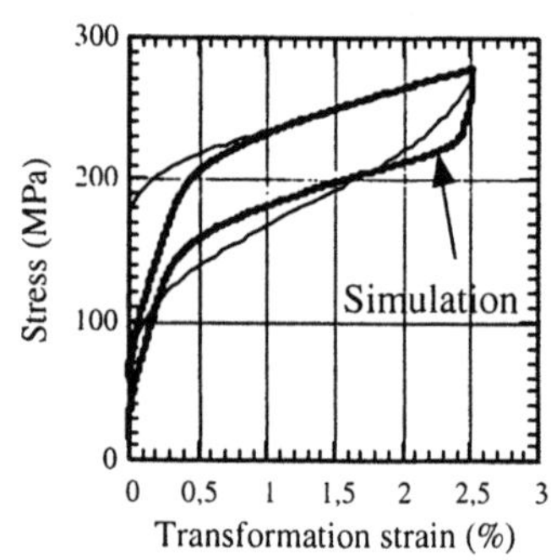

***Figure 1:** Comparison between simulation and experimental results performed on a polycrystalline CuAlBe SMA for a superelastic tensile test at T =60°C*

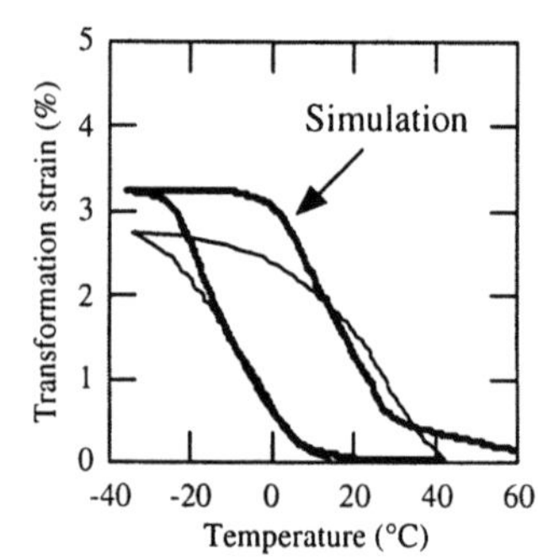

***Figure 2:** Comparison between simulation and experimental results performed on a polycrystalline CuAlBe SMA for a cooling-heating sequence under a constant applied tensile stress of 100MPa*

2.3 Thermomechanical behavior of TRIP steels

The remarkable properties of ductility and strength of TRIP steels arise on the one hand from the inelastic strain accompanying martensitic transformation, and on the other hand from the internal stress field emerging from this phase change, leading to an additional plastic flow known as the TRIP effect.

The calculation of the surface dissipation in (24) for any topology of S is complicated, and can be realized only through numerical procedures. However, in many situations (dislocation loops, martensitic plates, twinning) the moving boundaries S can be simulated by ellipsoidal inclusions. This allows to use Eshelby's relationships concerning inelastic ellipsoidal inclusions, if the inelastic strain field is piecewise uniform. Therefore, the jump $[\varepsilon^{\mathrm{tp}}]$ of the inelastic strain is assumed uniform along S. Furthermore, the stress σ^+ (not uniform along S) is linked to σ^- through the interface operator $Q(C,n)$

$$\sigma^+ = \sigma^- - Q(C,n)[\varepsilon^{\mathrm{tp}}] \tag{30}$$

and the jump $[\varepsilon^{\mathrm{tp}}]^I$ of the inelastic strain across the moving boundary S^I of a microdomain belonging to a martensitic variant I, is given by

$$[\varepsilon^{\mathrm{tp}}]^I = \varepsilon^{\mathrm{p}+} - \left(\varepsilon^{\mathrm{p}-} + \varepsilon^{\mathrm{tr}_I}\right) = -\varepsilon^{\mathrm{tr}_I} \tag{31}$$

where the plastic strains $\varepsilon^{\mathrm{p}+}$ and $\varepsilon^{\mathrm{p}-}$ on each side of the growing domain are taken to be equal, according to the instantaneous growth hypothesis (Cherkaoui et al., 1998).

By accounting for eqns (30) and (31) and according to the uniformity of σ^- and $[\varepsilon^{\mathrm{tp}}]^I$ along S^I, the surface dissipation becomes

$$D^{\mathrm{S}} = \left(\sigma^-_{ij}\varepsilon^{\mathrm{tr}_I}_{ij} - B\left(T - T^{\mathrm{o}}\right)\right)\frac{1}{V}\int_{S^I}\omega_\alpha n_\alpha d\Sigma + \frac{1}{2}\varepsilon^{\mathrm{tr}_I}_{ij}\left(\frac{1}{V}\int_{S^I}Q_{ijkl}\omega_\alpha n_\alpha d\Sigma\right)\varepsilon^{\mathrm{tr}_I}_{kl} \tag{32}$$

The ellipsoidal growing hypothesis allows us to obtain the following result:

$$\dot{f}^I = \frac{1}{V}\int_{S^I}\omega_\alpha n_\alpha dS, \quad \frac{1}{V}\int_{S^I}Q\omega_\alpha n_\alpha dS = C\left(I - \mathbf{S}^I\right)\dot{f}^I + C\dot{\mathbf{S}}^I f^I \tag{33}$$

where f^I is the volume fraction of the microdomain and $\mathbf{S}^I$ is the corresponding Eshelby tensor depending on the aspect ratio as well as on the orientation of the microdomain; $\dot{f}^I$ and $\dot{\mathbf{S}}^I$ correspond to the variation of f^I and $\mathbf{S}^I$ due to the ellipsoidal growth. With (33), (32) leads to

$$D^{\mathrm{S}} = \left(\sigma^-_{ij}\varepsilon^{\mathrm{tr}_I}_{ij} - B\left(T - T^{\mathrm{o}}\right)\right)\dot{f}^I + \frac{1}{2}\varepsilon^{\mathrm{tr}_I}_{ij}C_{ijmn}\left(I_{mnkl} - \mathbf{S}^I_{mnkl}\right)\varepsilon^{\mathrm{tr}_I}_{kl}\dot{f}^I + \frac{1}{2}\varepsilon^{\mathrm{tr}_I}_{ij}C_{ijmn}\dot{\mathbf{S}}^I_{mnkl}\varepsilon^{\mathrm{tr}_I}_{kl}f^I \tag{34}$$

The ellipsoidal growth hypothesis leads to a simple form of the surface dissipation, which allows one to choose the volume fraction of a martensitic domain belonging to each crystallographically possible martensitic variant as the internal variable describing the evolution of martensitic phase change. The thermodynamic force acting on this internal variable is deduced from (34) as follows:

$$F^I = \sigma^-_{ij}\varepsilon^{\mathrm{tr}_I}_{ij} - B\left(T - T^{\mathrm{o}}\right) + \frac{1}{2}\varepsilon^{\mathrm{tr}_I}_{ij}C_{ijmn}\left(I_{mnkl} - \mathbf{S}^I_{mnkl}\right)\varepsilon^{\mathrm{tr}_I}_{kl} + \frac{1}{2}\varepsilon^{\mathrm{tr}_I}_{ij}C_{ijmn}\frac{\dot{\mathbf{S}}^I_{mnkl}}{\dot{f}^I}\varepsilon^{\mathrm{tr}_I}_{kl}f^I, \tag{35}$$

where $\dot{\mathbf{S}}^I/\dot{f}^I$ are given for the most favorable growing mode.

Eqn (35) gives the thermodynamic driving force for nucleation and growth of martensitic microdomains belonging to different martensitic variants; it requires the knowledge of stress inside the growing domains. In addition to overall applied stress, this stress field contains several contributions due to different couplings between plasticity and phase transformation at the microscale. In the following, several ways for the determination of internal stresses are discussed from a micromechanics point of view:

a - The simplest way is to neglect any source of internal stresses. In such conditions, Patel and Cohen have formulated the energy term resulting only from the interaction of applied stress with transformation strains. This corresponds in (35) to the term $\Sigma : \varepsilon^{\mathrm{tr}_I} - B(T - T^{\circ})$, where σ^{-} is assumed equal to the applied stress Σ, and all the other contributions are neglected.

b - It is possible to assume an ellipsoidal growth of a martensitic microdomain inside a homogeneous stress field corresponding to the average stress over the austenitic phase. At the current configuration of the RVE, if we denote by $\bar{\varepsilon}^{\mathrm{p_A}}$ the average plastic strain in the austenitic phase, and by $\bar{\varepsilon}^{M_I}$ the one over a martensitic variant which consists of a set of microdomains, the interactions between plasticity and martensitic phase transformation is taken into account through the average stress field σ^{A} in the austenitic phase given by (Cherkaoui et al., 1998):

$$\sigma^{\mathrm{A}} = \Sigma - \sum_{I=1}^{N} f^{I} C\left(I - \mathbf{S}^{I}\right)\left(\bar{\varepsilon}^{\mathrm{p_A}} - \varepsilon^{\mathrm{tr}_I} - \bar{\varepsilon}^{\mathrm{M}_I}\right) \tag{36}$$

where N is the number of martensitic variants active at the current configuration of the RVE and f^{I} are the corresponding volume fractions.

With the instantaneous growth hypothesis, the stress inside a growing microdomain belonging to a martensitic variant I is related to σ^{A} by the following simple form:

$$\sigma^{-} = \sigma^{\mathrm{A}} - C\left(I - \mathbf{S}^{I}\right)\varepsilon^{\mathrm{tr}_I} \tag{37}$$

With (36) and (37), (35) leads to

$$F^{I} = \sigma_{ij}^{\mathrm{A}} \varepsilon_{ij}^{\mathrm{tr}_I} - B\left(T - T^{\circ}\right) - \frac{1}{2} \varepsilon_{ij}^{\mathrm{tr}_I} C_{ijmn}\left(I_{mnkl} - \mathbf{S}_{mnkl}^{I}\right)\varepsilon_{kl}^{\mathrm{tr}_I} + \frac{1}{2} \varepsilon_{ij}^{\mathrm{tr}_I} C_{ijmn} \frac{\dot{\mathbf{S}}_{mnkl}^{I}}{\dot{f}^{I}} \varepsilon_{kl}^{\mathrm{tr}_I} f^{I}. \tag{38}$$

In the thermodynamic force (38) one can distinguish two contributions:

- A long range internal stress effect through the term $\sigma^{A} : \varepsilon^{\mathrm{tr}_I} - B(T - T^{\circ})$. The stress σ^{A} contains the effects of plastic strains of both phases, as well as of transformation strains undergone by the pre-existing martensitic variants. Depending on applied and internal stresses, σ^{A} plays the role of variant selection (extended Magee effect),
- A self-internal stresses effect through the term

$$1/2\, \varepsilon^{\mathrm{tr}_I} : C : (I - \mathbf{S}^{I}) : \varepsilon^{\mathrm{tr}_I} + \frac{1}{2} \varepsilon^{\mathrm{tr}_I} : C : (\dot{\mathbf{S}}^{I} / \dot{f}^{I}) : \varepsilon^{\mathrm{tr}_I} f^{I}$$

depending essentially on the morphology of the growing microdomain

c - The effect of the internal stress field, emerging from the plastic flow, on the martensitic phase transformation is known as the strain induced martensitic phase transformation phenomenon. Classically, the plastic flow at the grain level is described by a homogeneous plastic strain through the plastic slip on crystallographic glide systems. The intragranular stresses arising from this description correspond to second order internal stresses. However, for ductile materials undergoing martensitic phase transformation, the role of plastic strain in the phase transition is more complicated. In fact, the martensitic plates nucleate at dislocation pile ups, dislocation dipole or intersection of slip bands. In such a situation, the description of the strain-induced martensitic transformation is insufficient through a homogeneous plastic strain. In other words, a third order stress field emerging from the heterogeneity of plastic strain has to be taken into account. Within a micromechanical model, one obtains the following interaction matrix describing the strain induced martensitic phase transformation through the third order internal stresses

$$H^{I_n} = R_{ij}^{I}\, C_{ijmn} \left(\mathbf{S}_{mnkl}^{I} - \mathbf{S}_{mnkl}^{\mathrm{p}_n} \right) P_{kl}^{\mathrm{A}_n} \qquad (39)$$

$\mathbf{S}^{\mathrm{p}}$ is the Eshelby tensor depending on the morphology of the plastic "defect" and corresponding to the initial shape of a martensitic nucleus; R^{I} is given in eqn (1), and the P^{A_n} are the Schmidt tensors associated with the slip systems in the austenitic phase. For the plastic flow in austenitic and martensitic phases, the associated driving forces are the resolved shear stresses on slip systems in austenite and martensitic variants given in a previous work (see Cherkaoui et al., 1998).

These driving forces are compared with critical forces describing the resistance to phase transformation and plastic flow. The expressions for critical forces are based on physical considerations, taking into account the hardening of austenite and martensite due to their own plastic strains, and the hardening of the martensitic phase due to the plastic strain inherited from the austenite. These results are combined with eqn (14) to derive the constitutive equation of an austenitic single crystal from which the overall behavior of polycrystalline TRIP materials is deduced by using the self-consistent scale transition method.

We present numerical results for the behavior of a polycrystalline TRIP steel obtained by this approach. The results are compared with the experimental data of Olson and Azrin performed on a Fe-Ni base alloy. For the numerical simulation, 100 grains having different crystallographic orientations describe the polycrystalline structure. The grains are assumed to be spherical in shape and identical in size. The elasticity of the material is considered as isotropic and homogeneous defined by the shear modulus ($\mu = 80\mathrm{GPa}$) and the Poisson's ratio ($\nu = 0.3$). In such conditions, we have calculated the stress-strain curve of the TRIP steel and martensitic volume fraction evolution at different temperatures (Fig. 3 (a) and Fig. 4 (a)). The corresponding experimental results of Olson and Azrin (1978) are plotted in Fig. 3 (b) and Fig. 4 (b). There is good agreement between experimental and theoretical results.

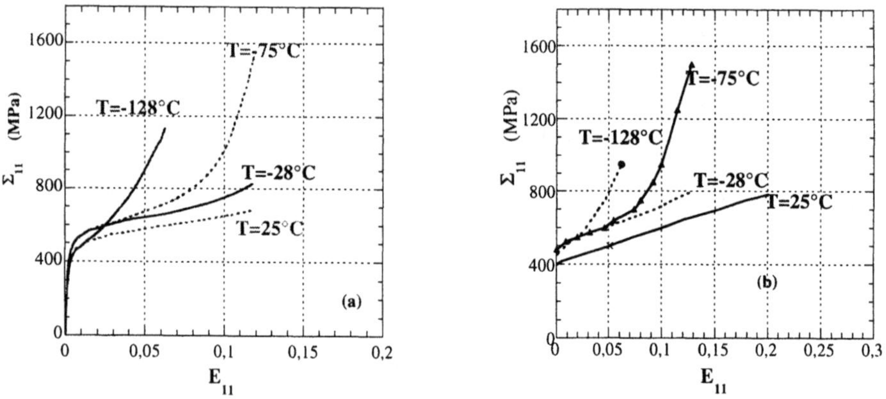

Figure.3 Tensile properties at different temperatures for Fe-Ni TRIP steel theoretical Results, (b) experimental Results

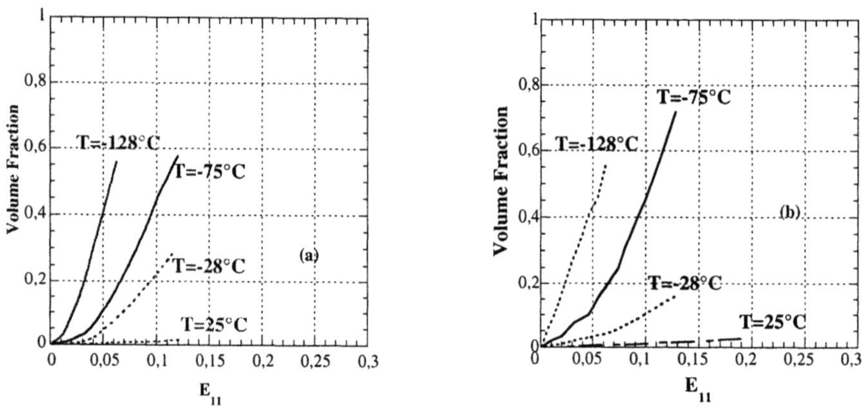

Figure.4 Transformation curves at different temperatures for Fe-Ni TRIP steel (a) theoretical Results, (b) experimental Results

REFERENCES

[Patoor et al., 1987] Patoor, E.; Eberhardt, A.; Berveiller, M. Potentiel pseudoelastique et plasticité de transformation martensitique dans les mono et polycristaux. *Acta Metall.* Vol. 35, pp. 2779-2789

[Cherkaoui et al., 1998] Cherkaoui, M.; Berveiller, M.; Sabar, H. "Micromechanical modelling of the martensitic transformation induced plasticity (TRIP) in austenitic single crystals. *Int. J. Plasticity*, Vol 14 pp. 597-626.

Modelling of coupled effects of damage by microcracking and friction in closed cracks

André Dragon and Damien Halm
LMPM (UMR 6617 CNRS), ENSMA, Poitiers, 1 Avenue Clément Ader, BP 40109
86961 Futuroscope-Chasseneuil Cedex
dragon@lmpm.ensma.fr

Abstract : The paper addresses crucial issues concerning inelastic behaviour of quasi-brittle solids, principally rock-like materials. Inelastic response for this class of solids results from the evolution of a large number of micro- and mesocracks accompanied with frictional effects regarding closed cracks for complex compression-dominated loading paths. Progressive microcracking and frictional blocking/sliding induce anisotropic behaviour, volumetric dilatancy and complex hysteretic effects due to crack opening/closure transition and plasticity-like sliding evolution. These problems are explored and modelled in the framework of rate-type constitutive theory with internal variables. The model is three-dimensional and micromechanically motivated in its essential ingredients. Meanwhile it is built to provide a tool for efficient structural analysis and, as such, represents a continuum damage approach coupled with a form of plasticity. The settlement between apparently conflicting requirements of physical pertinency on the one hand and of applicability of the model on the other hand, is attempted through relative simplicity of the approach (a small number of material constants to identify) and its modular character involving three increasing levels of complexity. The first, 'basic' level, concerns modelling of anisotropic degradation by multiple microcrack growth generating volumetric dilatancy and permanent strain. The second level consists in accounting for the 'normal' moduli recovery due to crack closure under predominantly compressive loads (unilateral effect). These two levels are outlined briefly at the beginning of the paper. The text focuses on the third level of modelling involving concomitant dissipative phenomena of damage by microcracking and frictional sliding leading to complex hysteretic effects such as inelastic unloading. The interaction of the two phenomena is successfully managed by the coupled model.

Keywords: Anisotropic damage, Microcracking, Frictional sliding, Coupled effects, 3D model.

1. INTRODUCTION AND SCOPE

Crucial issues concerning inelastic response of quasi-brittle solids, principally rock-like materials, are considered and modelled in this paper in the framework of rate-type constitutive theory with internal variables. This methodology has largely proved its efficiency to deal with essential features of structural rearrangements in engineering materials leading to non-linear (dissipative) models. Germain is among the principal precursors who established rigourous foundations for plasticity, viscoplasticity, damage and fracture modelling in this scope based on sound thermodynamic background. In the context related to the present study the reader can refer f.ex. to [Germain, 1986], [Germain et al., 1983].

G.A. Maugin et al. (eds.), Continuum Thermomechanics, 113-124.
© 2000 *Kluwer Academic Publishers.*

It is well confirmed nowadays that the non-linear behaviour of quasi-brittle, rock-like solids results from emergence and growth of multiple internal micro- and mesocrack-like defects under applied loading. The latter process generates progressive degradation of elastic moduli, induces secondary anisotropy effects, volumetric dilatancy and other events, see f.ex. [Dragon and Mroz, 1979], [Kachanov, 1982], [Horii and Nemat-Nasser, 1983], [Andrieux et al., 1986]. Opening and closure of microcrack-sets under varying loads lead to further complex phenomena like recovery of some degraded moduli in the direction perpendicular to that of closed crack systems and frictional blocking and/or sliding over internal crack surfaces. In particular, when considering cyclic loading with some crack-sets constrained to closure, the initial unloading process is frequently friction-locked, exhibiting a high apparent rigidity. Further unloading may be dissipative if reverse multistage frictional sliding (reverse with respect to a loading branch) becomes active. The inelastic unloading is just one specific effect generated by damage by microcracking and a form of plasticity linked to frictional sliding on closed microcracks. It has been, very recently, one of challenges for modelling involving necessarily a form of damage-plasticity coupling.

The problem of modelling received some attention in the past, see e.g. [Walsh, 1969], [Kachanov, 1982], [Horii and Nemat-Nasser, 1983], [Andrieux et al., 1986], [Ju, 1991], [Krajcinovic et al., 1994], [Gambarotta and Lagomarsino, 1993], [Fond and Berthaud, 1995], [Lawn and Marshall, 1998]. For the most part the texts cited represent pertinent micromechanical studies, resulting in pleasant models capable to cover a limited range of stress-strain paths (two-dimensional, axisymmetric, etc.). The purpose of this paper is to address, in a synthetic manner, basic issues of the 3D – modelling proposed by the present authors [Halm and Dragon, 1998], [Dragon et al., 1999] attempting to provide an efficient, macroscopic – whereas micromechanically motivated – approach suitable for treatment of boundary value problems involving joint process of anisotropic damage by microcracking and frictional sliding at closed cracks. As mentioned above, the approach presented employs an internal variable formalism founded on some micro-mechanical background.

The approach presented is based on an anisotropic damage model, the 'basic version', proposed by [Dragon et al., 1994] and extended by [Halm and Dragon, 1996] to include the unilateral effect concerning normal stiffness recovery with respect to a mesocrack system constrained to closure. This extended version is summarized in Section 2 ; the reader interested in more detailed account can refer to the quoted papers. The present study is focused on the third level of modelling adding micro- and mesocrack friction plasticity effects to the extended damage model by Halm and Dragon. The corresponding developments are given in Sections 3 and 4. First, crack-friction induced plasticity is introduced at any arbitrary damage state (Section 3). Further (Section 4) damage and frictional sliding interaction is examined. Specific hysteretic effects resulting from coupled non-linear behaviour are commented.

The crucial issue of the control of microcrack closure and opening is addressed through Section 2-4. In Sections 3 and 4 the friction-induced stiffness recovery enters into consideration next to 'normal' frictionless recovery considered before.

The central simplifying hypothesis, conveyed through the developments proposed, consists in reduction of any real microcrack-set configuration to an equivalent configuration of three mutually orthogonal systems of parallel cracks characterized by three eigenvectors $\mathbf{v}^k$ ($k = 1,2,3$) and three non-negative eigenvalues D_k of the second-order damage tensor $\mathbf{D}$. In such a manner the damage-induced anisotropy is systematically limited to a form of orthotropy, see also [Kachanov, 1992].

2. ANISOTROPIC DAMAGE AND NORMAL UNILATERAL EFFECT

This section summarizes some essential features of the anisotropic damage model by [Dragon et al., 1994], [Halm and Dragon, 1996], which forms the framework for further developments in Sections 3 and 4. An objective of the damage model outlined below is to describe – in a realistic manner applicable to structural analysis – the process of microcrack-induced anisotropic degradation and relative inelastic behaviour of a rock-like quasi-brittle solid. It thus stipulates evaluation of effective (degraded) elastic moduli and an adequate description of the damage growth. The emphasis upon an 'open' formulation of the model has been made to allow further extension and couplings. It is based on the hypotheses and developments ordered below in the items from (i) to (v) :

(i) A single damage internal variable is constituted by a symmetric, second-order tensor **D** indicating orientation of microcrack set(s) as well as the dissipative mechanism under consideration, namely generation and growth of decohesion microsurfaces :

$$\mathbf{D} = \sum_i d^i(s)\, \mathbf{n}^i \otimes \mathbf{n}^i \tag{1}$$

The scalar density $d^i(s)$ is proportional to the extent s of decohesion surface while the unit normal vector $\mathbf{n}^i$ describes orientation of the i-th set of parallel crack-like defects. The form (1) is motivated by micromechanical considerations (see e.g. [Kachanov, 1992]) but in the context further on here the density d(s) is reckoned as a macroscopic quantity. The expression (1) is in itself a guiding microstructural interpretation of damage-related internal variable **D**. Since **D** is a symmetric second-order tensor it has three positive eigenvalues D_k (k = 1, 2, 3) and three orthogonal eigenvectors $\mathbf{v}^k$. This means that any system of microcracks (1), decomposed into 1,...,i,...n of subsystems of parallel mesocracks can be reduced to three equivalent orthogonal sets of cracks characterized by densities D_k and normal vectors $\mathbf{v}^k$.

$$\mathbf{D} = \sum_{k=1}^{3} D_k \mathbf{v}^k \otimes \mathbf{v}^k = \sum_{k=1}^{3} \mathbf{D}^k \tag{2}$$

(ii) The expression of damage-dependent strain energy (free energy per unit volume) w(**ε**,**D**) generates a form of elastic orthotropy – in connection to the three eigensystems (2) – for $\mathbf{D} \neq \mathbf{0}$; w is assumed a linear function of **D** and in this way corresponding to non-interacting cracks hypothesis. On the other hand, it contains linear and quadratic terms in **ε**. A particular invariant form given below (formula (4)) comprises a single linear term reading g tr(**ε**.**D**), g = const, corresponding to damage-induced residual phenomena. The damage induced macroscopic residual stress for $\boldsymbol{\varepsilon} = \mathbf{0}$ is thus explicitly obtained equal to g**D**. Inversely, for $\boldsymbol{\sigma} = \mathbf{0}$, non-zero residual strain is induced.

(iii) Under predominantly compressive loading, favourably oriented cracks close leading to an elastic moduli recovery phenomenon in the direction normal to the closed cracks. It is called here normal unilateral effect – in the absence of frictional sliding (the latter, when accounted for later, will induce a 'shear' recovery effect as well) – and requires more involved damage characterization. In fact, for a set of cracks constrained against opening a fourth-order tensorial density is necessary for a rigorous, micromechanically motivated description. A compromise solution has been advanced in [Halm and Dragon, 1996] between micromechanical considerations imposing an additional fourth-order damage variable and macroscopic modelling efficiency. The formulation maintains the orthotropy of the effective, elastic properties – instead of eventual more general anisotropy induced by a new fourth-order damage tensor – and the complementary fourth-order entity $\hat{\mathbf{D}}$, necessary to account for the normal unilateral effect, is directly assembled with the eigenvalues and eigenvectors of $\mathbf{D}$ and cannot therefore be considered as a new damage internal variable :

$$\hat{\mathbf{D}} = \sum_{k=1}^{3} D_k \, v^k \otimes v^k \otimes v^k \otimes v^k \tag{3}$$

(iv) A single scalar simultaneous invariant of $\hat{\mathbf{D}}$ and ε , namely $\varepsilon : \hat{\mathbf{D}} : \varepsilon$, completes the expression of the free energy $w[\varepsilon, \mathbf{D}, \hat{\mathbf{D}}(\mathbf{D})]$ (thermodynamic potential), with no additional material constant with respect to the basic form $w(\varepsilon,\mathbf{D})$ postulated in (ii). Rigorous continuity analysis in the framework of multilinear elasticity (for a given damage state), recast in [Halm and Dragon, 1996], leads to a simple microcrack closure condition for an equivalent set, namely : $v^k.\varepsilon.v^k \leq 0$. This condition is equivalent to that postulated in [Chaboche, 1992]. The detailed expression of $w(\varepsilon,\mathbf{D})$ including the supplementary term allowing for normal unilateral effect is :

$$\begin{aligned} w(\varepsilon, \mathbf{D}) = {} & \frac{1}{2}\lambda(\mathrm{tr}\,\varepsilon)^2 + \mu \mathrm{tr}(\varepsilon.\varepsilon) + g\mathrm{tr}(\varepsilon.\mathbf{D}) + \alpha\, \mathrm{tr}\,\varepsilon\, \mathrm{tr}(\varepsilon.\mathbf{D}) + 2\beta\, \mathrm{tr}(\varepsilon.\varepsilon.\mathbf{D}) \\ & - (\alpha + 2\beta)\, \varepsilon : \left[\sum_{k=1}^{3} H(-v^k.\varepsilon.v^k)\, D_k v^k \otimes v^k \otimes v^k \otimes v^k \right] : \varepsilon, \end{aligned} \tag{4}$$

where H is the classical Heaviside function ; α, β are material constants related to modified elastic moduli for a given damage state. λ and μ are conventional Lamé constants for elastic (non damaged) solid matrix.

The corresponding damage-induced orthotropic elasticity representation $\sigma(\varepsilon,\mathbf{D})$ and the damage driving (thermodynamic) force $\mathbf{F}^D$ are determined by corresponding partial derivation :

$$\begin{aligned} \sigma = \frac{\partial w}{\partial \varepsilon} = {} & \lambda(\mathrm{tr}\varepsilon)\mathbf{1} + 2\mu\varepsilon + g\mathbf{D} + \alpha[\mathrm{tr}(\varepsilon.\mathbf{D})\mathbf{1} + (\mathrm{tr}\varepsilon)\mathbf{D}] + 2\beta\,(\varepsilon.\mathbf{D} + \mathbf{D}.\varepsilon) \\ & - 2\,(\alpha + 2\beta) \sum_{k=1}^{3} H(-v^k.\varepsilon.v^k)\, D_k (v^k.\varepsilon.v^k)\, v^k \otimes v^k \end{aligned} \tag{5}$$

$$\mathbf{F}^D = -\frac{\partial w}{\partial \mathbf{D}} = -g\varepsilon - \alpha(\mathrm{tr}\varepsilon)\varepsilon - 2\beta(\varepsilon.\varepsilon) + (\alpha + 2\beta)\sum_{k=1}^{3} H(-v^k.\varepsilon.v^k).(v^k.\varepsilon.v^k)^2 v^k \otimes v^k \tag{6}$$

In spite of the presence of the Heaviside function $H(-v^k.\varepsilon.v^k)$, w, σ and $\mathbf{F}^D$ remain continuous when passing from the open mesocracks configuration to the closed mesocracks configuration and vice versa.

(v) The evolution of $\mathbf{D}$, corresponding to the brittle, splitting-like crack kinetics, has been found to follow the normality rule with respect to a criterion in the space of components of the proper thermodynamic force (affinity) $\mathbf{F}^D$. The damage evolution is thus apparently following the principle of maximum (damage) dissipation, and is related here to tensile (positive) straining $\boldsymbol{\varepsilon}^+$ and to actual damage pattern. It should be stressed however that the particular damage criterion $f(\mathbf{F}^D,\mathbf{D}) \le 0$ is explicitly dependent on the part $\mathbf{F}^{D1+} = -\, g\boldsymbol{\varepsilon}^+ = \mathbf{F}^D - \mathbf{F}^{D2} - \mathbf{F}^{D1-}$ of the driving force $\mathbf{F}^D$. $\mathbf{F}^{D1}$ is the strain energy release rate term related to residual 'locked' effects : $\mathbf{F}^{D1} = -\, g\boldsymbol{\varepsilon}$, $\mathbf{F}^{D2}$ represents the remaining recoverable energy release rate. The former term is decomposed into the splitting part $\mathbf{F}^{D1+} = -\, g\boldsymbol{\varepsilon}^+$, $\boldsymbol{\varepsilon}^+ = \mathbf{P}^+:\boldsymbol{\varepsilon}$, with $\mathbf{P}^+$ a positive fourth-order projection operator selecting positive eigenvalues from strain, and the non-splitting part $\mathbf{F}^{D1-} = -\, g(\boldsymbol{\varepsilon} - \boldsymbol{\varepsilon}^+)$. The damage criterion and rate-independent damage evolution law are thus as follows :

$$
\begin{aligned}
f\left(\mathbf{F}^D - \mathbf{F}^{D2} - \mathbf{F}^{D1-};\mathbf{D}\right) = &\sqrt{\frac{1}{2}\mathrm{tr}\left[\left(\mathbf{F}^D - \mathbf{F}^{D2} - \mathbf{F}^{D1-}\right).\left(\mathbf{F}^D - \mathbf{F}^{D2} - \mathbf{F}^{D1-}\right)\right]} \\
&+ B\,\mathrm{tr}\left[\left(\mathbf{F}^D - \mathbf{F}^{D2} - \mathbf{F}^{D1-}\right).\mathbf{D}\right] - \left(C_o + C_1\mathrm{tr}\mathbf{D}\right) \le 0
\end{aligned}
\tag{7}
$$

$$
\dot{\mathbf{D}} = \Lambda_D \frac{\partial f}{\partial \mathbf{F}^D} = \begin{cases} 0 & \text{if } f < 0 \text{ or } f = 0,\ \dot{f} < 0 \\[2ex] \Lambda_D\left[\dfrac{\boldsymbol{\varepsilon}^+}{\sqrt{2\mathrm{tr}(\boldsymbol{\varepsilon}^+.\boldsymbol{\varepsilon}^+)}} + B\mathbf{D}\right] & \text{if } f = 0 \text{ and } \dot{f} = 0\ ;\ \Lambda_D \ge 0 \end{cases}
\tag{8}
$$

Remarks :

The damage criterion (7) transferred to stress-space (see [Dragon et al., 1994]) represents a strongly pressure-sensitive surface with a marked dissymmetry of traction vs compression limits.

The fourth-order tensor $\hat{\mathbf{D}}$ depends entirely on $\mathbf{D}$ (see definition (3)) ; it does not require a separate evolution law.

In numerical calculations any loading path is considered as a collection of **D**-proportional segments. The form of Eqn (6) is valid for such a segment, i.e. for a given configuration of principal directions of **D**. Otherwise it should be completed to account for a novel configuration of the tensor $\hat{\mathbf{D}}$.

A purely implicit numerical integration scheme has been applied for the model at stake. Indeed it appears perfectly compatible with the latter. Not only the implicit scheme ensures unconditionally stability properties of an algorithm but the value of Λ_D can be determined at each step as the solution of a linear equation (equation enforcing incremental damage consistency) and does not require iterative procedure unlike for most of elastoplastic models, see [Dragon et al., 2000].

The above non-linear and anisotropic model contains eight material constants : λ, μ, α, β, g, B, C_o and C_1, which can be relatively easily determined. This problem as well as the predictive capacities of the model are illustrated and commented in [Dragon et al., 1994], [Halm and Dragon, 1996], [Dragon et al., 2000]. It is shown in particular that volumetric dilatancy effects resulting from pronounced damage are well evidenced and conformable to experimental data. Furthermore, when examining loading programmes involving a compressive stage leading to microcracks closure (e.g. the lateral overloading sequence following axisymmetric 'triaxial' compression test) one observes orderly stiffness recovery effects, [Halm and Dragon, 1996].

3. MESOCRACK FRICTION INDUCED PLASTICITY

The unilateral normal effect included in the model summarized in Section 2 allows a moduli recovery in the direction normal to the closed mesocracks. It fails to capture a shear moduli recovery in the direction parallel to the crack plane, resulting from some blocking of mesocrack lips displacement due to roughness and corresponding friction phenomena. Experimental data involving loading-unloading cycles for specimens subjected to torsion and hydrostatic compression for instance show hysteretic effects generated by such a blocking and subsequent frictional sliding on closed mesocrack lips. The beginning of unloading is characterized by a quasi-vertical curve while further decreasing slope is linked to progressive sliding, see for example [Pecqueur, 1995]. Some attempts of micromechanical modelling of the phenomena deserve attention. However they are not directly operational for an efficient structural analysis. Earlier attempts [Kachanov, 1982], [Horii and Nemat-Nasser, 1983] consider the influence of friction on effective moduli but do not provide satisfactory thermodynamic interpretation of sliding evolution. Most of existing approaches are limited to two-dimensional analyses, as e.g. [Andrieux et al., 1986], with the notable exception of the more recent work by [Gambarotta and Lagomarsino, 1993].

3.1 Elastic-damage-and-friction response

Accounting for the global strain expression for a representative volume of a solid with microcracks and, in particular, for the specific contribution of the system (i) of parallel cracks, the sliding variable, for the set (i) at stake, is chosen in the form :

$$\gamma^i = \frac{s^i \xi^i}{V} \operatorname{sym}(\mathbf{n} \otimes \mathbf{g})^i, \tag{9}$$

the symmetrisation being operated for the expression in parentheses. ξ^i represents the amount of sliding in the direction $\mathbf{g}^i, \mathbf{g}^i \perp \mathbf{n}^i$. The similarity with (1) is striking : as for $\mathbf{D}$, the form of γ is motivated by micromechanics ; as for d(s) the quantity $s^i \xi^i / V$ cannot be explicitly calculated in the framework of a macroscopic model. Moreover, as any system of microcracks represented by $\mathbf{D}$ reduces to three equivalent sets according to (2), the sliding tensor γ can be written in the analogous manner :

$$\gamma = \sum_{k=1}^{3} \frac{s^k \xi^k}{V} \operatorname{sym}(\mathbf{v} \otimes \mathbf{g})^k = \gamma_k \operatorname{sym}(\mathbf{v} \otimes \mathbf{g})^k = \sum_{k=1}^{3} \gamma^k$$

where $\mathbf{v}^k$, k = 1,2,3 are $\mathbf{D}$-eigenvectors.

The objective here is to argue for an enlarged form of the free-energy function $w(\varepsilon;\mathbf{D},\gamma)$ accounting for the frictional blocking and sliding effects for closed crack sets.

Let us consider the transition from open cracks to closed ones, assuming friction resistant lips when in contact. The crack-open form of (4), applies for the former case. When the cracks are closed and blocked by friction resistance at a given γ, the shear modulus μ is recovered and this should be properly reflected in the new modified expression $w(\varepsilon;\mathbf{D},\gamma)$. The β-term governing the shear moduli degradation should be counterbalanced in this expression. The α-term, having no influence on shear moduli, enters as before. Additional invariants including γ can be only simultaneous $(\gamma,\mathbf{D})$-invariants as there is no sliding on crack lips in the absence of damage. As from (9) one infers tr $\gamma = 0$ and tr $(\gamma.\mathbf{D}) = 0$ (for conservative damage axes), only two simultaneous invariants of ε, γ and $\mathbf{D}$ convey useful information. They are : tr $(\varepsilon.\gamma.\mathbf{D})$ and tr $(\gamma.\gamma.\mathbf{D})$. The argument for the quantity including β in the last term of (4) was to restitute the normal stiffness reduced by the term 2βtr $(\varepsilon.\varepsilon.\mathbf{D})$ in the first line of (4), but since this latter term is going to be counterbalanced, the former quantity has to disappear from w. Doing so allows one to write the expression $w(\varepsilon;\mathbf{D},\gamma)$ for closed friction-resistant crack lips in the form (for a single crack system) :

$$\begin{aligned} w(\varepsilon,\mathbf{D},\gamma) &= \frac{1}{2}\lambda(\mathrm{tr}\varepsilon)^2 + \mu\,\mathrm{tr}(\varepsilon.\varepsilon) + g\,\mathrm{tr}(\varepsilon.\mathbf{D}) + \alpha\,\mathrm{tr}\varepsilon\,\mathrm{tr}(\varepsilon.\mathbf{D}) \\ &- \alpha(\varepsilon:\hat{\mathbf{D}}:\varepsilon) + \eta_1\mathrm{tr}(\varepsilon.\gamma.\mathbf{D}) + 2\eta_2\mathrm{tr}(\gamma.\gamma.\mathbf{D}) \end{aligned} \tag{10}$$

where η_1 and η_2 are material constants to be identified.

From the micromechanics viewpoint there are infinity of crack-closure paths possible (straight, slantwise, mixed,...). The macroscopic model continuity requires continuity for expressions of $w(\varepsilon;\mathbf{D},\gamma)$ and $\boldsymbol{\sigma}$ for crack opening-to-closure (and reverse) transition. This leads to the following condition at the closure-point :

$$\begin{cases} \varepsilon.\mathbf{D} = \gamma.\mathbf{D} \\ \mathbf{D}.\varepsilon = \mathbf{D}.\gamma \end{cases} \quad \Leftrightarrow \quad \gamma_{ij} = \mathrm{sym}(\varepsilon_{ik}v_k v_j) \;\text{ at closure-point} \tag{11}$$

The latter formula constitutes an initialization for the sliding variable γ and can be explained as follows : at closure point, the sliding quantity γ is equal to the strain ε in the crack plane, the matrix transmits its deformation to the crack.

According to the continuity conditions for multilinear elasticity [Wesolowski, 1969], [Curnier et al., 1995] already employed in [Halm and Dragon, 1996] in the context of unilateral normal effect, see eqns (4)-(6) Sect. 2, the jump of effective elastic stiffness $[[\mathbf{C}^*]]$ between open cracks (the corresponding energy is designated by w° below) and closed crack respective configurations should be a singular operator. It is sufficient that its all second-order determinants be equal to zero.

In the present context – eqn (10) at the very closure point, taking into account (11) – $[[\mathbf{C}^*]]$ is given as follows :

$$[[\mathbf{C}^*]] = \left.\frac{\partial^2 w}{\partial\varepsilon\partial\varepsilon}\right|_{\gamma,\mathbf{D}} - \left.\frac{\partial^2 w^\circ}{\partial\varepsilon\partial\varepsilon}\right|_{\mathbf{D}};$$

$$[[C^*_{ijkl}]] = \left(\frac{1}{2}\eta_1 + \eta_2 - \beta\right)\left(\delta_{ik}D_{jl} + \delta_{jl}D_{ik} + \delta_{il}D_{jk} + \delta_{jk}D_{il}\right) - 2\alpha\hat{D}_{ijkl}$$

The above-mentioned singularity requirement and the additional stronger condition allowing no stress jump in the strain space across the surface $\mathbf{v}.\varepsilon.\mathbf{v} = 0$, applied by [Halm and Dragon, 1998] (in the way similar as in [Halm and Dragon, 1996]) lead respectively to :

$$\begin{cases} \frac{1}{2}\eta_1 + \eta_2 - \beta = 0 \\ \eta_1 = 4\beta \end{cases} \tag{12}$$

The free-energy $w(\varepsilon;\mathbf{D},\gamma)$ can now be written as follows (for either open or closed cracks) :

$$
\begin{aligned}
w(\varepsilon;\mathbf{D},\gamma) &= \frac{1}{2}\lambda(\mathrm{tr}\varepsilon)^2 + \mu \mathrm{tr}(\varepsilon.\varepsilon) + g\mathrm{tr}(\varepsilon.\mathbf{D}) + \alpha \mathrm{tr}\varepsilon \mathrm{tr}(\varepsilon.\mathbf{D}) + 2\beta \mathrm{tr}(\varepsilon.\varepsilon.\mathbf{D}) \\
&+ H(-\mathbf{v}.\varepsilon.\mathbf{v})\left[-\alpha\varepsilon : \hat{\mathbf{D}} : \varepsilon - 2\beta \mathrm{tr}(\varepsilon.\varepsilon.\mathbf{D}) + 4\beta \mathrm{tr}(\varepsilon.\gamma.\mathbf{D}) - 2\beta \mathrm{tr}(\gamma.\gamma.\mathbf{D})\right]
\end{aligned}
\tag{13}
$$

The expression (13) can be generalized to three non-interactive equivalent crack sets represented by eigenvectors $\mathbf{v}^k$ associated with the principal components D_k, k=1,2,3. One can select the k-th set using the following projection operator $\mathbf{L}^k$:

$$
\begin{aligned}
\mathbf{L}^k &= \mathbf{v}^k \otimes \mathbf{v}^k \otimes \mathbf{v}^k \otimes \mathbf{v}^k \\
\mathbf{D}^k &= D_k \mathbf{v}^k \otimes \mathbf{v}^k = \mathbf{L}^k : \mathbf{D}
\end{aligned}
\tag{14}
$$

This allows to write counterpart equations of (4)-(6) independently for each equivalent set, all possible configurations being included (open or closed, sliding or non sliding sets) :

$$
\begin{aligned}
w(\varepsilon,\mathbf{D},\gamma) &= \frac{1}{2}\lambda(\mathrm{tr}\varepsilon)^2 + \mu \mathrm{tr}(\varepsilon.\varepsilon) + g\mathrm{tr}(\varepsilon.\mathbf{D}) + \alpha \mathrm{tr}\varepsilon \mathrm{tr}(\varepsilon.\mathbf{D}) + 2\beta \mathrm{tr}(\varepsilon.\varepsilon.\mathbf{D}) \\
&+ \sum_{k=1}^{3} H(-\mathbf{v}^k.\varepsilon.\mathbf{v}^k)\left[-\alpha\varepsilon : (D_k \mathbf{L}^k) : \varepsilon - 2\beta \mathrm{tr}(\varepsilon.\varepsilon.\mathbf{D}^k) + 4\beta \mathrm{tr}(\varepsilon.\gamma^k.\mathbf{D}^k) - 2\beta \mathrm{tr}(\gamma^k.\gamma^k.\mathbf{D}^k)\right]
\end{aligned}
\tag{15}
$$

$$
\begin{aligned}
\sigma = \frac{\partial w}{\partial \varepsilon} &= \lambda(\mathrm{tr}\varepsilon)\,\mathbf{1} + 2\mu\varepsilon + g\mathbf{D} + \alpha\left[\mathrm{tr}(\varepsilon.\mathbf{D})\,\mathbf{1} + (\mathrm{tr}\varepsilon)\mathbf{D}\right] + 2\beta(\varepsilon.\mathbf{D} + \mathbf{D}.\varepsilon) \\
&+ \sum_{k=1}^{3} H(-\mathbf{v}^k.\varepsilon.\mathbf{v}^k)\left[-2\alpha D_k (\mathbf{v}^k.\varepsilon.\mathbf{v}^k)\mathbf{v}^k \otimes \mathbf{v}^k - 2\beta(\varepsilon.\mathbf{D}^k + \mathbf{D}^k.\varepsilon) + 2\beta(\gamma^k.\mathbf{D}^k + \mathbf{D}^k.\gamma^k)\right]
\end{aligned}
\tag{16}
$$

$$
\begin{aligned}
\mathbf{F}^D = -\frac{\partial w}{\partial \mathbf{D}} &= -g\varepsilon - \alpha(\mathrm{tr}\varepsilon)\varepsilon - 2\beta(\varepsilon.\varepsilon) \\
&+ \sum_{k=1}^{3} H(-\mathbf{v}^k.\varepsilon.\mathbf{v}^k)\left[\alpha(\mathbf{v}^k.\varepsilon.\mathbf{v}^k)^2 \mathbf{v}^k \otimes \mathbf{v}^k + 2\beta\mathbf{L}^k : (\varepsilon.\varepsilon) - 4\beta\mathbf{L}^k : (\varepsilon.\gamma^k) + 2\beta\mathbf{L}^k : (\gamma^k.\gamma^k)\right]
\end{aligned}
\tag{17}
$$

As each equivalent set of the normal $\mathbf{v}^k$ is to be considered independently, the corresponding affinity (thermodynamic force) is :

$$
\mathbf{F}^{\gamma k} = -\frac{\partial w}{\partial \gamma^k} = H(-\mathbf{v}^k.\varepsilon.\mathbf{v}^k)\left[-2\beta(\varepsilon.\mathbf{D}^k + \mathbf{D}^k.\varepsilon) + 2\beta(\gamma^k.\mathbf{D}^k + \mathbf{D}^k.\gamma^k)\right]
\tag{18}
$$

The remark concerning eqn (6), Section 2, stating its validity for a **D**-proportional segment, i.e. for a given configuration of principal direction of **D**, is still in force for eqn (17).

3.2. Sliding criterion and evolution

The model herein considers frictional non-sliding/sliding phenomena on mesocrack lips on a macroscopic scale, by an approach similar to that to damage, notwithstanding the micromechanical background and interpretations of **D** and γ. So, the Coulomb criterion form, function of the corresponding shear and normal tractions on a crack lip, employed in micromechanical models ([Horii and Nemat-Nasser, 1983], [Andrieux et al., 1986], [Gambarotta and Lagomarsino, 1983]), is methodologically less suitable in the present context. The pertinent thermodynamic affinity governing frictional sliding on an equivalent system k, (k=1,2,3) is the entity $\mathbf{F}^{\gamma k}$ defined above as the strain energy release-rate with respect to γ^k.

The frictional non sliding/sliding complementary law is based on the hypotheses as follows :

(i) The sliding criterion depends explicitly on the norm of the tangential part $\mathbf{F}^{\gamma Tk}$ of the "force" $\mathbf{F}^{\gamma k}$ and on the normal strain $\mathbf{v}^k.\boldsymbol{\varepsilon}.\mathbf{v}^k$ consecutively to the strain-related representation of the energy w and the crack-closure criterion at stake ($\mathbf{v}^k.\boldsymbol{\varepsilon}.\mathbf{v}^k \leq 0$).

(ii) Contrarily to inconsistencies relative to the normality rule in the classical Coulomb framework affected by appearance of a normal separating velocity (cf. for example [Curnier, 1984]) a standard scheme in the space of forces conjugate to γ^k keeps physical pertinence. The normality rule appears to relate the frictional sliding rate to the tangential force $\mathbf{F}^{\gamma Tk}$ indicating its leaning to the crack plane (for a $\mathbf{D}^k$-proportional loading segment).

Consequently, the corresponding convex reversibility domain $h^k \leq 0$ can be written as :

$$h^k\left(\mathbf{F}^{\gamma k}-\mathbf{F}^{\gamma Nk},\mathbf{v}^k.\boldsymbol{\varepsilon}.\mathbf{v}^k\right)=\sqrt{\frac{1}{2}\mathrm{tr}\left[\left(\mathbf{F}^{\gamma k}-\mathbf{F}^{\gamma Nk}\right)\left(\mathbf{F}^{\gamma k}-\mathbf{F}^{\gamma Nk}\right)\right]}+\rho\mathbf{v}^k.\boldsymbol{\varepsilon}.\mathbf{v}^k\leq 0 \quad \text{if } \mathbf{v}^k.\boldsymbol{\varepsilon}.\mathbf{v}^k\leq 0 \tag{19}$$

where ρ is a material constant, a strain-related friction coefficient in the space ($\mathbf{F}^{\gamma k}$,$\boldsymbol{\varepsilon}$) and

$$\mathbf{F}^{\gamma k}=\mathbf{F}^{\gamma Tk}+\mathbf{F}^{\gamma Nk} \quad ; \quad \mathbf{F}^{\gamma Tk}=\mathbf{F}^{\gamma k}-(\mathbf{v}^k.\mathbf{F}^{\gamma k}.\mathbf{v}^k)\,\mathbf{v}^k\otimes\mathbf{v}^k$$
$$\mathbf{F}^{\gamma Nk}=(\mathbf{v}^k.\mathbf{F}^{\gamma k}.\mathbf{v}^k)\,\mathbf{v}^k\otimes\mathbf{v}^k \tag{20}$$

The normality rule for $\dot{\gamma}^k$ is then

$$\dot{\gamma}^k=\Lambda^k_\gamma\frac{\partial h^k\left(\mathbf{F}^{\gamma k}-\mathbf{F}^{\gamma Nk},\mathbf{v}^k.\boldsymbol{\varepsilon}.\mathbf{v}^k\right)}{\partial\mathbf{F}^{\gamma k}}=\begin{cases}0 & \text{if } h^k<0 \text{ or } h^k=0,\dot{h}^k<0\\ \Lambda^k_\gamma\dfrac{\mathbf{F}^{\gamma Tk}}{\sqrt{2\mathrm{tr}\left(\mathbf{F}^{\gamma Tk}.\mathbf{F}^{\gamma Tk}\right)}} & \text{if } h^k=0 \text{ and } \dot{h}^k=0\ ;\ \Lambda^k_\gamma\geq 0\end{cases} \tag{21}$$

Detailed comments on salient aspects of the criterion $h^k = 0$ in the strain space and in the stress space are given by [Halm and Dragon, 1998].

By examining the complete set (15)-(21) of the equations of the model, one can see that the frictional sliding does not sweep away the relative simplicity of the enlarged model (see the end of Section 2). Only one additional constant ρ adds to eight material constants (λ,μ,α,β,C_o,C_1,g,B).

4. DAMAGE AND FRICTIONAL SLIDING INTERACTION. COUPLED MODEL

The model completed in Section 3 incorporating friction-induced blocking and sliding on equivalent mesocrack-sets is valid for a given ('freezed') damage state or for conservative damage evolution ($\mathbf{D}^k$-proportional loading paths). It has proved conclusive in representing multistage loading-and-unloading dissipative cycles due to blocking-and-sliding sequences, see [Halm and Dragon, 1998] for illustrations.

The splitting-like damage kinetics considered in Section 2 is approximately valid for closed sliding mesocracks even when some branching occurs, see for example [Horii and Nemat-Nasser, 1985] for some experimental insight. This type of kinetics will be still considered as the predominant mechanism furthest for $\mathbf{D}^k$-non-proportional loading. This means the complementary damage law (7)-(8) being reconducted for more complex stress-strain paths involving varying $\mathbf{D}^k$ orientations. However, as the frictional blocking-and-sliding is inevitably affecting the stress-strain response, so the stress threshold corresponding to damage criterion f=0 is subsequently affected.

This is mostly the sliding complementary rule (19)-(21) which needs to be perfected to describe fairly the $\mathbf{D}^k$-non-proportional loading paths. If the principal axes of $\mathbf{D}^k$ rotate the orthogonality $\gamma^k : \mathbf{D}^k = 0$ is no longer true and discontinuities may arise, especially for crack closure-opening transition. So, an enhanced form of $h^k \le 0$ needs to account for the $\mathbf{D}^k$-axes rotation. The form (19), depending on $\mathbf{F}^{\gamma Tk}$, produced – via normality rule (21) – sliding $\dot{\gamma}^k$ in the mesocrack plane. A judicious modification of this basic assumption should be compatible with sliding and damage departure from the actual mesocrack equivalent plane. This is achieved by means of the following partition of $\mathbf{F}^{\gamma k}$, given below for a single crack set of normal $\mathbf{v}^k$:

$$\mathbf{F}^{\gamma k} = \mathbf{F}^{\gamma Tk} + \mathbf{F}^{\gamma Nk} = \underbrace{\mathbf{F}^{\gamma Tk} + 4\beta(\gamma^k : \mathbf{D}^k)\mathbf{v}^k \otimes \mathbf{v}^k}_{} - 4\beta(\varepsilon : \mathbf{D}^k)\mathbf{v}^k \otimes \mathbf{v}^k \tag{22}$$

$$= \qquad \mathbf{F}^k \qquad - 4\beta(\varepsilon : \mathbf{D}^k)\mathbf{v}^k \otimes \mathbf{v}^k$$

$\mathbf{F}^k$ is the appropriate part of $\mathbf{F}^{\gamma k}$ to enter the more general expression of $h^k \le 0$ suitable for the model including $\mathbf{D}^k$-axes rotation. First, one obtains that for $\mathbf{D}^k$-proportional loading $\mathbf{F}^k$ reduces to $\mathbf{F}^{\gamma Tk}$ (as $\gamma^k : \mathbf{D}^k = 0$) and the new representation $h^k(\mathbf{F}^k, \mathbf{v}^k.\varepsilon.\mathbf{v}^k)$ reduces to (19). Secondly, the above-mentioned, crucial stress continuity problem is effectively dealt with. In fact, comparing (11),(18),(20),(22), one can see that the closure-opening transition point for sliding crack-set can be alternatively defined as :

$$\mathbf{F}^k = \mathbf{F}^{\gamma k} = \mathbf{0} \qquad \Leftrightarrow \qquad \varepsilon.\mathbf{D}^k + \mathbf{D}^k.\varepsilon = \gamma^k.\mathbf{D}^k + \mathbf{D}^k.\gamma^k \tag{23}$$

Despite the fact that the above equation represents weaker condition than (11), it allows to verify the singularity requirement for [[C*]] (cf. Section 3.1) leading to the stress continuity.

It can be remarked that though $\gamma^k : \mathbf{D}^k \ne 0$ as equivalent crack-axes rotate, no additional invariants are necessary in the strain energy expression (15). They are not required by the continuity considerations (see above) and bring neither significant information. For example, introducing $\mathrm{tr}\,\varepsilon\ \mathrm{tr}(\gamma^k.\mathbf{D}^k)$, $\mathrm{tr}\,\gamma^k\ \mathrm{tr}(\gamma^k.\mathbf{D}^k)$ and $\gamma^k : D_k\, \mathbf{L}^k : \gamma^k$ contributes to no more record on shear moduli degradation than existing invariants $\mathrm{tr}(\varepsilon.\gamma^k.\mathbf{D}^k)$ and $\mathrm{tr}(\gamma^k.\gamma^k.\mathbf{D}^k)$.

The above considerations lead to the following improved expression for the sliding complementary rule :

$$h^k(\mathbf{F}^k, v^k.\varepsilon.v^k) = \sqrt{\frac{1}{2}\mathrm{tr}(\mathbf{F}^k.\mathbf{F}^k)} + \rho v^k.\varepsilon.v^k \leq 0 \tag{24}$$

$$\dot{\gamma}^k = \Lambda_\gamma^k \frac{\partial h(\mathbf{F}^k, v^k.\varepsilon.v^k)}{\partial \mathbf{F}^{\gamma k}} = \begin{cases} 0 & \text{if } h^k < 0 \text{ or } h^k = 0, \dot{h}^k < 0 \\ \Lambda_\gamma^k \dfrac{\mathbf{F}^k}{\sqrt{2\mathrm{tr}(\mathbf{F}^k.\mathbf{F}^k)}} & \text{if } h^k = 0 \text{ and } \dot{h}^k = 0\,;\ \Lambda_\gamma^k \geq 0 \end{cases} \tag{25}$$

The direction of $\dot{\gamma}^k$ is thus allowed to leave the equivalent crack-set plane consecutively to the rotation of the latter. The incremental procedure leading to numerical integration of the equations of the coupled model is summarized in [Dragon et al., 2000]

5. CONCLUSION

Two dissipative mechanisms, namely damage by microcracking and frictional sliding over internal crack surfaces, are considered in the framework of the three-dimensional model presented, applicable for quasi-brittle rock-like solids. The internal variable formalism employed for the joint process under consideration is based on second-order tensorial representation of damage and sliding effects respectively. The critical issue of the control of microcrack closure and opening phenomena is dealt with in the spirit of continuum damage approach associated with rigorous analysis of the stiffness recovery for closed microcracks. The basic simplifying assumption, carried on through the proposed modelling, consists in contraction of any real microcrack-set configuration (containing a number of sets) to an equivalent configuration of three mutually orthogonal systems. The analysis has been restricted furthermore to non-interacting crack systems regarding the energy expression even if the microcrack related damage evolution depends on current damage state.

The model constitutes an attempt to construct a tool applicable for efficient structural analysis through its relatively easy identifiability and algorithmic feasibility. In the same time a strong micromechanical motivation is being preserved in corresponding developments. It can be situated as an intermediary modelling approach between genuine micromechanical studies and pure phenomenological ones.

REFERENCES

[Andrieux et al., 1986] Andrieux, S.; Bamberger, Y.; Marigo, J.-J.; Un modèle de matériau microfissuré pour les bétons et les roches; *J. Méc. Théor. Appl.* 5, 471-513

[Chaboche, 1992] Chaboche, J.-L.; Une nouvelle condition unilatérale pour décrire le comportement des matériaux avec dommage anisotrope; *C.R. Acad. Sci. Paris*, t. 314, II, 1395-1401

[Curnier, 1984] Curnier, A.; A theory of friction; *Int. J. Solids Struct.* 20, 637-647

[Curnier et al., 1995] Curnier, A.; He, Q.; Zysset, P.; Conewise linear elastic materials; *J. Elasticity*, 37, 1-38

[Dragon and Mroz, 1979] Dragon, A.; Mroz, Z.; A continuum model for plastic brittle-behaviour of rock and concrete; *Int. J. Engng Sci.*, 17, 121-137

[Dragon et al., 1994] Dragon, A.; Cormery, F.; Désoyer, T.; Halm, D.; Localised failure analysis using damage models; In: *Localisation and Bifurcation Theory for Soils and Rocks*, Chambon, Desrues, Vardoulakis (eds), pp. 127-140; Balkema; ISBN 90 5411 9

[Dragon et al., 2000] Dragon, A.; Halm, D.; Désoyer, T.; Anisotropic damage in quasi-brittle solids : modelling, computational issues and applications; *Comput. Methods Appl. Mech. Engng.* [in press]

[Fond and Berthaud, 1995] Fond, C.; Berthaud, Y.; Extensions of the pseudo-tractions technique for friction in cracks, circular cavities and external boundaries; effects of the interactions on the homogenized stiffness; *Int. J. Fracture* 74, 1-28

[Gambarotta and Lagomarsino, 1993] Gambarotta, L.; Lagomarsino, S.; A microcrack damage model for brittle materials; *Int. J. Solids Struct.*, 30, 177-198

[Germain et al., 1983] Germain, P.; Nguyen, Q.S.; Suquet, P.; Continuum thermodynamics, *Trans. ASME, J. Appl. Mechanics*, 50, 1010-1020

[Germain, 1986] Germain, P.; Sur quelques concepts fondamentaux de la Mécanique; In; *Large Deformation of Solids : Physical Basis and Mathematical Modelling*, Gittus, Zarka, Nemat-Nasser (eds), pp. 3-25; Elsevier; ISBN 1-85166-016-X

[Halm and Dragon, 1996] Halm, D.; Dragon, A.; A model of anisotropic damage by mesocrack growth; unilateral effect; *Int. J. Damage Mech.*; 5, 384-402

[Halm and Dragon, 1998] Halm, D.; Dragon, A.; An anisotropic model of damage and frictional sliding for brittle materials; *Eur. J. Mech. A/Solids*, 17, 439-460

[Horii and Nemat-Nasser, 1983] Horii, H.; Nemat-Nasser, S.; Overall moduli of solids with microcracks : load-induced anisotropy; *J. Mech. Phys. Solids*, 31, 151-171

[Horii and Nemat-Nasser, 1985] Horii, H.; Nemat-Nasser, S.; Compression-induced microcrack growth in brittle solids : axial splitting and shear failure; *J. Geophys. Rech.* 90, 3105-3125

[Ju, 1991] Ju, J.W.; On two-dimensional self-consistent micromechanical damage models for brittle solids; *Int. J. Solids Struct.*, 27, 227-258

[Kachanov, 1982] Kachanov, M.; A microcrack model of rock inelasticity – part I : frictional sliding on microcracks; *Mech. Mater.* 1, 19-27

[Kachanov, 1992] Kachanov, M.; Effective elastic properties of cracked solids : critical review of some basic concepts; *Appl. Mech. Rev.* 45, 304-335

[Krajcinovic et al., 1994] Krajcinovic, D.; Basista, M.; Sumarac, D.; Basic principles; In: *Damage Mechanics of Composite Materials*, Talreja (ed), pp. 1-51; Elsevier; ISBN 0-444-42525-X

[Lawn and Marshall, 1998] Lawn, B.R.; Marshall, D.B.; Nonlinear stress-strain curves for solids containing closed cracks with friction; *J. Mech. Phys. Solids*, 46, 85-113

[Pecqueur, 1995] Pecqueur, G.; Etude expérimentale et modélisation du comportement d'une craie et d'un grès en torsion, Doctoral Thesis, Univ. Lille I

[Walsh, 1969] Walsh, J.B.; The effect of cracks on the uniaxial elastic compression of rocks; *J. Geophys. Rech.*, 70, 399-411

[Wesolowski, 1969] Wesolowski Z.; Elastic material with different elastic constants in two regions of variability of deformation; *Arch. Mech.*, 21, 449-468

Structural plastic microbuckling and compressive strength of long-fibre composite materials

Drapier S., Grandidier J.-C., Jochum C. and Potier-Ferry M.
Laboratoire de Physique et Mécanique des Matériaux,ISGMP
Île du Saulcy, 57045 Metz cedex 02, France
potier-ferry@lpmm.univ-metz.fr

Abstract : It is now well established that the axial compressive strength of organic long-fibre composites is limited by the development of a local instability of the fibres. This so-called microbuckling depends strongly on the coupling of the fibre initial geometric imperfection with the resin non-linear shear behaviour. In addition, it has been demonstrated experimentally that failure strains also depend on some structural parameters at the ply scale, the effect of which is less well known. We have been investigating this structure effect for several years from both experimental and theoretical points of view. In the present paper a model is proposed that is able to account for both local and structural parameters, with reasonable computation amounts. Then, by comparing predictions from this model with experiments, we demonstrate the necessity of accounting for the structure effect when designing composites against compression. Eventually, the last parameter still to be determined is the fibre initial imperfection which is thought to result from a fibre instability occuring during cure. A viscoelastic micromechanical model is presented here which very well captures the microbuckling of a single fibre induced by the resin thermosetting shrinkage.
Keywords: compressive strength, plastic microbuckling, structure effect, geometric imperfection, cure of composites.

1. INTRODUCTION

Predicting the mechanical behaviour of organic long fibre composites is in essence a difficult task. Especially, their compressive strength raises concerns both from experimental and theoretical points of view. Indeed, although this problem has been extensively studied for more than 30 years (since [Rosen, 1964]), predictions can rarely agree with experimental results.

The compressive strength is usually determined using direct compression tests that are very unstable and yield poor estimates. Conversely, bending fixtures yield higher strength while ensuring controlled fractures. Through the use of bending-compression devices, the influence of the structure at mesoscopic scale on the compressive failure of UD plies has been clearly established. As an example, under a bending loading, T300/914 unidirectional

G.A. Maugin et al. (eds.), Continuum Thermomechanics, 125-136.
© 2000 *Kluwer Academic Publishers.*

plies can withstand compressive strains greater than 2 %, whereas under a pure compression loading this strength is lower than 1.2 % ([t'Hart et al., 1991]). Furthermore, [Wisnom, 1991] showed that thickening the specimen lowers the strength, while [Grandidier, 1991] established that a high gradient of loading across the specimen thickness increases the compressive strength. [Wisnom, 1997] and [Rahier, 1998] recently confirmed these results, using a pin-ended buckling device in which hinge rotations can be restrained. Other experimental works reported in the literature show that, in laminates, the larger the number of consecutive plies in the loading direction, the lower their strength ([Grandsire-Vinçon, 1993]).

From a theoretical point of view, early research traced the origin of failure to the development of local plastic microbuckling ([Argon, 1972] [Budiansky, 1983]). The effect of the structure at ply scale on the failure mechanism of laminated composites has been studied only recently ([Grandidier, 1991] [Grandidier et al., 1992]). Since then, we have been extensively investigating such structure effects from both an experimental ([Grandidier, 1991] [Grandidier et al., 1992] [Rahier, 1998]) and a theoretical point of view ([Grandidier et al., 1992] [Drapier et al., 1996] [Drapier et al., 1999]). Here we aim at comparing predictions from our plastic microbuckling model including structure effect, and experimental results.

Eventually, in the scope of quantitative predictions the imperfection must be properly assessed, either measured or predicted. Since measurements on real composites are difficult to carry out, a viscoelastic micromechanical model is presented as a first attempt to predict the fibre imperfection initiated during the cure of the material.

2. STRUCTURAL MODEL FOR PLASTIC MICROBUCKLING

Although experimentally the structure effect has been demonstrated several times, it is very seldom, or at the best incompletely, tackled in the literature. In order to account for the influence of the structure on the microscopic instability, a non-linear microbuckling model is set at the mesoscopic scale. It aims at describing the failure mechanism with low computation amounts but accounting for every influential parameter ([Drapier et al., 1999]): size and shape of the fibre initial geometric imperfection, drop of stiffness induced by the plastic response of the matrix, and structural data across the laminate thickness.

2.1 Formulation of the mesoscopic problem

A bidimensional representation of a laminate is used (Figure 1) where $\mathbf{e_1}$ corresponds to the loading direction (0°). Displacement along $\mathbf{e_1}$ is $u(\mathbf{x})$ and displacement along $\mathbf{e_2}$ is $v(\mathbf{x})$. Stresses (second Piola-Kirchhoff tensor) are denoted S, and Green-Lagrange's strain tensor is γ. Based on works of [Grandidier, 1991] and [Grandidier et al., 1992], a formulation of the plastic microbuckling problem can be proposed. The corresponding equilibrium is presented here under the form of the virtual powers principle written for any virtual displacement field $\delta\mathbf{u}$ (1).

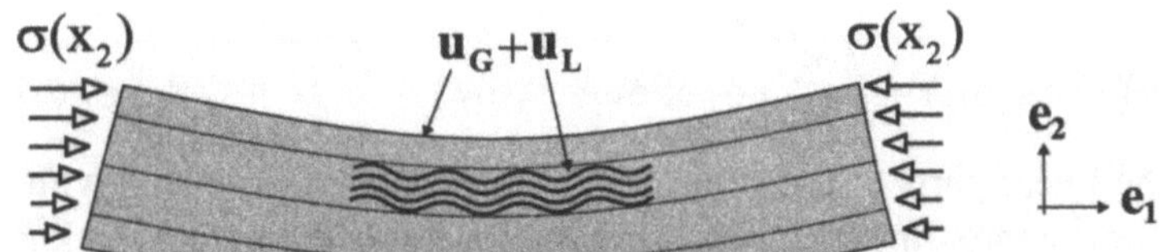

Figure 1 ; Bidimensional domain ω and splitting of the displacement field.

$$\int_\omega \left\{ f E_f r_{gf}^2 v" \delta v" + \mathbf{S}.\delta\gamma \right\} d\omega + < \mathbf{F}, \delta\mathbf{u} >= 0, \quad \forall \delta\mathbf{u} \tag{1}$$

where f is the fibre volume fraction, E_f is the fibre Young's modulus, r_{gf} is the fibre gyration radius and where $\mathbf{F}$ represents the external loading. The constitutive law (2) is anisotropic, the secant modulus tensor $\mathbf{L}$ being obtained from an explicit homogenisation formula based upon the properties of the constituents ([Gardin and Potier-Ferry, 1992]). Only the unreinforced material is assumed to be non-linear such that the changes (softening) in the constituve law induced by fibre microbuckling are simply described. It follows an isotropic law of J2 deformation type that yields good predictions of plastic buckling [Hutchinson, 1974].

$$\mathbf{S}(\gamma) = \mathbf{L}(\gamma).\gamma \tag{2}$$

This medium is not classical, due to the first term of (1) which represents the fibre bending stiffness and which is essential for predicting the effect of the structural data ([Gardin and Potier-Ferry, 1992] [Drapier et al., 1999]). The presence of this bending term has been justified by a homogenisation study using the multi- scale method ([Gardin and Potier-Ferry, 1992]) and also by comparing modes and buckling loads from this approach with micro-heterogeneous modelling results ([Drapier et al., 1996]).

Considering mainly uniaxial loadings and representing the initial fibre misalignment with a 'deflection field' $\mathbf{v_0}(\mathbf{x})$, we reduce the non-linear part of the strain tensor (3) to a single axial component :

$$\gamma(\mathbf{u}) = \frac{1}{2}\left(\nabla\mathbf{u} + \nabla^t\mathbf{u}\right) + \left(\frac{1}{2}\left(v'\right)^2 + v'v_0'\right) \mathbf{e_1} \otimes \mathbf{e_1} \tag{3}$$

where X' stands for the first derivative of any function X with respect to x_1, and similarly X'' is the corresponding second derivative.

2.2 Mesoscopic approach

Usually a numerical approach can be deduced from the continuous formulation through an adequate discretisation. In the present case, a further refinement is introduced which leads to a tractable model to represent this local short-wavelength phenomenon at the ply scale. The solution of the microbuckling problem is sought in the form of a displacement field (Figure 1) evolving at the scale of the structure (denoted $\mathbf{u_G}$), very locally modulated by a displacement field evolving at the ply scale (denoted $\mathbf{u_L}$) : $\mathbf{u}(\mathbf{x}) = \mathbf{u_G}(\mathbf{x}) + \mathbf{u_L}(\mathbf{x})$. With the hypothesis of rapid variations of $\mathbf{u_L}$ and slow variations of $\mathbf{u_G}$, the strain tensor can be simplified (4) and reads :

$$\gamma(\mathbf{u}) = \epsilon_{\mathbf{G}}(\mathbf{u_G}) + \left(\frac{v_G^{2'}}{2}\right)\mathbf{e_1}\otimes\mathbf{e_1} + \epsilon_{\mathbf{L}}(\mathbf{u_L}) + \left(\frac{v_L^{'2}}{2} + v_L' v_0'\right)\mathbf{e_1}\otimes\mathbf{e_1} \qquad (4)$$

with ϵ_G, ϵ_L the linear parts of the strain tensor, respectively associated with $\mathbf{u_G}$ and $\mathbf{u_L}$. Considering these approximations and assuming that displacement $\mathbf{u_G}$ is a known solution of the global equilibrium (1), we obtain the variational equation (5) describing the mesoscopic equilibrium, the solution of which is $\mathbf{u_L}(\mathbf{x})$:

$$\int_\omega \left\{ fE_f r_{gf}^2 v"_L \delta v"_L + \mathbf{S_L}(\gamma_\mathbf{G},\gamma_\mathbf{G}).\delta\gamma_\mathbf{L} + S_{11}(\gamma_\mathbf{G})(v_L' v_0')\delta v_L' \right\} d\omega = 0, \ \forall \delta\mathbf{u_L} \qquad (5)$$

where $\mathbf{S_L}(\gamma_\mathbf{G},\gamma_\mathbf{L}) = \mathbf{S}(\gamma_\mathbf{G}+\gamma_\mathbf{L}) - \mathbf{S}(\gamma_\mathbf{G})$. In this mesoscopic formulation the external loading appears through the global strain tensor which here is limited to its axial component sufficient to represent compression or bending-compression states : $\gamma_\mathbf{G}(\mathbf{u_G}) = \gamma_{G11}\mathbf{e_1}\otimes\mathbf{e_1}$.

2.3 Displacement approximation

In the framework of cellular instabilities, the displacement field approximation is chosen as a product of amplitudes across the ply thickness with a few Ritz basis functions in the fibre direction (6). This hypothesis also allows reduction of the bidimensional domain to a single wavelength in the fibre direction, greatly reducing computations. Ritz basis functions are selected so that both microbuckling elastic modes obtained in [Drapier et al., 1996] can be reproduced, and that a quasi-constant buckling stress can be obtained.

$$\mathbf{u}(\mathbf{x}) = \begin{cases} U_1(x_2)cos(kx_1) + U_2(x_2)sin(2kx_1) \\ V_1(x_2)sin(kx_1) + V_2(x_2)sin(3kx_1) \end{cases} \qquad (6)$$

where k is the wavenumber, and functions $U_i(x_2)$, $V_i(x_2)$ are the magnitudes of the displacement field which are discretised by three-noded finite element of Lagrange type. The imperfection $v_0(\mathbf{x})$ is represented similarly to $v(\mathbf{x})$. There is thus no limitation to any strain and imperfection distribution across the laminate thickness. This is important in order to account for the influence of structural parameters. Conversely, the Ritz approximation in the axial direction is more restrictive and this limits proper a representation of the post-microbuckling behaviour. Consequently we focus on the response up to the maximum load corresponding to the instability occurrence and associated with failure.

3. COMPARISON OF PREDICTIONS VERSUS MEASUREMENTS

This section is concerned with the prediction of microbuckling development in laminates. The full laminate thickness is discretised and free-edge conditions are prescribed on both top and bottom faces. The influence of imperfection distribution and structure effect on the instability development are studied separately for the sake of simplicity. According to the measurements from [Paluch, 1994], initial geometric imperfections are assumed to be of waviness form along $\mathbf{e_1}$ with a wavelength of 0.9 mm and angles from 0.5 ° to 2.5 °. The only unknown parameter is the imperfection distribution across the thickness which has never been measured. Following the understanding of imperfection presented in section 4, a parabolic distribution is chosen with a maximum amplitude at mid-thickness. For quantitative comparisons with experimental results, realistic constitutive laws are used for the constituents.

3.1 Loading

The effect of loading that is known from an experimental point of view ([t'Hart et al., 1991], [Grandidier, 1991], [Wisnom, 1991]) is demonstrated theoretically on a 3.2 mm thick UD modelled, first under constant loading across the thickness (pure compression) and secondly under a linearly varying loading (pure bending). Figure 2 shows that, for both loadings, the larger the imperfection angle, the lower the compressive strength. Complementary investigations also showed that the imperfection wavelength could be as influential as the imperfection angle in predicting the compressive strength. But the main result is that for any imperfection the compressive strength under flexural loading is larger than that for compression loading. With an imperfection angle close to 0.5° the predicted strength agrees very well with experimental measurements : 1.95% for bending and 1.2% for compression([t'Hart et al., 1991]). Predictions of kink-band models (here [Budiansky and Fleck, 1993]) are close to values commonly obtained with pure compression test fixtures.

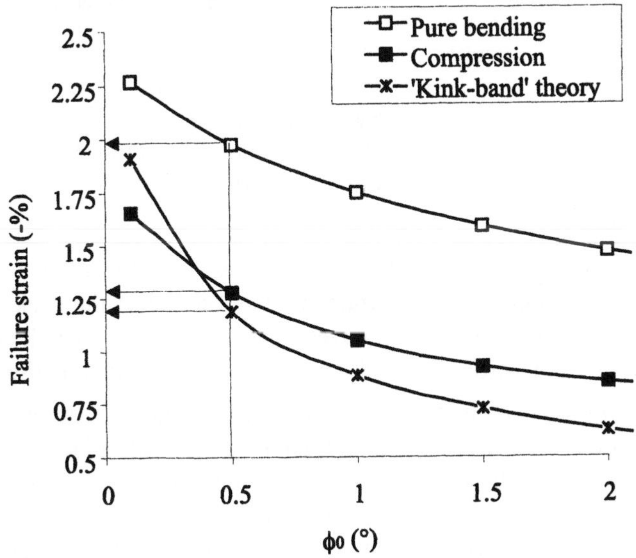

Figure 2 ; Failure strains vs. imperfection angle for compression, bending and the kink-band model.

The difference in strength for the two types of loading is almost constant whatever the imperfection angle considered. This is due to stress and strain distributions across the thickness that are characteristic of each loading and independent of the imperfection amplitude. Under pure compressive loading, strain and stress distributions across the tickness are homogeneous. Conversely under bending loading the plastic flow area is reduced to a third of the total thickness, in the most compressed zone : the gradient of loading sets the transverse characteristic length upon which microbuckling develops.

3.2 Thickness / Gradient of loading

The effect of the ply thickness on the characteristic transverse length, and thus on failure strains, is characterised on UDs of various thickness. Both compression and bending loadings are considered, along with several imperfection angles (Figure 3).

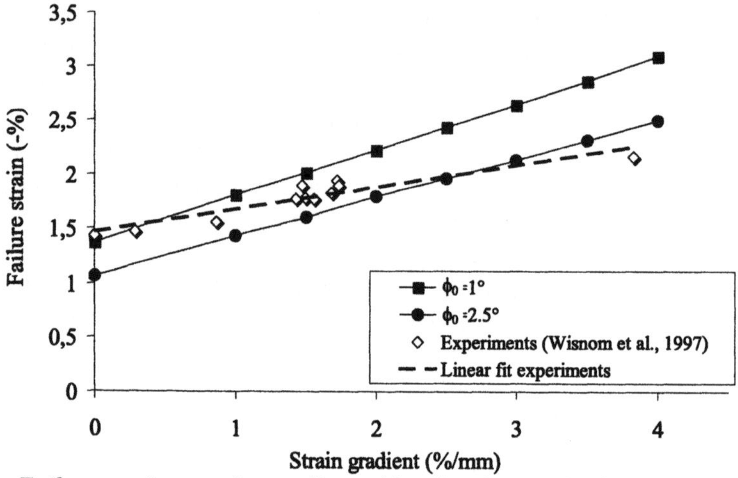

***Figure 3** ; Failure strains vs. the gradient of loading (prescribed strain), both predicted and measured.*

Under pure compression loading, the thickness has very little influence on the compressive strength since constant stress and strain distributions across the thickness cannot determine the transverse characteristic length over which microbuckling will develop. Only the imperfection distribution across the thickness can induce a structural effect.

Conversely, under bending loading an increase in strength is observed for decreasing thickness, or similarly for increasing gradients of loading. This influence of the loading gradient on the compressive strength discussed by [Grandidier, 1991] has recently been confirmed experimentally by [Wisnom et al., 1997] and [Rahier, 1998], who used a specific test fixture. In Figure 3 predictions from our approach appear to correlate well with experimental data from [Wisnom et al., 1997], who established on T800/924 material that as a first approximation, the compressive strength is linearly related to the gradient of loading.

3.3 Stacking sequence

Here, the influence of both stacking sequence and transverse plies stiffness (90° and 45°) is investigated. Five laminates made up of 16 T300/914 plies are considered ($[0_{16}]$, $[0_2, 45_2]_{2S}$, $[0_2, 90_2]_{2S}$, $[0, 90_3]_{2S}$, $[0, 90]_{4S}$) which can be gathered into three families depending on their number of 0° consecutive plies: 16, 2 and 1 ply. Whatever the loading, the thicker the 0° consecutive plies, the lower the laminate strength (Figure 4). The presence of 90° or 45°

cross plies determines the characteristic transverse length by clamping transverse displacements close to the interface with 0° plies. Orientation of these cross plies has a similar negligible influence on the instability since $[0_2, 45_2]_{2S}$ and $[0_2, 90_2]_{2S}$ strength are very close. In contrast the thickness of the transverse plies has a strong influence on the mechanism. Although very few data are available, predictions appear to be of the order of magnitude as the tests results from [Grandsire-Vinçon, 1993].

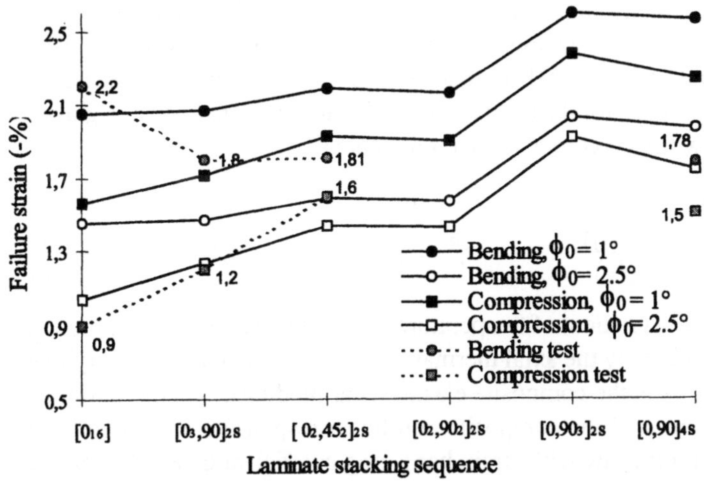

Figure 4 ; Failure strain both predicted (for 2 imperfections) and measured for bending and compression loadings, vs the laminate stacking sequence.

4. FIBRE IMPERFECTION GROWTH DURING CURE

Predictions obtained through the previous model have been shown to depend strongly on the fibre initial imperfections. Among the very few studies devoted to this phenomenon, only [Paluch, 1994] traced this imperfection by a succession of cross cuttings for a volume of $1x0.1x0.07mm^3$. But predicting this key parameter seems to be more realistic than measuring it. Indeed, not only the waviness must be known, but any variation of the waviness amplitude through the ply thickness must also be characterised since it will induce a structure effect as stated previously. Based on a simple single-fibre model, we demonstrate that the hypothesis commonly admitted of an imperfection induced during the cooling stage fails and we propose a new approach.

4.1 Microbuckling instability during the cooling ?

([Grandsire-Vinçon, 1993]) assumed that the instability appears during the final cooling phase. Theoretically, the mechanism which generates the instability may be due to the difference between axial thermal dilatation coefficients of epoxy matrices (positive) and carbon fibres (negative). According to the consistency of the fibre-matrix interface, a compressive loading is thus generated on the fibre during the cooling stage. As this phenomenon begins for hot temperatures, the support provided by the matrix against fibre microbuckling is not sufficient and leads to the wavy imperfection formation.

Here, the thermally-induced stress that acts on the fibre is compared with the critical microbuckling stress, taking into account the change in matrix behaviour during cooling. The applied thermally-induced stress (7-a) is evaluated by using a simple model in which the laminate is seen as a stacking of stiff (fibre) and soft (matrix) layers. [Rosen, 1964] and [Budiansky and Fleck, 1993] approaches (7 b-c) are then used to describe the fibre critical microbuckling stress, for the matrix in either a rubbery state (7-b) or a glassy state (7-c), *i.e.* above or below the glass transition temperature (Tg).

$$\overset{(a)}{\sigma_{fibre}^{th} = \frac{\Delta T E_f(\alpha_m - \alpha_f)}{1 + \frac{E_f}{E_m}\left(\frac{f}{1-f}\right)}} \qquad \overset{(b)}{\sigma_{cr}^{R} = \frac{G_m}{f(1-f)}} \qquad \overset{(c)}{\sigma_{cr}^{B-F} = \frac{G}{1 + n\left(\frac{3}{7}\right)^{\frac{1}{n}}\left(\frac{\bar{\phi}/\gamma_{ys}}{n-1}\right)^{\frac{n-1}{n}}}} \tag{7}$$

In (7-a) ΔT is the gradient of temperature, E_m is the matrix Young's modulus, and α_m and α_f denote respectively the axial thermal dilatation coefficients of epoxy matrix and carbon fibre. In (7-c), G is the composite elastic shear modulus, $\bar{\phi}$ is the initial inclination angle of the kink-band, and the composite non-linear response is modelled through a Ramber-Osgood constitutive law with strain hardening coefficient denoted by n and yield strain in shear γ_{ys}. Matrix modulus change and thermal expansion were assessed during the cooling by way of a Dynamical and Mechanical Thermal Analysis (DMTA).

Results show (Figure 5) that a misalignment angle of at least 2° would be necessary for the thermally-induced stress acting on the fibre to reach the critical microbuckling stress (7 b-c). The impossibility of microbuckling occurring during the cooling stage has been confirmed by single fibre cure tests carried out with different cure cycles, but having the same final cooling stage. It was demonstrated that hot-stage microbuckling occurs only for fast cure cycles, independently of cooling stages (*see also* [Ahlstrom, 1991]).

4.2 Microbuckling instability during the thermosetting reaction ?

Previous results lead us to consider that the microbuckling instability may be generated during the resin thermosetting reaction. Indeed, the epoxy resin shrinkage observed during the gelation (liquid/rubber phase change) reaction could induce a compressive loading on fibres which receive very little support from the matrix in its viscous state. When studying such a phenomenon, two main concerns arise. On the one hand, mechanical characteristics of the resin change strongly during chemical reaction, and on the other hand the thickness thermal gradient generated by the exothermic aspect of the reaction induces different heterogeneities in the conversion rates. Thus, the thermosetting reaction was investigated from a physical and mechanical point of views to search for microbuckling possibilities.

The physical aspects of the epoxy resin thermosetting reaction were characterized by DMTA analyses. Both elastic and viscous effects were characterised. The fundamental aspects of the physics involved are studied by examining a single carbon fibre embedded in

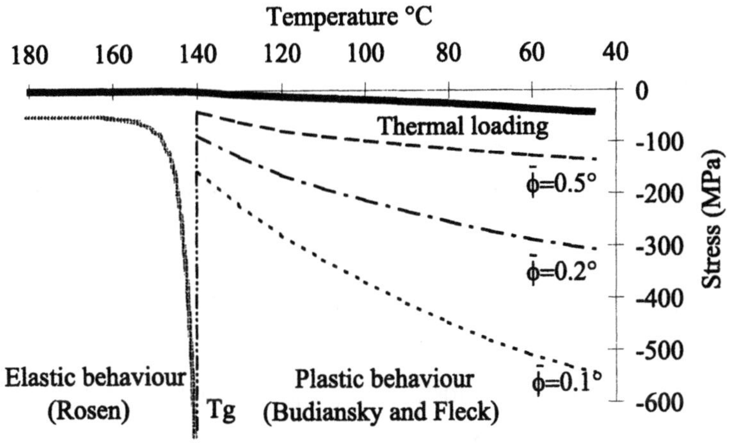

***Figure 5** ; Stress required to initiate microbuckling: thermally induced and calculated through relations (7a-b).*

an infinite matrix.

Experimental Cole-Cole plots were established during the cure cycle. After [Cazeneuve and Prepin, 1984] a bi-parabolic Zener model can be used to represent the viscoelastic behavior of fully cured epoxy resins. In our case, this result is extended to the expoy resin behaviour during the thermosetting reactions, but in order to get a tractable model, a classical Zener model is used. It is modified to account for the cylindrical area of matrix surrounding the fibre and which is affected by the occurrence of microbuckling. This area was observed experimentally through a polarized light, but it is quite complicated to follow during the cure process. Therefore, the characteritic radius of this zone was computed by assuming that the fibre axial strain can be approached through a Zener viscoelastic response involving a material with very low viscosity.

Except fibre properties that are supposed to remain constant during classical cure cycles (around 200°C), all terms of the model governing equations depend on time, temperature and degree of cure. A fourth order Runge-Kutta scheme is employed to determine the viscoelastic matrix strain change. Then, the fibre axial strain and stress (through Hooke's law) can be estimated which eventually yields the load $P(t)$ thus generated. Then the microbuckling wavelength is estimated from [Bhalerao and Moon, 1996] who considered a perfectly slipping fibre in an infinite viscoelastic matrix :

$$\lambda = 2\pi h \sqrt{\frac{S_f E_f}{4P(t)(1-\nu_f^2)}} \tag{8}$$

where S_f denotes the fibre cross-sectional area; ν_f is the fibre Poisson's ratio. To simplify this approach, we focused on the 120°C isothermal case. Results for this cure cycle are plotted in Figure 6.

It can be postulated that after vitrification, the process of imperfection formation is almost fixed as far as the wavelength is concerned. Then, from Figure 6, before cooling the

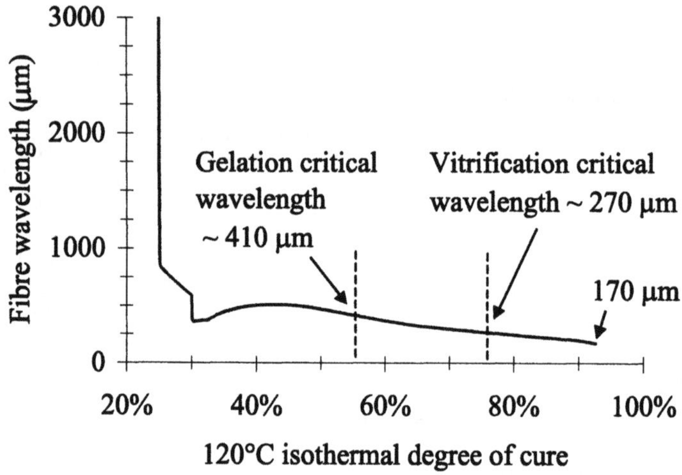

Figure 6 *; Critical wavelength vs. the degree of conversion of the resin.*

wavelength should not be lower than $170\mu m$. This result correlates well with the wavelengths observed experimentally on both single fibre specimens cured during 2h at 100°C and 2h at 140°C (Direct 120°C isothermal cure is too much an exothermic reaction; it induces epoxy darkennig which does not allow any optical observations). The experimental wavelength measured is about $210\mu m$ with an amplitude around $10\mu m$; this validates the present approach.

Results show clearly that the microbuckling instability is induced by the shrinkage property of the thermosetting reaction. Of course, a post buckling analysis is necessary to determine fibre waviness amplitude. However, this work sets the basis for the mechanical model that will be extended to fibre distribution in order to predict fibre waviness imperfections remaining after cure. Finally, the fibre imperfection distribution which is thermally induced is very likely to follow the thermal distributions observed in composites during cure which is parabolic across laminates. The microbuckling approach can thus be considered as being a good candidate for describing the residual stress state and thus demonstrates the structural aspect of fibre imperfection distribution in composites.

5. CONCLUSION

A modelling approach of compressive failure has been established which allows for quantitative predictions with very little computational effort. It was established that whatever the loading considered, the effect of the imperfection on the compressive strength is of the same order. Results for UDs provide a reliable explanation of the high strength obtained under bending loading. The characteristic length is the central parameter that controls the influence of the structural effects by determining the distribution of plastic microbuckling across the thickness.

This transverse dimension can be influenced by a combination of loading and thickness, which result in changing the gradient of loading. This effect was demonstrated through quantitative comparisons with experiments. For cross-plies laminates the number of consecutive plies in the loading direction is the key parameter. More precisely, the total thickness of these consecutive plies sets the characteristic transverse length.

The mechanism of failure in laminates under compression or bending is perfectly described provided more information can be obtained on the geometrical imperfection. A single fibre viscoelastic model was validated which was shown to describe properly the expoxy resin shrinkage that initiates microbuckling. Relying on DMTA measurements, we showed that predictions of this model compare well with experimental measurements made on single fibre specimens.

REFERENCES

[Ahlstrom, 1991] Ahlstrom, C.; Interface fibre de verre / matrice polyepoxy - Introduction d'une interphase à propriétés contrôlées; *PhD thesis, INSA Lyon*

[Argon, 1972] Argon, A.S.; Fracture of composites; In : *Treatise of Materials Science and Technology, Academic Press, New-York* **1**

[Bhalerao and Moon, 1996] Bhalerao, M. and Moon, T.; On the growth-of-waviness in fiber-reinforced polymer composites : viscoelastic bifurcation and imperfection sensitivity; *J. Appl. Mech.* **63**, pp. 460–466

[Budiansky, 1983] Budiansky, B.; Micromechanics; *Comp. Struct.* , pp. 3–12

[Budiansky and Fleck, 1993] Budiansky, B.; Fleck, N.A.; Compressive failure of fiber composites; *J. Mech. Phys. Solids* **41**, pp. 183–211

[Cazeneuve and Prepin, 1984] Cazeneuve, B.; Prepin A; Analyse des contraintes résiduelles de cuisson dans les composites carbone-epoxy *à* plis croisés; In: *Proceedings JNC4, Paris 11-13 september Pluralis*

[Drapier et al., 1996] Drapier, S.; Gardin, C.; Grandidier, J.-C.; Potier-Ferry, M.; Microbuckling and structure effect; *Comp. Sci. Techn.* **56**, pp. 861–867

[Drapier et al., 1999] Drapier, S.; Grandidier, J.-C.; Potier-Ferry, M.; Towards a numerical model of the compressive strength for long-fibre composites; *Eur. J. Mech/A* **18(1)**, pp. 1159–1164

[Grandsire-Vinçon, 1993] Grandsire-Vinçon, I.; Compression des composites unidirectionnels : méthodes d'essais et approche micromécanique; *Ph.D. Thesis ENS Cachan (France)*

[Grandidier, 1991] Grandidier, J.-C.; Compression et microflambage dans les matériaux composites *à* fibres longues; *Ph.D. Thesis University of Metz (France)*

[Gardin and Potier-Ferry, 1992] Gardin, C.; Potier-Ferry, M.; Microflambage des fibres dans un matériau composite *à* fibres longues : analyse asymptotique 2-D; *C. Rend. Acad. Sc. Paris* **t315 Série II**, pp. 1159–1164

[Grandidier et al., 1992] Grandidier, J.-C.; Potier-Ferry, M.; Ferron, G.; Microbuckling and strength in long- fiber-composites : theory and experiments; *Int. J. Solids Struct.* **29(14/15)**, pp. 1753–1761

[Hutchinson, 1974] Hutchinson, J.W.; Plastic buckling; *Adv. Appl. Mech.* **14**, pp. 67–144

[Kyriakides et al., 1995] Kyriakides, S.; Arseculeratne, R.; Perry, E.J.; Liechti, K.M.; On the compressive failure of fiber reinforced composites; *Int. J. Solids Struct.* **32(6/7)**,pp. 689–738

[Paluch, 1994] Paluch, B.; Analyse des imperfections géométriques affectant les fibres dans un matériau composite *à* renfort unidirectionel; *La Recherche Aéronautique* **6**, pp. 431–448

[Rahier, 1998] Rahier, O.; Grandidier, J.C.; Daridon, L; Pure compression testing of advanced fibre composites; *Comp. Sc. Techn.* **58(5)**, pp. 735-740

[Rosen, 1964] Rosen, B.W.; Mechanics of composite strengthening; In : *Fiber Composite Materials, Am. Soc. Metals Seminar, Metal Parks, Ohio.*, pp. 37–75

[t'Hart et al., 1991] t'Hart, W.J.G.; Aoki, R.; Bookholt, H.; Curtis, P.T.; Krober, I.; Marks, N.; Sigety, P.; The utilisation of advanced composites in military aircraft; In: *AGARD Report 785, 73rd Meeting of the AGARD Structures and Materials Panel, San-Diego, 7th-11th October 1991*

[Wisnom, 1991] Wisnom, M.R.; The effect of the specimen size on the bending strength of unidirectional carbon fibre-epoxy; *Comp. Struct.* **18**, pp. 47–63

[Wisnom, 1997] Wisnom, M.R.; Atkinson, J.W.; Jones, M.I.; Reduction in compressive strain to failure of unidirectional carbon fibre-epoxy with increasing specimen size in pin-ended buckling tests; *Comp. Sci. Techn.* **57(9/10)** , pp. 1303–1308

A new method of optimisation for composite structures

Georges DUVAUT
LM2S, Université Pierre et Marie Curie
8 rue du Capitaine Scott, 75015 Paris
duvaut@ccr.jussieu.fr
Ghyslaine TERREL
ONERA / LILLE et LM2S

Abstract: We present a new method for optimizing the rate and direction of fibers at each point in a composite structure for minimizing the compliance of the loaded structure. We give anexistence theorem and a numerical algorithm. We show an application to a two dimensionalelastic structure. The method can also be applied to bending plates and shells.
Keywords: Composite structure, optimisation, numerical methods.

1 INTRODUCTION

Most composite structures are reinforced by using fibers or woven glass fabrics, but their resulting mechanical behaviour is greatly modifiedd by varying the proportion of fibers and their orientation at each point of the structure. Traditional design considers these parameters piecewise constant, in the structure; but the geometry of the structure and the type of loads may impose that they vary, and recent progress in manufacturing techniques make it possible to fabricate composite structures that have curvilinear fibers with spatially variable volume fraction. In these situations one main difficulty is to determine the optimal direction and volume fraction of fibers at each point of the structure; the answer can be given by experience, but this way is not a scientific one-which is not necessarily a drawback-, but it is not easily transmissible.

The method which is proposed here is systematic and can be operated by engineers, even with little experience. It consists of minimizing a functional which includes the compliance of the structure, the price and the weight of the structure, and perhaps other parameters. It relies on variational formulations and theorems associated with the potential energy. It is presented here for elastic structures, but can be extended to other types of materials. Other authors ([6],[7],[8]) have been working on this type of problems, but the numerical methods that they use lead to very long computational times.

G.A. Maugin et al. (eds.), Continuum Thermomechanics, 137-151.
© 2000 *Kluwer Academic Publishers.*

The success of our algorithm, inspired by topological optimisation[1], comes from the fact that it consists of two alternating minimisations: a local one and a global one. The first one is to improve, at each point of the structure, the volume fraction and the direction of the fibers; the second one is for solving a homogenized structure problem when the homogenized constitutive relations are known.; some other parameters can also be introduced, such as the type of fibers. The algorithm is convergent because, at each step, the functional, which is positive, decreases. The method quickly leads to an optimal composite structure. The optimal fibre partition which is found creates a good stress distribution, in the critical regions of the structure, in the sense that the Hill-Tsai criterion is greatly reduced. The theory is given here for linear three-dimensional elasticity but it can be extended to other types of constitutive relations. The method has a very wide field of applications: aeronautics, naval, automotive, civil engineering. Several examples are given for two-dimensional elasticity structures, with or without holes, subjected to different loadings. The method has also been developed for shells (see G. Terrel thesis[11]),and applications are in progress in collaboration with aerospace industry.

2 PROBLEM STATEMENT

Consider an elastic structure in a region $\Omega \subset \mathbf{R}^3$, whose boundary $\partial\Omega$ is composed of Γ_0 where the displacements are zero, and $\Gamma_1$where we apply surface forces $F($ $\partial\Omega = \Gamma_0 \cup \Gamma_1$, $\Gamma_0 \cap \Gamma_1 = \phi$). Body forces f are given in Ω. If the elastic coefficients a_{ijkh} (or A_{ijkh}) satisfy the usual hypotheses [2], we know that there exists a unique solution of the elasticity problem: displacements $u = (u_1, u_2, u_3)$ and stresses $\sigma = (\sigma_{ij})$.

In order to reinforce this structure, we introduce fibers, whose direction $\alpha_1(x)$ $(0 \leq \alpha_1(x) \leq 2\pi)$ and volume fraction $\alpha_2(x)$ $(0 \leq \alpha_2(x) \leq \alpha_{2\max} < 1)$ may depend on the point $x \in \Omega$, in order to minimize the compliance plus some cost function: $G(\alpha) = SOUPL(\alpha) + COUT(\alpha)$, where

$$\begin{aligned} SOUPL(\alpha) &= \int_\Omega f.u \, dx + \int_{\Gamma_1} F.u \, dS = \int_\Omega A_{ijkl}(\alpha)\sigma_{ij}\sigma_{kl}dx, \\ COUT(\alpha) &= \int_\Omega cout(\alpha)dx. \end{aligned}$$

We have set $\alpha = (\alpha_1, \alpha_2)$. The scalar value function is the local cost implied by the introduction of fibers; it depends on α,and may take in account of the increase of mass of the structure.

The function $G(\alpha)$ is strictly positive; this implies that it has a minimum G_{MIN}. We want to show that this minimum is reached for some $\alpha(x) = (\alpha_1, \alpha_2)$ In section

3 we study the theoretical aspect of the problem and prove that, in an appropriate function space, the minimum is reached for some field $\alpha(x) = (\alpha_1, \alpha_2)$ .In section 4 we give a numerical algorithm that leads to a numerical solution.

3 EQUATIONS AND VARIATIONAL FORMULATIONS

We present the method in a three-dimensional composite elastic structure that occupies the open set $\Omega \in R^3$ described in the previous paragraph.

The equations and boundary conditions lead to the following system of equations and relations,

$$\frac{\partial \sigma_{ij}}{\partial x_j} + f_i = 0 \text{ , on } \Omega. \tag{1}$$

$$\sigma_{ij} = a_{ijkh}\epsilon_{kh}(u), \text{ where } \epsilon_{kh}(u) = \frac{1}{2}(\partial u_i/\partial x_j + \partial u_j/\partial x_i) \tag{2}$$

$$u = 0 \quad \text{on } \Gamma_0 \tag{3}$$

$$\sigma_{ij}n_j = F_i \text{ , on } \Gamma_1. \tag{4}$$

where (u, σ), are the displacement and stress tensor field solution, $n = (n_1, n_2, n_3)$ is the outer unit normal to the boundary $\partial\Omega$, $a_{ijkh}(..)$ are the elastic coefficients which depend on $x = (x_1, x_2, x_3) \in \Omega$, and on some other parameters : $\alpha_1, \alpha_2...$; the parameter $\alpha_1 = \theta(x)$ defines the direction of the fibers at point $x \in \Omega$; $\alpha_2 = \alpha(x)$ is the volume fraction of the fibers at x; other parameters may also refer, for example, to the type of fibers. For convenience let us state that $\alpha = (\alpha_1, \alpha_2, ...)$ is a vector function on Ω. We shall make the following assumptions on the coefficients $a_{ijkh}(x, \alpha)$: there exist two constants C_1 and C_2, strictly positive, such that

$$C_1|e|^2 \leq a_{ijkh}e_{ij}e_{kh} \leq C_2|e|^2, \ \forall e = (e_{ij}), \ e_{ij} = e_{ji}, \ |e|^2 = \sum_{i,j}(e_{ij})^2. \tag{5}$$

This implies also that the inverse matrix (A_{ijkl}), defined by

$$\epsilon_{kh}(u) = A_{ijhk}\sigma_{hk} \tag{6}$$

satisfies a similar set of inequalities, say,

$$D_1|e|^2 \leq A_{ijkh}e_{ij}e_{kh} \leq D_2|e|^2, \ \ \forall e = (e_{ij}), \ e_{ij} = e_{ji}, \ |e|^2 = \sum_{i,j}(e_{ij})^2. \tag{7}$$

The problem that we state is the following: what are the functions $\alpha_i(x)$ such that the loaded structure shows the highest stiffness (or the lower compliance), at the smallest cost. The compliance is defined by

$$E(\alpha) = \int_{\Omega} f.udx + \int_{\Gamma_1} F.udS \tag{8}$$

It appears to be the work done by the exterior forces under the solution displacement. We shall see, from the variational formulation (or the principle of virtual work), that we have

$$E(\alpha) = \int_{\Omega} \sigma_{ij}\epsilon_{ij}(u)dx = \int_{\Omega} a_{ijhk}\epsilon_{hk}(u)\epsilon_{ij}(u)dx = \int_{\Omega} A_{ijhk}\sigma_{hk}\sigma_{ij}dx \tag{9}$$

which shows that $E(\alpha)$ is also a measure of the global strain or stress solution.

In order to obtain a variational formulation of the problem (see [4] for details), let us state the set of kinematically admissible displacement fields:

$$V = \{v | \ v = (v_1, v_2, v_3), v_i \ smooth, \ v_i = 0 \ on \ \Gamma_0, \ i = 1, ...3.\} \tag{10}$$

If u is the displacement field solution , then $u \in V$.

Take any $v \in V$, multiply by (1) and integrate over Ω ; after integration by parts and using (2), we get

$$u \in V, \ \ a(u, v) = L(v), \ \ \forall v \in V \tag{11}$$

where

$$a(u, v) = \int_{\Omega} a_{ijhk}\epsilon_{hk}(u)\epsilon_{ij}(v)dx, \ \ L(v) = \int_{\Omega} f_i v_i dx + \int_{\Gamma_1} F_i v_i dS \tag{12}$$

Choosing $v = u$ in (11), we get

$$E(\alpha) = L(u) = a(u, u) \tag{13}$$

which shows that $E(\alpha)$ is always positive. Furthermore (11) is also equivalent to (14) (see [2] for a proof),

$$I(v) = \frac{1}{2}a(v, v) - L(v) \leq I(u), \ \forall v \in V. \tag{14}$$

We can also get a variational formulation in term of the stress. Let us introduce the set of statically admissible stresses:

$$\Sigma_{ad} = \left\{ \tau \mid \tau = (\tau_{ij}), \tau_{ij} = \tau_{ji}, \frac{\partial \tau_{ij}}{\partial x_j} + f_i = 0 \;\; on \;\; \Omega; \tau_{ij} n_j = F_i \;\; on \;\; \Gamma_1 \right\} \tag{15}$$

If σ is the stress field solution, we have

$$\sigma \in \Sigma_{ad}.$$

Let us now write (2) in term of the strain to get

$$\epsilon_{ij}(u) = A_{ijhk} \sigma_{hk} \tag{16}$$

Now, for any $\tau \in \Sigma_{ad}$, multiply (16) by $\tau_{ij} - \sigma_{ij}$ to get

$$\sigma \in \Sigma_{ad}, \;\; A(\sigma, \tau - \sigma) = 0, \;\; \forall \, \tau \in \Sigma_{ad} \tag{17}$$

where

$$A(\sigma, \tau) = \int_{\Omega} A_{ijhk} \sigma_{hk} \tau_{ij} dx \tag{18}$$

We can prove that (17) is equivalent to

$$\sigma \in \Sigma_{ad}, \;\; J(\tau) \geq J(\sigma), \; \forall \, \tau \in \Sigma_{ad} \tag{19}$$

where

$$J(\tau) = \frac{1}{2} A(\tau, \tau) \tag{20}$$

From (14) and (19) and the fact that u and σ are respectively the displacement and stress fields solution, we get the double inequality (see [2], for a detailed proof),

$$\begin{aligned} \sigma &\in \Sigma_{ad}, \qquad \sigma \in \Sigma_{ad}, \\ I(v) &\geq I(u) = -J(\sigma) \geq -J(\tau), \quad \forall \, v \in V, \quad \forall \, \tau \in \Sigma_{ad} \end{aligned} \tag{21}$$

Using (13) , we obtain

$$-2I(v) \leq -2I(u) = E(\alpha) = 2J(\sigma) \leq 2J(\tau), \quad \forall \, v \in V, \;\; \forall \, \tau \in \Sigma_{ad} \tag{22}$$

From this double inequality we shall get an algorithm.

4 NUMERICAL ALGORITHM

We want to minimize $E(\alpha)$ or $E(\alpha)$ plus some cost function which depends on the functions $\alpha(x)$, say

$$\int_{\Omega} cost(\alpha)dx$$

Consequently, if we set

$$G(\alpha) = \int_{\Omega} A_{ijhk}(\alpha)\sigma_{hk}\sigma_{ij}dx + \int_{\Omega} cost(\alpha)dx$$

we have to find

$$\underset{\alpha}{Inf} G(\alpha) \tag{23}$$

where α appears in $A_{ijhk}(\alpha)$ and in the stress field solution $\sigma = (\sigma_{ij})$.

From this remark, we can construct an algorithm : assume that we start with an initial value of α, say $\alpha^{(0)}$. We compute the stress field solution corresponding to $\alpha^{(0)}$, say $\sigma^{(0)}$. In a second step, and for any $x \in \Omega$ we get $\alpha^{(1)}(x)$ which realises locally

$$\underset{\alpha}{Inf} \left[A_{ijhk}(\alpha)\sigma_{hk}^{(0)}\sigma_{ij}^{(0)} + cost(\alpha)\right] = A_{ijhk}(\alpha^{(1)})\sigma_{hk}^{(0)}\sigma_{ij}^{(0)} + cost(\alpha^{(1)}). \tag{24}$$

With this value $\alpha^{(1)}(x)$ we compute the corresponding stress field solution, say $\sigma^{(1)}$, which, according to (21), satisfies

$$\int_{\Omega} A_{ijhk}(\alpha^{(1)})\sigma_{hk}^{(1)}\sigma_{ij}^{(1)}dx \leq \int_{\Omega} A_{ijhk}(\alpha^{(1)})\sigma_{hk}^{(0)}\sigma_{ij}^{(0)}dx$$

and, consequently,

$$G(\alpha^{(1)}) \leq G(\alpha^{(0)}) \tag{25}$$

We can iterate these calculations : as in (24) , we obtain $\alpha^{(2)}(x)$ by

$$\underset{\alpha}{Inf} \left[A_{ijhk}(\alpha)\sigma_{hk}^{(1)}\sigma_{ij}^{(1)} + cost(\alpha)\right] = A_{ijhk}(\alpha^{(2)})\sigma_{hk}^{(1)}\sigma_{ij}^{(1)} + cost(\alpha^{(2)}).$$

and then we compute the stress field solution $\sigma^{(2)}$ corresponding to $\alpha^{(2)}(x)$,which will satisty, as in (25)

$$G(\alpha^{(2)}) \leq G(\alpha^{(1)})$$

and so on; at each step the functional $G(\alpha)$ decreases; as it is positive it must converge to a finite positive value, say G_{MIN}. This numerical algorithm will be used in section 6, for numerical applications; it converges very rapidly.

5 THEORETICAL RESULTS

5.1 Preliminary Remark

Since $G(\alpha)$ is positive, there exists G_{MIN} such that,

$$\underset{\alpha\in(L^\infty(\Omega))^2}{Inf} \quad G(\alpha) = G_{MIN} > 0$$

and a minimizing sequence $\alpha_n(x) \in (L^\infty(\Omega))^2$

This sequence is bounded in $(L^\infty(\Omega))^2$ because $\alpha_1(x)$ and $\alpha_2(x)$ are bounded for all points in Ω . Consequently this sequence α_n has a subsequence, still denoted by α_n, which converges in $(L^\infty(\Omega))^2$, weak dual of $(L^1(\Omega))^2$. We deduce easily that the corresponding sequences u_n and σ_n are bounded respectively in V and $(L^2(\Omega))^6$ then converge weakly to u and σ respectively, such that

$$u \in V, \quad \sigma \in \Sigma_{ad}.$$

Unhappily we do not know how to pass to the limit in

$$\sigma_{ij}^{(n)} = a_{ijkh}(\alpha_n)\epsilon(u^n).$$

We stand in a situation mentioned in [9](Remarque 7.6, page 95).

5.2 Reduction of the problem

In order to avoid the previous difficulty, we shall look for a solution α in a subset of $(L^\infty(\Omega))^2$ of functions that are piecewise constant . For that we make a partition (P) of Ω into N sub-domains ω_i , $i = 1...N$. such that

$$\bigcup_{i=1...N} \overline{\omega}_i = \overline{\Omega}, \quad \omega_j \cap \omega_k = \emptyset, \; si \; j \neq k$$

and we introduce the space of functions on Ω,whose restriction to each ω_k is constant, with values in R^2. Such a function $\alpha = (\alpha_1, \alpha_2)$ is nothing but an element of R^{2N}, and, since α_1, α_2 are bounded, α stays in a compact set $K(P)$ of R^{2N}. We then easily obtain the following result:

Theorem 1 *The restriction of the function $G(\alpha)$ to the subset $K(P)$ takes its minimum value for $\widetilde{\alpha} \in K(P)$, $G_{MIN} = G(\widetilde{\alpha})$ and the associated u, σ will satisfy*

$$u \in V, \quad \sigma \in \Sigma_{ad}, \quad \sigma_{ij} = a_{ijkh}(\widetilde{\alpha})\epsilon(u).$$

Of course $\widetilde{\alpha}$ depends on (P), say $\widetilde{\alpha} = \alpha_P$.

Proof. Let us set

$$\underset{\alpha \in K(P)}{Inf}\, G(\alpha) = G(\alpha_P) = G_P,$$

and $\alpha_n \in K(P)$ a minimising sequence; it has a sub-sequence, still denoted by α_n, which converges in $K(P)$, to $\widetilde{\alpha} \in K(P)$.The correspondingsequences u_n and σ_n are bounded respectively in V and $(L^2(\Omega))^6$ and converge weakly to u and σ respectively, such that

$$u \in V, \quad \sigma \in \Sigma_{ad}.$$

In addition we can pass to the limit in the sense of distributions in

$$\sigma_{ij}^{(n)} = a_{ijkh}(\alpha_n)\epsilon_{kh}(u^n)$$

because the convergence in $K(P)$ implies pointwise convergence in Ω of the sequence $\alpha_n(x)$; for any $\varphi \in D(\Omega)$, we state that

$$\int_\Omega (a_{ijkh}(\alpha_n)\epsilon_{kh}(u^n) - a_{ijkh}(\widetilde{\alpha})\epsilon_{kh}(u))\varphi dx = X_1 + X_2.$$

$$X_1 = \int_\Omega (a_{ijkh}(\alpha_n) - a_{ijkh}(\widetilde{\alpha}))\epsilon_{kh}(u^n))\varphi dx$$

tend to zero, for $\epsilon_{kh}(u^n)$ is bounded in $L^2(\Omega)$, and $[(a_{ijkh}(\alpha_n) - a_{ijkh}(\widetilde{\alpha}))\varphi]^2$ is bounded by a constant and tends to zero in each point, which implies that X_1 tends to zero (Lebesgues theorem).

$$X_2 = \int_\Omega a_{ijkh}(\widetilde{\alpha})\varphi\epsilon_{kh}(u^n - u)dx$$

also tends to zero, because it is a sum of terms which are scalar products in $L^2(\Omega)$ of a fixed element $a_{ijkh}(\widetilde{\alpha})\varphi \in L^2(\Omega)$ and $\epsilon_{kh}(u^n - u)$ which tends weakly to zero in $L^2(\Omega)$.

Furthermore we easily get

$$\lim_{n\to\infty} G(\alpha_n) = \lim_{n\to\infty} \left[\int_\Omega f.u^n\, dx + \int_{\Gamma_1} F.u^n\, dS + \int_\Omega cout(\alpha_n)dx\right] = G(\widetilde{\alpha}).$$

which ends the proof of the theorem. ■

5.3 Remark 1

The solution obtained here is of the same type as the one that we get numerically by using finite elements.

5.4 Remark 2

The difficulty mentioned in [9] (Remark 7.6, page 95) can also be avoided by choosing the control in a space of piecewise constant functions; in fact it solves the problem from a practical point of view, but it does not explain the mathematical difficulty encountered. A similar kind of problem appears in [10].

6 NUMERICAL APPLICATIONS

We have applied the method explained in section 3 for finding the optimal fiber directions and volume fraction of fibres at each point in a rectangular plate in plane deformation. The plate occupies the following region in R^2,

$$0 < x < b, \quad -a < y < a.$$

It is fixed along $x = 0$, free on $y = \pm a$,and loaded on $x = b$. The plate is composed of matrix and long fibers. The matrix is assumed isotropic, with Young modulus E^m , Poisson coefficient ν^m.Fibers are othotropic and transversly isotropic, with

Young's moduli:	E_l^f , E_t^f,
Poisson's ratio:	ν_{lt}^f , ν_t^f,
shear moduli:	G_{lt}^f , G_t^f,
related by	$E_t^f = 2G_t^f(1+\nu_t^f)$.

The composite plate, where the volume fraction of fibers is α, is transversly isotropic; the coefficients may be obtained either by homogenisation, which gives a very precise but not explicit result, or by approximate formulas which are the following:

Longitudinal Young's modulus	$E_l = \alpha E_l^f + (1-\alpha)E^m$
Tranverse Young's modulus	$\frac{1}{E_t} = \frac{\alpha}{E_t^f} + \frac{1-\alpha}{E^m}$
Shear modulus	$\frac{1}{G_{lt}} = \frac{\alpha}{G_{lt}^f} + \frac{1-\alpha}{G^m}$
Poisson's ratio	$\nu_{lt} = \alpha\nu_l^f + (1-\alpha)\nu^m$

The two-dimentional constitutive relation written in an orthotropy frame is given by

$$\begin{bmatrix} \epsilon_{11} \\ \epsilon_{22} \\ 2\epsilon_{12} \end{bmatrix} = \begin{vmatrix} \frac{1}{E_l} & -\frac{\nu_{lt}}{E_t} & 0 \\ -\frac{\nu_{lt}}{E_t} & \frac{1}{E_t} & 0 \\ 0 & 0 & \frac{1}{G_{lt}} \end{vmatrix} \cdot \begin{bmatrix} \sigma_{11} \\ \sigma_{22} \\ \sigma_{12} \end{bmatrix}.$$

The criterion to minimise is given by

$$G(\alpha) = \int_\Omega \left[\frac{1}{E_l}(\sigma_{11})^2 - 2\frac{\nu_{lt}}{E_t}\sigma_{11}\sigma_{22} + \frac{1}{E_t}(\sigma_{22})^2 + \frac{1}{G_{lt}}(\sigma_{12})^2\right] dxdy + \int_\Omega cout(\alpha)dxdy$$

where

$$cout(\alpha) = C\alpha_2,\ C = const,$$

is proportional to fiber volume representing both the finantial cost of the fibers and the increase of mass due to the fibers.

Four applications will be presented corresponding to
i) shear forces on a square plate,
ii) shear forces on a rectangular plate,
iii) oblique traction upward on a square plate with two holes,
iv) oblique traction downward on a square plate with two holes.

For each case, we indicate
i) the system of forces,
ii) the decrease of G(α) versus the number of iterations, when avalable,
iii) the Hill-Tsai criteria defined by

$$C_{rupt} = \frac{(\sigma_{ll})^2}{(\sigma_{ll})^2_{rupture}} + \frac{(\sigma_{tt})^2}{(\sigma_{tt})^2_{rupture}} - \frac{\sigma_{ll}.\sigma_{tt}}{(\sigma_{ll})^2_{rupture}} + \frac{(\sigma_{lt})^2}{(\sigma_{lt})^2_{rupture}},$$

which must stay lower than one,
iv) the picture showing the optimal fiber orientations and volume fractions in (Ω).

For each case the convergence of the numerical algorithm has been obtained in less than twenty iterations, and the computational time is less than sixty seconds on these simple examples. Of course for industrial problem the situation will be much different, but will remain very reasonable.

6.1 Shear forces on a square (b=a)

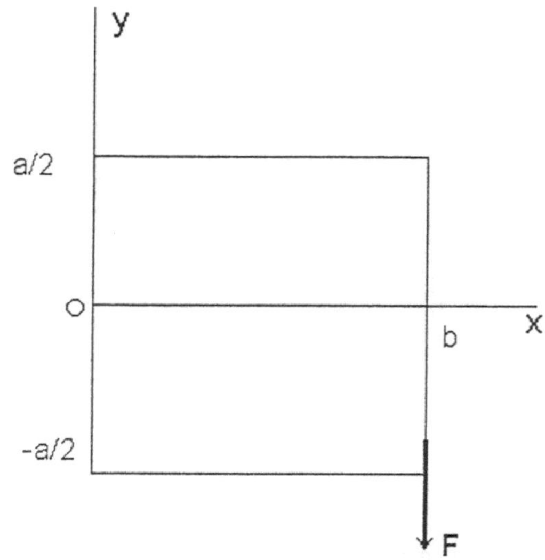

Iteration numbers	G(α)	C_{rupt} max.
Initial state	140.90	28.00
1	21.61	1.80
2	19.72	1.30
3	19.10	1.22
4	18.80	1.18
5	18.60	1.14
6	18.46	1.10
7	18.39	1.07
8	18.36	1.03
9	18.30	1.00
10	18.28	0.96
11	18.27	0.93
12	18.23	0.91
13	18.22	0.89
14	18.22	0.89

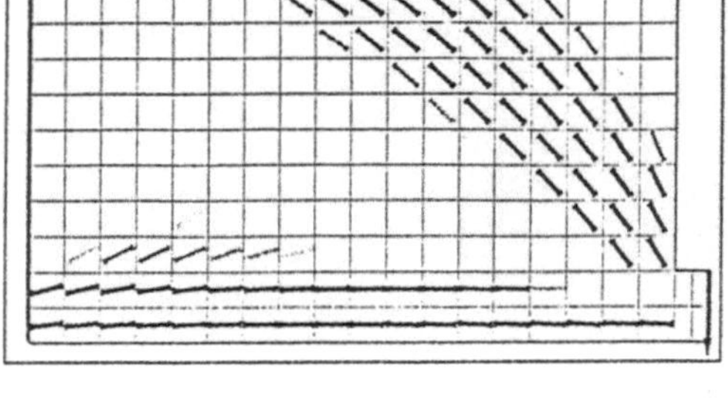

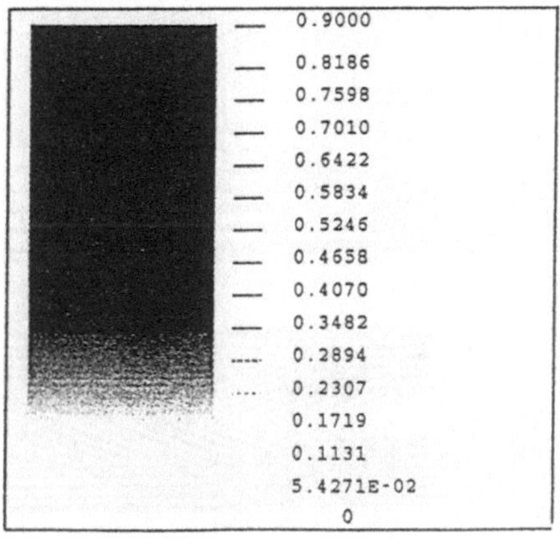

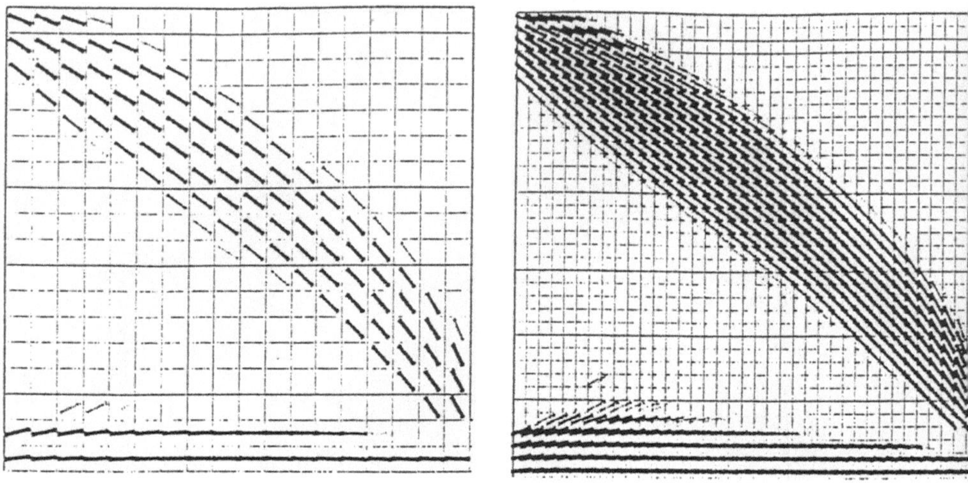

Influence of the size of the mesh (weak)

6.2 Shear forces on a rectangle (b=2a and b=4a).

Orientations and fibers volume ratio

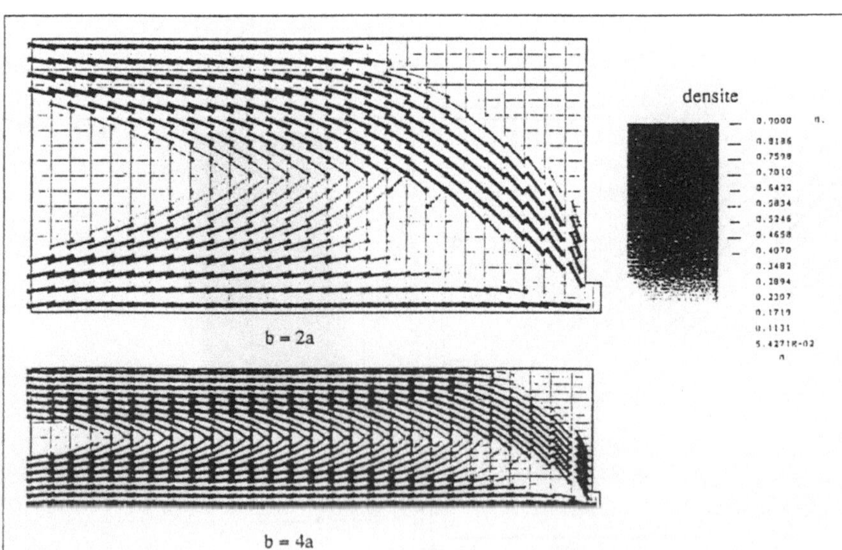

6.3 Oblique forces (topward) on a square with two holes.

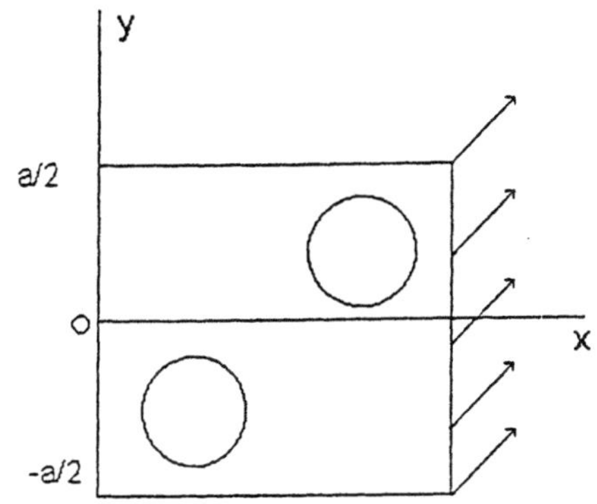

Iteration numbers	G(α)	C_{rupt} max
Initial state	13.3	8.058
1	4.28	1.590
2	4.18	1.211
3	4.00	0.912
4	3.95	0.909
5	3.90	0.907
6	3.89	0.906
7	3.88	0.906
8	3.87	0.906
9	3.87	0.906
10	3.87	0.906

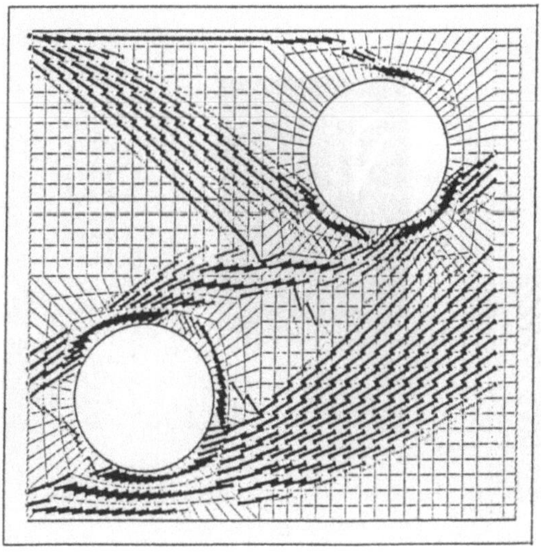

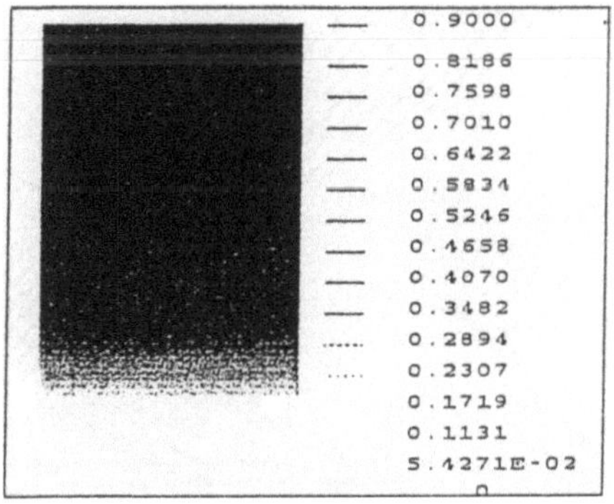

6.4 Oblique forces (downward) on a square with two holes.

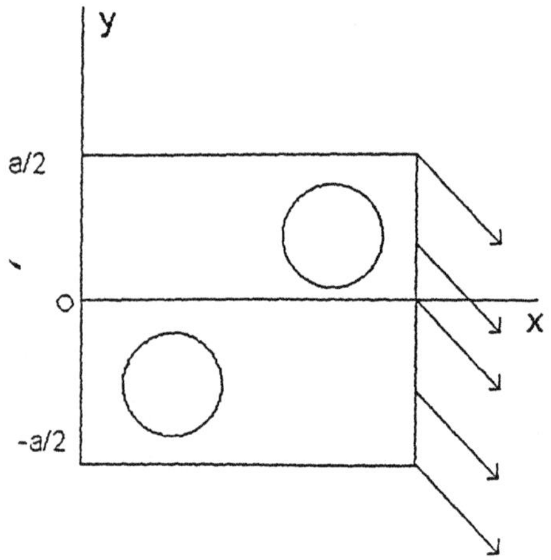

Iteration numbers	G(α)	C_{rupt} max
Initial state	12.60	6.574
1	3.98	1.104
2	3.86	0.880
3	3.76	0.812
4	3.67	0.790
5	3.57	0.781
6	3.56	0.774
7	3.55	0.769
8	3.55	0.765
9	3.55	0.763
10	3.55	0.763

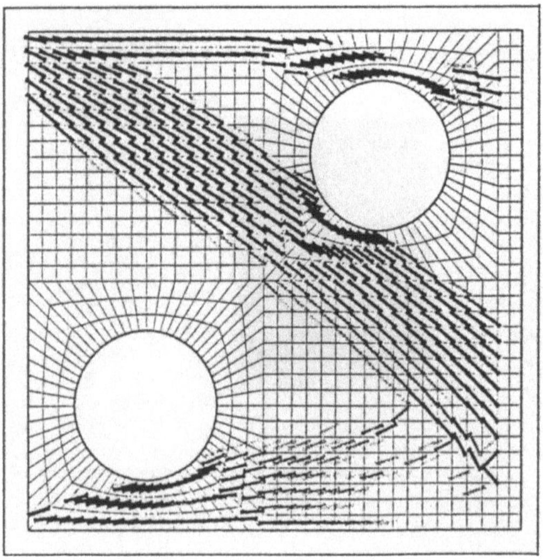

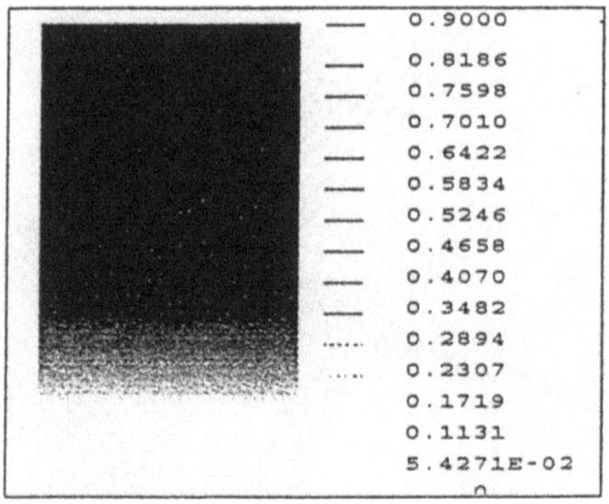

Orientations and fibers volume ratio (iteration 10)

7 CONCLUSION

The method presented does not need long computations, but nevertherless, at least on the relatively simple examples treated, leads to interesting results, which appear in good agreement with what can be estimated intuitively.

In addition, the optimal fibre partition which is found leads to a stress field which presents a maximum Hill-Tsai criterion that is much lower that an initial one corresponding to a single fiber orientation in the structure. One can give an intuitive explanation by noting that the optimal fiber orientations reduces the shear stresses. It also shows the importance of using curvilinear fibers to get the maximal benefit of using composite materials in the design of structures. This fact has not escaped to the attention of aerospace engineers, and industrial applications are in progress.

8 References:

[1] Allaire G. and Kohn R.V., Topology optimization and optimal shape using homogenization, *Proceedings of Topology design of Structures, pp.207-218, 1993, Kluwer Academic publishers, Netherlands.*

[2] Duvaut G., *Mécanique des Milieux Continus*, Masson, Paris, 1990.

[3] Duvaut G., Optimisation of Composite Structures, Keynote lecture, *Proceedings of ICCE/6, 27 june-3 July 1999, Orlando, Florida, USA, pp. B17-B20.*

[4] Duvaut G. et Lions J.L., *Les Inéquations en Mécanique et en Physique*, Dunod, Paris, 1972.

[5] Duvaut G. et Terrel G., Optimisation of fiber reinforced composites, *Proceedings of the 2d International Conference on Composite Sience and Technology, 9-11 June 1998, Durban, South Africa.*

[6] Gurdal Z. and Olmedo R., In-plane response of laminates with spatially varying fiber orientations : variable stiffness concept, *AIAA J., 1993, vol. 31, pp. 751-758.*

[7] Hayer M.W. and Charette R.F., The use of curvilinear fiber format in composite structure design, *Proceedings of the of the 33th AIAA, ASME, ASCE, AHS, Structural Dynamics and Material Conference, April 1989, pp.2137-2145.*

[8] Hayer M.W. and Lee H.H., The use of curvilinear fiber format to improve buckling resistance of composite plates with central circular holes, *Composite Structures, 1991, vol 18, pp.239-261.*

[9] Lions J.L., *Contrôle optimal des Systèmes Gouvernés par des Equations aux Dérivées Partielles*, Dunod-Gauthier-Villars, Paris, *1968.*

[10] Murat F. et Tartar L., Calcul des variations et homogénéisation, *Coll. de la Direction des Etudes et Recherches de Electricité de France, Eyrolles, 1985, pp. 27-68.*

Material evolution in plasticity and growth

***Marcelo Epstein*[1] *and Gérard A. Maugin*[2]**
[1] *Department of Mechanical and Manufacturing Engineering*
The University of Calgary, Calgary, Alberta T2N 1N4, Canada
epstein@enme.ucalgary.ca
[2] *Laboratoire de Modélisation en Mécanique*
Université Pierre et Marie Curie (Paris VI)
Paris, France

Abstract: The theories of plasticity and growth, as well as other theories of anelastic behaviour, are shown to share the same formal structure. The unifying concept is that of material evolution, rooted in the theory of material inhomogeneities. Particular attention is given to models based on first and second-grade elastic prototypes. A detailed treatment of the formal restrictions to be imposed on any possible evolution law is followed by the formulation of balance equations in a volumetrically growing body. A short final section is devoted to the discussion of thermodynamic inequalities and their consequences.
Keywords: Inhomogeneities, anelasticity, growth, evolution.

1. WHAT IS MATERIAL EVOLUTION?

In many anelastic material phenomena, such as plasticity, viscoelasticity and growth [Maugin, 1992, 1995; Rodriguez et al., 1994; Taber, 1995], it can be observed, or assumed, that, in spite of the radical structural changes taking place in the body, the material points "remain the same" in some sense as far as their constitutive identity is concerned. Specifically, in plasticity it can be assumed that the massive motion of dislocations leaves the underlying *local* properties of the crystallographic lattice unchanged. Similarly, in a theory of volumetric growth one may assume that new material *of the same kind* is somehow squeezed in, thus possibly causing internal stresses, but leaving the material properties essentially unchanged. What *does* then change or evolve?

The simplest possible theory of a material evolution of the kind just described is the following [Epstein and Maugin, 1996]. The basic, unchangeable, material properties are specified by means of the *purely elastic* response of a fixed "reference crystal" (Figure 1) with a given constitutive equation:

$$\boldsymbol{t} = \boldsymbol{t}_c(\boldsymbol{H}) \tag{1}$$

G.A. Maugin et al. (eds.), Continuum Thermomechanics, 153-162.
© 2000 *Kluwer Academic Publishers*.

where $\boldsymbol{H}$ is a variable non-singular matrix representing the deformation of the reference crystal, $\boldsymbol{t}$ is the Cauchy stress, and $\boldsymbol{t}_c$ is the specific function describing the elastic response of the reference crystal. This response function will remain unchanged, so as to represent the intuitive idea that the material itself remains essentially the same throughout the process of deformation and evolution. What *may* change with time is the way in which this prototypical material element is inserted in the body at some fixed reference configuration. This implant operation will be denoted by $\boldsymbol{P}$. If the body is uniform (i.e., made of the same material at all of its points), we can use the same reference crystal for each and every point. This implant will naturally depend both on $\boldsymbol{X}$ (point in reference configuration) and t (time), and will accordingly be written as $\boldsymbol{P} = \boldsymbol{P}(\boldsymbol{X}, t)$. This is then the essence of material evolution. The model is still elastic, but, because of the possible time evolution of $\boldsymbol{P}$, the functional response of the body to a deformation gradient $\boldsymbol{F}$ is time dependent, namely:

$$\boldsymbol{t}(\boldsymbol{F},\boldsymbol{X}) = \boldsymbol{t}_c(\boldsymbol{F}\boldsymbol{P}(\boldsymbol{X},t)) \tag{2}$$

as can be seen from Figure 1.

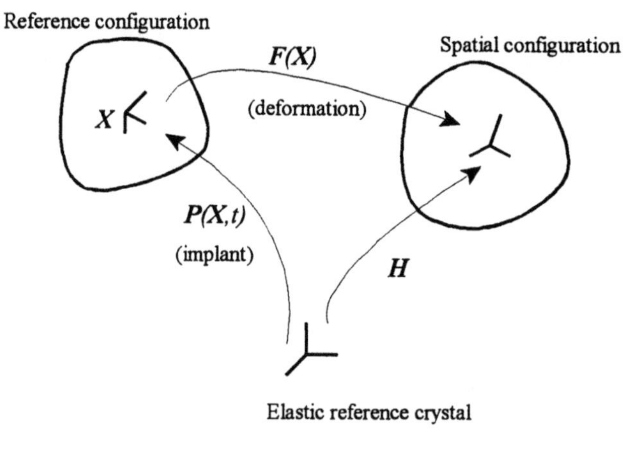

***Figure 1**: The interplay between the reference crystal, the body and the deformation*

Denoting by ρ_c the fixed density of the reference crystal, the density field of the reference configuration at time t is given by

$$\rho_0 = J_P^{-1}\rho_c \tag{3}$$

where J_P is the determinant of $P(X, t)$. It follows that, unless J_P is constant in time for each point X, there will be material growth or resorption. One may then conclude that, at least from the purely formal point of view, plasticity and growth differ only by the fact that the condition

$$\dot{J}_P = 0 \tag{4}$$

is satisfied by the former and violated by the latter. Other interpretations of this condition are, of course, possible. In this presentation, however, we identify the volumetric growth μ (net mass increase per unit volume and per unit time) as:

$$\mu = \dot{J}_P^{-1}\rho_c = -J_P^{-1}tr(\boldsymbol{P}^{-1}\dot{\boldsymbol{P}})\rho_c = -\rho_0\, tr\boldsymbol{L}_{P_C} \tag{5}$$

where a superimposed dot stands for the material time derivative, and *tr* is the trace operator. The quantity $\boldsymbol{L}_{P_C} = \boldsymbol{P}^{-1}\dot{\boldsymbol{P}}$ can be legitimately called the gradient of the velocity of evolution (or the inhomogeneity velocity gradient) at the reference crystal level. Its counterpart at the reference configuration level is $\boldsymbol{L}_P = \dot{\boldsymbol{P}}\boldsymbol{P}^{-1}$, both tensors having the same trace.

2. FORMAL RESTRICTIONS ON EVOLUTION LAWS

A complete constitutive theory in the context of material evolution will consist of two parts, namely: (i) the specification of the constitutive functionals for the reference crystal (which is elastic, in the above treatment), and (ii) the specification of a time evolution law for the material implant operation $\boldsymbol{P}$. In this section we will discuss some aspects of the second item, since it is not included in standards treatments of continuum mechanics.

The simplest evolution law is a differential equation of the first order for $\boldsymbol{P}$, of the form:

$$\dot{\boldsymbol{P}} = f(\boldsymbol{P}, \boldsymbol{a}) \tag{6}$$

where $\boldsymbol{a}$ represents a list of arguments (such as the stress tensor, the Eshelby stress, the deformation gradient, etc.). It is not difficult to show that the function f governing the evolution cannot be completely arbitrary, but must be subjected to a number of formal restrictions arising from physical considerations.

The first obvious restriction stems from the assumed uniformity of the body. If the material, although evolving differently at different points, is assumed to remain essentially unique, we must conclude that the list of arguments $\boldsymbol{a}$ cannot include $\boldsymbol{X}$ explicitly.

The second type of restriction to be imposed on the evolution law is of a more subtle nature. It arises from the fact that the evolution law must, in some precise sense, be independent of the reference configuration chosen for the body. Upon a change of reference configuration $\lambda = \lambda(\boldsymbol{X})$, the implant maps $\boldsymbol{P}(\boldsymbol{X})$ change by composition to

$$\hat{\boldsymbol{P}}(\lambda(\boldsymbol{X})) = \nabla\lambda(\boldsymbol{X})\ \boldsymbol{P}(\boldsymbol{X}) \tag{7}$$

and, similarly, the time derivative changes to

$$\frac{D\hat{\boldsymbol{P}}(\lambda(\boldsymbol{X}))}{Dt} = \nabla\lambda(\boldsymbol{X})\ \dot{\boldsymbol{P}}(\boldsymbol{X}) \tag{8}$$

Analogously, the list of arguments $\boldsymbol{a}$ will change in some specific way to a transformed list which we symbolically indicate by $\hat{\boldsymbol{a}}_{\nabla\lambda}$. (For example, $\hat{\boldsymbol{F}} = \boldsymbol{F}\,\nabla\lambda^{-1}$). Since the reference crystal itself has not been altered, we demand now that the evolution law be valid for the hatted quantities *with the same function* f, namely:

$$\frac{D\hat{\boldsymbol{P}}(\lambda(\boldsymbol{X}))}{Dt} = f(\hat{\boldsymbol{P}}, \hat{\boldsymbol{a}}_{\nabla\lambda}) \tag{9}$$

or, according to Equations (7) and (8),

$$\nabla\lambda(\boldsymbol{X})\ \dot{\boldsymbol{P}}(\boldsymbol{X}) = f(\nabla\lambda\,\boldsymbol{P}, \hat{\boldsymbol{a}}_{\nabla\lambda}) \tag{10}$$

identically for all possible changes of reference configuration λ. Choosing, therefore, instantaneously at an arbitrary point $\nabla\lambda = \boldsymbol{P}^{-1}$, we conclude that the evolution law must be of the form:

$$\boldsymbol{L}_{P_C} = f(\hat{\boldsymbol{a}}_{P^{-1}}) \tag{11}$$

where f is an arbitrary tensor function. In other words, the evolution law can only prescribe the evolution of the inhomogeneity velocity gradient at the reference crystal level in terms of the list of arguments pulled instantaneously back to the reference crystal.

Two further restrictions of the possible form of the evolution law stem from considerations of material symmetry. Indeed, let $\mathcal{G}_c$ be the symmetry group of the reference crystal. Then, the collection $\mathcal{P}(X, t)$ of all implant maps equivalent to a given implant map $\boldsymbol{P}(X, t)$ is given by $\mathcal{P}(X, t) = \boldsymbol{P}(X, t)\, \mathcal{G}_c$. If the evolution law is to prescribe a true structural change, therefore, we must make sure that $\boldsymbol{P}$ does not move instantaneously within $\mathcal{P}$, making use of the degree of freedom afforded by the symmetry group. This is the same as saying ("principle of actual evolution" [Epstein and Maugin, 1996; Epstein, 1999]) that the values of the function f must not be attained within the Lie algebra of the symmetry group $\mathcal{G}_c$. For example, if the material happens to be a fully isotropic solid (for which the symmetry group is the full orthogonal group, whose Lie algebra is the collection of all skew-symmetric matrices), the evolution law should be such that the symmetric part of $\boldsymbol{L}_{P_C}$, as prescribed by Equation (11), never vanishes. In other words, for a fully isotropic solid the "spin" part of the inhomogeneity velocity gradient is irrelevant. For other materials, however, the spin part of $\boldsymbol{L}_{P_C}$ can be meaningfully prescribed.

The final restriction to be mentioned here arises from symmetry considerations too. Indeed, symmetry transformations can be imposed on the list of arguments $\hat{\boldsymbol{a}}_{P^{-1}}$ to derive identities for the function f, as done in the usual constitutive theory.

So far, we have presented the simplest example of material evolution, based on a prototypical first-grade elastic reference crystal. A similar treatment can be extended to second-grade elastic materials [Epstein, 1999], a feature that may be necessary in theories of growth to account for mass diffusion. The main difference with the first-grade theory just presented is that the implant maps consist not just of a 2nd-order tensor $\boldsymbol{P}$, but also of an independent 3rd-order symmetric tensor $\boldsymbol{Q}$ representing the second-gradient behaviour. The evolution equations are now of the form:

$$\dot{\boldsymbol{P}} = f(\boldsymbol{P}, \boldsymbol{Q}, \boldsymbol{a}) \tag{12}$$

and

$$\dot{\boldsymbol{Q}} = g(\boldsymbol{P}, \boldsymbol{Q}, \boldsymbol{a}) \tag{13}$$

A treatment analogous to the previous one shows that these equations can be reduced to the form

$$\boldsymbol{L}_{P_C} = f(\hat{\boldsymbol{a}}_{(P,Q)^{-1}}) \tag{14}$$

and

$$\boldsymbol{L}_{Q_C} = \boldsymbol{g}(\hat{\boldsymbol{a}}_{(P,Q)^{-1}}) \tag{15}$$

where the second-grade part of the inhomogeneity velocity gradient is given by

$$\boldsymbol{L}_{Q_C} = \boldsymbol{P}^{-1}(\dot{\boldsymbol{Q}} - \boldsymbol{Q}\boldsymbol{L}_{P_C} - (\boldsymbol{Q}\boldsymbol{L}_{P_C})^T) \tag{16}$$

The transposition operation is naturally to be effected on the lower indices only.

As to the "actual evolution" restriction, it turns out to be exactly the same as before, except that the Lie algebra is now to be understood in the more involved sense of the extended material symmetry group of a second-grade elastic point [Elzanowski and Epstein, 1992; Epstein and de León, 1996]. A case of particular importance for the theory of growth is that in which the second-grade character of the material is represented by just a dependence on the density gradient. It is then possible to show that the symmetry group associated with this property consists of all third-order symmetric tensors with zero trace (the trace being taken by contracting the upper index of the tensor with either of the lower indices). This is an Abelian group under addition, so it coincides with its own Lie algebra. The principle of actual evolution in this case necessitates, therefore, that the function $\boldsymbol{g}$ have a non-zero trace.

3. BALANCE EQUATIONS

Let Ω be a fixed volume in a given reference configuration with density $\rho_0 = \rho_0(\boldsymbol{X}, t)$, which is allowed to vary with time. If Π denotes a smooth volumetric mass source and M is an assumed mass flux through the boundary, we may write the following non-standard global equation of mass balance:

$$\frac{D}{Dt}\int_\Omega \rho_0 d\Omega = \int_\Omega \Pi d\Omega + \int_{\partial\Omega} M dS \tag{17}$$

By Cauchy's tetrahedron argument, there exists a vectorial mass flux $\boldsymbol{M}$ such that, on an area with unit normal N, the scalar mass flux is given by $M = \boldsymbol{M} . N$. The local Lagrangian form of mass balance is then

$$\dot{\rho}_0 = \mu = -\rho_0 \, tr\, \boldsymbol{L}_{P_c} = \Pi + Div\boldsymbol{M} \tag{18}$$

where *Div* is taken in the reference configuration. The corresponding Eulerian form is

$$\dot{\rho} = \pi + div\, \boldsymbol{m} - \rho\, div\, \boldsymbol{v} \tag{19}$$

where *div* is taken in space and where $\boldsymbol{v}$ is the velocity field. The Eulerian production and flux quantities are related to their Lagrangian counterparts by the classical formulas involving the deformation gradient $\boldsymbol{F}$ and its determinant J.

Because of the new mass generated during the process of growth, new sources of momentum, energy and entropy need to be included in the respective balance equations. We shall divide these sources into two classes. The first type (which somewhat imprecisely may be called "reversible") consists of those inputs carried by the entering mass assumed to make its appearance with the same velocity, specific energy and specific entropy as the preexisting local substratum. Those sources may then be written explicitly. Other sources, which may be termed "irreversible", and can easily be lumped as extra terms, as done in [Cowin and Hegedus, 1976; Epstein and Maugin, in press], will be ignored here for the sake of brevity. Under these conditions, the Lagrangian form of the balance of linear momentum reads:

$$\rho_0 \dot{\boldsymbol{v}} = \boldsymbol{f}_b + Div\, \boldsymbol{T} + (Grad\, \boldsymbol{v})^T \boldsymbol{M} \tag{20}$$

where *Grad* is taken in the reference configuration and where $\boldsymbol{f}_b$ is the volumetric body force, and $\boldsymbol{T}$ is the first Piola-Kirchhoff stress. We omit the Eulerian version, which can be obtained by means of the standard transformation of variables.

The balance of angular momentum yields the symmetry of the modified Cauchy stress tensor given by:

$$\tilde{\boldsymbol{t}} = \boldsymbol{t} + \boldsymbol{m} \otimes \boldsymbol{v} \tag{21}$$

Alternatively, in terms of the corresponding modified first Piola-Kirchhoff stress

$$\tilde{\boldsymbol{T}} = \boldsymbol{T} + \boldsymbol{M} \otimes \boldsymbol{v} \tag{22}$$

the balance of angular momentum implies the symmetry of $\boldsymbol{F}\tilde{\boldsymbol{T}}$.

The Lagrangian local form of the energy balance is obtained as:

$$\rho_0\, \dot{\epsilon}_0 = tr\,(\boldsymbol{T}\ (Grad\, \boldsymbol{v})^T) - Div\, \boldsymbol{Q}_h + \boldsymbol{M} \cdot Grad\, \epsilon_0 + r_b \tag{23}$$

where ϵ_0 is the internal energy per unit mass, Q_h is the referential heat flux vector, and r_b is the referential volumetric heat supply, if any. Note the presence of a "diffusive" term involving the mass flux vector acting on the material gradient of the energy density..

The Clausius-Duhem inequality results in the following local form:

$$\rho_0 \dot{\eta}_0 \geq \frac{r_0}{\theta} - Div\left(\frac{\boldsymbol{Q}_h}{\theta}\right) + \boldsymbol{M} \cdot Grad\, \eta_0 \qquad \textbf{(24)}$$

where η_0 is the entropy per unit mass, and θ is the absolute temperature. Following the standard procedure, Equation (24) can be combined with (23) to eliminate the volumetric heat supply term and to express the second law of thermodynamics in terms of the free energy density $\psi_0 = \epsilon_0 - \theta\, \eta_0$.

4. FURTHER CONSIDERATIONS

We have already established the need to specify a set of evolution equations for the implant maps as an integral part of a theory of anelasticity or growth, and we have determined a number of formal restrictions that those evolution laws must satisfy. To obtain a complete theory it is necessary to specify also constitutive equations for the stress, the free energy, the entropy, the heat flux and the mass flux. We will now see that, within the framework of a set of adopted constitutive equation, some more restrictions impose themselves.

In a first-grade theory, the list of independent variables might consist of the deformation gradient, the temperature and the temperature gradient. According to the same philosophy which led us to write Equation (2), we will have similar formulas relating each of the constitutive quantities at a point in the reference configuration with the prototypical constitutive equation of the reference crystal. Thus, for example, the free energy per unit mass is given by:

$$\psi_0(\boldsymbol{F}, \theta) = \psi_c(\boldsymbol{F}\boldsymbol{P}, \theta) \qquad \textbf{(25)}$$

where ψ_c is the free energy density of the reference crystal. Following the standard exploitation of the Clausius-Duhem inequality, it is not difficult to show that the free energy, in addition to being independent of the temperature gradient, acts as a potential for the stress and the entropy. The residual dissipation inequality turns out to be the following [Epstein and Maugin, in press]:

$$tr(\boldsymbol{B}\boldsymbol{L}_P{}^T) - \frac{\boldsymbol{Q}_h \cdot Grad\theta}{\theta} - \phi_M \geq 0 \qquad \textbf{(26)}$$

where $\boldsymbol{B}$ is the Mandel stress [Mandel, 1971; Lubliner, 1990; Cleja-Tigoiu and Maugin, i.p.]:

$$\boldsymbol{B} = \boldsymbol{F}^T \boldsymbol{T} \tag{27}$$

It is interesting to note that the Mandel stress appears naturally as a result of differentiating the free energy density with respect to the implant maps, as required by the chain rule of differentiation coupled with Equation (25). The term ϕ_M represents the dissipation associated with the mass flux. It is explicitly given by the following expression:

$$\rho_0 \, \phi_M = F^i_{K;I} \, M^I \, T^K_i \tag{28}$$

where indices have been used to avoid any ambiguity. The semicolon indicates covariant differentiation with respect to the *material connection* [Noll, 1967], whose Christoffel symbols are:

$$\Gamma^K_{IJ} = - \, (P^{-1})^\alpha_I \, \frac{\partial P^K_\alpha}{\partial X^J} \tag{29}$$

Since in a first-grade material the second gradient does not appear in the list of independent variables, the inequality (26) implies that $\boldsymbol{M}$ must vanish, so that there can be no diffusive effects. These effects are, however, possible in a second-grade theory, in which the second gradient of the deformation is included ab-initio in the list of independent variables. In a second-grade theory, a careful use of the chain rule of differentiation and the standard use of the Clausius-Duhem inequality would yield a residual dissipation involving also Mandel hyperstresses producing dissipation through the time derivative of the second-order implant $\boldsymbol{Q}$. In any case, we see that if the list of independent arguments in the evolution law is made to include the Mandel or the Eshelby stresses, then the dissipation inequality imposes further restrictions upon the evolution functionals.

ACKNOWLEDGEMENTS

This work has been partially supported by a NATO collaborative grant and by the Natural Sciences and Engineering Research Council of Canada.

REFERENCES

[Cleja-Tigoiu and Maugin, in press] Cleja-Tigoiu, S., Maugin, G.A., "Eshelby's stress tensors in finite elastoplasticity", Acta Mechanica.

[Cowin and Hegedus, 1976] Cowin, C.S.; Hegedus, H.D.; "Bone remodeling I: Theory of adaptive elasticity", J Elasticity, **6**, pp. 313-326.

[Elzanowski and Epstein, 1992] Elzanowski, M; Epstein, M.; "On the symmetry group of second-grade materials", Int J Nonlinear Mech, **27**, pp. 635-638.

[Epstein, 1999] Epstein, M.; "Toward a complete second-grade evolution law", Math and Mech of Solids, **4**, pp. 251-266.

[Epstein and de León, 1996] Epstein, M.; de León, M; "The geometry of uniformity in second-grade elasticity", Acta Mechanica, **83**, pp. 127-133.

[Epstein and Maugin, 1996] Epstein, M; Maugin, M.A.; "On the geometrical material structure of anelasticity", Acta Mechanica, **115**, pp. 19-131

[Epstein and Maugin, in press]. Epstein, M.; Maugin, G.A.; "Thermomechanics of volumetric growth in uniform bodies", Int J Plasticity.

[Lubliner, 1990] Lubliner, J.; "Plasticity theory", McMillan, New York.

[Mandel, 1971] Mandel J.; 'Plasticité classique et viscoplasticité", (CISM Lecture Notes, Udine, Italy), Springer-Verlag, Vienna.

[Maugin, 1992] Maugin, G.A.; "Thermomechanics of plasticity and fracture", Cambridge University Press, U.K.

[Maugin, 1995] Maugin, G.A.; "Material forces: Concepts and applications", ASME Appl Mech Rev, **48**, pp. 213-245.

[Noll, 1967] Noll, W.; "Materially uniform simple bodies with inhomogeneities", Arch Rational Mech Analysis, **27**, pp. 1-32.

[Rodriguez et al., 1994] Rodriguez, E.K.; Hoger, A.; McCulloch, A.D.; "Stress-dependent growth in soft elastic tissues", J Biomechanics, **27**, pp. 455-467.

[Taber, 1995] Taber, L.A.; "Biomechanics of growth, remodeling and morphogenesis", ASME Appl Mech Rev, **48**, pp. 487-545.

Thermoelasticity of second–grade media

Samuel Forest, Jean-Marc Cardona
Ecole des Mines de Paris / CNRS
Centre des Matériaux / UMR 7633
BP 87 - 91003 Evry France
Samuel.Forest@mat.ensmp.fr

Rainer Sievert
Bundesanstalt für Materialforschung und -prüfung (BAM)
Unter den Eichen 87, D-12200 Berlin, Germany

Abstract : The method of virtual power and continuum thermodynamics are used to incorporate temperature and temperature gradients into the theory of second grade solids settled by [Germain, 1973] in the isothermal case. In a second part, it is shown that heterogeneous classical materials submitted to slowly-varying mean fields can be replaced by a homogeneous equivalent medium including higher order gradients of displacement and temperature. For that purpose, an asymptotic analysis of thermoelastic heterogeneous periodic materials is performed. The form of the derived effective properties are compared to the previous phenomenological framework.
Keywords : second gradient, Cattaneo equation, virtual power, homogenization theory, asymptotic method.

1. INTRODUCTION AND NOTATIONS

Although the mechanical and constitutive framework of second gradient theory has been settled by Mindlin, the work of Germain [Germain, 1973] shows that the principle of virtual power is a powerful and elegant tool to derive the balance equations and boundary conditions for a medium modelled by the first and second gradients of the displacement field.

In several works including recent ones like [Boutin, 1996] and [Triantafyllidis, Bardenhagen, 1996], the need for such a theory arises in the mechanics of heterogeneous materials and in particular homogenization theory. The usual assumption that the size of the heterogenities is much smaller than the size of the considered structure inevitably leads to a classical first gradient Cauchy medium to model a homogeneous equivalent medium (HEM). As soon as slowly-varying mean fields over the heterogenities are possible (due to strong deformation gradients in a structure for instance), the HEM, if it exists, must be regarded as a generalized continuum [Pideri, Seppecher, 1997] [Forest, 1998]. In the case of periodic media, asymptotic methods can be used to derive the effective properties of a second grade medium [Boutin, 1996] [Triantafyllidis, Bardenhagen, 1996].

G.A. Maugin et al. (eds.), Continuum Thermomechanics, 163-176.
© 2000 *Kluwer Academic Publishers.*

The need for higher grade theories appears also in the thermomechanical framework. For instance, strong stress gradients can develop in a structure made of a heterogeneous material as a result of strong thermal gradients. Accordingly, a thermoelasticity theory must be developed for second grade media. This has been undertaken in [Cardona et al., 1998], where motivations and examples based on the mechanics of heterogeneous materials are given. Part 2 of the present work shows how the principle of virtual power combined with the thermodynamical principles can be used to settle a thermoelasticity theory of second grade media. The linear case is presented in part 3. In the last part, the asymptotic analysis of periodic heterogeneous classical media of [Boutin, 1996] is extended to the thermomechanical case, in order to show that the additional constitutive tensors arising in the pheneomenological theory of part 2 can be computed explicitely as functions of the classical thermomechanical properties of the constituents. In this sense, the mechanics of heterogeneous materials provides an example of thermoelastic second grade effective medium, for which balance and constitutive equations have the form predicted by the basic principles of continuum thermomechanics.

An invariant notation is used throughout this work. First, second, third and fourth order tensors are respectively denoted by $\underline{\mathbf{a}}, \underset{\sim}{\mathbf{a}}, \underset{\approx}{\underline{\mathbf{a}}}$ (or $\underset{\approx}{\mathbf{a}}$) and $\underset{\approx}{\mathbf{a}}$. The gradient and divergence operators are defined as follows

$$\nabla \underline{\mathbf{U}} = U_{i,j} \underline{\mathbf{e}}_i \otimes \underline{\mathbf{e}}_j, \quad \nabla\nabla \underline{\mathbf{U}} = U_{i,jk} \underline{\mathbf{e}}_i \otimes \underline{\mathbf{e}}_j \otimes \underline{\mathbf{e}}_k \tag{1}$$

$$\nabla . \underset{\sim}{\mathbf{\Sigma}} = \Sigma_{ij,j} \underline{\mathbf{e}}_i, \quad \nabla . \underset{\approx}{\underline{\mathbf{S}}} = S_{ijk,k} \underline{\mathbf{e}}_i \otimes \underline{\mathbf{e}}_j \tag{2}$$

where the nabla operator and an orthonormal basis $(\underline{\mathbf{e}}_i)_{i=1,3}$ have been introduced : $\nabla = \frac{\partial .}{\partial x_i} \underline{\mathbf{e}}_i$. The notations for the contraction of tensors are :

$$\underset{\sim}{\mathbf{A}} : \underset{\sim}{\mathbf{B}} = A_{ij} B_{ij}, \quad \underset{\approx}{\underline{\mathbf{K}}} \vdots \underset{\approx}{\underline{\mathbf{L}}} = K_{ijk} L_{ijk} \tag{3}$$

The gradient operator can be decomposed into its normal and tangent parts D_n and D_t, $\underline{\mathbf{n}}$ being the normal to a surface :

$$\nabla \underline{\mathbf{u}} \;=\; D_n \underline{\mathbf{u}} + D_t \underline{\mathbf{u}} \quad \text{with} \quad D_n \underline{\mathbf{u}} \;=\; (\nabla \underline{\mathbf{u}}).\underline{\mathbf{n}} \tag{4}$$

The gradient of the velocity field $\underline{\dot{\mathbf{U}}}$ on a solid Ω can be decomposed into a symmetric and a skew–symmetric part :

$$\nabla \underline{\dot{\mathbf{U}}} = \underset{\sim}{\mathbf{D}} + \underset{\sim}{\mathbf{W}} \tag{5}$$

For simplicity, the small perturbation framework is adopted, so that Lagrangian and Eulerian configurations are not distinguished. That is why tensor $\underset{\sim}{\mathbf{D}}$ will be replaced by $\underset{\sim}{\dot{\mathbf{E}}} = (\nabla \underline{\dot{\mathbf{U}}} + (\nabla \underline{\dot{\mathbf{U}}})^T)/2$. The boundary $\partial\Omega$ of body Ω will be regarded as twice differentiable and thus admits at each point a unique mean curvature R. The reader is referred to [Germain, 1973] for the treatment of edges and corners.

2. MECHANICS AND THERMODYNAMICS OF SECOND GRADE MEDIA

2.1 Principle of virtual power

The method of virtual power has proved to be an efficient tool for deriving balance equations and boundary conditions and has been applied to several coupled mechanical problems. Following [Maugin, 1980], the first step consists in defining the set $\mathcal{V}^\circ$ of virtual motions relevant for the considered physical situation. Within the present thermomechanical framework, velocity fields $\dot{\underline{\mathbf{U}}}^*$ and temperature rates $\dot{T}^*$ are regarded as generalized virtual motions. It means that temperature is treated formally as an additional degree of freedom. The set $\mathcal{V}$ then contains the variables that have to be introduced in a second grade theory :

$$\mathcal{V}^\circ = \{\dot{\underline{\mathbf{U}}},\ \dot{T}\},\quad \mathcal{V} = \{\dot{\underline{\mathbf{U}}}\ ,\ \nabla\dot{\underline{\mathbf{U}}}\ ,\ \nabla\nabla\dot{\underline{\mathbf{U}}}\ ,\ \dot{T},\ \nabla\dot{T}\} \tag{6}$$

The latter may be restricted to objective virtual fields $\mathcal{V}^{\text{obj}}$ and an additional set $\mathcal{V}^c$ is defined :

$$\mathcal{V}^{\text{obj}} = \{\underset{\sim}{\mathbf{D}}, \nabla\underset{\sim}{\mathbf{D}}, \dot{T}, \nabla\dot{T}\},\quad \mathcal{V}^c = \{\dot{\underline{\mathbf{U}}}, D_n\dot{\underline{\mathbf{U}}}, \dot{T}\} \tag{7}$$

Since the small perturbation framework is used in this work, $\underset{\sim}{\mathbf{D}}$ will be replace by $\dot{\underset{\sim}{\mathbf{E}}}$ in the following. The principle of vitual power states that, in a Galilean frame, the virtual power of inertial forces balances the virtual power of internal and external acting forces, for all generalized virtual motions ϑ^* and for all subdomain $\mathcal{D}$ of body Ω :

$$\mathcal{P}^{(i)}(\vartheta^* \in \mathcal{V}^{\text{obj}}) + \mathcal{P}^{(d)}(\vartheta^* \in \mathcal{V}) + \mathcal{P}^{(c)}(\vartheta^* \in \mathcal{V}^c) = \mathcal{P}^{(a)}(\vartheta^* \in \mathcal{V}^\circ) \tag{8}$$

The virtual power of internal (i), volume (d), contact (c) and inertial forces (a) are supposed to admit densities according to :

$$\mathcal{P}^{(i)} = -\int_{\mathcal{D}} p^{(i)}\,dV,\quad \mathcal{P}^{(c)} = \int_{\partial\mathcal{D}} p^{(c)}\,dS,\quad \mathcal{P}^{(d,a)} = \int_{\mathcal{D}} p^{(d,a)}\,dV \tag{9}$$

The densities are then taken as linear forms on the appropriate set of generalized virtual motions :

$$p^{(i)}(\dot{\underline{\mathbf{U}}}^*, \dot{T}^*) = \underset{\sim}{\mathbf{\Sigma}} : \dot{\underset{\sim}{\mathbf{E}}}^* \;+\; \underset{\approx}{\mathbf{S}} \vdots \nabla\dot{\underset{\sim}{\mathbf{E}}}^* \;+\; a^{(i)}\,\dot{T}^* \;+\; \underline{\mathbf{b}}^{(i)}.\nabla\dot{T}^* \tag{10}$$

$$p^{(d)}(\dot{\underline{\mathbf{U}}}^*, \dot{T}^*) = \underline{\mathbf{f}}.\dot{\underline{\mathbf{U}}}^* \;+\; \underset{\sim}{\mathbf{C}} : \underset{\sim}{\mathbf{W}}^* \;+\; \underset{\sim}{\mathbf{F}} : \dot{\underset{\sim}{\mathbf{E}}}^* \;+\; \underset{\approx}{\mathbf{P}} \vdots \nabla\dot{\underset{\sim}{\mathbf{E}}}^* \;+\; a^{(d)}\,\dot{T}^* \;+\; \underline{\mathbf{b}}^{(d)}.\nabla\dot{T}^* \tag{11}$$

$$p^{(c)}(\dot{\underline{\mathbf{U}}}^*, \dot{T}^*) = \underline{\mathbf{T}}.\dot{\underline{\mathbf{U}}}^* \;+\; \underline{\mathbf{M}}.(D_n\underline{\mathbf{U}}^*) \;+\; a^{(c)}\,\dot{T}^*;\quad p^{(a)}(\dot{\underline{\mathbf{U}}}^*) = \rho\ddot{\underline{\mathbf{U}}}.\dot{\underline{\mathbf{U}}}^* \tag{12}$$

The quantities dual to the strain rate and strain rate gradient in the power density of internal forces are the symmetric stress tensor $\underset{\sim}{\mathbf{\Sigma}}$ and the hyperstress tensor $\underset{\approx}{\mathbf{S}}$ $(S_{ijk} = S_{jik})$. Volume forces $\underline{\mathbf{f}}$, volume couples $\underset{\sim}{\mathbf{C}}$ and volume double and triple forces $\underset{\sim}{\mathbf{F}}$ and $\underset{\approx}{\mathbf{P}}$ may exist in general. A traction vector $\underline{\mathbf{T}}$ and a normal surface double force $\underline{\mathbf{M}}$ may act on the domain surface. The acceleration at a given material point reads $\ddot{\underline{\mathbf{U}}}$. For the sake of generality, additional terms associated with virtual temperature rates and their gradients have been systematically introduced. Note that a similar term should in principle appear in $(12)_2$. However it will be assumed that, if such additional terms exist in the power of acceleration forces and in the expression of the kinetic energy $\mathcal{K}$, they are still such that :

$$\mathcal{P}^{(a)} = \dot{\mathcal{K}} \tag{13}$$

Note that the homogenization procedure developed in section 4 and restricted to the static case, should be applied to the full dynamical case in order to justify the existence or not of additional terms in $\mathcal{P}^{(a)}$ and $\mathcal{K}$.

The application of the principle of virtual power and the successive use of Stokes theorem for volumes and surfaces as shown in [Germain, 1973] and in [Cardona et al., 1998] lead to the balance equations

$$\nabla . \underset{\sim}{\mathcal{T}} + \underline{\mathbf{f}} = \rho \underline{\ddot{\mathbf{U}}} \quad \text{with} \quad \underset{\sim}{\mathcal{T}} = \underset{\sim}{\Sigma} - \underset{\sim}{\mathbf{F}} - \underset{\sim}{\mathbf{C}} - \nabla . \underset{\approx}{\mathbf{S}} + \nabla . \underset{\approx}{\mathbf{P}} \tag{14}$$

$$\nabla . \underline{\mathbf{b}} - a = 0 \quad \text{with} \quad a = a^{(i)} - a^{(e)} \text{ and } \underline{\mathbf{b}} = \underline{\mathbf{b}}^{(i)} - \underline{\mathbf{b}}^{(e)} \tag{15}$$

and associated boundary conditions :

$$\underline{\mathbf{T}} = \underset{\sim}{\mathcal{T}} . \underline{\mathbf{n}} + 2R \underset{\approx}{\mathbf{S}} : (\underline{\mathbf{n}} \otimes \underline{\mathbf{n}}) - D_t \, (\underset{\approx}{\mathbf{S}} . \underline{\mathbf{n}}); \; \underline{\mathbf{M}} = \underset{\approx}{\mathbf{S}} : (\underline{\mathbf{n}} \otimes \underline{\mathbf{n}}); \; a^{(c)} = \underline{\mathbf{b}} . \underline{\mathbf{n}} \tag{16}$$

Note that the equations (15) and $(16)_3$ can be regarded as a definition of a and $a^{(c)}$ depending on $\underline{\mathbf{b}}$, and thus the formulation does not require any additional partial differential equation to be solved. In the classical first gradient theory, the term $a^{(i)}\dot{T}$ could also be introduced in the power density of internal forces, but, for vanishing $a^{(e)}$, it would have no counterpart $\underline{\mathbf{b}}^{(i)}$ so that the application of the principle finally implies $a^{(i)} = 0$. As a result, the proposed thermomechanical framework is relevant for a second grade theory and reduces to the classical case if higher order gradients are excluded. Similarily, the subsequent developments will show that a constitutive dependence on temperature gradient that seems to be necessary for instance in the modelling on some thermal treatments, can be introduced only within the framework of a second grade theory for the displacement.

2.2 Energy and entropy principles

The global form of the energy balance equation on Ω takes the forms :

$$\dot{\mathcal{E}} + \dot{\mathcal{K}} = \mathcal{P}^{(d)} + \mathcal{P}^{(c)} + \mathcal{Q}_e; \quad \dot{\mathcal{E}} = -\mathcal{P}^{(i)} + \mathcal{Q}_e \tag{17}$$

where $\mathcal{E}$ is the internal energy of the system and $\mathcal{Q}_e$ the total heat supply. In the second form, the kinetic energy theorem (13) and the principle of virtual power (8) have been applied. Introducing the specific internal energy ϵ, the heat flux vector $\underline{\mathbf{Q}}$, and an inner heat production rate r,

$$\dot{E} = \int_\Omega \rho \dot{\epsilon} \, dV \; ; \; \mathcal{Q} = -\int_{\partial\Omega} \underline{\mathbf{Q}} . \underline{\mathbf{n}} \, dS + \int_\Omega r \, dV \tag{18}$$

a local form of the energy balance is obtained :

$$\rho \dot{e} = p^{(i)} - \nabla . \underline{\mathbf{Q}} + r \tag{19}$$

where the expression (10) of the power density of internal forces is to be substituted.

The global form of the second principle reads : $\dot{\mathcal{S}} \geq \mathcal{Q}_s$, where $\mathcal{S}$ is the global entropy of the system and $\mathcal{Q}_s$ is the total flux of entropy. Introducing the local specific entropy s, the following relations are assumed to hold :

$$\mathcal{S} = \int_\Omega \rho s \, dV, \quad \mathcal{Q}_s = - \int_{\partial\Omega} \underline{\mathbf{J}}_s . \underline{\mathbf{n}} \, dS + \int_\Omega \frac{r}{T} \, dV \quad \text{and} \quad \underline{\mathbf{J}}_s = \frac{\underline{\mathbf{Q}}}{T} \tag{20}$$

where $\underline{\mathbf{J}}_s$ is the entropy flux vector. A local form of the entropy imbalance is adopted :

$$\rho \dot{s} + \nabla . (\frac{\underline{\mathbf{Q}}}{T}) - \frac{r}{T} \geq 0 \tag{21}$$

Combining (19) and (21) and introducing the Helmholtz free energy $\Psi = \epsilon - Ts$, we get the Clausius-Duhem inequality :

$$-\rho(\dot{\Psi} + s\dot{T}) + p^{(i)} - \frac{1}{T}\underline{\mathbf{Q}}.\nabla T \geq 0 \tag{22}$$

In the case of hyperelastic material behaviour, the specific free energy is a function of $(\underset{\sim}{\mathbf{E}}, \underset{\approx}{\mathbf{K}} = \nabla \underset{\sim}{\mathbf{E}}, T, \nabla T)$. The Clausius-Duhem inequality (22) can then be expanded as follows :

$$\begin{aligned}(\underset{\sim}{\Sigma} - \rho \frac{\partial \Psi}{\partial \underset{\sim}{\mathbf{E}}}) : \dot{\underset{\sim}{\mathbf{E}}} &+ (\underset{\approx}{\mathbf{S}} - \rho \frac{\partial \Psi}{\partial \underset{\approx}{\mathbf{K}}}) \vdots \dot{\underset{\approx}{\mathbf{K}}} + (a^{(i)} - \rho s - \rho \frac{\partial \Psi}{\partial T}) \dot{T} \\ &+ (\underline{\mathbf{b}}^{(i)} - \rho \frac{\partial \Psi}{\partial \nabla T}).(\nabla \dot{T}) - \frac{\underline{\mathbf{Q}}}{T}.(T\underline{\nabla}) \geq 0\end{aligned} \tag{23}$$

from which the state laws can be deduced :

$$\underset{\sim}{\Sigma} = \rho \frac{\partial \Psi}{\partial \underset{\sim}{\mathbf{E}}}; \quad \underset{\approx}{\mathbf{S}} = \rho \frac{\partial \Psi}{\partial \underset{\approx}{\mathbf{K}}}; \quad \underline{\mathbf{b}}^{(i)} = \rho \frac{\partial \Psi}{\partial (\nabla T)}; \quad \rho s = -\rho \frac{\partial \Psi}{\partial T} + a^{(i)} \tag{24}$$

When compared to the isothermal second grade theory in [Germain, 1973], the proposed thermodynamical framework takes full account of the introduction of variable ∇T in the constitutive modelling. The main consequence is the modification of the entropy by the term $a^{(i)}$ which, according to the balance equation (15), is nothing but the divergence of the generalized thermodynamical force $\underline{\mathbf{b}}$ associated with the temperature gradient. As a result, dissipation is reduced to its thermal part :

$$D = -\frac{\underline{\mathbf{Q}}}{T}.\nabla T \tag{25}$$

2.4 Alternative formulations

Alternative thermodynamical formulations of second grade thermoelasticity exist and some of them have been described in [Cardona *et al.*, 1998]. The common feature of the alternative formulations is to avoid the introduction of additional terms in the power density of internal forces (10) and the modifications may then appear in the energy balance (thus extending the treatment of [Dunn and Serrin, 1985] in the isothermal case) or in the entropy flux as recommended in [Maugin, 1990] for the introduction of gradients of internal variables in the constitutive framework. According to the latter procedure, the $\dot{T}$ and $\nabla\dot{T}$ terms can be dropped in (10) and an extra-entropy flux must be defined :

$$\underline{\mathbf{J}}_s = \frac{\underline{\mathbf{Q}}}{T} + \underline{\mathbf{k}}, \quad \text{with} \quad \underline{\mathbf{k}} = \frac{\dot{T}}{T}\underline{\mathbf{b}}, \quad \text{and} \quad \underline{\mathbf{b}} = \rho\partial\Psi/\partial(\nabla T) \tag{26}$$

This framework however leads to the same state laws as (24), but the classical heat conduction inequality is changed leading to a non-classical Fourier's law modified by the extra-entropy flux $\underline{\mathbf{k}}$.

3. SECOND GRADE LINEAR THERMOELASTICITY

The previous general framework is explicited in the case of linear thermoelasticity in the static case and the associated constitutive properties are derived. The additional terms arising then in the heat equation are discussed.

3.1 Linearized constitutive equations

A reference thermomechanical state $(\underset{\sim}{\mathbf{E}}_0 = 0, \underset{\approx}{\mathbf{K}}_0 = 0, T_0, (\nabla T)_0 = 0)$ is considered and the kinematic, balance and constitutive equations are linearized with respect to this reference state. The small perturbation framework requires sufficiently small strains, strain gradients, temperature changes and temperature gradients. The relevant variables then are :

$$\underset{\sim}{\mathbf{E}} = e(\underline{\mathbf{U}}), \quad \underset{\approx}{\mathbf{K}} = \nabla\underset{\sim}{\mathbf{E}}, \quad \Delta = T - T_0, \quad \nabla T \tag{27}$$

where the symmetric gradient operator e means $e(\underline{\mathbf{U}}) = (U_{i,j} + U_{j,i})/2\, \underline{\mathbf{e}}_i \otimes \underline{\mathbf{e}}_j$. The free energy is then taken as a quadratic form in all these variables :

$$\begin{aligned}\rho\Psi &= \frac{1}{2}\underset{\sim}{\mathbf{E}} : \underset{\approx}{\mathbf{C}} : \underset{\sim}{\mathbf{E}} - \Delta\underset{\sim}{\mathbf{E}} : \underset{\approx}{\mathbf{C}} : \underset{\sim}{\boldsymbol{\alpha}} - \frac{1}{2}\beta\Delta^2 + \frac{1}{2}\underset{\approx}{\mathbf{K}} \vdots \underset{\approx}{\mathbf{A}} \vdots \underset{\approx}{\mathbf{K}} + \underset{\sim}{\mathbf{E}} : \underset{\approx}{\mathbf{M}} \vdots \underset{\approx}{\mathbf{K}} + \Delta\underset{\approx}{\mathbf{K}} \vdots \underset{\approx}{\mathbf{H}} \\ &+ \frac{1}{2}\nabla T.\underset{\sim}{\mathbf{B}}.\nabla T + \Delta\underline{\mathbf{F}}.\nabla T - \underset{\approx}{\mathbf{K}} \vdots \left((\underset{\approx}{\mathbf{A}} : \underset{\sim}{\mathbf{P}}).\nabla T\right) + \underset{\sim}{\mathbf{E}} : \underset{\approx}{\mathbf{N}}.\nabla T\end{aligned} \tag{28}$$

The state laws follow from (24) :

$$\underset{\sim}{\boldsymbol{\Sigma}} = \underset{\approx}{\mathbf{C}} : (\underset{\sim}{\mathbf{E}} - \Delta\,\underset{\sim}{\boldsymbol{\alpha}}) + \underset{\approx}{\mathbf{M}} \vdots \underset{\approx}{\mathbf{K}} + \underset{\approx}{\mathbf{N}}.\,\nabla T \tag{29}$$

$$\underset{\approx}{\mathbf{S}} = \underset{\approx}{\mathbf{A}} \vdots (\underset{\approx}{\mathbf{K}} - \underset{\sim}{\mathbf{P}} \otimes \nabla T) + \underset{\approx}{\mathbf{M}} \vdots \underset{\sim}{\mathbf{E}} + \Delta\,\underset{\approx}{\mathbf{H}} \tag{30}$$

$$\underline{\mathbf{b}}^{(i)} = \underset{\approx}{\mathbf{N}} : \underset{\sim}{\mathbf{E}} - (\underset{\approx}{\mathbf{A}} : \underset{\sim}{\mathbf{P}}) \vdots \underset{\approx}{\mathbf{K}} + \Delta\,\underline{\mathbf{F}} + \underset{\sim}{\mathbf{B}}\,.\,\nabla T \tag{31}$$

$$\rho s = (\underset{\approx}{\mathbf{C}} : \underset{\sim}{\boldsymbol{\alpha}}) : \underset{\sim}{\mathbf{E}} - \underset{\approx}{\mathbf{H}} \vdots \underset{\approx}{\mathbf{K}} + \beta\,\Delta - \underline{\mathbf{F}}.\nabla T + a^{(i)} \tag{32}$$

in which classical and additional terms can be recognized. In particular, there exist a thermal eigenstrain $\Delta\underset{\sim}{\alpha}$ and, in the same way, an eigen-(strain gradient) $\underset{\sim}{\mathbf{P}} \otimes \nabla T$, the interpretation of which is given in [Cardona et al., 1998]. In the linearized scheme, the introduced constitutive properties are independent from temperature. The constitutive equations can be written as a single relation linking the effective stress tensor $\underset{\sim}{\tau} = \underset{\sim}{\Sigma} - \nabla.\underset{\approx}{\mathbf{S}}$ and strain and temperature gradients :

$$\underset{\sim}{\tau} = \underset{\approx}{\mathbf{C}} : (\underset{\sim}{\mathbf{E}} - \Delta\underset{\sim}{\alpha}) + (\underset{\approx}{\mathbf{M}} - \underset{\approx}{\mathbf{M}}^*) \vdots \underset{\approx}{\mathbf{K}} + (\underset{\approx}{\mathbf{N}} - \underset{\approx}{\mathbf{H}}).\nabla T - \underset{\approx}{\mathbf{A}} \vdots \left(\nabla \underset{\approx}{\mathbf{K}} - \underset{\sim}{\mathbf{P}} \otimes (\nabla\nabla T)\right) \quad (33)$$

where $M^*_{ijpqr} = M_{pqijr}$. In this expression, second order strain gradients and second order temperature gradients necessarily appear.

3.2 Linearized heat equation

The heat equation is deduced from the energy equation (19) and takes the form :

$$\rho T \dot{s} \;=\; -\nabla.\underline{\mathbf{Q}} + r \quad (34)$$

Substituting the linear state laws in (34) and taking the fact that $a^{(i)}$ is nothing but the divergence of $\underline{\mathbf{b}}^{(i)}$ (setting $\underline{\mathbf{b}}^{(e)}$ to zero without loss of generality) into account, the linearized heat equation is obtained :

$$T\beta\dot{T} = r - \underline{\mathbf{Q}}.\underline{\nabla} - T\dot{\underset{\sim}{\mathbf{E}}} : \underset{\approx}{\mathbf{C}} : \underset{\sim}{\alpha} + T(\underset{\approx}{\mathbf{A}} : \underset{\sim}{\mathbf{P}}) :: \nabla\dot{\underset{\approx}{\mathbf{K}}} - T\underset{\sim}{\mathbf{B}} : (\nabla\nabla\dot{T}) - T(\underset{\approx}{\mathbf{N}} - \underset{\approx}{\mathbf{H}}) \vdots \dot{\underset{\approx}{\mathbf{K}}} \quad (35)$$

In the isotropic case, all odd order constitutive tensors vanish so that the last term disappears. It turns out that the proposed approach leads to an additional thermo-mechanical coupling term $(\underset{\approx}{\mathbf{A}} : \underset{\sim}{\mathbf{P}}) :: \nabla\dot{\underset{\approx}{\mathbf{K}}}$ in addition to the classical term $\dot{\underset{\sim}{\mathbf{E}}} : \underset{\approx}{\mathbf{C}} : \underset{\sim}{\alpha}$.

It can be noted that a generalized specific heat can be defined [Cardona et al., 1998], the positivity of which is preserved at least in the present linearized case. In the purely thermal case, the heat equation is also modified as indicated below. Introducing a linear relationship between the heat flux vector and the temperature gradient :

$$\underline{\mathbf{Q}} = -\underset{\sim}{\kappa}.\nabla T \quad (36)$$

that still identically fulfills the positive thermal dissipation requirement (25). The purely thermal part of the heat equation reduces, in the isotropic case, to :

$$\beta\dot{T} = \kappa\nabla.\nabla T \;-\; B\nabla.\nabla\dot{T} \quad (37)$$

where $\nabla.\nabla$ simply is the Laplacian operator. It is interesting to notice that this equation is identical to the first Cattaneo equation presented for instance in [Müller and Ruggeri, 1993]. Cattaneo's argument is rather based on a modification of Fourier's law :

$$\underline{\mathbf{Q}} = -\kappa\;\nabla T + B\;\nabla\dot{T} \quad (38)$$

which may lead to up-hill heat diffusion, contrary to the present formulation. However Cattaneo proposed a second modification of Fourier's law aiming at restoring the hyperbolic character of the heat equation. This modification falls into the framework of extended thermodynamics and is not contained in our formulation.

4. ASYMPTOTIC ANALYSIS OF HETEROGENEOUS MATERIALS

The aim of this section is to show that the additional constitutive properties associated with higher order strain and temperature gradients that have been introduced in a purely phenomenological manner in the previous sections, can be deduced from an analysis of the effective properties of a heterogeneous material subjected to slowly-varying mean fields. For simplicity, classical heterogeneous materials with periodic microstructure are considered, in the static case. In a first analysis, the local temperature field is assumed to be known and the fully coupled thermomechanical problem is treated at the end. The main features of the constitutive framework for thermoelastic second grade media proposed in section 2 and 3 are confirmed by the following analysis but additional ones are pointed out in the last section.

4.1 Field equations at the local scale

The linear thermomechanical properties of a heterogeneous classical material, adequately represented by a Cauchy continuum, are considered. The aim is to deduce the global properties of an homogeneous equivalent medium from the local properties. The local and global (effective) variables (free energy, deformation, stress, temperature, temperature change, specific entropy and heat flux) must be distinguished :

$$\textit{local variables} \quad : \quad \psi, \quad \underset{\sim}{\varepsilon}, \quad \underset{\sim}{\sigma}, \quad \theta, \quad \delta = \theta - \theta_{ref}, \quad \eta, \quad \underline{\mathbf{q}} \tag{39}$$

$$\textit{global variables} \quad : \quad \Psi, \quad \underset{\sim}{\mathbf{E}}, \quad \underset{\sim}{\Sigma}, \quad T, \quad \Delta = T - T_0, \quad s, \quad \underline{\mathbf{Q}} \tag{40}$$

The heterogeneous material is described locally by the following free energy with constitutive properties varying in space but independent from temperature (linearized formulation) :

$$\rho\,\psi(\underset{\sim}{\varepsilon},\theta) = \frac{1}{2}\,\underset{\sim}{\varepsilon} : \underset{\approx}{\mathbf{C}} : \underset{\sim}{\varepsilon} \;-\; \delta\,\underset{\sim}{\varepsilon} : \underset{\approx}{\mathbf{c}} : \underset{\sim}{\alpha} \;-\; \frac{1}{2}\,\beta\,\delta^2 \tag{41}$$

The local specific heat for constant deformation is related to parameter β through :

$$\rho\, C_\varepsilon \;=\; \theta\,\rho\,\left(\frac{\partial \eta}{\partial \theta}\right)_{\underset{\sim}{\varepsilon}} \;=\; \theta\,\beta \;=\; -\,\theta\,\frac{\partial^2 \psi}{\partial \theta^2} \tag{42}$$

The equations of the linear coupled thermomechanical problem $\mathcal{P}$ to be solved on the heterogenous material are Hooke's and Fourier's laws and the balance of momentum and energy :

$$\underset{\sim}{\sigma} = \underset{\approx}{\mathbf{c}} : (\underset{\sim}{\varepsilon} - \delta\,\underset{\sim}{\alpha}); \quad \nabla.\underset{\sim}{\sigma} + \underline{\mathbf{f}} = 0 \tag{43}$$

$$\underline{\mathbf{q}} = -\underset{\sim}{\lambda}.\nabla T; \quad \beta\dot{\theta} = -\nabla.\underline{\mathbf{q}} - \theta\,\underset{\sim}{\alpha} : \underset{\approx}{\mathbf{c}} : \underset{\sim}{\varepsilon} + r \tag{44}$$

where $\underline{q}$ stands in fact for $\underline{q}/\theta_{ref}$, r for r/θ_{ref} and $\underset{\sim}{\lambda}$ for Fourier's heat conduction tensor divided by θ_{ref}. This convention holds for the remaining of this part. The unknowns fields are the local displacement $\underline{u}(\underline{x})$ and temperature $\theta(\underline{x})$.

In this part, the problem is restricted to an inifinite body, so that the additional problems associated with boundary conditions are not addressed. Initial conditions for the evolution problem close the formulation of $\mathcal{P}$.

4.2 Dimensional analysis and asymptotic developments

For simplicity, it is assumed that the heterogeneous material admits a periodic microstructure that can be described by a unit cell Y^l of characteristic size l. The macroscopic scale is characterized by a typical wave length L_ω of variation of the mean fields like overall stress and strain. In a finite body, L_ω may be of the order of magnitude of its size L. The dimension analysis performed below provides the small parameters involved in the thermomechanical problem, to be taken into account in a homogenization procedure.

Dimensionless space coordinates, displacements, time and temperature, several operators and constitutive properties can be defined :

$$\underline{x}^* = \underline{x}/L_\omega, \quad \underline{u}^* = \underline{u}/L_\omega, \quad t^* = t/\bar{t}, \quad \theta^* = \theta/\theta_{ref} \tag{45}$$

$$e(\underline{u}) = e^*(\underline{u}^*), \quad \nabla^* = L_\omega \nabla, \quad \nabla\theta = \frac{\theta_{ref}}{L_\omega}\nabla^*\theta^* \tag{46}$$

$$\underset{\approx}{c} = \bar{c}\,\underset{\approx}{c}^*, \quad \underset{\sim}{\alpha} = \bar{\alpha}\,\underset{\sim}{\alpha}^*, \quad \underset{\sim}{\lambda} = \frac{\bar{c}\,\bar{\alpha}}{\theta_{ref}\bar{t}}\, l_\lambda^2\, \underset{\sim}{\lambda}^*, \quad \beta = \frac{\bar{c}\,\bar{\alpha}}{\theta_{ref}}\left(\frac{l_\lambda}{L_\omega}\right)^2 \frac{t_\beta}{\bar{t}}\,\beta^* \tag{47}$$

where reference time $\bar{t}$ and characteristic time t_β and length l_λ have been introduced. A dimensionless formulation $\mathcal{P}^*$ of the thermomechanical problem $\mathcal{P}$ can then be expressed on a dimensionless unit cell Y^* :

$$\nabla^*.\underset{\sim}{\sigma}^* + \underline{f}^* = 0 \quad \text{with} \quad \underset{\sim}{\sigma}^* = \underset{\sim}{\sigma}/\bar{c} = \underset{\approx}{c}^* : (e^*(\underline{u}^*) - \bar{\alpha}\theta_{ref}\delta^*\underset{\sim}{\alpha}^*), \quad \underline{f}^* = \frac{L_\omega}{\bar{c}}\underline{f} \tag{48}$$

$$-\nabla^*.\underline{q}^* - \frac{\partial e^*(\underline{u}^*)}{\partial t^*} : \underset{\approx}{c}^* : \underset{\sim}{\alpha}^* = \varepsilon_\beta \varepsilon_\lambda^2 \beta^* \frac{\partial \theta^*}{\partial t^*}, \quad \text{with} \quad \underline{q}^* = \frac{\bar{c}\,\bar{\alpha} L_\omega}{\bar{t}}\underline{q}^* \tag{49}$$

The characteristic numbers of the thermomechanical problem then are :

$$\varepsilon = \frac{l}{L_\omega}, \quad \varepsilon_\lambda = \frac{l_\lambda}{L_\omega}, \quad \varepsilon_\beta = \frac{t_\beta}{\bar{t}} \tag{50}$$

Note that another characteristic length appears if boundary conditions like heat convection are introduced but this is not treated here.

In the following a homogenization procedure is considered for which ε is the relevant small parameter whereas ε_λ and ε_β are regarded as constant and independent from ε. For that purpose and similarly to classical multiscale asymptotic methods used in periodic homogenization [Sanchez-Palencia, 1980], a series of problems

$(\mathcal{P}_\varepsilon)_{\varepsilon>0}$ is considered. Once the small parameter of the problem has been chosen after considering the dimensionless problem $\mathcal{P}^*$, it is possible to define each $\mathcal{P}_\varepsilon$ on Y

$$Y = \{\underline{\mathbf{y}} \,|\, \underline{\mathbf{y}} = \underline{\mathbf{x}}/\varepsilon,\ \underline{\mathbf{x}} \in Y^l\} \tag{51}$$

and the equations of $\mathcal{P}_\varepsilon$ are chosen to be the equations (43) and (44) in which displacements, temperature, stress and heat flux have to be replaced by $\underline{\mathbf{u}}^\varepsilon, \theta^\varepsilon, \underset{\sim}{\sigma}^\varepsilon$ and $\underline{\mathbf{q}}^\varepsilon$. Each constitive tensor a appearing in these equations is to be replaced by a^ε such that

$$a^\varepsilon(\underline{\mathbf{y}}) = a(\varepsilon\,\underline{\mathbf{y}}) \tag{52}$$

The limiting case obtained for $\varepsilon \to 0$ gives the balance and constitutive equations of a homogeneous equivalent medium. This is recalled in the next section where the corrections for non-vanishing ε are also investigated since they are relevant when the macroscopic mean fields are not strictly constant but slowly varying. All fields are regarded as functions of the two variables $\underline{\mathbf{x}}$ and $\underline{\mathbf{y}}$, that can be expanded in a series of powers of small parameter ε :

$$\underline{\mathbf{u}}^\varepsilon(\underline{\mathbf{x}}) = \underline{\mathbf{u}}_0(\underline{\mathbf{x}},\underline{\mathbf{y}}) + \varepsilon\,\underline{\mathbf{u}}_1(\underline{\mathbf{x}},\underline{\mathbf{y}}) + \varepsilon^2\,\underline{\mathbf{u}}_2(\underline{\mathbf{x}},\underline{\mathbf{y}}) + \varepsilon^3\,\underline{\mathbf{u}}_3(\underline{\mathbf{x}},\underline{\mathbf{y}}) + \ldots \tag{53}$$

$$\theta^\varepsilon(\underline{\mathbf{x}}) = \theta_0(\underline{\mathbf{x}},\underline{\mathbf{y}}) + \varepsilon\,\theta_1(\underline{\mathbf{x}},\underline{\mathbf{y}}) + \varepsilon^2\,\theta_2(\underline{\mathbf{x}},\underline{\mathbf{y}}) + \varepsilon^3\,\theta_3(\underline{\mathbf{x}},\underline{\mathbf{y}}) + \ldots \tag{54}$$

$$\underset{\sim}{\sigma}^\varepsilon(\underline{\mathbf{x}}) = \underset{\sim}{\sigma}_0(\underline{\mathbf{x}},\underline{\mathbf{y}}) + \varepsilon\,\underset{\sim}{\sigma}_1(\underline{\mathbf{x}},\underline{\mathbf{y}}) + \varepsilon^2\,\underset{\sim}{\sigma}_2(\underline{\mathbf{x}},\underline{\mathbf{y}}) + \varepsilon^3\,\underset{\sim}{\sigma}_3(\underline{\mathbf{x}},\underline{\mathbf{y}}) + \ldots \tag{55}$$

$$\underline{\mathbf{q}}^\varepsilon(\underline{\mathbf{x}}) = \underline{\mathbf{q}}_0(\underline{\mathbf{x}},\underline{\mathbf{y}}) + \varepsilon\,\underline{\mathbf{q}}_1(\underline{\mathbf{x}},\underline{\mathbf{y}}) + \varepsilon^2\,\underline{\mathbf{q}}_2(\underline{\mathbf{x}},\underline{\mathbf{y}}) + \varepsilon^3\,\underline{\mathbf{q}}_3(\underline{\mathbf{x}},\underline{\mathbf{y}}) + \ldots \tag{56}$$

where the terms of the series are Y-periodic with respect to the second variable. In order to put these expressions into the balance and constitutive equations of $\mathcal{P}_\varepsilon$, the gradient operator can be split into partial derivatives with respect to $\underline{\mathbf{x}}$ and $\underline{\mathbf{y}}$:

$$\nabla = \nabla_x + \frac{1}{\varepsilon}\nabla_y, \quad e(.) = e_x(.) + \frac{1}{\varepsilon}e_y(.) \tag{57}$$

4.3 Derivation of the effective balance and constitutive equations

The expansion of stress and heat flux are introduced into the equations of balance of momentum and energy. The terms can be ordered with respect to the powers of ε. Identifying the terms of same order, the different contributions in the expansions (53) to (56) can be shown to be solution of the following auxiliary problems on Y :
Problem $\mathcal{A}_0$:

$$e_y(\underline{\mathbf{u}}_0) = 0;\ \nabla_y\theta_0 = 0 \tag{58}$$

Problem $\mathcal{A}_1$:

$$\underset{\sim}{\sigma}_0 = \underset{\approx}{\mathbf{c}}^\varepsilon : (e_x(\underline{\mathbf{u}}_0) + e_y(\underline{\mathbf{u}}_1) - \underset{\sim}{\alpha}^\varepsilon\,\delta_0)\,;\ \nabla_y.\underset{\sim}{\sigma}_0 = 0 \tag{59}$$

$$\underline{\mathbf{q}}_0 = -\lambda^\varepsilon(\nabla_x\theta_0 + \nabla_y\theta_1);\ \nabla_y.\underline{\mathbf{q}}_0 = 0 \tag{60}$$

Problem $\mathcal{A}_2$:

$$\underset{\sim}{\sigma}_1 = \underset{\approx}{\mathbf{c}}^\varepsilon : (e_x(\underline{\mathbf{u}}_1) + e_y(\underline{\mathbf{u}}_2) - \underset{\sim}{\alpha}^\varepsilon\,\delta_1)\,;\ \nabla_x.\underset{\sim}{\sigma}_0 + \nabla_y.\underset{\sim}{\sigma}_1 + \underline{\mathbf{f}}^\varepsilon = 0 \tag{61}$$

$$\underline{q}_1 = -\lambda^\varepsilon(\nabla_x\theta_1 + \nabla_y.\theta_2);\ -\nabla_x.\underline{q}_0 - \nabla_y.\underline{q}_1 - \underset{\approx}{c}^\varepsilon : \underset{\sim}{\alpha}^\varepsilon : (e_x(\dot{\underline{u}}_0) + e_y(\dot{\underline{u}}_1)) + r^\varepsilon = \beta^\varepsilon\dot{\theta}_0 \tag{62}$$

Problem $\mathcal{A}_3$:

$$\underset{\sim}{\sigma}_2 = \underset{\approx}{c}^\varepsilon : (e_x(\underline{u}_2) + e_y(\underline{u}_3) - \underset{\sim}{\alpha}^\varepsilon\ \delta_2)\ ;\ \nabla_x.\underset{\sim}{\sigma}_1 + \nabla_y.\underset{\sim}{\sigma}_2 = 0 \tag{63}$$

$$\underline{q}_2 = -\lambda^\varepsilon(\nabla_x\theta_2 + \nabla_y\theta_3);\quad -\nabla_x.\underline{q}_1 - \nabla_y.\underline{q}_2 - \underset{\approx}{c}^\varepsilon : \underset{\sim}{\alpha}^\varepsilon : (e_x(\dot{\underline{u}}_1) + e_y(\dot{\underline{u}}_2)) = \beta^\varepsilon\dot{\theta}_1 \tag{64}$$

The solutions of problem $\mathcal{A}_i$ are $(\underline{u}_i, \underset{\sim}{\sigma}_{i-1}, \theta_i, \underline{q}_{i-1})$, for $i > 0$. Problem $\mathcal{A}_0$ gives $\underline{u}_0$ and θ_0. The resolution of $\mathcal{A}_0$ and $\mathcal{A}_1$ corresponds to classical homogenization theory applied to coupled thermoelasticity. The reader is referred to [Francfort, 1983] and [Brahim-Otsmane et al., 1992] for a detailed solution. We simply give the form of the solution :

$$\underline{u}_0(\underline{x}, \underline{y}) = \underline{U}_0(\underline{x});\ \theta_0(\underline{x}, \underline{y}) = \Theta_0(\underline{x}) \tag{65}$$

$$\underline{u}_1(\underline{x}, \underline{y}) = \underline{U}_1(\underline{x}) + \underset{\approx}{X}(\underline{y}) : e_x(\underline{U}_0) + \underline{X}'(\underline{y})\delta_0;\ \theta_1(\underline{x}, \underline{y}) = \Theta_1(\underline{x}) + \underline{X}''(\underline{y}).\nabla_x\theta_0 \tag{66}$$

where $\underset{\approx}{X}, \underline{X}', \underline{X}''$ are concentration tensors, the existence of which is ensured by the linearity of the problem.

4.4 Links with second gradient theory

Following the technique used for instance in [Boutin, 1996], the problems $\mathcal{A}_2$ and $\mathcal{A}_3$ dealing with the correcting terms $\underline{u}_2$ and $\underline{u}_3$ can be solved in order to evidence some links with second gradient theory. In this section and for simplicity, the analysis is restricted to the thermomechanical problem for a given temperature field $\theta^\varepsilon(\underline{x}) = \Theta_0(\underline{x})$. The problem associated with the two first equations of $\mathcal{A}_2$ (resp. $\mathcal{A}_3$) can then be interpreted as an elasticity problem with fictitious body forces that are linear in $e_x(\underline{U}_1)$, $\nabla_x e_x(\underline{U}_0)$ and $\nabla_x\theta_0$ (resp. $e_x(\underline{U}_2), \nabla_x e_x(\underline{U}_1), \nabla_x\nabla_x e_x(\underline{U}_0)$ and $\nabla_x\nabla_x\theta_0$). Accordingly, there exist 4 additional concentration tensors such that :

$$\underline{u}_2(\underline{x}, \underline{y}) = \underline{U}_2(\underline{x}) + \underset{\approx}{X}(\underline{y}) : e_x(\underline{U}_1) + \underset{\approx}{Y}(\underline{y}) \vdots (\nabla_x e_x(\underline{U}_0)) + \underset{\sim}{Y}'(\underline{y}).\nabla_x\theta_0 \tag{67}$$

$$\begin{aligned}\underline{u}_3(\underline{x}, \underline{y}) &= \underline{U}_3(\underline{x}) + \underset{\approx}{X}(\underline{y}) : e_x(\underline{U}_2) + \underset{\approx}{Y}(\underline{y}) \vdots (\nabla_x e_x(\underline{U}_1)) \\ &+ \underset{\approx}{Z}(\underline{y}) :: (\nabla_x\nabla_x e_x(\underline{U}_0)) + \underset{\approx}{Z}'(\underline{y}) : (\nabla_x\nabla_x\theta_0)\end{aligned} \tag{68}$$

where the $\underline{U}_i$ are constant translation terms. The concentration tensors can be taken such that their average over the unit cell vanishes. These expressions are now used to compute the local strain fields with the correctors :

$$\begin{aligned}e(\underline{u}^\varepsilon) &= e_x(\underline{U}_0) + \hat{\underset{\approx}{X}}(\underline{y}) : e_x(\underline{U}_0) + \hat{\underset{\sim}{X}}'(\underline{y})\ \delta_0 - \underset{\sim}{\alpha}^\varepsilon\ \delta_0 \\ &+ \varepsilon\left(\underset{\approx}{X}(\underline{y}) \vdots (\nabla_x e_x(\underline{U}_0)) + \hat{\underset{\approx}{Y}}(\underline{y}) \vdots (\nabla_x e_x(\underline{U}_0)) + \underset{\sim}{X}'(\underline{y}).\nabla_x\theta_0 + \hat{\underset{\sim}{Y}}'(\underline{y}).\nabla_x\theta_0\right) \\ &+ \varepsilon^2\left(\underset{\approx}{Y}(\underline{y}) \vdots (\nabla_x\nabla_x e_x(\underline{U}_0)) + \hat{\underset{\approx}{Z}}(\underline{y}) :: (\nabla_x\nabla_x e_x(\underline{U}_0))\right) \\ &+ \varepsilon^2\left(\underset{\sim}{Y}'(\underline{y}).(\nabla_x\nabla_x\theta_0) + \hat{\underset{\approx}{Z}}'(\underline{y}) : \nabla_x\nabla_x\theta_0\right)\end{aligned} \tag{69}$$

where concentration tensors modified by application of the symmetric gradient operator have been introduced and the translation terms have been put into the first term $\underline{\mathbf{U}}_0$ for simplicity. The mean stress is obtained by averaging the local stress over the unit cell :

$$\begin{aligned} < \underset{\sim}{\boldsymbol{\sigma}}^\varepsilon > &= < \underset{\approx}{\mathbf{c}}^\varepsilon : (\underset{\sim}{\mathbf{e}}(\underline{\mathbf{u}}^\varepsilon) - \delta^\varepsilon \underset{\sim}{\boldsymbol{\alpha}}^\varepsilon) >= \underset{\approx}{\mathbf{C}}^0 : \left(e_x(\underline{\mathbf{U}}_0) - \delta_0 \underset{\sim}{\boldsymbol{\alpha}}^0 \right) \\ &+ \varepsilon \underset{\approx}{\mathbf{C}}^1 \vdots (\nabla e_x(\underline{\mathbf{U}}_0) - \underset{\sim}{\boldsymbol{\alpha}}^1 . \nabla \theta_0) + \varepsilon^2 \underset{\approx}{\mathbf{C}}^2 :: (\nabla \nabla e_x(\underline{\mathbf{U}}_0) - \underset{\approx}{\boldsymbol{\alpha}}^2 : \nabla \nabla \theta_0) \end{aligned} \quad (70)$$

where $< . >$ denotes the averaging process. This relation can be interpreted as the effective constitutive equation for the homogenization problem up to order 3. The involved constitutive tensors are related to the concentration tensors as follows :

$$\underset{\approx}{\mathbf{C}}^0 =< \underset{\approx}{\mathbf{c}}^\varepsilon : \hat{\underset{\approx}{\mathbf{X}}} >; \ \underset{\approx}{\mathbf{C}}^0 : \underset{\sim}{\boldsymbol{\alpha}}^0 =< \underset{\approx}{\mathbf{c}}^\varepsilon : (\hat{\underset{\sim}{\mathbf{X}}}' - \underset{\sim}{\boldsymbol{\alpha}}^\varepsilon) >; \ \underset{\approx}{\mathbf{C}}^1 =< \underset{\approx}{\mathbf{c}}^\varepsilon : (\underset{\approx}{\mathbf{X}} + \hat{\underset{\approx}{\mathbf{Y}}}) >; \ ... \quad (71)$$

It appears that the overall constitutive equations involve higher order gradients of the overal strain and temperature fields. They are found to be identical to the constitutive equations (33) of a linear thermoelastic medium, providing that the mean stress is interpreted not as an overall Cauchy stress tensor but rather as the "effective" stress $\underset{\sim}{\boldsymbol{\tau}}$ of a second grade medium defined by (33). The mean stress can be shown to fulfill the following effective balance equation :

$$< \nabla . \underset{\sim}{\boldsymbol{\sigma}}^\varepsilon > + < \underline{\mathbf{f}}^\varepsilon >= 0. \quad (72)$$

which actually is the balance of momentum equation (14) satisfied by the "effective" stress $\underset{\sim}{\boldsymbol{\tau}}$ of a second grade medium. The identification of the homogeneous equivalent medium with a second grade thermoelastic material should be closed by the statement of the boundary conditions of the boundary value problem for the body endowed with the properties of the HEM.

4.5 Coupled thermomechanical problem

The resolution of the fully coupled thermomechanical problem of heterogeneous elasticity goes through the resolution in cascade of problems $\mathcal{A}_0$ to $\mathcal{A}_3$ for the displacement end temperature fields $\underline{\mathbf{u}}_i$ and θ_i. Note that, in each case, the thermal and mechanical problems are decoupled, since the coupling term in the heat equation of problem $\mathcal{A}_{i+1}$ involves the solution of problem $\mathcal{A}_i$ only. The mechanical part of the problem has been solved in the previous section. The partial differential equation of each thermal problem in $\mathcal{A}_i$ can be interpreted as a heat equation with fictitious heat source terms linear in $\dot{\theta}_0, e_x(\dot{\underline{\mathbf{U}}}_0), \nabla\dot{\theta}_0, \nabla e_x(\dot{\underline{\mathbf{U}}}_0)$, so that the solutions for temperature can be proved to have the form :

$$\theta_2(\underline{\mathbf{x}}, \underline{\mathbf{y}}) = \underset{\sim}{\mathbf{Y}}''(\underline{\mathbf{y}}) : (\nabla_x \nabla_x \theta_0) + \underset{\sim}{\mathbf{Y}}^{(3)}(\underline{\mathbf{y}}) : e_x(\dot{\underline{\mathbf{U}}}_0) + Y^{(4)}(\underline{\mathbf{y}}) \, \dot{\theta}_0 \quad (73)$$

$$\theta_3(\underline{\mathbf{x}}, \underline{\mathbf{y}}) = \underset{\approx}{\mathbf{Z}}^{(4)}(\underline{\mathbf{y}}) \vdots (\nabla_x \nabla_x \nabla_x \theta_0) + \underset{\approx}{\mathbf{Z}}^{(5)}(\underline{\mathbf{y}}) \vdots (\nabla e_x(\dot{\underline{\mathbf{U}}}_0)) + \underline{\mathbf{Z}}^{(6)}(\underline{\mathbf{y}}) . \nabla \dot{\theta}_0 \quad (74)$$

800 μm. These macrocracks are sometimes the result of the coalescence of many microcracks. The microcracking may be intergranular (figure 2a), transgranular or mixed (figure 2b).

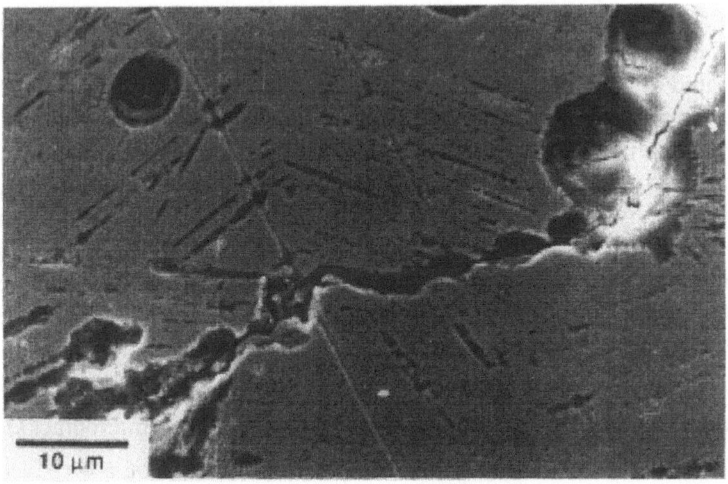

Figure 1. *Initiation of surface defects and slip planes attack for a strain rate of* $6x10^{-6}s^{-1}$ *[Touzet et al., 1994].*

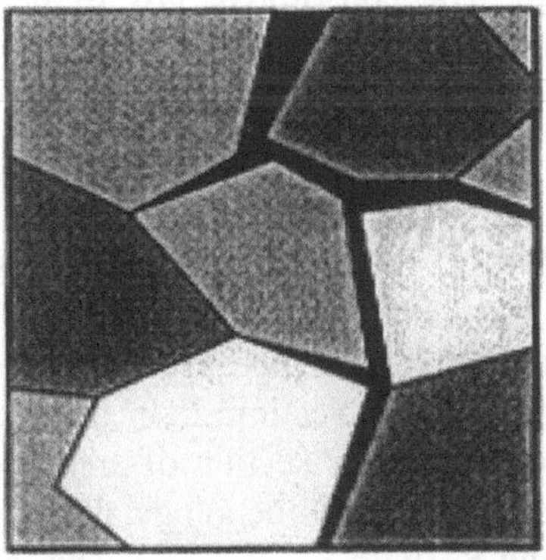

Figure 2a. *Intergranular microcrack.*

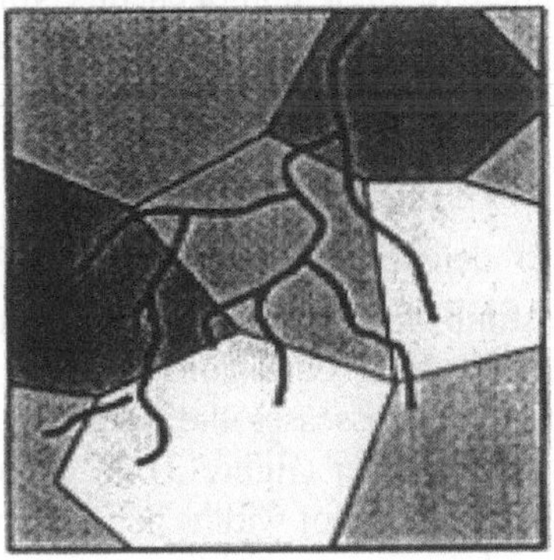

Figure 2b. *Transgranular microcrack.*

The general direction of micro or macrocracks propagation is perpendicular to the stress direction. The features of rupture show a behaviour of the fragile type, even in the case of very ductile materials such as stainless steel.

3. THEORETICAL ASPECT

Continuum damage mechanics is classically interested in the rise and growth of microcracks or more generally various decohesion processes within the material, what leads to the rupture of the volume element (in the sense of continuum mechanics). This corresponds to the consideration of discontinuities of which the dimensions are typically small compared to the representative volume element of the material [Maugin., 1999].

The theory of the effective stress has been introducted by Kachanov (1958). It is based on the notion of effective resisting area. In the phenomenological development of the effective stress concept, the term damage is used to denote a loss of effective area. However, in the physical sense, one may choose to interpret damage in terms of some other convenient macroscopic measure, such as the depth of microcrack penetration (intergranular or granular stress corrosion cracking) and relate it to the phenomenological description. This physical interpretation is adopted in the following without loss of generality and without introducing any new notation for damage. Given S a plane section area of the virgin material let $\tilde{S}$ the effective resisting area of the damaged material ($\tilde{S} < S$). The total area of the set of defects (microcracks) taking into account stress concentration effects and possible interaction is $S_c = S - \tilde{S}$. The damage internal parameter is then defined by $D = S_c/S$. Thus the parameter D characterizes a damage which is generally supposed to affect quasi uniformly all plane sections of the damaged material. For the tridimensionnal case considered here, this definition of D is valid for all plane sections perpendicular to the load. Then it is an internal variable representative of a homogeneous volume damage. The constitutive law of damage coupled with elasticity is written in a classical form (hypothesis of equivalence in strain, cf. [Lemaître et al., 1990]):

$$\sigma = E_0(1-D)\varepsilon_e \tag{1}$$

where σ denotes the stress, E_0 is the Young modulus of the virgin material, ε_e is the elastic strain, and $D \in [0, 1[$ is the internal damage parameter.

In the stress corrosion case, the law (1) is not relevant because the effective resisting area possesses the particular features described in Section 2. In the present study, we consider (figure 3) an element of plane rectangular section of area S (half thickness e by unit of width) perpendicular to the load. The phenomenology allows us to distinguish two areas of different nature noted S_0 and S_1 (with $S = S_0 + S_1$). The former S_0 (thickness ℓ_0 by unit of width) is located in the middle of the material and is not affected by the microcracks (virgin material). The second part S_1 (width ℓ_1) near to the surface of the sample corresponds to the area of gradual saturation of the density of microcracks.

As soon as a length of microcracks is greather than ℓ_1 (we are then in the presence of a macrocrack) we observe a quick propagation of these. The length ℓ_1 (critical length of defects) is therefore, by definition, the boundary which separates the field of continuum damage mechanics from that of fracture mechanics. For a given level of damage, we note δ the average depth of microcracks; the surface by unit of width of these microcracks is therefore $S_{cl} = \delta \times 1$. According to the classical definition of damage, which here is contained only near to the surface of material, it is necessary to define the parameter of damage of stress corrosion D_s by $D_s = S_{cl}/S_l = \delta/\ell_1$. Moreover, we set $\ell_o = (n - 1)\ell_1$, thus defining kind of a plane rectangular section of the sample so that the half thickness is equal to $e = \ell_0 + \ell_1 = n\ell_1$. It results from this that the concept of effective resisting area $\tilde{S} = S_0 + S_l - S_{cl}$ leads to :

$$\tilde{S} = S(1 - \frac{D_S}{n}) \quad (2).$$

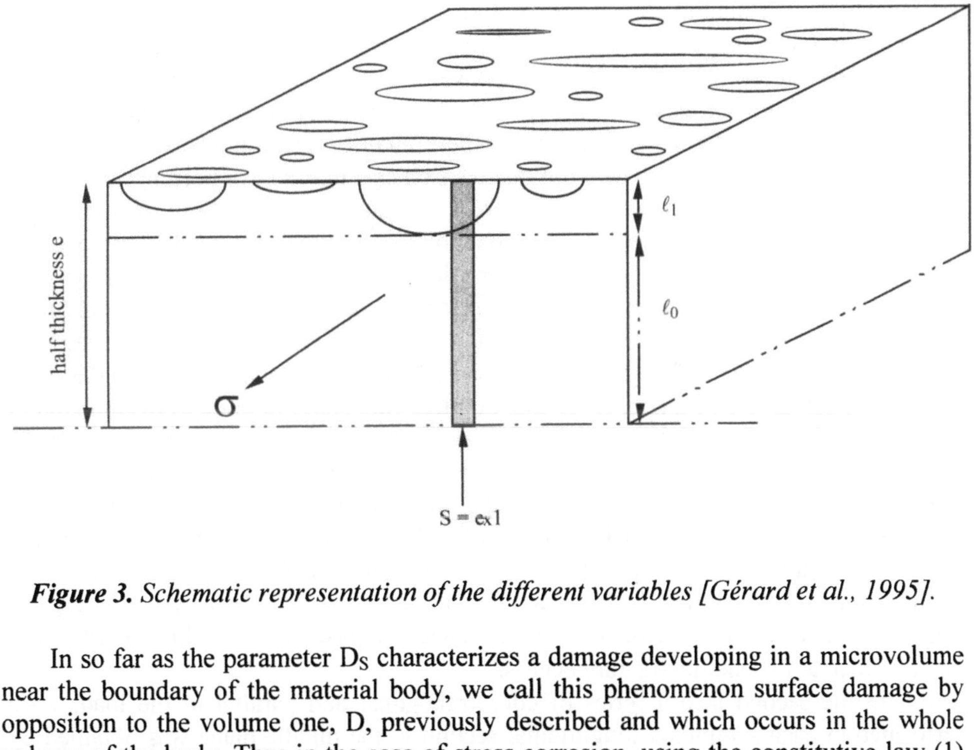

Figure 3. *Schematic representation of the different variables [Gérard et al., 1995].*

In so far as the parameter D_S characterizes a damage developing in a microvolume near the boundary of the material body, we call this phenomenon surface damage by opposition to the volume one, D, previously described and which occurs in the whole volume of the body. Thus in the case of stress corrosion, using the constitutive law (1) leads to considering an "apparent" volume damage which is not achieved in reality. Taking into account the relation (2) and the effective stress concept, it is more convenient to write the elastic constitutive law coupled to damage, for stress corrosion, in the form :

$$\sigma = E_0(1 - \frac{D_S}{n})\varepsilon_e \quad (3).$$

4. EXPERIMENTAL RESULTS

The macroscopic observations described in the previous section show that for a stainless steel Z3 CN 18-10 immersed in Scheil reagent ($Mg\ Cl_2$ at 153° Celsius) ℓ_1 has a rough estimate of 200 μm (ℓ_1 = 200 μm). Numerous creep tests at 200 MPa were performed on plate samples by laser cutting in form of "I" of useful length L=30 mm, retangular section S = 10 mm^2, and thickness 2e = 2 mm between heads ($\ell_0 + \ell_1$ = 1 mm). Samples were annealed at 1050° C for 30 minutes and then quenched. All the samples had identical surface finishing e.g. an electrolytic polishing followed by an acid etching. With these previous values we obtain the coefficient n = 5 in relation (3) then D_S = 5 D. These tests are interrupted at various level of damage.

In a first step, the samples were equipped with extensometric gauges and put in a test machine under neutral atmosphere and kept at room temperature. Specimens were immersed in Scheil reagent for 2 hours before slow strain rate tests were performed at a strain rate of 6 x 10^{-6} s^{-1}. The elastic response of each sample was investigated. The analysis of load-unload elastic cycles gives via the constitutive law (1) the values of damage D_S shown in figure 4. These values are obtained according to the creep level reached. Interrupted constant extension rate tests on specimens have been performed for different strain rates. For each point in figure 4, 10 tests were carried out. At the elastic range and beginning of the plastic range (first point on figure 4, D_S = 0,15) microscopic observation shows attacks of slip planes and initiation of defects on the surface. No crack traces can be seen over the longitudinal section. At the next point (D_S = 0,26) the first crack traces appear. A stereoscopic study of these surface cracks show that their depth does not exceed 5 μm. The localization of corrosion occurred and for the great majority of cracks are initiated. After this point, it was possible to observe (third point D_S = 0,4) a saturation of the number of total crack traces. Their size remains small, lower than 200 μm. The size of the maximum crack increases very slowly. At the beginning of the fourth point (D_S = 0,59) the total number of crack traces remains the same. A shift of the size of crack traces can be observed corresponding to the necking phenomenon; the accomodation of the plastic deformation at the tip of a few cracks (which leads to the failure of the specimen) certainly caused the arrest of the great majority of cracks. The size of the longest crack increases very quickly, reaches 400 μm, and then at the end the fracture occurs.

In the second step a series of cuts, orthogonal and parallel to the load, were performed in each sample. A longitudinal section in the middle plane was obtained by mechanical polishing. From the micrographs and microscopic observations the effective resisting area was evaluated. So was the average depth of microcracks. The size of the smallest crack observed was about 5μm. Crack traces were counted on the two faces all long the calibrated length L. The final distribution is the average of the two distributions on the two faces of the specimens. We establish a good correlation between the experimental measurements of Ds (□) obtained by strain gauges and the values of this parameter given directly by the micrograph estimate of the average depth of microcracks (•).

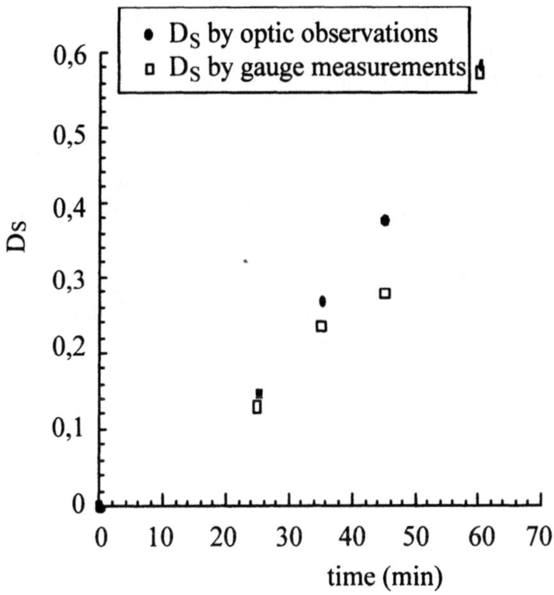

Figure 4. *Damage versus time of relaxation test [Gérard et al., 1995].*

5. CONCLUSION

In this work we have shown that the parameter D_S depends on the density of cracks that have reached a given size. We have seen that the use of two related techniques to characterize the damage in stress corrosion studies gives very interesting reults.

Following the surface internal parameter of damage D_S provides first simple predictive precursor of the failure of materials by stress corrosion.

It is possible to see when mechanical and electrochemical processes are preponderant [Touzet., 1990]. Also a thermodynamic formulation of this important problem SCC taking into account coupling between mechanical and electrochemical process is the object of further studies.

REFERENCES

[Desjardins et al., 1992] Desjardins D.; Oltra R.; "Corrosion sous contrainte : phénoménologie et mécanismes", *Editions de Physique*, pp. 1-17, 1992.

[Gérard et al., 1995] Gérard A.; Touzet M.; Puiggali M.; "Un modèle d'endommagement continu par corrosion sous contrainte", C. R. Acad. Sci. Paris, t. 320, Série II b, pp. 429-432, 1995.

[Kachanov, 1958] Kachanov L. M.; "Time of rupture process under deep conditions", Izv. Akad. Nauk. SSSR, **8**, pp. 26, 1958.

[Lemaître et al., 1990] Lemaître J.; Chaboche J. L.; "Mechanics of materials", *Cambrige University Press*, Cambridge, 1990.

[Maiya, 1985] Maiya P. S. ; Corrosion NACE **41**, pp. 630-642, 1985.

[Maugin, 1999] Maugin G. A. ; "The thermomechanics of nonlinear irreversible behaviors. An introduction", *World Scientific Publishing*, Series on Nonlinear Science, Series A, **27**, 1999 ; ISBN 981-02-3375-2.

[Touzet et al., 1994] Touzet M.; Puiggali M.; Cid M.; Desjardins D.; "Characterization of damage in SCC by an electrochemical impedance and statistical study of multiple cracking", Cor. Sci., **36**, pp. 815-825, 1994.

MODELLING WETTING BEHAVIOUR

H. Lanchon-Ducauquis, F. Biguenet,
T. Liraud, E.Csapo and Y. Guemouri
LEMTA-CNRS, UMR 7563, 2, avenue de la Forêt de Haye, BP 160,
54504 VANDOEUVRE Cedex, France. E-mail : dsimonigh@ensem.u-nancy.fr
November, 6th 1999

Abstract

Our aim is to improve the macroscopic description of the wetting property. Wettability characterises a capillary behaviour implying three partners : a liquid $F^{(1)}$, a solid S and a complementary one which is the surrounding fluid $F^{(2)}$. "What are the parameters adequate to describe such behaviour ?" A first simple approach, lying on the statics of fluids and Laplace's law, produces a mathematical model for the equilibrium of a $F^{(1)}/F^{(2)}$ interface, in the presence of a solid wall. That model induces a complementary experimental approach for obtaining a phenomenological relation between three unavoidable parameters intervening in the mathematical model : the spreading parameter η, the contact angle θ, and the contact mean curvature μ. At the same time, a thermomechanical research is implemented to obtain a general framework of capillary behaviour ; it lies on the *Cahn-Hilliard Fluid Theory*, and leads to the notion of *Fluid with internal wettability.*

Keywords : Capillarity, wetting, identification, second gradient, ODE.

1 INTRODUCTION

Wettability can be defined as a property shared by a solid and a liquid, which reflects the ability of the liquid to spread out on the solid in the presence of a second surrounding fluid. This phenomenon, which reveals in fact a *behaviour*, plays a decisive part in numerous industrial processes : oil assisted recovery, coating, napkin-diaper design, detergent efficiency, etc.

The macroscopic mechanical models usually proposed by engineers to take into account the capillary situations, are based on Laplace's and Young's laws, which involve the notions of interfacial tensions and contact angle ; this framework is usually insufficient for obtaining well-posed problems.

Our objective is to improve the description of the wetting property. Several approaches can be considered :

G.A. Maugin et al. (eds.), Continuum Thermomechanics, 197-208.
© 2000 *Kluwer Academic Publishers.*

• Identify the macroscopic parameters that are necessary for the completion of a mathematical model (based on the general laws of the classical theory of continuous media) ; these identifications usually need both a rigorous mathematical analysis and adequate experimentation. This method is somewhat indirect.

• Direct research of the mutual behaviour ; the starting point is usually the second principle of Thermodynamics, expressed with more or less subtlety, this allows one to include some dissipation mechanism, which fits the physical phenomena which are studied.

The three following sections of the paper relate to the *indirect method* which was the initial approach of our group ; the **second** section proposes an inventory of the static mechanical relations which are the starting points of the present study [Marron, 1987][Jacquot, 1998]; the **third** one describes the experiments[Jacquot, 1998], [Csapo and al, 1999], directly inspired by the preceding relations ; they relate to axisymmetric sessile drops ; their objective is to propose a realistic law, in order to be able to identify the nature of the wetting dissipation ; another interest of the experimental results is to find a phenomenological relation between the above-mentionned parameters η, θ and μ. The **fourth** section evokes some theoretical results obtained about general cylindrical and axisymmetrical interfaces, which ones can be completely characterised by a simple profile [Liraud, 1998]; from there, it is proposed a realistic strategy, based on simple experimental measures, to obtain relatively precises values of the parameters θ and μ.

The **fifth** section relates a more fundamental and *direct* approach of wetting behaviour ; it is based on the second gradient theory of the Cahn-Hilliard Fluid and explains, how the corresponding law has to be modified in order to take into account specific solid-fluid interactions near the wall boundary. A model of fluid with internal wettability is built [Biguenet, 1998].

Lastly, in the **conclusion**, the complementary aspects of the precedent studies are pointed out, and their common objective is emphasized.

2 BASIC STATIC MODELS

2.1 The general Bashforth and Adams relation

To identify the macroscopic parameters adequate for characterising the wetting, it is useful to have a static model of the liquid / fluid interface Σ. The non-dimensional BASHFORTH and ADAMS relation consists of a combination of the static and the Laplace's law governing fluids [Bashforth et al, 1892]; it leads to :

$$2H(M) - s\varepsilon x = a\,\text{constant} \qquad \forall M \in \Sigma \tag{1}$$

$$\varepsilon = \frac{L^2}{L_c^2} \quad , \quad L_c^2 = \frac{\sigma}{|\rho^{(1)} - \rho^{(2)}|\,g} \quad , \quad s = \frac{\rho^{(1)} - \rho^{(2)}}{|\rho^{(1)} - \rho^{(2)}|} \quad , \tag{2}$$

where $H(M)$ and x represent respectively the mean curvature and the altitude (measured on a vertical upward axis) at the considered point M of the interface Σ ; σ and $\rho^{(\alpha)}g$ are the interfacial tension and volume weight, relative to each fluid $F^{(\alpha)}$ (the sign of $H(M)$ and the numbering α depend on the orientation of the normal vector) ; the BOND number ε is small, as the characteristic length L is weak compared to the capillary one's L_c. In the absence of gravity , $\varepsilon = 0$; in this situation (1) shows that the mean curvature of Σ is a constant. If Γ is an *a priori* unknown contact line, intersection of Σ with a solid wall, the relation (1) implies, as a particular consequence:

$$2H(M) = 2H(C) + s\varepsilon(x - x_c) \qquad \forall M \in \Sigma \text{ and, } C \in \Gamma(\text{ of altitude} \, x_c) \qquad (3)$$

When M describes Σ, the relation (3) leads to a universal, second-order partial-differential equation, one solution of which defines the shape of the $F^{(1)}/F^{(2)}$ interface Σ. Each particular imposed geometry has to be specified with the help of adequate boundary conditions on Γ. Note that Γ is precisely the contact line, in the vicinity of which the wetting comes to light. It is also interesting to notice that the connection between Σ and the wall is necessarily characterised by the shape of Γ (i.e. the spreading $\eta(C)$ of the liquid) and the contact angle $\theta(C)$ of Σ with the wall. Lastly, to get rid of $H(C)$, an unknown parameter in the equation (3), it is reasonable to introduce the new "local altitude":

$$z = \frac{2H(C)}{\varepsilon} + s(x - x_c), \qquad (4)$$

which points to the importance of the parameter $H(C)/\varepsilon$. So, the formulation of adequate boundary conditions $\eta(C)$ and $\theta(C)$ for $z = \frac{2H(C)}{\varepsilon}$, when C describes Γ, allows one to identify the wetting parameters.

2.2 Application to axisymmetric sessile drops

An axisymmetric drop can be characterised by its cross-sectionnal profile $\mathcal{P}$; it is sessile if its apparent weight (here $(\rho^{(1)} - \rho^{(2)})g$) is oriented towards the horizontal solid support; we have then $s = 1$.

In this situation, it is reasonable to use the representation $r = \overline{\varphi}(x)$ for the profile $\mathcal{P}$; we have here $x_c = 0$. If we introduce the notation :

$$\mu = -2H(C) \qquad , \qquad (5)$$

the change of variable (4) (if $\varepsilon \neq 0$) :

$$z = x - \frac{\mu}{\varepsilon} \qquad , \qquad \varphi(z) = \overline{\varphi}(z + \frac{\mu}{\varepsilon}) \qquad , \qquad (6)$$

leads (after an analytic writing of $H(M)$ for $M = \left\{r, z + \frac{\mu}{\varepsilon}\right\} \in \mathcal{P}$), to the differential equation with the initial conditions

$$\left\{\begin{array}{ll} \varphi" = \varepsilon z(1+\varphi'^2)^{\frac{3}{2}} + \frac{1+\varphi'^2}{\varphi} & \\ \varphi(-\frac{\mu}{\varepsilon}) = \eta \quad ; & \varphi'(-\frac{\mu}{\varepsilon}) = \lambda \stackrel{def}{=} -cotg(\theta) \end{array}\right. \tag{7}$$

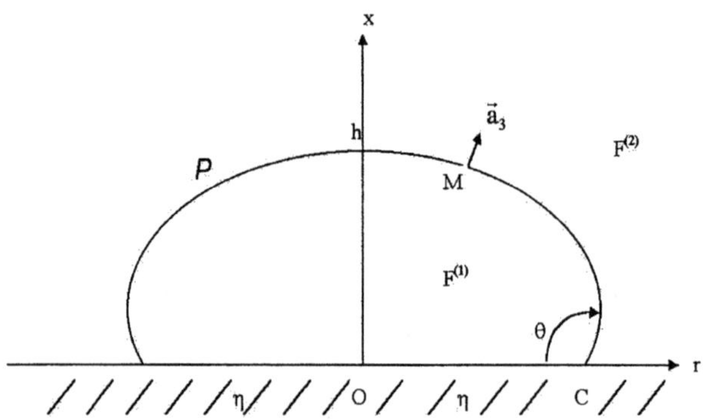

Figure 1: *Profile of an axisymmetric drop*

Remark 1 we recognize a Cauchy problem in R^2 (outside of the axis $r = 0$),which , for any initial conditions specified by the parameters $\left\{-\frac{\mu}{\varepsilon}; \eta, \lambda\right\}$ with $\eta \neq 0$, admits a unique local solution. However, in order to be realistic for the profile of a drop, the function φ has to satisfy the following conditions :
i) $\exists\, h$, a bounded positive number, such that :

$$\varphi(h - \frac{\mu}{\varepsilon}) = 0 \quad ; \quad \varphi'(h - \frac{\mu}{\varepsilon}) = -\infty \tag{8}$$

(that expresses the fact that the drop is bounded by the height h and is axisymmetric).
ii) The drop has given finite volume V, whence :

$$\pi \int_{-\frac{\mu}{\varepsilon}}^{h-\frac{\mu}{\varepsilon}} \varphi^2(z)dz = V \tag{9}$$

These two constraints imply two relationships between the three terms $\frac{\mu}{\varepsilon}, \eta, \lambda$ *of the initial conditions in (7)*; the first one has to guarantee the closing of the curve $r = \varphi(z)$ on the axis $r = 0$, which is singular for the problem (7) ; the second relationship is :

$$V = \frac{\pi\eta^2}{\varepsilon}\left(\mu - \frac{2sin(\theta)}{\eta}\right) \tag{10}$$

it comes from a first integral of (7).

An important consequence of *i)* an *ii)* is the necessary strict convexity of the sessile drop ; this assertion is equivalent to $\varphi'' < 0$. Others consequences are the following useful inequalities:

$$h < \frac{\mu}{\varepsilon} \qquad \text{and} \qquad 0 \leq \frac{2}{\eta\sqrt{1+\lambda^2}} \leq \mu \tag{11}$$

Remark 2 In the absence of gravity ($\epsilon = 0$), the change of variable (6) is no longer valid, and the relationship (10) implies

$$\mu = \frac{2sin(\theta)}{\eta} \quad , \tag{12}$$

which is, in that particular situation, the closing condition described earlier. It is possible to determine $\overline{\varphi}(x)$, and then, it appears that the interface Σ is a portion of a sphere ; the volume is now :

$$V = \pi\eta^3 G(\theta) \tag{13}$$

where G is an explicit, strictly increasing function of θ.

The expressions (10) of V is a source of inspiration for experiments and for physical interpretation of the fundamental parameters μ, η and θ. In fact (because wettability is defined as the ability of the drop to spread out on the solid) it seems natural from (10) to measure the evolution of the contact area, when incremental variations of the volume are imposed , it is then possible to deduce the link between $\mathcal{A} = \pi\eta^2$ and V, let :

$$\mathcal{A} = \mathcal{F}(V) \tag{14}$$

The *a priori* aim of this experimental study is to obtain information both about the wetting dissipation and, [from (10)], a third phenomenological relation between μ, η and θ ; this last one,

$$\mu - \frac{2sin(\theta)}{\eta} = \frac{\varepsilon V}{\pi\eta^2} = \frac{\varepsilon V}{\mathcal{F}(V)} \tag{15}$$

together with the other two relations, is supposed to achieve the determination of φ via (7), when the history of V is given. Some experimental results about the law (14) are presented in section 3.

3 FIRST EXPERIMENTAL APPROACH AND ACTUAL RESULTS

As mentioned in the section 2.2, the principle of the experiments which have already been carried out, is simple; it consists in measuring $\mathcal{A} = \pi\eta^2$ (contact area), while controling incremental variations of the drop volume. First the volume is increased from V_0 to V_M, then decreased from V_M to V_m ; cf. Fig.2. It is usually admitted that an hysteretic phenomenon is observed, for η as well as θ; however, only the constant phases of η are clearly seen: first when V decreases from V_M to a threshold phenomenological value V_w (after which η decreases with V), and again when V increases from V_m to a new threshold value V_a (after which η increases with V). The respective loci (withdrawal $\mathcal{C}_w$ and advanced $\mathcal{C}_a$ curves) of V_w and V_a (as V_M and V_m are modified by the experimentalists) are never specified ; it is only said that θ is constant (equal to θ_a and θ_w along respectivly $\mathcal{C}_a$ and $\mathcal{C}_w$). We ourselves have observed, with good repeatability, the following type of complete cycles shown in Fig. 2 for all the triplets $F^{(1)}$, $F^{(2)}$ and S of fluids and solids we have studied.

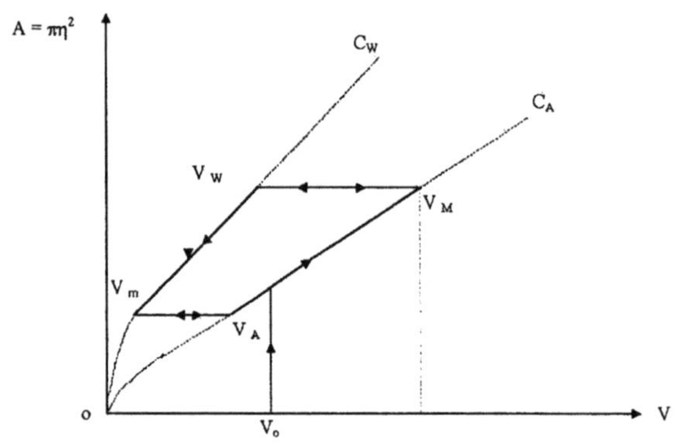

Figure 2: *Variations of the contact area with the volume (with reversible and irreversible paths)*

After several attempts to identify linear or simple power laws for $\mathcal{C}_a$ and $\mathcal{C}_w$, we are now able to conclude that two explicit behaviour laws are admissible. The first, proposed by Jacquot / Lanchon-Ducauquis [Jacquot, 1999], although continuously derivable, admits a threshold value V_S ; its dimensionless, normalized version (for $L = V_s^{\frac{1}{3}}$) is

$$A = J(V) = \begin{cases} V^b & \text{for } V \leq 1 \\ bV + 1 - b & \text{for} V \geq 1 \end{cases} \tag{16}$$

with $$A = \frac{b}{a}\,\mathcal{A}, \qquad a = \alpha V_s^{\frac{1}{3}}, \qquad b = \left(1 + \frac{\beta}{\alpha\, V_s}\right)^{-1};$$

α and β are respectively the dimensional slope and, ordinate at $V = 0$, of the linear phase.

The second law, proposed by Cunat and Csapo [not yet published, 1999], has a single expression; with the same normalization as (16) it can be written

$$A = C(V) = CV + D\, V^P \tag{17}$$

Both these laws, depend on three phenomenological coefficients (either V_s, a and b ; or C, D and P)

Remark 3 In the absence of gravity ($\varepsilon = 0$), the theoretical relationship between $\mathcal{A}$ and V obtained from (13) is of the type :

$$\pi\eta^2 = K(\theta)\, V^{\frac{2}{3}} \tag{18}$$

However, in the laboratory ($\varepsilon \neq 0$), it has been empirically observed that the powers b and P in (16) and (17) are not equal but relatively close to $\frac{2}{3}$.

Remark 4 The confrontations between theoretical and experimental approaches described above, although already well engaged, are not yet conclusive.

4 GENERAL CYLINDRICAL AND AXISYMMETRICAL INTERFACES

In the situations in which interfaces can be characterised by a simple profile, the mathematical model is reduced to a second order differential equation with either initial or limit conditions.

For general cylindrical interfaces, an analysis in the hodograph plane is possible and the equation can be explicitly integrated with the help of elliptic functions. In this favourable situation, there are three categories of solutions , depending on the value of the reduced mean curvature in a particular point M_0 (not necessarily on the profile, but well characterised by the theory), compared to the numbers $\pm\sqrt{\varepsilon}$.The different types are characterised by *i)*$H(M_0) = \pm\sqrt{\varepsilon}$, *ii)*$|H(M_0)| > \sqrt{\varepsilon}$, *iii)*$|H(M_0)| < \sqrt{\varepsilon}$.

• The interfaces produced by a Whilelmy blade belongs to the second type.

• Those generated, either between two parallel vertical plates, or by a sessile "cylindrical" drop (limit situations of liquid in a dihedron, when the angle tends to π) are always in the third category.

• The interfaces relative to non sessile "cylindrical" drops are either of the first type, if there are not convexe, or of the third type, if they are convexe.

For general axisymmetrical interfaces, when $\varepsilon \neq 0$, neither the hodograph analysis nor the direct integration of the differential equation is possible, but the preceding study allows one, by analogy, to foresee some interesting qualitive properties. Anyway, **for both situations**, the main objective, which is to find sufficient relations between the parameters : η, θ, μ and ε, is not yet rigourous even for the cylindrical interfaces; in fact, in this last situation the elliptic functions which appear in the explicit expression of the solution are not really adapted to the present objective ; A general strategy has then to be brought into play, in order to build adequate relations ; this strategy can be briefly described by the following necessary steps :

1. The choice of a characteristic length : adequate to make of ε a, more or less small, variable parameter, but also to temporarily neutralize one of the other parameters.
2. The choice in each situation of a simple experiment, the control could be one of the following : the spacing between two plates (for cylindrical interfaces), the diameter of capillary tubes or the volume of a drop ; the response to the control variation has to be easy to measure, as for instance the rising height along the plates and the tubes, or the spreading η of a drop on the solid horizontal support.
3. A numerical computation, as rigourous as possible (either from explicit solution or directly from the differential system), likely to provide a reference for the behaviour (when θ is fixed) of different parameters in the function of ε ; a *log-log* representation of this behaviour gives in fact a good idea of the order of these parameters when ε tends either to zero or to infinity.
4. Each parameter of interest has an explicit approximation in powers of ε, with coefficients depending implicitly of θ ; the appropriate highest power of ε being deduced from the preceding *log-log* study.
5. Asymptotic expansions of the same parameters in either ε or $\frac{1}{\varepsilon}$, depending on the initial or limit conditions of the differential system ; these expansions are used for the identification of the coefficients in terms of θ, in the approximating functions proposed in the previous step.
6. Finally, the computations of the differences between the values of the parameters obtained, on the one hand by the preceding approximations (of orders 0, 1 or 2), on the other hand by the reference numerical solutions.

With the strategy described above, several significative results could be obtained: a good first example is a clear **improvement of Jurin's** law giving the rising height of the liquid in a capillary tube as a function of its diameter d, the contact angle θ and the capillary length L_c ; **another still more convincing example** is illustrated by the following table which exhibits from the experimental data and measures η, V, ε (for axisymmetric sessile drops), four sequences of values obtained for the contact angle and the mean curvatures θ and H_c ; first by the numerical computations, then by the approximating functions of orders one and two, and finally by the experiments ; the percentage differences between these and the reference numerical values emphasize the efficiency of the approximations, but also the extreme difficulty of obtaining realistic values of the contact mean curvature H_c by experiments.

Measurements and computations for a sessile drop of water on glass in presence of air **($L_c = 2.74$ *mm* , $L = \eta$)**

η (mm)	Volume (mm^3)	ϵ	θ (num) (rad)	2H$_c$ (num)
1.54433	2.64	0.3117	0.825	0.9982
1.61959	3.18	0.3494	0.856	0.98403
1.90442	5.18	0.4831	0.863	0.8585
2.58863	12.8	0.8926	0.869	0.6710
2.86766	16.2	1.0954	0.827	0.5967
3.08122	19.8	1.2646	0.824	0.5648
3.18418	22.6	1.3505	0.850	0.5664

θ (order 1) (rad)	error (%)	2H$_c$ (order 1)	error (%)	θ (order 2) (rad)	error (%)	2H$_c$ (order 2)	error (%)
0.828	0.364	0.9990	0.080	0.826	0.121	0.9991	0.090
0.858	0.234	0.9860	0.203	0.856	0.000	0.9840	0.000
0.866	0.348	0.8608	0.268	0.863	0.000	0.8586	0.012
0.875	0.690	0.6740	0.447	0.870	0.115	0.6714	0.060
0.835	0.967	0.6005	0.637	0.828	0.121	0.5974	0.117
0.832	0.971	0.5683	0.620	0.825	0.121	0.5651	0.053
0.859	1.059	0.5701	0.653	0.852	0.235	0.5671	0.124

θ (exp) (rad)	error (%)	2H$_c$ (exp) (mm^{-1})	error (%)
0.8456	2.497	1.3816	38.41
0.8855	3.446	1.2952	31.62
0.9264	7.346	1.4173	65.09
0.9172	5.547	0.9634	43.57
0.9122	10.30	1.1012	84.55
0.8568	3.981	0.8614	52.51
0.9094	6.988	0.8821	53.73

5 THERMOMECHANICAL APPROACH TO WETTING BEHAVIOUR

We now embark on a more careful study on the physical phenomena involved in wetting. We use the *fluid with internal capillarity* theory, also known as Cahn-Hilliard's fluid Theory [Cahn-Hilliard, 1958] [Casal - Gouin, 1985], developed in this volume by P. Seppecher. An interface is considered as a transition layer, of finite thickness, through which the fluid properties vary rapidly but continuously. The equations obtained with this model and formulated at the fluid-fluid interface, lead to a limit relation (when the layer thickness tends to zero), known as Laplace's law. To obtain the property of wettability as a limit behaviour (when the distance from the wall tends to zero), we introduce the notion of *internal wettability* in the preceding theory ; this situation allows us to study the fluid-solid interfacial zone. For simplicity, we consider only the case of a liquid in its vapor.

In a first step, we recall that the theory of internal capillarity can be rigorously introduced by the thermomechanics of continuous media ; the point of departure is the notion that interfacial phenomena are due to non balanced intermolecular forces acting on particles of the transition zone ; this leads to the idea of a local internal energy e depending on the particle distribution near the considered point; this dependency is expressed through the density gradient $\vec{\nabla}\rho$. More precisely, Gibbs' equation is written as :

$$de = Tds + \mu d\rho + \alpha d\beta \qquad \text{with} \qquad \beta = \frac{1}{2}\left|\vec{\nabla}\rho\right|^2 \quad , \tag{19}$$

where T is the temperature, s the entropy, μ the chemical potential and α the capillary coefficient. The constitutive law is built with the local state assumption and the Thermomechanics of Irreversible Processes method ; as a consequence, the use of a second gradient theory appears as a necessity; the Cahn-Hilliard theory generates then a finer, but more complex, description of the internal forces (represented by two stress tensors: one of second order $\underline{\underline{\sigma}}$ and the other $\underline{\underline{\underline{\lambda}}}$ of third order) ; it requires a new formulation of the general laws. So, the fundamental principle of dynamics becomes

$$\rho\frac{d\vec{V}}{dt} = div\underline{\underline{\tau}} + \vec{f} \qquad \text{where } \underline{\underline{\tau}} = \underline{\underline{\sigma}} + div\underline{\underline{\underline{\lambda}}}. \tag{20}$$

For a non dissipative fluid, the constitutive law is :

$$\underline{\underline{\sigma}} = -\left(p + \frac{1}{2}\alpha\left|\vec{\nabla}\rho\right|^2\right)\underline{\underline{1}} - \alpha\vec{\nabla}\rho \otimes \vec{\nabla}\rho \qquad ; \underline{\underline{\underline{\lambda}}} = \alpha\rho\underline{\underline{1}} \otimes \vec{\nabla}\rho \quad , \tag{21}$$

where p is the usual pressure term and $\underline{\underline{1}}$ is the unit tensor of second order. Finally, to complete this model, it is necessary to propose two state laws (for p and α).

In this second step, we extend the preceding model in order to take into account the solid surface effects. Near the wall, a privileged direction $\vec{N}$ appears in the fluid, associated with the position of the wall. The internal energy of the fluid loses its *isotropic* property, in such a way that two privileged directions exist. Since the internal energy e is a scalar function, it cannot depend on the directions of these vectors, but only on their magnitudes and scalar product (which characterizes their relative directions). Consequently, following the precedent construction and this remark, we obtain the *fluid with internal wettability* model. For simplicity, we consider only a plane solid surface; $\overrightarrow{N}$ is then the constant unit normal to the wall, and we have

$$de = Tds + \mu d\rho + \alpha d\beta + K d\varpi \quad , \qquad \text{with } \varpi = \overrightarrow{\nabla}\rho \bullet \overrightarrow{N} \quad . \tag{22}$$

In the simplified situation of a non dissipative fluid, the constitutive law becomes

$$\begin{cases} \underline{\underline{\sigma}} = -\left(p + \frac{1}{2}\alpha\left|\overrightarrow{\nabla}\rho\right|^2\right)\underline{\underline{1}} - \alpha\overrightarrow{\nabla}\rho \otimes \overrightarrow{\nabla}\rho - K\overrightarrow{\nabla}\rho \otimes \overrightarrow{N}, \\ \underline{\underline{\underline{\lambda}}} = \rho\underline{\underline{1}} \otimes \left(\alpha\overrightarrow{\nabla}\rho + K\overrightarrow{N}\right) \end{cases} \tag{23}$$

This model requires three state laws which could be the expressions of the conjugated thermostatic terms p, α and K as functions of the principal ones s, ρ, β, $\overline{\omega}$; a simplified molecular theory [Gouin, 1997] can be used to provide these expressions. Our first application of this model consists in obtaining a limit law of wettability, which could be integrated into a macroscopic model; however, this theory can also be used to obtain further information. In fact, the substitution of (23) in (20), and the state laws lead to a mathematical problem, where the unknown terms are the density ρ and the velocity $\overrightarrow{V}$ fields. By determining these unknowns, we can deduce the pressure field, as well as the network of isodensity surfaces (which localize precisely the transition zones). However, to solve such a problem, it is first necessary to write down sufficient boundary conditions (compatible with the second gradient theory and accounting for the solid effects) and then to use a numerical treatment.

In the particular case of a static axisymmetric drop, we obtain through this limit procedure, far from the solid wall, the Laplace law. Moreover, the asymptotic matched expansions of the equations near the triple zone seem to confirm that the contact angle is a characteristic parameter of the wettability and that the influence of the mean curvature term vanishes near the wall. But the limit procedure has not yet given a law as simple as we have expected.

6 CONCLUSION

So many questions are raised from the two kinds of analysis of the wetting phenomenon, that it is better to conclude with some basic remarks.

First of all, we have reinforced our *a priori* conviction that the contact angle θ alone cannot represent the property of wetting ; it is in fact intimately coupled with the contact mean curvature H_c (represented by μ) and, like this second parameter, it is really difficult to measure it by experiments. We point out that the spreading length η, is a real perceptible manifestation of wetting, while the mean curvature of the $F^{(1)}/F^{(2)}$ interface, is a real perceptible manifestation of capillarity, even far from a solid boundary. It is then natural to look to characterize the wetting behaviour for a given triplet $\left\{F^{(1)}, F^{(2)}, S\right\}$, by an invariant grouping of the three introduced parameters, in particular along an hysterese curve.

Another important remark concerns the perturbative part of the parameter ε ; this parameter represents the competition between capillarity and gravity; it is as small as, the gravity is weak, or the characteristic length is small ; it is omnipresent in usual laboratory experiments but cannot be present in the expression of the wetting behaviour ; in fact, this expression, in so far as a constitutive law is concerned, has to be independent of the choice of any reference frame, and in consequence independent of gravity.

Finally, another interesting remark concerns the common objective of all the various approaches evoked above : the first method, combined conjugates mathematical arguments and experimental measurements; it is based on relatively simple ideas; however, until now, it has been exclusively concerned with geometrical parameters. The second method follows the usual way of characterizing behaviour, by the choice of appropriate thermodynamical variables and the description of the type of dissipation; it is indispensable for providing some physical consistency to the three parameters, and considered as crucial for the wetting description. Much work has still to be done to achieve agreement between these two approaches.

REFERENCES

[**Bashforth and Adams, 1882**] Bashforth, F.; Adams, J.C.; An Attempt to Test the Theory of Capillary Action, *Cambridge University Press and Deighton, Bell and Co.*

[**Biguenet, 1998**] Biguenet, F.;Contribution à l'identification de paramètres pour la mouillabilité à partir de modèles énergétiques de milieux continus, Thèse INPL, Nancy.

[**Cahn and Hilliard, 1958**] Cahn, J.W.; Hilliard, J.E.; Free energy of a non uniform system : Interface free energy, *Journal of Chemical Physics*, Vol.28, N°2, pp. 258-267.

[**Casal and Gouin, 1985**] Casal, P.; Gouin, H.; Relation entre l'équation de l'énergie et l'équation du mouvement en théorie de Korteweg de la capillarité, *CR.Acad sci Paris* , t. 300, série II, N°7.

[**Csapo et al, 1999**] Csapo, E.; Jacquot, F.; Lanchon-Ducauquis, H; Lois non classiques pour l'étalement d'une goutte axisymmetrique, *14ème Congrès Français de Mécanique Toulouse*, Ref 274, ISBN : 2-84088-040-7.

[**Gouin, 1997**] Gouin, H.; Energy of interaction between solid surfaces and liquids, *The Journal of Physical Chemistry B*, Vol 102, pp.1212-1218.

[**Jacquot, 1998**] Jacquot, F.; Contributions expérimentale et théorique à une nouvelle approche de la mouillabilité, *Thèse INPL*, Nancy..

[**Liraud, 1998**] Liraud, T.; Interfaces capillaires possédant certaines symétries, *Thèse INPL*, Nancy.

[**Marron, 1997**] Marron, F.; Contribution mathématique à l'identification des paramètres adéquats pour la modélisation de la mouillabilité, *Thèse INPL*, Nancy.

Thermodynamics and phenomenology

Jean Lemaitre - Ahmed Benallal – René Billardon – Didier Marquis
Laboratoire de Mécanique et Technologie
Ecole Normale Supérieure de Cachan – CNRS – Université Paris 6
61, avenue du Président Wilson 94235 Cachan Cedex, France

Abstract: The general way to derive constitutive equations for different phenomena, coupled or not, is briefly described with an emphasis on the choice of state variables and analytical expressions for the state potential and the (pseudo) potential of dissipation by qualitative experiments. Quantitative experiments are used to determine the numerical values of the materials dependent parameters. Applications are given for anisotropic damage coupled to stress-strain behaviour of materials, for aging coupled to plasticity and for magneto-mechanical couplings. A brief «state of the art» on localisation phenomena is given as a conclusion.

1. INTRODUCTION

1975! When «L.M.T.-Cachan» just started, it was clear that experiments should play a major role in the development of mechanics of materials, but how to bridge the gap between esoteric theories and pragmatic measurements? Up to that time, the worlds of theoreticians of thermodynamics and the world of experimentalists were completely disconnected, both writing for their own world without almost any mutual understanding. «Useless theories» against «empiricism»!
Fortunately we were the children of Paul Germain; he spent much time making basic thermodynamics clear in order to teach it in an understandable way to graduate students (Germain, 1973). It became clear that any material behaviour can be stated as a mathematical model, provided the second principle of thermodynamics is fulfilled and provided a proper choice of state variables, a proper choice of an analytical expression of a state potential Ψ, and a proper choice of another analytical expression of a dissipative (pseudo-)potential F are made (Lemaitre and Chaboche, 1994). In this sentence the word «choice» is used three times. This means that Thermodynamics gives the framework for building models, but not all the keys. Fortunately, human sensibility still plays a role:

- The choice of variables depends upon the purpose of modelling: which phenomena should be modelled, under which conditions, and which predictions should

G.A. Maugin et al. (eds.), Continuum Thermomechanics, 209-223.
© 2000 *Kluwer Academic Publishers.*

be made. The observable variables and internal variables V_i define the state of a Representative Volume Element of any material. Each of them represents a mechanism to be identified by observations either at a meso-scale or at a micro-scale and by quantitative experiments. This is often a difficult task, since hidden variables correspond to hidden mechanisms: elastic and plastic strains ε_e and ε_p, isotropic and kinematic hardenings r and α, damage D, phase change a^*, anhysteretic and internal magnetic fields H_{an}, H_i.;

- The choice of an analytical expression for the state potential taken as the specific Helmholtz free energy Ψ, (or its dual the Gibbs specific enthalpy if more convenient), a function of all the state variables, is the second key to establishing the state laws. It gives the definition of the state couplings between phenomena and allows one to calculate the associated forces A_k such that the product $A_k\,\dot{V}_k$ is the dissipated power involved by the mechanism described by V_k within the hypothesis of small strains: $\Psi(\varepsilon_e, T, V_k)$

$$
\begin{aligned}
&\frac{\partial\rho\Psi}{\partial\varepsilon} = \sigma \rightarrow \text{ law of elasticity}\\
&-\frac{\partial\Psi}{\partial T} = s \rightarrow \text{ entropy density}\\
&\frac{\partial\rho\Psi}{\partial V_k} = A_k \rightarrow \text{ associated variables}
\end{aligned}
\tag{1}
$$

Only experiments can help to make this choice.

- The choice of an analytical expression for the dissipation potential is the third key to writing the kinetic laws of evolution of the dissipative phenomena: $F(A_k..;V_k)$

$$
\dot{\lambda}\frac{\partial F}{\partial A_k} \text{ or } \frac{\partial F}{\partial A_k} = \dot{V}_k \rightarrow \text{kinetic laws} \tag{2}
$$

Here again experiments are needed in order to define the function $F(A_k;V_k)$. Let us add that the intellectual dexterity of the « modeller » is also an important factor in obtaining models easy to use. The choice of the analytical form may also depend on the way the problem is numerically solved.

2. QUALITATIVE AND QUANTITATIVE IDENTIFICATION

A Physical model is meant to describe observed phenomena and to be predictive for situations that were not included in its construction and identification process. The few experiments needed for the identification must be carefully chosen to represent basic situations. The choice of the variables and the choice of the analytical expressions of the potentials require a sensitive analysis that indicates trends and do not require a good accuracy of the test results: this is qualitative identification. Conversely, the numerical values of the material parameters need precise results for a good quantitative identification.

For qualitative identification, the number of variables necessary to describe observed phenomena should be a minimum and consistent with the continuum description if one expects to use the models for design. Therefore, the route to the choice of these variables is the identification of all the physical mechanisms involved in the situation. This is based on experiments carried out on the Representative Volume Element but it may be complemented by physical observations and/or by homogenisation procedures whenever possible. Once all the mechanisms are exhibited, a variable is affected by each of them. The state potential will be then dependent on all these variables. The (pseudo-) dissipation potential will be dependent on forces A_k but also on the variables V_k. The first step in the construction of these potentials is the analysis of the interactions between all the defined mechanisms. Experiments can be designed to check if a given mechanism has or has not an influence on another, and what type of influence it is.

A state coupling is said to exist between mechanisms i and j if a change in state variable V_i induces a change in the associated force A_j, i.e. if: $\dfrac{\partial^2\Psi}{\partial V_i \partial V_j} \neq 0$. If such a coupling does not exist, then the state potential contains two independent terms in V_i and V_j. Along the same lines, a kinetic coupling between two mechanisms i and j exists when a change in V_i implies a change in the rate of V_j (Lemaitre and Marquis, 1992).

For quantitative identification, since the determination of the parameters is in general a very difficult task (though many efficient optimisation techniques exist), the number of parameters entering a model should also be kept to a minimum. Here again, well-defined experiments on the Representative Volume Element (homogeneous behaviour) should be used in order to be able to measure all the variables properly, directly or indirectly. However, in many instances it is not possible to design experiments where a significant part of the specimen is in a homogeneous state; in this case, quantitative identification also requires the use of inverse procedures and corresponding algorithms.

3. APPLICATION TO DAMAGE MODELLING

Let us consider how to make the three choices introduced in Section 1 when modelling the progressive mechanical degradation of materials whatever their nature: metals, polymers, concrete, wood, ceramics.

3.1. Choice of the variable

Physically, damage corresponds to the creation of new micro free surfaces inside a Representative Volume Element at a meso-scale. A natural choice of the variable is the relative density of these defects: volume fraction of cavities for ductile damage (Gurson, 1977), crack density for brittle damage and fatigue (Kachanov, 1958). To link the two mechanisms in only one variable let us consider the maximum surface density $D_{(\vec{n})}$ of « defects » in any plane of normal $\vec{n}$ of the R.V.E.

If damage is isotropic, the intrinsic variable can be chosen as a scalar: D ; if not it is a tensor: a fourth order tensor in the most general case (Krajcinovic, 1985), a second order tensor if damage may be considered as orthotropic (Murakami, 1988). When damage of metals is observed under a microscope it is always, or almost always, associated with plasticity either at a meso-scale or at a micro-scale. Then, if damage is governed by plasticity described by a second order tensor, damage can also be represented by a second order tensor. Furthermore, the micro-mechanical analysis of an array of randomly distributed cracks shows that the elastic strain energy is represented by three terms that are dependent upon the crack pattern. The first term is a scalar; the second a second order tensor; and the third a fourth order tensor, negligible with respect to the other two terms (Kachanov, 1994). Thus, at least for two main reasons, the choice of a second order tensor D_{ij} is reasonable (but not rigorous!).

3.2. Choice of the state potential

When modelling linear isotropic elasticity, and isotropic and kinematic hardening plasticity, the free energy Ψ is a quadratic function of the elastic strain ε_{ij}^e and the sum of two non linear functions of accumulated plastic strain p and back strain α_{ij}.

Many experiments have shown a decrease of Young's modulus as damage progresses either in ductile creep or low-cycle fatigue phenomena. Coupling damage with elasticity is imposed by experiments, but how? One hypothesis, that is still a principle since it has not yet been demonstrated in the general case, solves the problem (Lemaitre, 1971; Lemaitre, 1978). The *principle of strain equivalence* states that « any strain constitutive equation of a damaged material is derived in the same way as for the virgin material if the stress is replaced by the effective stress ». This effective stress $\tilde{\sigma}$ is the stress acting on the unit surface diminished by the surface of defects. Due to the definition of the damage variable, in one dimension the effective stress is defined as: $\tilde{\sigma} = \sigma/(1-D)$.

In three dimensions with anisotropic damage it is more complicated because the simple extension of the above statement does not yield the existence of an elastic potential (Cordebois and Sidoroff, 1982). This difficulty is solved if the elastic energy is divided into two parts, distortion and hydrostatic. By using the Gibbs energy Ψ^* with the second order damage tensor D_{ij} of mean value $D_H = 1/3D_{kk}$ (Lemaitre et al, 1999), we find

$$\rho\Psi^* = \frac{1+\nu}{2E} H_{ij}\sigma^D_{jk}H_{kl}\sigma^D_{li} + \frac{3(1-2\nu)}{2E}\frac{\sigma_H^2}{1-\eta D_H} + \Psi_{\rho(r,\alpha)} \quad (3)$$

where,

$$(\delta_{ij} - D_{ij})^{1/2} \text{ and } \sigma^D_{ij} = \sigma_{ij} - \sigma_H\delta_{ij} \quad \sigma_H = \frac{1}{3}\sigma_{kk} \quad (4)$$

and where ρ denotes the mass density, E and ν the elastic constants of the virgin material and η a parameter that is required for a correct representation of experiments concerning the variation of the contraction ratio. For several metals $\eta \cong 3$, and $\eta = 1$ corresponds to isotropic damage. This is the price to pay for phenomenology; the parameter η has to be measured to evaluate the components of damage, but it does not appear in the damage kinetic law.
The law of elasticity derives from the Gibbs potential, so that

$$\varepsilon^e_{ij} = \rho\frac{\partial\Psi^*}{\partial\sigma_{ij}} = \frac{1+\nu}{E}\tilde{\sigma}_{ij} - \frac{\nu}{E}\tilde{\sigma}_{kk}\delta_{ij} \quad (5)$$

and the effective stress is defined as

$$\tilde{\sigma}_{ij} = (H_{ik}\sigma^d_{kl}H_{li})^D + \frac{\sigma_H}{1-\eta D_H}\delta_{ij} \quad (6)$$

The force associated with damage variable is the strain energy density release rate tensor Y:

$$Y_{ij} = \rho\frac{\partial\Psi^*}{\partial D_{ij}} \quad (7)$$

Since it is easier to introduce the coupling between damage and plasticity through the dissipation potential, the plastic part Ψ^*_p of the state potential Ψ^* is not affected by damage.

3.3. Choice of the dissipation potential

There is much more freedom to choose a «good» dissipation potential. This is the reason why so many models of kinetic laws have been proposed!
It has been observed that:

- Ductile damage is governed by plasticity ;
- Fracture and ductile damage are influenced by stress triaxiality, which is related to the ratio of the hydrostatic elastic energy to the distortion energy;

- Damage measurements by means of elasticity changes show that ductile damage grows approximately proportionally to cumulative plastic strain, whereas fatigue damage evolution evolves non linearly with stress or elastic strain.

Hence, the (pseudo-)dissipation potential F has to be a function of the corresponding variables. Furthermore, since plastic constitutive equations must also be derived from this potential F, it must include the yield criterion f. The coupling with damage is introduced by means of the effective stress associated with the principle of strain equivalence, so that

$$F(\tilde{\sigma}, R, X, Y) = f(\tilde{\sigma}, R, X) + F_D(Y, \varepsilon^p) \tag{8}$$

Here f denotes the von Mises function giving the plastic multiplier $\dot{\lambda}$ by the consistency condition $\dot{f} = 0$: $\dot{\lambda} = \dot{r}, r$ is the variable associated with isotropic hardening R.

For isotropic damage the following choice of F_D has given a damage law of evolution extensively used with success for more than 15 years for ductile, quasi-brittle and fatigue damage (Lemaitre, 1983):

$$F_D = \frac{S}{s+1}(\frac{Y}{S})^{s+1} \rightarrow \dot{D} = \dot{\lambda}\frac{\partial F}{\partial Y} = (\frac{Y}{S})^s \dot{p} \tag{9}$$

where $\dot{p}$ denotes the cumulative plastic strain rate: $\dot{p} = (\frac{2}{3}\dot{\varepsilon}^p_{ij}\dot{\varepsilon}^p_{ij})^{1/2}$.

Its extension to anisotropic damage imposes the choice: $F_D = \left(\frac{\overline{Y}(\varepsilon^e)}{S}\right)^s Y_{ij}\left|\frac{d\varepsilon^p}{dr}\right|_{ij}$,

where S and s denote material parameters, variable $\overline{Y}$ denotes the effective elastic strain energy such that: $\overline{Y} = \frac{1}{2}E_{ijkl}\varepsilon^e_{kl}\varepsilon^e_{ij}$, and operator $|\circ|$ applied to a tensor means the absolute value of the principal values. Finally, the damage evolution law appears as:

$$\dot{D}_{ij} = \frac{\partial F}{\partial Y_{ij}}\dot{\lambda} = \left(\frac{\overline{Y}}{S}\right)^s |\dot{\varepsilon}^p|_{ij} \tag{10}$$

Two other material-dependent parameters must be introduced in order to fit experimental observations:

- A damage threshold p_D is introduced such that $D_{ij} = 0$ if $p \leq p_D$;
- A critical value of the damage parameters is written in accordance with the principle of strain equivalence on the largest principal value of the damage tensor. This critical value is introduced such that

$$\underset{I}{Max}\ D_I = D_c \rightarrow \text{mesocrack intitiation.}$$

4. APPLICATION TO AGING AND PLASTICITY COUPLINGS

Among the many applications of this formalism to the modelling of the interactions between the mechanical behaviour and the chemical and physical transformations of materials, we present the modelling of an aluminium-copper alloy subjected to plastic deformation during the aging process. Aging of these alloys takes place after a dissolution heat treatment (1 hour at 500°C) followed by quenching in water at 20°C. After this heat treatment, the copper atoms are in solution in the aluminium matrix, and this configuration is not stable. Precipitates (Al_2Cu) appear, and lead to a strong hardening of the material. This aging process takes about 100 hours at 20°C. If the material experiences some plastic deformation during the aging process, it is assumed for sake of simplicity that the main interactions between the dislocations and the precipitates lead to a modification of the aging kinetics. The set of constitutive equations given hereafter is written for the isothermal case.

The state variables are the classical state variables of elasto-plasticity, and an aging variable denoted by a^*. The limiting values for a^* are 0 (absence of precipitates) and 1 (complete precipitation). To determine the state potential, the state couplings or uncouplings are analysed. The only state coupling is the coupling between aging and isotropic hardening due to the dislocations and precipitates interactions. The following expression of the free energy is deduced:

$$\psi(\varepsilon^e,\alpha,r,a)=\frac{1}{2}A:\varepsilon^e:\varepsilon^e+\frac{1}{2}Btr(\alpha^2)+W(r)+a^*(Cr-L)+L \tag{11}$$

The last terms (where C and L are material dependent parameters) lead to the state coupling:

$$R=W'(r)+Ca^* \tag{12}$$

This expression shows that the higher the aging value, the higher the isotropic hardening (or the yield stress).

The kinetic coupling of the plastic strain with the ageing mechanism is deduced from the interaction between the diffusion rate of copper atoms and the density of dislocations. The following expression of the potential of dissipation is deduced from specific experiments (Marquis and Costa Mattos 1991)

$$\varphi=I_f(\sigma,X,R)+\frac{1}{\tau}<a^*_\infty-a^*>Z \tag{13}$$

where τ and a^*_∞ denote functions of r and ε^p, and $\langle x\rangle=xH(x)$ where H is the Heaviside function.

From these two potentials, the following set of constitutive equations is obtained:

$$\varepsilon = \varepsilon^e + \varepsilon^p; \varepsilon^e = \frac{1+v}{E}\sigma - \frac{v}{E}tr(\sigma)I; \dot{\varepsilon}^p = \dot{\lambda}\frac{\partial f}{\partial \sigma'} \tag{14}$$

$$\dot{X} = B^{\infty}\varepsilon^p; \dot{R} = W''(r)\dot{r} + C\dot{a}^* \tag{15}$$

$$\dot{a}^* = \frac{1}{\tau} < a_{\infty}^* - a^* > \tag{16}$$

where f denotes the von Mises yield function with kinematic and isotropic hardenings such that

$$f = \left[\frac{3}{2}tr(\sigma^D - X)^2\right]^{1/2} - R - \sigma_y \tag{17}$$

where σ_y denotes the initial yield stress of the material, and σ^D the stress deviator.

The plastic multiplier $\dot{\lambda}$ is calculated from the consistency condition: $\dot{f} = 0$, and the internal variable r associated with the isotropic hardening is the cumulative plastic strain.
From specific experiments on the material at different levels of plastic strain and aging, the following expressions for the functions τ and a^* are identified:

$$\frac{1}{\tau} = [\alpha - \beta\exp(-\gamma' r)] \tag{18}$$

$$a_{\infty}^* = A + (1 - A)\exp\left[-k(\varepsilon_{II}^p)M\right] \tag{19}$$

where $\alpha, \beta, \gamma', A$ and k denote material parameters, and $(\varepsilon_{II}^p)_M$ denotes the maximum equivalent plastic strain over the time interval, so that $(\varepsilon_{II}^p)_M = \max\left[\frac{2}{3}tr(\varepsilon^p)^2\right]^{1/2}$.
The expression of the strain hardening function $W''(r)$ is chosen to ensure the limitation of the size of the elastic domain:

$$W''(r) = b\gamma\exp(-\gamma r) \tag{20}$$

where b and γ denote material parameters.
The set of all numerical values of the material parameters is given below for the 2024 aluminium alloy considered in the application:

$E = 72500$ Mpa, $v = 0.33$, $\mathrm{B} = 30000$ Mpa, $\sigma_y = 75 Mpa$,

$\mathrm{b} = 40.5$ Mpa, $\gamma = 5$, $\alpha = 0.00167\mathrm{s}^{-1}$,

$\beta = 0.00153\mathrm{s}^{-1}$, $\gamma' = 1$, $\mathrm{A} = 0.5$, $\mathrm{k}' = 28$, $\mathrm{C} = 130$Mpa.

5. APPLICATION TO MAGNETO-ELASTIC COUPLINGS

This section is devoted to some aspects of the complex couplings between the different phenomena that are involved in real thermo-magneto-elastic behaviour of soft ferromagnetic materials. The global (structural) couplings [Germain et al 1983, Maugin, 1980-1988] that involve the structural conservation laws viz. the heat equation, the mechanical conservation equation and the Maxwell equations, are not discussed here.
The phenomena that are discussed here refer to the fact that the magnetic behaviour is affected by stresses and conversely, to the fact that changes in the magnetisation generate so-called magnetostrictive strains. One of the practical consequences of these phenomena is the following: the magnetic behaviour of electrical machines is dependent not only on the behaviour of the raw materials they are made of, but also on the stresses induced in these machines by their manufacturing process and the in-service loadings.

A possible approach to developing coupled magneto-elastic constitutive equations consists in using so-called homogenisation techniques [Buiron et al, 1999]. More phenomenological models have also been proposed [see for instance: Maugin, 1991; Jiles, 1995; Sabir, 1995]. The approach used herein is also phenomenological though based on continuum thermodynamics.

5.1. Phenomenology at the mascroscopic scale

When subjected to a cyclic magnetic field, industrial materials exhibit a hysteretic behaviour. However, by superimposing a slowly decaying alternative field on a magnetic field of given amplitude, it is possible to reach the state of lowest internal energy that corresponds to one point of the so-called anhysteretic curve. This experimental procedure has been applied to an industrial non-oriented Fe-3%Si alloy commonly used in electrical power engineering. During the tests, the 0.5mm thick laminations were subjected to collinear uniaxial mechanical and magnetic excitations. The experimental results presented in Fig.1 show the influence of stresses on both the magnetostriction strain measured in the direction of the applied field, and on the magnetic response.

5.2. Microscopic behaviour

The magnetisation process can be sketched as follows. The material comprises numerous microscopic magnetic domains. In each domain, the magnetisation is uniform and oriented according to the local orientation of the crystal axes. The boundaries between the domains are so-called domain walls and correspond to a few atomic distances. At the macroscopic level, if no external field is applied, the average magnetisation tends to zero. This demagnetised state is often taken as a reference for the strain state of the material.
When an external magnetic field is applied, the domain configuration evolves: domains oriented in the direction of the external field grow, whereas other domains tend to shrink, so that both the average magnetisation and the average strain state change. Moreover, after a certain level of mean magnetisation has been reached, the magnetisation in each domain tends to rotate from the crystallographic axes to the

direction of the applied magnetic field. The final stage of this evolution corresponds to magnetic saturation: in Fig.1a this microscopically approximately uniform state has been chosen as a reference for the strain state. During the magnetisation process, the domain walls motions are restrained by all kinds of defects in the material (impurities, grain boundaries, inclusions, local stresses, etc); these interactions lead to repeated pinning, bulging and unpinning of the walls; this explains the hysteretic character of the magnetic behaviour.
The anhysteretic behaviour corresponds to the behaviour that would be observed for a material with no defect. Real materials may exhibit this ideal behaviour, which corresponds to the state of minimum energy, if the domain walls are «shaken» for instance by an alternating magnetic field so that an additional energy overcomes the local energy barriers created by the various obstacles.

5.3. A magneto-elastic model (Hirsinger, 1994; Gourdin, 1998)

5.3.1 State variables

To model these mechanisms, and as sketched on Fig.2, it is proposed to make a partition of the magnetic field into two contributions: the anhysteretic «effective» field $\mathbf{H}_{an}$ corresponds to the reversible behaviour and the internal variable $\mathbf{H}_i$ corresponds to the dissipative phenomena. Further partitioning of the internal field enables us to take account of the reversible bending of the pinned walls and irreversible jump to the next obstacle that are respectively described by fields $\mathbf{H}_{rev}$ and $\mathbf{H}_{irr}$:

$$\mathbf{H}_{an} = \mathbf{H} - \mathbf{H}_i \tag{21}$$

$$\mathbf{H}_i = \mathbf{H}_{rev} + \mathbf{H}_{irr} \tag{22}$$

$$\mathbf{M}(\mathbf{H}) = \mathbf{M}_{an}(\mathbf{H}_{an}) = \mathbf{M}_{an}(\mathbf{H} - \mathbf{H}_i) \tag{23}$$

Besides, it is assumed that total strains ε are the sum of pure mechanical strains ε^m, pure magnetostriction strains ε^μ and thermal expansion strains ε^{th}, so that when small strain and linear elasticity assumptions are made, the following strain decomposition can be used:

$$\varepsilon = \varepsilon^m + \varepsilon^\mu + \varepsilon^{th} = \mathbf{E}^{-1} : \sigma + \varepsilon^\mu(\sigma, \mathbf{H}, \mathbf{H}_{rev}, \mathbf{H}_{irr}, T) + \varepsilon^{th}(T) \tag{24}$$

where $\mathbf{E}$, σ and T respectively denote the elasticity moduli tensor, the stress tensor and temperature. This relation is based on the following uncoupling hypothesis: the elastic and thermal expansion moduli are unaffected by the magnetic field.

5.3.2 State potential

The specific free enthalpy Ψ is used as a state potential. This global state potential is divided into several uncoupled contributions:

$$\rho\Psi = \rho\Psi(\sigma, \mathbf{H}, \mathbf{H}_{rev}, \mathbf{H}_{irr}, T) \tag{25}$$

$$\rho\Psi = \rho\Psi^{\mu m}(\sigma, \mathbf{H}, \mathbf{H}_{rev}, \mathbf{H}_{irr}, T) + \rho\Psi^{\mu an}(\mathbf{H}-\mathbf{H}_{rev}-\mathbf{H}_{irr}, T) + \rho\Psi^{\mu i}(\mathbf{H}, \mathrm{H}_{rev}, \mathbf{H}_{irr}, T) \tag{26}$$

where ρ denotes the mass density.

The first term $\Psi^{\mu m}$ represents the state coupling via the magnetostriction strain; its expression is postulated as

$$\rho\Psi^{\mu m}(\sigma,\mathbf{H},\mathbf{H}_{rev},\mathbf{H}_{irr},T)=\frac{1}{2}\sigma:\mathbf{E}^{-1}:\sigma+\int_0^\sigma \varepsilon^{\mu}(\sigma,\mathbf{H},\mathbf{H}_{rev},\mathbf{H}_{irr},T):d\sigma+\varepsilon^{th}(T):\sigma \quad (27)$$

The second term $\rho\Psi^{\mu an}$ is linked to the anhysteretic behaviour of the material, its expression is postulated as

$$\rho\Psi^{\mu an}(\mathbf{H}-\mathbf{H}_{rev}\text{-}\mathbf{H}_{irr},T)=\int_0^{\mathbf{H}\text{-}\mathbf{H}_{irr}-\mathbf{H}_{rev}}\mu_0\mathbf{M}_{an0}(\mathbf{h},T)\mathbf{dh} \quad (28)$$

where subscript 0 refers to $\sigma = \mathbf{0}$ and μ_0 denotes the vacuum permeability.

The last term $\rho\Psi^{\mu i}$ accounts for the hysteretic part of magnetic behaviour.

5.3.2 State laws

The mechanical and magnetic state laws are derived from the state potential:

$$\varepsilon=\frac{\partial(\rho\Psi^{\mu m})}{\partial\sigma}=\mathbf{E}^{-1}:\sigma+\varepsilon^{\mu}(\sigma,\mathbf{H},\mathbf{H}_{rev},\mathbf{H}_{irr},T)+\varepsilon^{th}(T) \quad (29)$$

$$\mu_0\,\mathbf{M}_l=\frac{\partial(\rho\Psi^{\mu an}+\rho\Psi^{\mu i})}{\partial\mathbf{H}}+\int_0^\sigma\frac{\partial\varepsilon^{\mu}}{\partial\mathbf{H}}(\sigma,\mathbf{H},\mathbf{H}_{rev},\mathbf{H}_{irr},\mathbf{T}):\mathbf{d}\sigma \quad (30)$$

$$\mu_0\,\mathbf{M}_{rev}=\frac{\partial(\rho\Psi)}{\partial\mathbf{H}_{rev}} \quad (31)$$

$$\mu_0\,\mathbf{M}_{irr}=\frac{\partial(\rho\Psi)}{\partial\mathbf{H}_{irr}} \quad (32)$$

where $\mathbf{M}_l$, $\mathbf{M}_{rev}$ and $\mathbf{M}_{irr}$ respectively denote the forces associated to $\mathbf{H}$, $\mathbf{H}_{rev}$ and $\mathbf{H}_{irr}$.

5.4 Experimental validation of the state (un)couplings

To complete the model, evolution laws for the internal variables must be postulated and experimentally identified. These laws must satisfy the Clausius-Duhem inequality:

$$D=\varepsilon:\dot{\sigma}+\mu_0\,\mathbf{M}\,.\,\dot{\mathbf{H}}-\rho\,(\dot{\Psi}+s\,\dot{T})\geq 0 \quad (33)$$

where s denotes the entropy.

However, before proceeding towards this last step it is worthwhile -when possible- to validate the choices made when writing the state potential.

When written for anhysteretic evolutions, the Clausius-Duhem inequality shows that

$$\mu_0\,\mathbf{M}_l=\mu_0\,\mathbf{M}_{an0}(\mathbf{H}-\mathbf{H}_{rev}-\mathbf{H}_{irr},T)+\int_0^\sigma\frac{\partial\varepsilon^{\mu}}{\partial\mathbf{H}}(\sigma,\mathbf{H}-\mathbf{H}_{rev}-\mathbf{H}_{irr},\mathbf{T}):\mathbf{d}\sigma$$

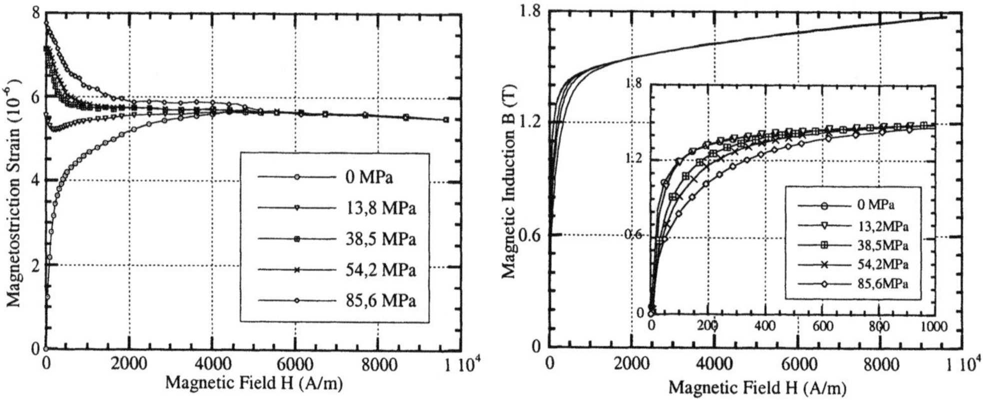

a. b.

Figure 1. Effect of an elastic tensile stress on the anhysteretic magnetostriction strain measured in the direction of applied field (**a**) and on the magnetic anhysteretic curve (**b**) for 0.5 mm thick laminations made of N.O. Fe-3%Si.

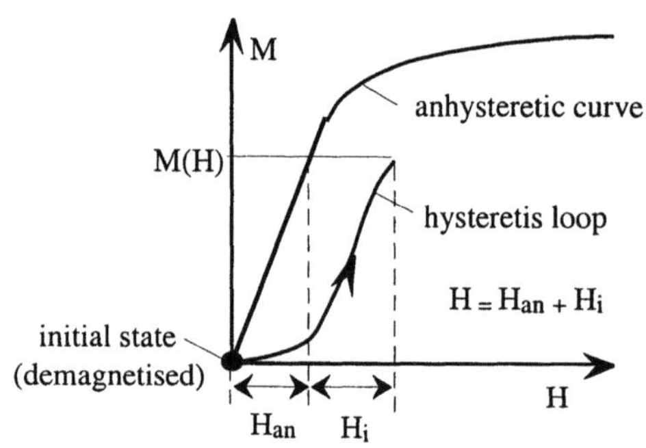

Figure 2. Definition of the state variables

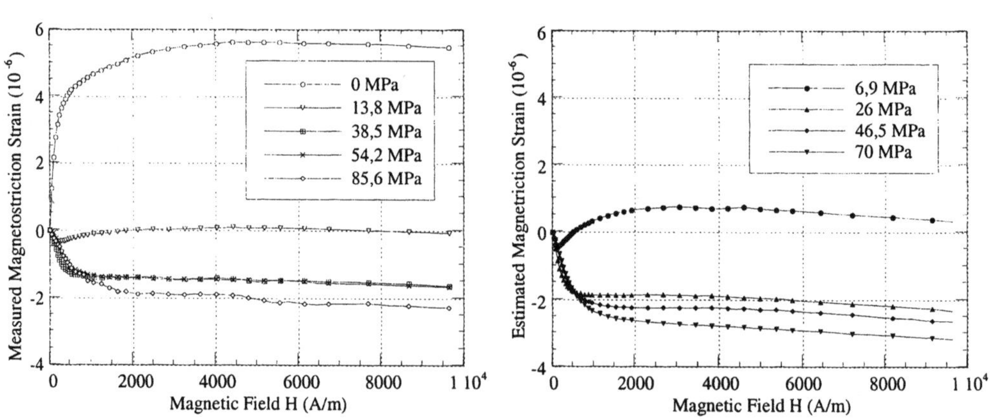

a. b.

Figure 3. Validation of the magnetoelastic state couplings: comparison between the measured (**a**) and the predicted (**b**) relative anhysteretic curves ($\varepsilon^{\mu an}_{//}(\mathbf{H},\sigma) - \varepsilon^{\mu an}_{//}(\mathbf{H}=\mathbf{0},\sigma)$).

$$=\mu_0 \, \mathbf{M}_{an} \, (\mathbf{H}_{an}, \sigma, T) \tag{34}$$

Differentiation of this relation with respect to stress σ, followed by integration with respect to $\mathbf{H}$ leads to

$$\mu_0 \frac{\partial \mathbf{M}_{an}}{\partial \sigma}(\mathbf{H},\sigma) = \frac{\partial \varepsilon^{\mu an}}{\partial \mathbf{H}}(\mathbf{H},\sigma) \tag{35}$$

$$\varepsilon^{\mu an}(\mathbf{H}, \sigma) - \varepsilon^{\mu an}(\mathbf{H}=0, \sigma) = \mu_0 \int_0^{\mathbf{H}} \frac{\partial \mathbf{M}_{an}}{\partial \sigma}(\mathbf{h},\sigma).\mathbf{dh} = \int_0^{\mathbf{H}} \frac{\partial \varepsilon^{\mu an}}{\partial \mathbf{H}}(\mathbf{h},\sigma).\mathbf{dh} \tag{36}$$

For isotropic materials, it is reasonable to assume that the magnetostriction strain tensor takes the following form

$$\varepsilon^{\mu}(\mathbf{H}, \sigma) = \begin{pmatrix} \varepsilon^{\mu}_{//}(\mathbf{H},\sigma) & 0 & 0 \\ 0 & \varepsilon^{\mu}_{\perp}(\mathbf{H},\sigma) & 0 \\ 0 & 0 & \varepsilon^{\mu}_{\perp}(\mathbf{H},\sigma) \end{pmatrix}_{(\mathbf{u}_{//},\mathbf{v}_{\perp},\mathbf{w}_{\perp})} \tag{37}$$

where $\mathbf{u}_{//}$, $\varepsilon^{\mu}_{//}(\mathbf{H},\sigma)$ and $\varepsilon^{\mu}_{\perp}(\mathbf{H},\sigma)$ respectively denote a unit vector parallel to the magnetisation $\mathbf{M}$, magnetostriction strain in the direction of magnetisation, and magnetostriction strain in perpendicular directions.

The results that are presented in Fig.3.a., correspond to the identification of $(\varepsilon^{\mu an}_{//}(\mathbf{H},\sigma) - \varepsilon^{\mu an}_{//}(\mathbf{H} = \mathbf{0}, \sigma))$ through a straightforward treatment of the experimental results presented in Fig.1.a. The results that are presented in Fig.3.b are directly derived from the experimental results presented in Fig.1.b; a finite difference scheme has been used to compute the term $\frac{\partial \mathbf{M}_{an}}{\partial \sigma}(\mathbf{H},\sigma)$. It is worth noticing that both plots 3.a and 3.b are directly derived from experiments and the expression of the state potential. In other words, this procedure –comparison between the direct measurements of magnetostrictive strains and the prediction of these strains from the magnetic response of the material under different stress states– provides an example of the choices made in writing the state potential.

6. LOCALISATION PHENOMENON AND ITS CONSEQUENCE

If, as mentioned above, Paul Germain was at the beginning of the story about the use of thermodynamics in the modelling of the mechanical behaviour, we find him also at the end. Indeed, the boundary value problems arising in non-linear solid mechanics (with the type of constitutive equations sketched above, damage mechanics and non-associated plasticity) may change their type locally during a loading process. It is well established and understood now that various types of constitutive equations lead to a rate-boundary value problem that may become hyperbolic at a certain stage of a loading process, while it was elliptic at the beginning. This loss of ellipticity suggests the appearance of discontinuity surfaces, and happens for example in the presence of softening or for certain non-associative behaviours. Note however that none of these features is necessary or sufficient. It is this type of change, this loss of ellipticity that is used to describe localisation phenomena for rate-independent solids, phenomena that are precursors to rupture. This sets the transition from a diffuse deformation behaviour to a very localized one.

Most of the time this boundary value problem becomes actually mixed (elliptic and hyperbolic at the same time). There are both an elliptic zone and a hyperbolic zone in the structure just like in fluid mechanics and in the context of transonic compressible flows where a subsonic zone coexists with a supersonic one, a field to which Paul Germain has significantly contributed.

It is also known that this loss of ellipticity, when it occurs, leads to some kind of ill-posedness in the sense that the boundary value problems may have an infinite number of linearly independent solutions, and that these solutions may not depend continuously on the data. This has led these last years to the reappraisal of generalised continuum theories such as higher gradient theories, Cosserat continua and nonlocal models. Paul Germain was also concerned by these theories. This reappraisal is meant to solve some physical problems that the classical local continuum framework fails to describe. All these theories bring in a way or another an internal length in the continuum that allows for instance one to fix the wavelength of certain bifurcation modes and a thickness to shear bands. The loss of ellipticity manifests itself in many ways: loss of ellipticity of the field equations but also through a compatibility between these field equations and boundary conditions or interfacial conditions (when the material is heterogeneous). See [Benallal et al, 1993] for a detailed discussion.

REFERENCES

[Benallal et al, 1993] Benallal, A.; Billardon, R.; Geymonat, G.; Bifurcation and localisation in rate-independent materials. Some general considerations, CISM Lecture Notes N° 327 on *Bifurcation and Stability of Dissipative Systems*, pp. 1-44, edited by Nguyen Quoc Son, Springer Verlag.

[Buiron et al, 1999] Buiron, N.; Hirsinger, L.; Billardon, R.; A multiscale model for magneto-elastic couplings, *J. Phys. IV France 9*, pp. 187-196.

[Cordebois et al, 1982] Cordebois, J.P.; Sidoroff, F.; Endommagement anisotrope en élasticité et plasticité, In *JMTA* Numéro spécial, pp. 45-60.

[Germain, P.; 1973] *Cours de Mécanique des Milieux Continus*, Tome 1, Masson Ed. Paris (France).

[Germain et al, 1983] Germain, P.; Nguyen, Q.S.; Suquet, P.; Continuum Thermodynamics, *J. Appl. Mech.,* 50, pp. 1010-1020.

[Gourdin, C., 1998] Identification et modélisation du comportement électromagnétique de structures ferromagnétiques", *Thèse de Doctorat, Université Paris 6.*

[Gurson, A.L., 1977] Continuum theory of ductile rupture by voids nucleation and growth, *J. Eng. Mat. Tech*, 99, pp. 2-15.

[Hirsinger, L., 1994] Etude des déformations magnéto-élastiques dans les matériaux ferromagnétiques doux, *Thèse de doctorat, Université Paris 6.*

[Jiles, D.C., 1995] Theory of the magnetomechanical effect, *J. Phys. D: Applied Physics*, 28, pp. 1537-1546.

[Kachanov, M., 1994] Elastic solids with many cracks and related problems, *Advances in Appl. Mech.*, Vol. 30, Acad Press Ed.

[Kachanov, M., 1958] Time of the Rupture Process under creep conditions, *T. V.Z. Akad. Nauk S.S.R. Otd. Tech.* Nauk, 8.

[Krajcinovic, D., 1985] Continuous damage mechanics revisited. Basic concepts and definitions, *J. Appl. Mech.*, 52, pp. 829-834.

[Lemaitre, J. ; 1971] Evaluation of dissipation and damage in metals, in *Proc. Int. Conf. on Mechanical Behavior of Materials*, Kyoto (Japan).

[Lemaitre, J. ; 1978] Théorie mécanique de l'endommagement isotrope appliquée à la fatigue des métaux, *Actes Matériaux et structures sous chargement cyclique*, pp. 133-144, edited by Halphen B. and Nguyen Q.S., Assoc. Amicale des Ingénieurs Anciens Elèves de l'E.NP.C.

[Lemaitre et al, 1994] Lemaitre, J.; Chaboche, J.L.; *Mechanics of Solid Materials*, Cambridge University Press, Cambridge (UK).

[Lemaitre, J., 1992] *A Course on Damage Mechanics*, Springer Verlag, Berlin (Germany).

[Lemaitre et al, 1999] Lemaitre, J.; Desmorat, R.; Sauzay, M.; Anisotropic damage law of evolution, *European J. Mech../ A Solids* (accepted).

[Lemaitre, J., 1983] Un modèle linéaire d'endommagement plastique ductile, *C.R.A.S. Paris*, t. 296 II, pp. 1359-1364.

[Lemaitre et al, 1992] Lemaitre, J.; Marquis, D.; Modeling Complex behavior of metals by the «state-kinetic coupling theory», *J. Eng. Mat. Tech,* 114, pp. 250-254.

[Marquis et al, 1991] Marquis, D.; Costa Mattos, H.; Modeling plasticity and aging as coupled phenomena, *Int. J. Plasticity*, 7, pp. 865-877.

[Maugin, G.A., 1980] The method of virtual power in continuum mechanics. Application to coupled fields, *Acta Mechanica*, 35, pp. 1-70.

[Maugin, G.A., 1988] *Continuum Mechanics of Electromagnetic Solids*, North-Holland.

[Maugin, G.A., 1991] Compatibility of magnetic hysteresis with thermodynamics, *Int. J. Appl. Electromag. Mat.*, 2, pp.7-19.

[Murakami, S., 1988] Mechanical modeling of material damage, *J. Appl. Mech.*, vol. 55, pp. 280-286.

[Sabir, M., 1995] Constitutive relations for magnetomechanical hysteresis in ferromagnetic materials, *Int. J. Engng. Sci.*, 33, 9, pp.1233-1249.

On the thermomechanical modelling of shape memory alloys

Ch. Lexcellent and M. L. Boubakar
Laboratoire de Mécanique Appliquée Raymond Chaléat,
UMR-CNRS 6604, Université de Franche-Comté,
24 chemin de l'Epitaphe, 25030 Besançon, France.
christian.lexcellent@univ-fcomte.fr
lamine.boubakar@univ-fcomte.fr

Abstract: On the one hand, at the scale of the crystal, a very "smart" mathematical theory of martensitic transformation is described. On the other hand, two more classical ones, the first at the mesoscopic scale ("meso-macro" self consistent integration), the second at the macroscopic scale, based on the thermodynamics of irreversible processes, are formulated. The basic foundations of these meso-macro and phenomenological models are recalled with special attention devoted to the discrepancies between them. This paper constitutes a challenge to reducing the gap between the mathematical theory and the classical ones.
Keywords: Shape memory alloys, Martensitic phase transition, Thermodynamics of solids, Finite strains.

1. INTRODUCTION

Over the past decade, decisive progress has been made in the theoretical research on the martensitic phase transformation in shape memory materials [James and Hane, 2000]. An important step has been made from the very well known crystalographic theory on martensitic transformation due to Weschler, Lieberman and Read [Weschler et al., 1953] to the actual mathematical one [Ball and James, 1987, 1992]. These authors formulate a free energy function that would produce the austenite-martensite interface by energy minimisation, and relate it to crystal structure. One principal result of this approach is the recognition that some of the common microstructures in shape memory materials are possible (as energy minimising microstructures) only with exceedingly special lattice parameters. However, their investigation is, at this day, restricted to the "reversible" behaviour (i.e. without hysteresis and dissipation) of single crystals or polycrystals [Bhattacharya and Kohn, 1996] under external stress free state conditions. The more classical approaches such as the "meso-macro" self consistent modelling [Patoor et al., 1993] [Sun and Hwang, 1993] or the macroscopical modelling within the thermodynamical approach, although less rigorous (in the mathematical sense), seems more predictive. Hence, important work on polycrystals [Siredey et al., 1999][Qidway and Lagoudas, 2000], the geometrical linear theory [Roytburd and Slutsker, 1999],

G.A. Maugin et al. (eds.), Continuum Thermomechanics, 225-236.
© 2000 *Kluwer Academic Publishers.*

hysteresis [Huo and Muller, 1993], constitutive equations [Raniecki and Lexcellent, 1998][Leclercq and Lexcellent, 1996][Abeyratne and Knowles, 1993], and dissipation [Peyroux et al., 1996] are reported in this general investigation.

This paper consists first part in a brief synthesis of the main results obtained from the mathematical theory of the martensitic transformation as compared to the experiments. In a second part, the basic foundations of the phenomenological models are described with special attention to the discrepancies between them.

Tensors will be denoted by underlined capitals in direct notation. Their juxtaposition implies the usual summation operation over two adjacent indices, while double dots indicate the summation product. A superposed dot indicates the rate, a superposed T the transpose, a superposed S the symmetric part and a superposed A the skew-symmetric part. The tensor $\underline{I}$ is the second-order identity tensor.

2. CONFRONTATION OF THE MATHEMATICAL THEORY OF THE MARTENSITIC TRANSFORMATION WITH THE EXPERIMENTS

The observations of Chu and James [Chu and James, 1996] on a Cu-14Al-3.9Ni (W^t %) single crystal revealed a transformed region consisting alternately of two different variants of martensite. These variants form an internally twinned martensite region which is needed to have an undeformed interface between austenite and martensite (Figure 1).

***Figure 1**; Austenite-twinned martensite microstructures obtained at stress free state [Chu, 1993]*

In this situation, this Cu-Al-Ni single crystal undergoes a martensitic transformation from a cubic structure β_1 to an orthorhombic structure γ_1' (2H). Six variant of martensite are defined as :

$$\underline{U}_1 = \begin{pmatrix} \beta & 0 & 0 \\ 0 & \frac{\alpha+\gamma}{2} & \frac{\alpha-\gamma}{2} \\ 0 & \frac{\alpha-\gamma}{2} & \frac{\alpha+\gamma}{2} \end{pmatrix} \qquad \underline{U}_2 = \begin{pmatrix} \beta & 0 & 0 \\ 0 & \frac{\alpha+\gamma}{2} & \frac{\gamma-\alpha}{2} \\ 0 & \frac{\gamma-\alpha}{2} & \frac{\alpha+\gamma}{2} \end{pmatrix}$$

$$\underline{U}_3 = \begin{pmatrix} \frac{\alpha+\gamma}{2} & 0 & \frac{\alpha-\gamma}{2} \\ 0 & \beta & 0 \\ \frac{\alpha-\gamma}{2} & 0 & \frac{\alpha+\gamma}{2} \end{pmatrix} \qquad \underline{U}_4 = \begin{pmatrix} \frac{\alpha+\gamma}{2} & 0 & \frac{\gamma-\alpha}{2} \\ 0 & \beta & 0 \\ \frac{\gamma-\alpha}{2} & 0 & \frac{\alpha+\gamma}{2} \end{pmatrix} \qquad (1)$$

$$\underline{U}_5 = \begin{pmatrix} \frac{\alpha+\gamma}{2} & \frac{\alpha-\gamma}{2} & 0 \\ \frac{\alpha-\gamma}{2} & \frac{\alpha+\gamma}{2} & 0 \\ 0 & 0 & \beta \end{pmatrix} \qquad \underline{U}_6 = \begin{pmatrix} \frac{\alpha+\gamma}{2} & \frac{\gamma-\alpha}{2} & 0 \\ \frac{\gamma-\alpha}{2} & \frac{\alpha+\gamma}{2} & 0 \\ 0 & 0 & \beta \end{pmatrix}$$

If a_0; a, b, c are the lattices parameters respectively of cubic and orthorhombic phases, the non-dimensional constants α, β, γ can be written as :

$$\alpha = \frac{a}{a_0\sqrt{2}} \qquad \beta = \frac{b}{a_0} \qquad \gamma = \frac{c}{a_0\sqrt{2}} \qquad (2)$$

The interface compatibility equation is :

$$\tilde{\underline{R}}\left[\lambda \underline{R}\underline{U}_i + (1-\lambda)\underline{U}_j\right] - \underline{I} = \vec{b} \otimes \vec{m} \qquad (3)$$

where $\underline{R}$ is the twin rotation, $\tilde{\underline{R}} \in SO_3$, $\vec{b}$ is the shape strain vector and $\vec{m}$ the unit normal to the austenite phase; the thickness of the two variants $\underline{R}\underline{U}_i$ and $\underline{U}_j$ are proportional respectively to λ and (1-λ).

On the other hand, it is observed that in a Cu-25.63Zn-4.2Al (W^t %) single crystal [Vivet and Lexcellent, 1999] the martensitic region does not consist of twins, but of an "untwinned" martensite, as explained by Hane [Hane, 1999].

For these transformations, an undeformed interface exists between austenite and a single variant of martensite. The single variant of martensite created in the mother phase with an "exact" austenite-martensite interface is obtained under a radial loading ($\sigma_1/\sigma_2 = \text{Constant}$) for a biaxial tensile test [Vivet, 1999] (Figure 2). With each loading axis, the variant which appears among the twelve possible ones corresponds to the maximum of shearing. The theoretical analysis of Hane [Hane, 1999] is useful in the following sense: if the lattice parameter a_0 of the DO_3 austenite and the lattice parameters (a, b, c, θ) of the monoclinic martensite are measured, for this Cu-Zn-Al (and also some Cu-Al-Ni, Cu-Zn-Ga, ...), the shape strain and the normal vector to the habit plane can be predicted in agreement with those obtained with the Weschler, Liebermann and Read (WLR) theory. But, it is no longer necessary to introduce a shear parameter k in order to achieve compatibility between austenite and a single variant of martensite. This shear k is a means by which the shuffles in the unit cells of the long-period stacking ordered structures can be obtained [Otsuka et al., 1976]. The "exact" interface is obtained between the classical DO_3 austenite cell and a new unit cell called $6M_1$. The term "untwinned martensites" which is used more often than "faulted martensites", has been proposed by Hane [Hane, 1999].
The simplest austenite-martensite microstructure is two adjacent regions, one with gradient on the austenite energy well and the other with gradient on a single energy well. The compatibility equation with the variant $\underline{U}_i$ is :

$$\underline{R}\underline{U}_i - \underline{I} = \vec{b} \otimes \vec{m} \qquad (4)$$

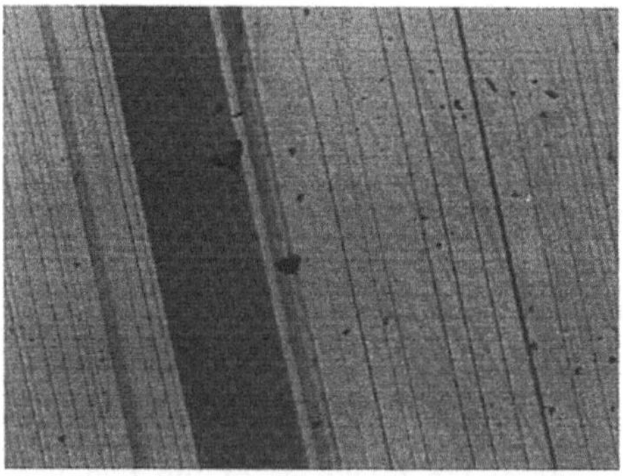

Figure 2; *Pseudoelastic biaxial tensile test on a Cu-Zn-Al single crystal with $\sigma_1/\sigma_2 = 7.25$ and $\sigma_1 = 110.6 MPa$ (T > Martensite start)[Vivet, 1999]*

The mathematical theory of martensite transformation is useful also for interpreting other observations, such as the presence of twin or untwinned martensites, the geometry of wedge triangles and diamond micro-structure interfaces. But if external loads are applied, the minimisation of the total free energy which is the starting point of that theory leads to a difficult non-linear elasticity problem [James and Hane, 2000]. An attempt to solve the case of a thin film constrained on its lateral boundaries has been performed by Shu and Bhattacharya [Shu and Bhattacharya, 1998]. These authors examine the link between the microstructure as "untwinned martensites" or the texture in a polycrystal and the ability to obtain efficient recovery strains. Certain microstructures can be built in thin film specimens used in the design of novel actuator devices. Finally, textures can be suggested for improved shape-memory effect.

3. PHENOMENOLOGICAL MODELLING

Thermomechanical behaviour is often viewed in the framework of generalised standard materials using the average volume fraction of martensite z for macroscopic scale on single crystals or polycrystals or volume fraction of martensite variants z_i; i=1,..., N for "meso-macro" approach on single crystal and the mean transformation strain over the volume of martensite as internal variables.
The expression of the specific Helmholtz global free energy is nearly the same in the different investigations

$$\varphi(\underline{\varepsilon}_e, T, z) = \varphi_{el} + \varphi_{th} + \varphi_{ch} + \varphi_{int} \qquad (5)$$

where φ_{el} represents the elastic part of φ. Its expression is:

$$\varphi_{el} = \frac{1}{2\rho} {}^T\underline{\varepsilon}_e \underline{\underline{L}} \underline{\varepsilon}_e \qquad (6)$$

This form is obtained by supposing that the elastic constants ($\underline{\underline{L}}$ is the generalised Hooke's tensor) and the mass density ρ of the austenite and the martensite are the same. Experimental measures show that the Young modulus of austenite is in fact between two or three times the one of the martensite but as the elastic deformation $\underline{\varepsilon}_e$ remains very small in comparison with the deformation associated with the martensitic transformation, this hypothesis appears to be reasonable.

The thermal part φ_{th} is expressed in the following classical way:

$$\varphi_{th} = c_v \left[(T - T_0) - T \ln \frac{T}{T_0} \right] \qquad (7)$$

The specific heat c_v is supposed to be independent of the phase state. T_0 is a reference temperature.

The chemical part φ_{ch} can be taken as (B is a material parameter):

$$\varphi_{ch} = B(T - T_0)z \quad (8)$$

The choice of the expression of the dissipation part φ_{int} remains a more opened problem. It depends on interface and multiple interaction effects between martensite platelets and the mother phase and also between the platelets themselves.

Within a macroscopic approach, the polycristalline microstructure is not taken into account because only the volume fraction of martensite z is considered. The distribution of the martensite variants inside a grain and the grain boundary interactions are not examined. The choice:

$$\varphi_{int} = z(1 - z)\varphi_{it} \qquad (9)$$

(i) with $\varphi_{it} > 0$ for stress induced phase transformation austenite <-> martensite in the case of single crystal pseudoelasticity when only one variant is created (see for instance the Cu-Zn-Al single crystal [Huo and Muller, 1993] and for the Ti-Ni or Cu-Zn-Al polycrystal [Raniecki et al., 1992][Raniecki and Lexcellent, 1994] (φ_{int} is non convex on z).

(ii) with $\varphi_{it} < 0$ in the special case of cubic to trigonal transformation austenite <->Rphase for Ti-Ni [Lexcellent et al., 1994] is efficient (φ is convex on z).

proves to be.

The following choice of the form of φ_{int} recently proposed by Qidway and Lagoudas [Qidway and Lagoudas, 2000]:

$$\varphi_{int} = \begin{cases} \dfrac{1}{3}\rho b_3^M z^3 + \dfrac{1}{2}\rho b_2^M z^2 + (\mu_1 + \mu_2)z & \text{if} \quad \dot{z} > 0 \\ \dfrac{1}{3}\rho b_3^A z^3 + \dfrac{1}{2}\rho b_2^A z^2 + (\mu_1 - \mu_2)z & \text{if} \quad \dot{z} < 0 \end{cases} \qquad (10)$$

where b_i^M, b_i^A (i = 2, 3) and μ_i (i = 1,2) are material parameters, is not consistent with the local state postulate and cannot be used. The choice of the free energy expression does not depend on the direction of the phase transformation as it is also made by Kamita and Matsuzaki [Kamita and Matsuzaki, 1998].

In order to extend the modelling to the case of non-isothermal behaviour i.e. the recovery strain or stress, the volume fraction of martensite z is split into two components: the self accommodated one z_T obtained under pure thermal action and the untwinned one z_σ obtained under mechanical loading [Brinson, 1993]: $z = z_\sigma + z_T$; $0 \leq z, z_\sigma, z_T \leq 1$. With these two quantities chosen as internal variables a new form of the energy component φ_{int} has been proposed by Leclercq and Lexcellent [Leclercq and Lexcellent, 1996]:

$$\varphi_{int} = z(1-z)\varphi_{it} + z_\sigma z_T \varphi_{it}^{m} \qquad (11)$$

Examining the behaviour of a grain, Siredey et al [Siredey et al., 1999] propose the following form of φ_{int} using the volume fraction z_i of each variant i of martensite as internal variables (meso-macro investigation):

$$\varphi_{int} = \frac{1}{2}\sum_n \sum_m H^{nm} z_n z_m \qquad (12)$$

where H^{nm} constitutes the interaction matrix. φ_{int} in this case is chosen convex in the internal variable z_i.

In a classical way, the shape memory material is considered as a "plastic" one but with different behaviours in loading and unloading. The process of stress induced phase transformation can be controlled by the addition to the specific free energy of two yield functions: $\psi^{(1)}$ for the forward transformation austenite -> martensite and $\psi^{(2)}$ for the reverse one martensite -> austenite with the consistency conditions $\dot{\psi}^{(1)} = 0$ and $\dot{\psi}^{(2)} = 0$.

For the process of martensite reorientation (self accommodated martensite M_T -> untwinned martensite M_σ), a very classical model for plasticity can be used [Roumagnac et al., 2000]

In another way, the examination of the intrinsic dissipation part D_i in the Clausius-Duhem inequality can be undertaken. In the case of pseudoelasticity (stress induced phase transformation) with only one internal variable z, we establish that :

$$D_i = \Pi^f \dot{z} \geq 0 \qquad (13)$$

Π^f and $\dot{z}$ form a couple of conjugate variables.

In a somewhat restrictive hypothesis Peyroux et al. [Peyroux et al., 1996] postulate that the intrinsic dissipation is negligible in regard to the latent heat, it means that $\Pi^f = 0$ when phase transformation occurs ($\dot{z} \neq 0$). One has to note that the slope in the stress-strain diagram of $\Pi^f = 0$ is positive. The hysteresis is taken into account by the local temperature effect calculated using the heat equation. (forward phase transformation austenite -> martensite is exothermal whereas reverse phase transformation martensite -> austenite is endothermal).

The choice of the form of the intrinsic dissipation D_i is opened. As an example, in the modelling of grain pseudoelastic behaviour within a meso-macro approach using z_1, ..., z_N as internal variables, it can be established that [Lexcellent et al., 1996]:

$$D_i = \sum_{s=1}^{N} \Pi_s^f \dot{z}_s = \sum_{s=1}^{N} \alpha^{tr} (z_s - z_s^m) \dot{z}_s \qquad (14)$$

and hence the criterion for forward and reverse phase transition of each variant s is finally defined.
α^{tr} is a coefficient and z_s^m is the memory volume fraction. z_s^m is defined as the value attained by z_s just before inversion of the loading.

In the frame of this modelling, a crucial point is the choice of the expression of the transformation strain rate. For a phenomenological approach at the macroscopic scale, a common expression is in the spirit of Von-Mises:

$$\underline{\dot{\varepsilon}}^{tr} = \underline{k}\dot{z} \qquad (15)$$

with:

$$\underline{k} = \frac{3}{2}\frac{\text{dev}\underline{\sigma}}{\bar{\sigma}}; \qquad \bar{\sigma} = \left(\frac{3}{2}\text{dev}\underline{\sigma} : \text{dev}\underline{\sigma}\right)^{1/2} \qquad (16)$$

In order to take into account the asymmetry between tension and compression the $\underline{k}$ expression must be changed:

$$\underline{k} = \frac{3}{2}\frac{\text{dev}\underline{\sigma}}{\bar{\sigma}} + \frac{a}{\bar{\sigma}^2}\left\{\bar{\sigma}\left[(\text{dev}\underline{\sigma})^2 - \frac{2}{9}\bar{\sigma}^2\underline{I}\right] - 3\det(\text{dev}\underline{\sigma})\text{dev}\underline{\sigma}\right\} \qquad (17)$$

Qidway and Lagoudas [Qidway and Lagoudas, 2000] add the eventual influence of the isotropic part of the stress tensor ($\text{tr}\underline{\sigma}\underline{I}$) on the phase transformation (i.e. the pressure effect). At this day, it constitutes an opened problem as the triaxial experiments of Gall et al [Gall and al., 1996] are not totally conclusive.

4. KINEMATICS FOR FINITE STRAIN ANALYSIS

It is well known that the pseudoelastic behaviour of shape memory materials is characterised by large recoverable strains. Even so, constitutive models based on small deformation theory have been developed and successfully used. Nevertheless, this approximation becomes inadequate when large displacements and rotations occur.

To built constitutive models for finite strain calculations, we introduce the following decomposition of the deformation gradient $\underline{F}$:

$$F = \underline{V}^e \underline{Q}\underline{F}^{tr} \quad (18)$$

The tensor $\underline{F}^{tr}$ represents the deformation gradient associated with the martensitic transformation. It connects a reference configuration to a particular configuration C_Q where a frame

attached to the material substructure preserves its initial orientation. It is the case, for example, if we choose a direction along a habit plane. During deformation, this frame is rotated by the rotation $\underline{Q}$ ($\underline{Q}\underline{Q}^T = \underline{I}; \det\underline{Q} = 1$) to an intermediate configuration from which a pure elastic strain $\underline{V}^e$ is finally applied. This decomposition is based on the fact that the martensite and the austenite have the same elastic parameters as supposed previously. In the case of small elastic strain ($\underline{V}^e = I + \underline{\varepsilon}_e; \|\underline{\varepsilon}_e\| \ll 1$), the velocity gradient $\underline{L} = \underline{\dot{F}}\underline{F}^{-1}$ leads to the following decomposition of the strain rate tensor $\underline{D}$ and the material spin tensor $\underline{W}$:

$$\begin{aligned} \underline{D} &= D_Q\underline{\varepsilon}_e + \underline{Q}[\underline{\dot{F}}^{tr}\underline{F}^{tr-1}]^S\underline{Q}^T \\ \underline{W} &= \underline{\dot{Q}}\underline{Q}^T + \underline{Q}[\underline{\dot{F}}^{tr}\underline{F}^{tr-1}]^A\underline{Q}^T \end{aligned} \quad (19)$$

$D_Q\underline{\varepsilon}_e$ is the rotational derivative of $\underline{\varepsilon}_e$ with respect to the rotation $\underline{Q}$.

Equation (15.a) expresses the additive decomposition of the strain rate tensor in an elastic part $D_Q\underline{\varepsilon}_e$ and a part $\underline{Q}[\underline{\dot{F}}^{tr}\underline{F}^{tr-1}]^S\underline{Q}^T$ associated with the martensitic transformation. This equation gives in the configuration C_Q:

$$(\underline{\dot{\varepsilon}}_e)_Q = \underline{d}_Q - [\underline{\dot{F}}^{tr}\underline{F}^{tr-1}]^S \quad (20)$$

with $(\dot{\underline{\varepsilon}}_e)_Q = d\left(\underline{Q}^T \underline{\varepsilon}_e \underline{Q}\right)/dt$ and $\dot{\underline{d}}_Q = \underline{Q}^T \underline{D} \underline{Q}$. $\underline{D}$ is defined as being the rotational derivative of the strain measure $\underline{d}$ with respect to $\underline{Q}$.

Equation (15.b) is a particularity of finite strains. If the material has no preferred orientations, $\underline{\dot{Q}Q}^T$ can be considered as the spin of a well chosen frame. The tensor $\underline{Q}[\dot{\underline{F}}^{tr}\underline{F}^{tr-1}]^A\underline{Q}^T$ represents then the velocity of the medium with respect to this frame. An objective frame family is given by ($\alpha \in [0,1]$):

$$\underline{\dot{Q}Q}^T = (1-\alpha)\underline{W} + \alpha\underline{\dot{R}R}^T \qquad (21)$$

Where $\underline{R}$ is the proper rotation derived from the polar decomposition of the deformation gradient $\underline{F}$.

Constitutive equations of the form of those developed under small deformation hypothesis can then be established in the configuration C_Q supposing that the state of the material is completely defined by the elastic strain $(\underline{\varepsilon}_e)_Q = \underline{Q}^T \underline{\varepsilon}_e \underline{Q}$, the volume fraction of martensite z and the temperature T [Boubakar et al., 1999].

5. CONCLUSION

Referring to the large number of publications, the thermomechanical modelling of shape memory alloys seems to be a very exciting challenge. On one hand, there is the very smart mathematical theory of martensite transformation with limited capacity for engineering applications. On the other hand, the modelling based on the thermodynamics of irreversible processes seems to be more predictive (one-way shape memory effect, pseudoelasticity, recovery strain or stress under anisothermal loading). But nowadays, the gap between the two approaches is obvious. Perhaps, the work of Roytburd and Slutsker [Roytburd and Slutsker, 1999] i.e. a geometrical linear theory, can constitute a link between the two investigations. However, some real problems such as the non- proportional loading paths modelling are not yet solved. In fact the coupling between the creation of martensite platelets and their reorientation under stress action, must be investigated. Moreover, modelling the training of shape memory alloys i. e. the two way memory effect constitutes an interesting goal.

Nota: The research field concerning shape memory alloys modelling is very dense. Many important papers are not quoted in this brief synthesis. The authors present their apologies in advance for this restriction.

REFERENCES

[Abeyaratne and Knowles, 1993] Abeyaratne, R.; Knowles, J. K.; "A continuum model of a thermoelastic solids capable of undergoing phase transitions"; In: *Journal of the Mechanics and Physics of Solids*, pp. 541-571, 41.

[Ball and James, 1987] Ball, J. M.; James, R. D.; "Fine phase mixtures as minimizers of energy"; In: *Archive for Rational Mechanics and Analysis*, pp. 13-52, 100.

[Ball and James, 1992] Ball, J. M.; James, R. D.; "Proposed experimental tests of a theory of fine microstructure and the two well problems"; In: *Philosophical Transaction of the Royal Society of London*, pp. 389-450, A338.

[Bhattacharya and Kohn, 1996] Bhattacharya, K.; Kohn, H. V.; "Symmetry, texture and recoverable strain of shape memory polycristals"; In: *Acta. Materialia and Metallurgica*, pp. 529-542, 44.

[Boubakar et al, 1999] Boubakar, L.; Moyne, S.; Lexcellent, Ch.; Boisse, Ph.; "SMA pseudoelastic finite strains. Theory and numerical application"; In: *J. of Eng. Mat. and Technology*, pp. 44-47, 121.

[Brinson, 1993] Brinson, L. C.; "One dimensional constitutive behaviour of shape memory alloys. Thermomechanical derivation with non consistent material function"; In: *J. of Intell. Mat. Syst. and Structures*, pp. 229-242, 4(2).

[Chu, 1993] Chu, C. H.; "Hysteresis and microstructures. A study of biaxial loading on compound twins of copper-aluminium-nickel single crystals"; PhD thesis of University of Minnesota, USA.

[Chu and James, 1995] Chu, C.; James, R. D.; "Analysis of microstructures in Cu 14.0% Al 3.9% Ni by energy minimisation"; In: *Journal de Physique III Colloque*, pp. 143-149, 5, C8.

[Gall et al, 1998,] Gall, K.; Sehitoglu, H.; Maier, H. J.; Jacobus, K.; "Stress induced martensitic phase transformation in polycristalline CuZnAl shape memory alloys under different stress states"; In: *Met. and Mat. Trans. A.*, pp. 765-773, 29A.

[Hane, 1999] Hane, K. F.; "Bulk and thin film microstructures in untwinned martensites"; In: *J. Mech. Phys. Solids*, pp. 1917-1939, 47.

[Huo and Muller, 1993] Huo, Y.; Muller, I.; "Non equilibrium thermodynamics of pseudoelasticity"; In: *Continum Mech. Thermodynamics*, pp. 163-204, 5.

[James and Hane, 1999] James, R. D.; Hane, K. F.; "Martensitic transformations and shape memory materials"; To appear in *Millenium Special Issue of Acta Materialia.*

[Kamita and Matsuzaki, 1998] Kamita, T.; Matsuzaki, Y.; "One dimensional pseudoelastic theory of shape memory alloys"; In: *Smart Mater. Struct.*, pp. 489-495, 7.

[Leclercq and Lexcellent, 1996] Leclercq, S.; Lexcellent, Ch.; "A general macroscopic description of the thermomechanical behaviour of shape memory alloys"; In: *J. Mech. Phys. Solids*, pp. 953-980, 49,6.

[Lexcellent et al., 1994] Lexcellent, Ch.; Tobushi, H.; Ziolkowski, A.; Tanaka, K.; "Thermodynamical model of reversible R-phase transformation in NiTi shape memory alloy"; In: *Journal of Pressure Vessels and Piping*, pp. 51-57, 58.

[Otsuka et al., 1993] Otsuka, K.; Ohba, T.; Tokonami, M.; Wayman, C. M.; "New description of long period stacking order structures of martensite in β-phase alloys"; In: *Scripta Metallurgica et Materialia*, pp. 1359-1364, 29.

[Patoor et al., 1993] Patoor, E.; Eberhadt, A.; Berveiller, M.; "Micromechanical modelling of superelasticity in shape memory alloys"; In: *Pitman Research Notes in Mathematical Series*, pp. 38-54, 296.

[Peyroux et al., 1996] Peyroux, R.; Chrysochos, A.; Licht, Ch.; Lobel, M.; "Phenomenological constitutive equation for numerical simulation of SMA's structures. Effects of thermomechanical couplings"; In: *J. de Physique IV*, pp. 347-356, 6, C1.

[Qidway and Lagoudas, 2000] Qidway, M. A.; Lagoudas, D. C.; "On thermomechanics and transformation surfaces of polycrystalline shape memory materials"; To appear in *Int. J. of Plasticity*.

[Raniecki et al., 1992] Raniecki, B.; Lexcellent, Ch.; Tanaka, K.; "Thermodynamic model of pseudoelastic behaviour of shape memory alloys"; In: *Arch. Mech.*, pp. 261-284, 44,3.

[Raniecki and Lexcellent, 1994] Raniecki, B.; Lexcellent, Ch.; "R_L models of pseudoelasticity and their specification for some shape memory solids"; In: *Eur. J. Mech. A/Solids*, pp. 21-50, 12, 1.

[Raniecki and Lexcellent, 1998] Raniecki, B.; Lexcellent, Ch.; "Thermodynamics of isotropic pseudoelasticity in shape memory alloys"; In: *Eur. J. Mech. A/Solids*, pp. 185-205, 17, 2.

[Roumagnac et al., 2000] Roumagnac, P.; Guillemer-Neel, C.; Saanouni, K.; Clavel, M.; "Mechanical behaviour and deformation mechanisms of NiTi shape memory alloys in tension"; To appear in *European Physical Journal of Applied Physics.*

[Roytburd and Slutsker, 1999] Roytburd, A. L.; Slutsker, J.; "Deformation of adaptive materials. Part I: Constrained deformation of polydomain crystals"; In: *J. Mech. Phys. Solids*, pp. 2299-2329, 47.

[Shu and Bhattacharya, 1998] Shu, Y. C.; Battacharya, K.; "Texture and recoverable strain in bulk and thin films shape memory polycristals"; In: *Proceedings of ICOMAT '98*, To appear in *Mat. Sci. And Eng.*

[Siredey et al., 1999] Siredey, N.; Patoor, E.; Berveiller, M.; Eberhadt, A.; "Constitutive equations for polycristalline thermoelastic shape memory alloys. Part I: Intergranular interaction and behaviour of the grain"; In: *Int. Jour. Of Solids and Structures*, pp. 4289-4315, 36.

[Sun et al., 1993] Sun, Q. P.; Hwang, K. C.; "Micromechanic modelling for the constitutive behaviour of polycristalline shape memory alloys", Part I and II; In: *J. Mech.Phys. Solids*, pp. 1-17 and 19-33, 41.

[Vivet and Lexcellent, 1999] Vivet, A.; Lexcellent, Ch.; "Observations and analysis of martensitic phase transformation on CuZnAl single crystal"; In: *Journal de Physique IV Colloque*, pp. 411-418, Pr9.

[Vivet, 1999] Vivet, A.; "Approche expérimentale du comportement pseudoélastique des alliages à mémoire de forme. Modélisation isotherme et anisotherme"; PhD Thesis of Franche-Comté University, France, n° 778.

Thermo-Mechanical Modelling of Nematic Polymers

Daniel Lhuillier
Laboratoire de Modélisation en Mécanique, UPMC-CNRS
8 rue du Capitaine Scott, case 162, 75015 Paris, France
lhuilier@ccr.jussieu.fr

Abstract : A new thermo-mechanical framework is presented, able to deal with internal state variables obeying a statistical distribution. That framework is used for modelling flowing nematic polymers. In fact, we extend the *Doi* model, allowing the mean-field potential acting on each molecule to depend on the gradients of orientation. The link with the *Ericksen-Frank-Leslie* model is established without ambiguities, and we stress on the differences of dynamic behaviour implied by the two models.

Keywords : Liquid Crystals, Polymer Melts, Smoluchovski Equation, Thermodynamics of Irreversible Processes.

1. INTRODUCTION

As high strength materials, liquid crystalline polymers have a considerable technological importance, but one still lacks of a theoretical model describing the flow of these complex fluids. The issue is to estimate the degree of orientation order, its modifications by various flows, and its feedback on the overall stress of the material. The well-known *Ericksen-Frank-Leslie* (EFL) model (De Gennes and Prost 1993) assumes a perfect and undisturbed order, and represents it by a unit vector, the so-called director giving the locally preferred orientation. That model is generally thought to hold when the orientation stays close to a uni-axial equilibrium distribution, and that means for relatively slow flows. The *Doi* model is more recent (Doi 1981, Doi and Edwards 1986) : instead of assuming a more or less frozen order, a probability distribution is introduced for the orientation and one considers its modifications in various flows. The *Doi* model also assumes that the interaction between neighbour molecules can be represented by a mean-field potential acting on each molecule and depending on its orientation. However, there is no equivalent of the Frank elastic energy, which means that the energy increase due to a spatially varying distribution of orientation is not taken into account. We propose to amend the *Doi* model by supposing the mean-field potential to depend not only on the orientation, but also on the orientation gradient. And we want to compare the predictions of that extended model with those of the EFL model. For that purpose, we will use a new thermo-mechanical approach, based on the well-known thermodynamics of irreversible processes, but enlarged so as to cope with internal variables of state distributed according to some probability law. General results

G.A. Maugin et al. (eds.), Continuum Thermomechanics, 237-246.
© 2000 *Kluwer Academic Publishers.*

for the entropy balance of liquid crystals are presented in section 2. The EFL model is presented in a somewhat unconventional way in section 3. The proposed extension is developed in section 4, followed by some concluding remarks and a comparison with previous work in the last section.

2. THE GENERAL ENTROPY BALANCE OF LIQUID CRYSTALS

Liquid crystals obey the usual conservation laws for mass, momentum and total energy,

$$d\rho/dt + \rho\, \nabla.\mathbf{v} = 0, \qquad \rho\, d\mathbf{v}/dt = \nabla.\sigma + \rho\mathbf{g}\,, \qquad \rho\, de^*/dt = \nabla.(\sigma.\mathbf{v} - \mathbf{Q}) + \rho\mathbf{g}.\mathbf{v}\ ,$$

where $d/dt = \partial/\partial t + \mathbf{v}.\nabla$ is the convective time-derivative. One of the peculiarities of liquid crystals is the presence of internal body couples and the absence of any intrinsic angular momentum. This means the stress tensor σ is not symmetric but the density of angular momentum is simply $\mathbf{r} \times \rho\mathbf{v}$ without any additional term, and the total energy per unit mass is $e^* = e + v^2/2$ where the internal energy e does not include any spin kinetic energy. For angular momentum to be conserved, one is led to introduce *couple stresses* represented by a third-order tensor **t**, anti-symmetric in the exchange of its last two indices, i.e. $t_{ijk} = - t_{ikj}$. These couple stresses enter the expressions of the stress tensor and the energy flux as

$$\sigma_{ij} = \sigma^s_{ij} - \partial t_{kij}/\partial x_k \qquad \text{and} \qquad Q_i = q_i - t_{ijk}\,\Omega_{kj}\ , \tag{2.1}$$

where σ^s is the symmetric part of the stress tensor, **q** is the heat flux and Ω is the fluid rotation rate. As a result, the balance of internal energy becomes

$$\rho\, de/dt = \sigma^s : \mathbf{D} + t_{kji}\,\partial\Omega_{ij}/\partial x_k - \nabla.\mathbf{q}\ ,$$

where **D** is the fluid strain rate ($\Omega_{ij} + D_{ij} = \partial v_i/\partial x_j$). The free enthalpy of a liquid crystal is a function of its pressure p, of its temperature T and of the extra variables describing the orientation order. The free enthalpy per unit mass is the chemical potential μ. The Gibbs relation of liquid crystals can be presented as

$$de = T\,ds - p\,d(1/\rho) + d\mu\,|_{p,T}\ ,$$

where s is the entropy per unit mass and $\mu\,|_{p,T}$ depends on the orientational variables only. That Gibbs relation holds not only at equilibrium but also out of equilibrium provided one considers it in the frame moving with velocity **v** (De Groot and Mazur 1962). As a consequence, one deduces the general form of the entropy balance of a liquid crystal

$$T\,(\rho\, ds/dt + \nabla.\mathbf{S}) = $$

$$(\sigma^s + pI) : \mathbf{D} + t_{kji}\, \partial\Omega_{ij}/\partial x_k - \mathbf{S}.\nabla T + \nabla.(T\mathbf{S} - \mathbf{q}) - \rho\, d\mu\,|_{p,T}/dt \;, \qquad (2.2)$$

where **S** is the entropy flux and ***I*** is the unit tensor. The right-hand side of this equation represents the entropy production rate (multiplied by T). For it to appear in the expected form of a sum of fluxes times thermodynamic forces, one needs a detailed expression for the time derivative of $\mu\,|_{p,T}$, in other words one needs to know what are the orientation variables entering $\mu\,|_{p,T}$, together with their evolution in time.

3. THE ERICKSEN- FRANK-LESLIE MODEL

The EFL model depicts the orientation order by a unit vector, the director **n**. A distortional energy F^d represents the energy penalty when distortions occur in the orientation field $\mathbf{n}(\mathbf{r},t)$. F^d depends on **n** itself, but is primarily a function of its gradient. It is related to the chemical potential and its time derivative by

$$\rho\,\mu\,|_{p,T} = F^d(n,\nabla n) \qquad \text{and} \qquad \rho d\mu\,|_{p,T}/dt = dF^d/dt + F^d\,\nabla.\mathbf{v}\,.$$

There is no difficulty when calculating dF^d/dt in terms of $d\mathbf{n}/dt$, but to present the result in a simple form one needs to introduce the following four definitions :

$$\pi^d_{ji} = \partial F^d/\partial(\partial n_i/\partial x_j)\;, \qquad \sigma^d_{ji} = -\,\pi^d_{jk}\,\partial n_k/\partial x_i\;, \qquad t^d_{kji} = (1/2)\,(\pi^d_{ki}n_j - \pi^d_{kj}n_i)\;,$$

$$h_i = -\,\partial F^d/\partial n_i + \partial\pi^d_{ji}/\partial x_j\;.$$

One then obtains

$$\rho d\mu\,|_{p,T}/dt = F^d\,\nabla.\mathbf{v} + \sigma^{ds} : \mathbf{D} + t^d_{kji}\,\partial\Omega_{ij}/\partial x_k$$

$$-\,\mathbf{h}.(d\mathbf{n}/dt - \Omega\times\mathbf{n}) + \nabla.[\,\pi^d.(d\mathbf{n}/dt - \Omega\times\mathbf{n})] \;, \qquad (3.1)$$

together with the rotational identity that must be satisfied by any scalar expression for F^d (see eq. 3.110 of De Gennes and Prost 1993)

$$(\sigma^d_{ji} - n_j h_i)^a + \partial t^d_{kji}/\partial x_k = 0\,. \qquad (3.2)$$

where the superscripts s and a stand for the symmetric and anti-symmetric part of a tensor respectively. Reporting the result (3.1) into the entropy balance (2.2), one deduces that the heat flux is related to the entropy flux by

$$\mathbf{q} = T\,\mathbf{S} - \pi^d.(d\mathbf{n}/dt - \Omega\times\mathbf{n}) \;, \qquad (3.3)$$

while the right-hand side of the entropy balance becomes

$$[\sigma^s - \sigma^{ds} + (p-F^d)I] : \mathbf{D} + \mathbf{h}.(d\mathbf{n}/dt - \Omega\times \mathbf{n}) + (t_{kji} - t^d_{kji})\partial\Omega_{ij}/\partial x_k - \mathbf{S}.\nabla T \ .$$

At this point, the consequences of choosing **n** as the orientation variable of state have been fully taken into account. What is needed now is the evolution equation for **n**. That equation is generally written as

$$d\mathbf{n}/dt - \Omega\times \mathbf{n} = (I - \mathbf{nn}).(\lambda\mathbf{D}.\mathbf{n} + \gamma\mathbf{h}) \tag{3.4}$$

where $I - \mathbf{nn}$ is the projector on a plane orthogonal to **n**, while λ and γ are two phenomenological coefficients. Taking that evolution equation into account, and defining the viscous stress σ^v and the viscous couple stress $\mathbf{t}^v$ by

$$\sigma^s = \sigma^v + \sigma^{ds} - (p-F^d)I - \lambda(\mathbf{nh})^s + \lambda\mathbf{nn}\,\mathbf{n}.\mathbf{h} \quad \text{and} \quad \mathbf{t} = \mathbf{t}^v + \mathbf{t}^d \ , \tag{3.5}$$

the final expression for the entropy balance appears as

$$T(\rho\, ds/dt + \nabla.\mathbf{S}) = \sigma^v : \mathbf{D} + t^v_{kji}\,\partial\Omega_{ij}/\partial x_k + \gamma(\mathbf{n}\times\mathbf{h})^2 - \mathbf{S}.\nabla T \ . \tag{3.6}$$

As a consequence, the coefficient γ must be positive. The overall stress tensor of the nematic liquid crystal is related to the viscous stress and viscous couple stress as

$$\sigma = \sigma^s - \nabla.\mathbf{t} = -(p-F^d)I + \sigma^d - (\mathbf{nh})^a - \lambda(\mathbf{nh})^s + \lambda\mathbf{nn}\,\mathbf{n}.\mathbf{h} + \sigma^v - \nabla.\mathbf{t}^v \tag{3.7}$$

where the invariance relationship (3.2) has been taken into account. The viscous couple stress $\mathbf{t}^v$ has always been neglected in previous studies. A detailed expression for σ^v, taking the uni-axial symmetry into account can be found in e.g. De Gennes and Prost 1993. It is noteworthy that many authors have included the non-dissipative stresses involving λ as well as the anti-symmetric stress $(\mathbf{nh})^a$ in their definition of the viscous stress. This is rather arbitrary and should be avoided. Moreover, the simultaneous presence of λ in results (3.4) and (3.7) is less a manifestation of the Onsager symmetry than a (more trivial) condition of elimination of λ from the entropy production.

4. THE EXTENDED DOI MODEL

As in the original Doi model [Doi 1981, Doi and Edwards 1986], a unit vector **u** depicts the orientation of an individual polymer molecule. The orientation order is characterized by the statistical distribution $N(r,t,u,\nabla u)$ giving the number density of macromolecules with an orientation **u** in a local environment described by $\nabla\mathbf{u}$. Note that N is not a probability as it was in the original Doi's model, but that it is related to the number density n of polymers as

$$n(r,t) = \int N\, dC \ ,$$

where $dC = d\mathbf{u}\ d\nabla\mathbf{u}$ is the elementary volume of the configuration state space. To simplify the issue, we will suppose the macromolecules to move all with the same velocity $\mathbf{v}$, and will write the conservation law for the density distribution as

$$\partial N/\partial t + \nabla . N\mathbf{v} + (\partial/\partial\mathbf{u}).(N d\mathbf{u}/dt) + \partial/\partial(\nabla\mathbf{u}):(N d\nabla\mathbf{u}/dt) = 0 \quad . \qquad (4.1)$$

Integrating over the configuration phase-space, one recovers the expected conservation law for the polymers number density

$$\partial n/\partial t + \nabla . n\mathbf{v} = 0 \ .$$

A mean-field energy is defined for each macromolecule and written as $U(u,\nabla u)/2$. The fluid energy depends on the mean value of U, but there is also an entropic contribution coming from the distribution N itself, and finally, the chemical potential of the liquid crystalline polymer appears as

$$\rho\, \mu\, |_{p,T} = \int [U/2 + kT \operatorname{Log} N]\ N\, dC \ , \qquad (4.2)$$

where k is the Boltzmann constant. The trick when calculating the time-derivative of μ (we henceforth delete the $|_{p,T}$ symbol) is to proceed in two steps. First, we consider μ as a function of $\mathbf{u}$ and ∇u, and we express $d\mu/dt$ as a function of $d\mathbf{u}/dt$ and of the statistical distribution N. In a second step, we use the evolution equation of $\mathbf{u}$ and get the final expression for the time-derivative appearing in the r.h.s. of (2.2).

Taking the conservation equation (4.1) into account and after integration by parts, one deduces from (4.2)

$$\rho\, d\mu/dt = -nkT\, \nabla.\mathbf{v} + \int N\, [\, \partial U^*/\partial\mathbf{u} \,.\, d\mathbf{u}/dt + \partial U^*/\partial(\nabla\mathbf{u}) \ :\ d(\nabla\mathbf{u})/dt\,]\, dC \ ,$$

with U^* defined as

$$U^* = U + kT \operatorname{Log} N \ . \qquad (4.3)$$

We now change the order of the derivatives d/dt and ∇ in the second term between brackets (a non-commutative operation !) and we replace everywhere $d\mathbf{u}/dt$ by the co-rotational derivative $d\mathbf{u}/dt - \boldsymbol{\Omega}\times\mathbf{u}$ because $d\mu/dt$ must be a Galilean invariant. We then get an expression for $\rho d\mu/dt$ which contains a contribution depending linearly on the rotation rate $\boldsymbol{\Omega}$. That contribution must disappear because $\rho\ d\mu/dt$ is a part of the entropy production rate which cannot display terms linear in $\boldsymbol{\Omega}$ (but terms proportional to the gradient of $\boldsymbol{\Omega}$ are allowed). As a result, we obtain the rotational identity

$$\int N\, (\Sigma^*_{ji} - u_j H^*_i)^a\, dC + \partial/\partial x_k \int N\, T^*_{kji}\, dC = 0 \quad , \qquad (4.4)$$

with the definitions

$$\Pi^*_{ji} = \partial U^*/\partial(\partial u_i/\partial x_j) \quad , \qquad H^*_i = -\,\partial U^*/\partial u_i + (1/N)\partial(N\Pi^*_{ji})/\partial x_j$$

and (4.5)

$$\Sigma^*_{ji} = -\,\Pi^*_{jk}\,\partial u_k/\partial x_i \;, \qquad T^*_{kji} = (1/2)\,(\Pi^*_{ki}u_j - \Pi^*_{kj}u_i) \;.$$

The rotational identity (4.4) being satisfied, $\rho d\mu/dt$ appears as

$$\rho\, d\mu/dt = -\,nkT\,\nabla.\mathbf{v} + \mathbf{D}:\int N\,(\Sigma^*)^s\,dC + \partial\Omega_{ij}/\partial x_k \int N\,T^*_{kji}\,dC$$
$$-\int N\,\mathbf{H}^*.(d\mathbf{u}/dt - \boldsymbol{\Omega}\times\mathbf{u})\,dC + \nabla.\left[\int N\;\Pi^*.(d\mathbf{u}/dt - \boldsymbol{\Omega}\times\mathbf{u})\,dC\right] \;, \tag{4.6}$$

It is now evident, when comparing (4.4) and (4.6) with (3.1) and (3.2) that there is a one-to-one correspondence between the *extended* Doi model and the EFL model. This is not the case of the original Doi model which assumes the mean-field potential to depend on the orientation only, and for which (4.6) simplifies into

$$\rho\, d\mu/dt = -\,nkT\,\nabla.\mathbf{v} + \int N(u)\,\partial U^*/\partial\mathbf{u}\,.(d\mathbf{u}/dt - \boldsymbol{\Omega}\times\mathbf{u})\,d\mathbf{u} \;,$$

while the rotational identity (4.4) (or its equivalent form (4.13)) amounts to the no-torque condition

$$\int N(u)\,\mathbf{u}\times(\partial U/\partial\mathbf{u})\,d\mathbf{u} = 0\,.$$

The second step of the calculation depends on the evolution equation for **u**. The original Doi model assumes that **u** behaves like the orientation of a long and thin rigid rod, acted upon by mean-field forces and Brownian diffusion forces

$$d\mathbf{u}/dt - \boldsymbol{\Omega}\times\mathbf{u} = (I - \mathbf{uu})\,.\,[\,\mathbf{D}.\mathbf{u} - (D_r/kT)\partial U^*/\partial\mathbf{u}\,] \tag{4.7}$$

where D_r is the rotational diffusion coefficient and U* is defined in (4.3). A more general evolution equation, which assumes no special molecular shape and which takes the ∇u-dependence of U* into account is

$$d\mathbf{u}/dt - \boldsymbol{\Omega}\times\mathbf{u} = (I - \mathbf{uu})\,.\,[\,\lambda(u):\mathbf{D} + (D_r/kT)\mathbf{H}^*\,] \quad , \tag{4.8}$$

where λ is a third-order tensor, characteristic of the polymer shape, depending on **u** only [Hinch and Leal, 1975], and symmetric in the exchange of its last two indices ($\lambda_{ijk} = \lambda_{ikj}$). The presence of $(D_r/kT)\mathbf{H}^*$ in (4.8) is in fact necessary to have a positive entropy production associated with the diffusion of orientation. When the above evolution of **u** is introduced into (4.6), one finds

$$\rho\, d\mu/dt \;=\; \sigma^{ED}:\mathbf{D} + \partial\Omega_{ij}/\partial x_k \int N\, T^*_{kji}\, dC \;-\; (D/kT)\int N\, (\mathbf{u}\times\mathbf{H}^*)^2\, dC$$
$$+\; \nabla.\left[\int N\;\; \Pi^*.(d\mathbf{u}/dt - \boldsymbol{\Omega}\times\mathbf{u})\, dC\right]\;, \qquad (4.9)$$

with the following definition of the non-dissipative stress in the extended Doi model

$$\sigma^{ED} \;=\; -\,nkT\,\boldsymbol{I} + \int N\,(\Sigma^*)^s\, dC \;-\; \int N\,(\mathbf{H}^* - \mathbf{u}.\mathbf{H}^*\mathbf{u}).\,\lambda(u)\, dC \qquad . \quad (4.10)$$

The last operation consists in reporting (4.9) into (2.2). The elimination of the terms under the divergence leads to (compare with (3.3))

$$\mathbf{q} \;=\; T\,\mathbf{S} \;-\; \int N\;\; \Pi^*.(d\mathbf{u}/dt - \boldsymbol{\Omega}\times\mathbf{u})\, dC \qquad ,$$

and after defining the viscous stress and viscous couple stress by

$$\sigma^s \;=\; \sigma^v - p\,\boldsymbol{I} + \sigma^{ED} \qquad \text{and} \qquad \mathbf{t} \;=\; \mathbf{t}^v + \int N\, \mathbf{T}^*\, dC \qquad , \quad (4.11)$$

the final expression of the entropy balance in the extended Doi model is

$$T\,(\,\rho\, ds/dt + \nabla.\mathbf{S}\,) \;=\; \sigma^v:\mathbf{D} \;+\; t^v_{kji}\,\partial\Omega_{ij}/\partial x_k \;+\; (D_r/kT)\int N\,(\mathbf{u}\times\mathbf{H}^*)^2\, dC \;-\; \mathbf{S}.\,\nabla T\;,$$

to be compared with (3.6). After taking results (4.10) and (4.11) into account, the total stress becomes

$$\sigma \;=\; -\,(p{+}nkT)\,\boldsymbol{I} \;+\int N\,(\Sigma^*)^s\, dC \;-\int N\,(\mathbf{H}^* - \mathbf{u}.\mathbf{H}^*\mathbf{u}).\lambda(u)\, dC - \nabla.\int N\, \mathbf{T}^*\, dC$$
$$+\; \sigma^v \;-\; \nabla.\mathbf{t}^v\;. \qquad (4.12)$$

Taking the identity (4.4) into account, one can present the total stress in a form easily comparable to (3.7). It is important to remember that Σ^*, $\mathbf{T}^*$ and $\mathbf{H}^*$ depend on U^* defined in (4.3). This means that each integral in (4.12) gives rise to *two* different kinds of integrals, the first one connected to the mean-field potential U, and the second one independent of it. This is the main difference between (4.12) and the EFL result (3.7). Let us call Π, Σ , T and $\mathbf{H}$ the quantities defined as in (4.5) but with U^* replaced by U. Then, it can be shown that the rotational identity (4.4) amounts to a condition to be satisfied by the mean-field potential $U(u, \nabla u)$ only

$$\int N\,[u_j\, \partial U/\partial u_i - \Pi_{jk}\, \partial u_k/\partial x_i \;+\; \Pi_{ki}\, \partial u_j/\partial x_k\,]^a\, dC \;=\; 0 \qquad . \quad (4.13)$$

And the above overall stress tensor can be rewritten as

$$\sigma = -(p-nkT)\,\boldsymbol{I} + \int N\,(\Sigma)^s\,dC - \int N\,(\mathbf{H}-\mathbf{u.Hu}).\lambda(u)\,dC - \nabla.\int N\,\mathbf{T}\,dC$$
$$- 4kT\int N\,[(\boldsymbol{I}-\mathbf{uu}):(\partial\lambda/\partial\mathbf{u}) - 2\,\mathbf{u}.\lambda(u)]\,dC + \sigma^v - \nabla.\mathbf{t}^v \quad . \tag{4.14}$$

To obtain more explicit expressions for the stress tensor, one needs an expression for the mean-field potential. We propose below the simplest potential containing a ∇u-dependence. It stems from an extension of the trick used by Maier and Saupe [see de Gennes and Prost, 1993, ch. II] to find the equilibrium distribution of orientation, and it reads

$$U(u,\nabla u) = -3/2\,kT\mathcal{U}\,[\,\langle u_iu_j\rangle\,u_iu_j - L^2\,\partial\langle u_iu_j\rangle/\partial x_k\,\partial u_iu_j/\partial x_k\,] \quad , \tag{4.15}$$

where $\mathcal{U}$ gives the strength of the nematic interaction and L is a parameter with the dimension of a length, presumably of the order of the polymer size. The mean-field potential is made of two terms, the first one favouring a non-isotropic configuration tensor $\langle u_iu_j\rangle$ and the second one penalising its spatial gradients. The symbol $\langle\ \rangle$ is defined as the average over the configuration space

$$\langle u_iu_j\rangle = (1/n)\int N\,u_iu_j\,dC \quad .$$

The above mean-field potential leads to the following distortion stress

$$\int N\,(\Sigma_{ji})^s\,dC = -3/2\,nkT\mathcal{U}\,L^2\,\partial\langle u_ku_l\rangle/\partial x_j\,\partial\langle u_ku_l\rangle/\partial x_i\,, \tag{4.16}$$

which is symmetric and depending on the gradients of the configuration tensor only. Moreover, one obtains for the molecular field **H**,

$$H_i - \mathbf{u.H}u_i = 3kT\mathcal{U}\,(\delta_{ij} - u_iu_j)u_k\,[\langle u_ju_k\rangle + L^2\,\nabla^2\langle u_ju_k\rangle] \quad .$$

It is remarkable that this molecular field can also be deduced (Feng et al., 1999) from the one-constant Marrucci-Greco (1991) potential

$$U(u) = -3/2\,kT\mathcal{U}\,[\langle u_iu_j\rangle + L^2\nabla^2\langle u_iu_j\rangle]\,u_iu_j \quad .$$

The above results have revealed the importance of the configuration tensor. Hence the need to know its evolution in time, as a part of the modelling of liquid-crystalline polymers. That equation is obtained by multiplying the i-component of (4.8) by u_j, then symmetrizing, and finally averaging. Separating the contribution of **H** to **H***,one obtains

$$\mathsf{D}_J\,\langle u_iu_j\rangle/\mathsf{D}t - \langle (u_j\delta_{lk} + u_l\delta_{jk} - 2u_iu_ju_k)\lambda_{kpq}\rangle\,D_{pq} - 6D_r(\langle u_iu_j\rangle - \delta_{ij}/3)$$
$$+ (D_r/kT)\,\langle u_jH_i + u_iH_j - 2u_iu_j\,\mathbf{u.H}\rangle \tag{4.17}$$

where D_J/Dt is the Jaumann or co-rotational derivative. The respective roles of the shape tensor λ and the molecular field **H** are clearly seen. The initial Doi model corresponds to the special shape factor $\lambda_{ijk} = \frac{1}{2}(\delta_{ij}u_k + \delta_{ik}u_j)$. In this case, the general results (4.14) and (4.16) give as a particular case the model equations proposed by Feng, Sgalari and Leal (1999). It is noteworthy that these authors have used the one-constant Marrucci-Greco potential, and they paid a lot of attention to a careful generalization of Doi's principle of virtual work to obtain a distortion stress. The advantage of the above thermodynamic approach is to deduce all the relevant quantities from a single ∇u - dependent mean-field potential.

5. Conclusions

The model proposed by Doi to describe the flow of liquid-crystalline polymers has played a major role for the last fifteen years, because of its ability to describe in a satisfactory way most of the experiments involving those complex liquids. Its only shortcoming is the absence of distortional energy which precludes its use in situations where the liquid-crystalline polymers display a spatially varying configuration. To solve that issue, Edwards, Beris and Grmela (1990) proposed a model in which the free-enthalpy $\rho\mu|_{p,T}$ was supposed to depend on $\langle u_i u_j \rangle$ and its gradients only. Their model lies "midway" between the EFL model and our extended Doi model. However, it is far from obvious that a mean-field potential exists such that (4.2) is expressible in terms of the second moment of the distribution N only. And while the distortion stress (4.16) obtained with the potential (4.15) comforts their assumption, the term **u.Hu** present in (4.14) and (4.17) has many chances to introduce higher-order moments. Marrucci and Greco (1991) have proposed an extension of the Doi-Maier-Saupe potential, and Feng, Sgalari and Leal (1999) have incorporated the simplest form of that new potential into the Doi theory to formulate a model containing distortion energy. Their approach is a nice alternative to ours but we have a hunch it will be difficult to extend it to anisotropic distortion energies. This is the reason why we prefer a mean-field potential depending on the orientation gradient. This was the starting point of the present work, and the use of a newly developed thermodynamic framework allowed us to deduce all its consequences concerning the macroscopic equations of motion.

Acknowledgements : This work was initiated when visiting the Chemical Engineering Department in U.C. Santa Barbara. I have benefited from numerous and stimulating discussions with Dr. Giorgia Sgalari and Prof. Gary Leal who introduced me to those odd but fascinating liquids.

REFERENCES

Doi M., Molecular dynamics and rheological properties of concentrated solutions of rod-like polymers in isotropic and liquid crystalline phases, *J. Polym Sci. Polym. Phys.* **19** (1981) pp 229-243.

Doi M. and Edwards S.F., *The Theory of Polymer Dynamics* (Oxford University Press, London, 1986).

Edwards B.J., Beris A.N. and Grmela M., Generalized constitutive equation for polymeric liquid crystals, Part I, *J. of Non-Newtonian. Fluid Mech.*, 35 (1990) 51-72.

Feng J.J., Sgalari G. and Leal L.G., A theory for flowing nematic polymers with orientational distortion (preprint 1999).

de Gennes P. G. and Prost J., *The Physics of Liquid Crystals*, 2nd ed. (Oxford University Press, New-York, 1993).

de Groot S.R. and Mazur P., *Non-Equilibrium Thermodynamics* (North-Holland Publishing Co. Amsterdam 1962) ch III.

Hinch E.J. and Leal L.G., Constitutive equations in suspension mechanics Part I : General formulation, *J. Fluid Mech.* 71 (1975) 481-495.

Marrucci G. and Greco F., The elastic constants of Maier-Saupe rod-like molecules nematics, *Mol. Cryst. Liq. Cryst.* **206** (1991) pp.17-30.

Multiscale thermomechanical approaches to SMA behaviour

LMGC *

Laboratoire de Mécanique et Génie Civil
Université Montpellier II,
Place Eugène Bataillon, 34095 Montpellier Cedex 05, FRANCE

Abstract : This paper presents, in a synthetic way, several works performed on shape memory alloys (SMAs). Three scales of description are used according to whether one seeks to numerically predict the possible microstructural configurations of phases in equilibrium or the behavior of a mono- and polycrystal of SMA during a phase transition.
Keywords: SMA, phase transition, thermomechanical coupling, multiscaling.

1. INTRODUCTION

Shape memory alloys (SMAs) may undergo remarkable microstructural transformations: they may change the structure of their crystallographic lattice under mechanical and/or thermal loading. This transformation, called martensitic transformation, is displacive in the sense that it corresponds to a collective displacement of atoms. From a thermodynamic point of view, it belongs to the family of first order phase transitions insofar as it is accompanied by a latent heat of phase change. SMAs are being closely studied at the present time, for various reasons. From a fundamental point of view, researchers have found in the SMAs an amazing example of solid-solid phase transition, the domain of which is generally close to room temperature and thus is relatively easy to observe. From a more practical point of view, the number of industrial applications, using SMAs properties, grows day by day.
Three length scales can be introduced to study the SMA behaviour. The *microscopic* length scale corresponds to the crystal lattice, and is about several times the size of the unit cell; it can also be related to the *representative volume element* (RVE) of the continuous medium that is associated with the lattice. As for the *mesoscopic* and *macroscopic* length scales, they correspond to the RVEs of mono and polycrystal, respectively.
Since this transformation is able to generate a large variety of phase mixtures, some authors have systematically studied the possible microstructural arrangements (twinning) or rearrangement mechanisms due to straining (variants reorientation) when the material is in mechanical and thermodynamic equilibrium [Ball *et al.*, 1987].

(*) *List of authors*: P. ALART, X. BALANDRAUD, A. CHRYSOCHOOS, C. LICHT, O. MAISONNEUVE, S. PAGANO, R. PEYROUX, B. WATTRISSE. name@lmgc.univ-montp2.fr.

G.A. Maugin et al. (eds.), Continuum Thermomechanics, 247-264.
© 2000 *Kluwer Academic Publishers.*

Others are interested in the phase transition itself and in its correlation with more macroscopic effects: pseudoelasiticy, self-accommodation, one-way or two-way shape memory effects, aging, etc. They are many experimental tests, mathematical modelling or numerical works on these matters. Many of these studies are directly related to engineering purposes and industrial applications, [Patoor *et al.*, 1994].
This paper gathers and tries to synthetize the researches, performed by several teams at LMGC, which concern the thermomechanical behaviour of SMAs. In section 2 we recall the thermodynamic concepts and results used in this paper, then we will present investigations into the three scales of description. In section 3, following works of Ball and James, we show numerically computed examples of twinned microstructures. At this scale of description, the symmetries of the crystal lattice and the material frame-indifference principle will be invoked to introduce non-convex potentials having multiple wells. Section 4 will be devoted to research performed at the mesoscopic scale. Two examples of "micro-meso" passage will be pointed out ; they will be based on a (quasi)convexification procedure of the thermodynamic potential. The goal here will be the deduction, from microstructural considerations of section 3, of the properties of mesoscopic (quasi-convex) potentials. A third example, more inspired by experimental results obtained on monocrystalline samples, will deal with the pseudoelastic behaviour of SMAs. We will underline the strong coupling that represents the mechanism of phase change, and the important role played by the temperature variations, particularly those induced by the latent heat. This model of monocrystal will be used, in section 5, to numerically predict the behaviour of a set of grains considered as a RVE for the polycrystal. Comparisons between calculations and experiments will be shown, considering at the same time the mechanical, thermal, and energy aspects of the phase transition.

2. A CONVENIENT THERMOMECHANICAL FRAMEWORK

Currently many physicists or chemists like to introduce phase transitions as unstable thermodynamic phenomena [Kondepudi *et al.*, 1998]. The instability is in the sense defined by the Gibbs-Duhem criterion [Glansdorff *et al.*, 1971] and is invoked in a wide range of situations [Papon *et al.*, 1999]. Deduced from an analysis of the entropy source, the thermodynamic instability is related to a change of the convexity of the thermodynamic potentials and may consequently imply a non-monotonic mechanical behaviour. In such a context, the reversibility assumption is necessary to derive the famous equal-area rule due to Maxwell which enables us to determine the phase diagram. These ingredients have been widely exploited for describing the martensitic transformation [Müller *et al.*, 1991], [Abeyaratne *et al.*, 1993].
In what follows, we have chosen other ways for determining the microstructures corresponding to the possible phase mixtures at equilibrium or for introducing the phase change mechanisms in the constitutive equations. We will first assume that the phase transition is a quasi-static process possibly accompanied by irreversibilities. A convenient thermomechanical framework is the *Continuum Thermodynamics* [Germain,

1973] that postulates the local state axiom. More particularly, we will use the formalism of generalised standard materials [Germain *et al.*, 1983] for which the constitutive equations can be derived from a thermodynamic potential and a dissipation potential. The interest of modelling phase transition into such a framework is at least twofold. Theoretically, it gives a convenient mathematically consistent background to predict phase change for any thermomechanical loading. In other words, it is able to include, in the same model phase diagram, properties and kinetics of phase change induced by stress and/or temperature variations. Of course, when only stable equilibrium states of phase mixture are considered, the dissipation potential is no longer necessary and a thermostatic framework is sufficient. Experimentally, energy balance construction facilitates the interpretation of thermal and calorimetric phenomena accompanying the phase change. Besides, it leads to a better understanding of certain pseudoelastic effects such as the hysteresis loops of the stress-strain curves [Chrysochoos *et al.*, 1993].

To describe the martensitic transformation at meso or macrolevel, the following variables are often chosen to characterise the thermodynamic state of each volume element : T the absolute temperature, ε a strain tensor and $x_1,..,x_n$, n variables describing the phase mixture.

If W denotes the specific Helmholtz free energy, the Clausius-Duhem inequality defines the intrinsic dissipation d_1 and the thermal dissipation d_2, both supposed to be separately positive :

$$\begin{cases} d_1 = \sigma : \mathrm{D} - \rho W_{,\varepsilon} : \dot{\varepsilon} - \rho W_{,x_i}\, \dot{x}_i \geq 0 \\ d_2 = -\dfrac{q}{T} . \mathrm{grad} T \geq 0 \end{cases},$$

where σ is the Cauchy stress tensor, D the Eulerian strain rate tensor, ρ the mass density, and q the heat influx vector. The dot stands for the material time derivative. The equality $d = d_1 + d_2 = 0$ characterises reversible thermodynamic processes.

The intrinsic dissipation per unit volume d_1 is the difference between the inelastic energy rate $w'_{in} = \sigma : \mathrm{D} - \rho W_{,\varepsilon} : \dot{\varepsilon}$ and the stored energy rate $w'_s = \rho W_{,x_i} \dot{x}_i$. Deduced from both principles of thermodynamics, the local heat conduction equation is

$$\rho C_{\varepsilon,x} \dot{T} + \mathrm{div}\, q = d_1 + \rho T\, W_{,T,\varepsilon} : \dot{\varepsilon} + \rho T\, W_{,T,x_i}\, \dot{x}_i + r_e ,$$

where $C_{\varepsilon,x}$ denotes the specific heat capacity at constant ε and $(x_i)_{i=1,n}$, while r_e stands for the external heat supply. The intrinsic dissipation d_1, the thermoelastic coupling term $\rho T\, W_{,T,\varepsilon} : \dot{\varepsilon}$ and the rate of latent heat $\rho T\, W_{,T,x_i}\, \dot{x}_i$ have been collected in the right hand member of the heat equation. Taking into account an isotropic conduction of heat ($q = -k\, \mathrm{grad} T$), we underline that the left hand side becomes a partial derivative operator applied to the temperature. This property has been experimentally used to deduce the distribution of heat sources w'_{ch} from temperature charts by the use of infrared techniques [Chrysochoos *et al.*, 2000]. The volume heat source is defined by the equation :

$$w'_{ch} = d_1 + \rho T W_{,T,\varepsilon} : \dot{\varepsilon} + \rho T\, W_{,T,x_i}\, \dot{x}_i \,.$$

We emphasise that the rate of latent heat is here defined as the heat source related to the thermomechanical couplings between temperature and state variables characterising the phase mixture. An interesting property can be directly derived from the energy balance when a mechanical hysteresis loop is associated with a thermodynamic cycle of duration C. In [Peyroux *et al.*, 1998], we showed that :

$$A_h = \oint_C \sigma : D\, dt = \oint_C w'_{ch}\, dt \,,$$

where A_h is the energy associated with the hysteresis area. This result shows that the presence of a mechanical hysteresis is not only due to dissipative effects, but also to thermomechanical coupling mechanisms. The experimental investigations confirmed the main role played by the temperature on the mechanical behaviour [Chrysochoos *et al.*, 1995]. In fact, even during quasi-static loading, the temperature variations induced by the latent heat are of the same order of magnitude as the transition domain "width". Consequently, we emphasized that these variations strongly modify the kinetics of an apparent stress-induced phase change and we claimed that the transformation is to be considered *a priori* as an anisothermal process. The thermomechanical couplings and the heat diffusion generate a time-dependence on SMA behaviour, to which we will return in the last sections of this paper.

3. ATTEMPTS AT MULTISCALE MODELLING

3.1. Microscopic scale

3.1.1. A variational model for microstructures

A neat description, which has its origins in the work of J. L. Ericksen, can be found in [Ball *et al.*, 1987], [Ball *et al.*, 1992], [James *et al.*, 1989] and roughly summarized as follows. Under the Cauchy and Born rule [Ericksen, 1984], an abstract continuous medium is associated with the crystalline lattice. This medium is assumed to be thermoelastic. In the framework of the finite transformations, the free energy density function W depends on the temperature T and on the gradient of deformation F. This density function inherits some properties of a crystalline lattice encountering phase transitions:

i) W is frame indifferent:

$$W(QF,T) = W(F,T), \forall (Q,F,T) \in O_3^+ \times M_3^+ \times R^+ ,$$

O_3^+ is the set of all rotations, M_3^+ the set of all square matrices of order 3 with positive determinant,

ii) W has a finite number of potential wells:

at each T, there exists a finite number $m(T)$ of symmetric, positive definite matrices ot order 3, $U_i(T), i = 1, m$, such that the minimizers of $W(.,T)$ are the orbits

$O_3^+U_i(T), i=1,m$. For instance « at each temperature greater than T_c, the minimizers of $W(\bullet,T)$ are $O_3^+U_A(T)$» corresponds to the existence of the austenite state at a temperature greater than the transition temperature T_c. While « at each temperature lower than T_c, the minimizers of $W(\bullet,T)$ are $O_3^+U_{Mi}(T),\ i=1,k$ » corresponds to the existence, at each temperature below T_c, of the martensite state which appears in k *variants*, k being the index of the symmetry group of the martensite lattice in the symmetry group of the austenite lattice. Of course, the minimizers of $W(\bullet,T_c)$ are the k+1 orbits $O_3^+U_A(T), O_3^+U_{Mi}(T),\ i=1,k$.

Finding stable equilibrium configurations of a piece of monocrystal at a *uniform* temperature *T*, with for instance a displacement boundary condition on its whole boundary, leads to the determination of the exact minimizers of the total energy

$$I(\varphi)=\int_\Omega W(\nabla\varphi(x))dx$$

on the reference configuration $\overline{\Omega}$ for all admissible deformations φ. Here, the temperature acts as a parameter, thus the variable *T* is, from now on, omitted in this section. In many situations, because of the numerous potential wells, the previous problem has no solutions. The (weak) limits of the minimizing sequences of I do not minimize I, and they have to develop finer and finer oscillations. The fact that these spatial oscillations correspond to the observed microstructures is the basic assumption of this theory.

Our own contribution is confined to the determination of microstructures by *numerical* energy minimization. The aim of our numerical experiments is to check if minimizers of the discrete problems, which always exist, account for the microstructures. We denote by *Y* the microscopic domain of observation, and *F* the mesoscopic gradient of deformation (i. e. the average on *Y* of the microscopic gradient of deformation).

3.1.2. Numerical experiments at microscale

In contrast with most previous computations [Collins, 1993] we do not use, in this part, either finite elements or gradient-like algorithms, but a method using a trigonometric interpolation and a decoupling of the deformation fields from their gradients. Saddle points of an augmented Lagrangian:

$$L(\varphi,\ q;\ \lambda)=\int_\Omega\left\{W(q)+\lambda:(\nabla\varphi-q)+\frac{r}{2}|\nabla\varphi-q|^2\right\}dx\ ,\qquad r>0$$

are obtained by an Uzawa algorithm:

$$L\left(\varphi^{n+1},\ q^{n+1};\ \lambda^n\right)\le L\left(\varphi,\ q;\ \lambda^n\right)\ \text{and}\ \lambda^{n+1}=\lambda^n+r\left(\nabla\varphi^{n+1}-q^{n+1}\right).$$

This method is worthwhile, since a step of a relaxation procedure to minimize $L(.,.;\lambda^n)$ involves a global linear problem of Laplace type, and *local* nonlinear problems of minimization of a quadratic perturbation of W in a suitable set of matrices. Moreover, to compute the microstructures, we take advantage of being free to choose the domain of computation and, to a certain extent, the boundary conditions. With Y cubic and a boundary condition of place, we can use a trigonometric interpolation on a grid of order N, rather than finite elements. This method avoids storing a stiffness matrix and the fast Fourier transform algorithm (FFT) is faster than solving the linear system involved by finite elements.

For our numerical two-dimensional experiments [Licht, 97] we have chosen an Ericksen-like energy density function [Collins, 93]:

$$W(F) = \tilde{W}(C),\ C = F^T F$$

$$\tilde{W}(C) = k_1\left|trC - 2 - \delta^2\right|^2 + k_2\left|C_{12}^2 - \delta^2\right|^2 + k_3\left|C_{11} - C_{22} - \delta^2\right|^2,\ k_1, k_2, k_3, \delta > 0.$$

If $F = xS^+ + (1-x)S^-, 0 < x < 1,\ S^\pm = I \pm \delta e_2 \otimes e_1$, the microstructure is *unique* [Ball *et al.*, 92] and is a single twinned laminate orthogonal to e_1 with phases corresponding to the wells S^+, S^- in proportions x, 1- x. Thus a *valuable* test of this numerical method is to check if minimizers φ_N of the discretized problem account for this microstructure.

Figures 1a (resp.1b) displays $\psi(\nabla\varphi_N)$ for the case x = 1/3 and N=16 (resp.N=32) where, as proposed by [Collins, 93],

$$\psi(F) = \left|F^T F - \left(U^+\right)^2\right| / \left(\left|F^T F - \left(U^+\right)^2\right| + \left|F^T F - \left(U^-\right)^2\right|\right).$$

Except of course in the vicinity of ∂Y, where φ_N must satisfy the boundary condition $\varphi_N(y) = Fy$, $\nabla\varphi_N$ essentially takes the values S^+ and S^- in vertical layers in proportions x and $(1-x)$. Whereas φ_N depends strongly on N, the distribution of the values of $\nabla\varphi_N$ does not.

Thus, our numerical method gives the microstructure on the scale of the grid, but these nice results were obtained through initial fields with oscillations of the same wavelength as the expected microstructure. Random initialization (Fig.1c) gives a less regular distribution of S^+ and S^-, but the horizontal proportions are still about x and $(1-x)$. Even with slowly decreasing random perturbation of the updating of the multiplier λ, we did not succeed in avoiding this kind of local minimum (Fig.1d), however the average proportions of the *vertical layers* are x and $(1-x)$. Figures 1e-1h concern the case in which x = 1/2 and $S^\pm$ are replaced by $S^\pm R$, $R \in O_3^+$. When $R^T e_1$ is compatible with the grid, twinning is once more obtained.

Nevertheless trigonometric interpolation seems more flexible than piecewise affine finite elements for capturing slanting oscillations, because of the Hadamard jump condition. But as in [Collins, 93], we never succeeded in trapping laminates of order two (layers within layers)!

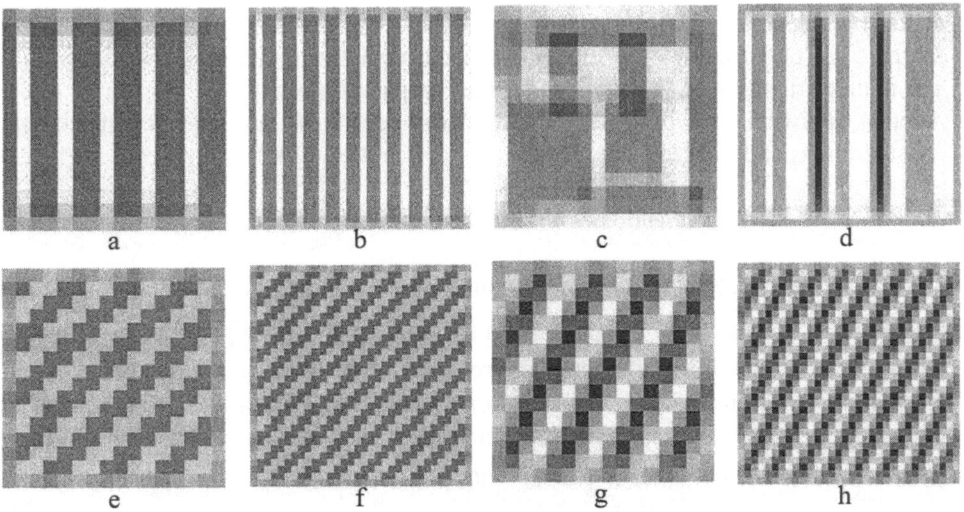

Figure 1 *; Examples of microstructures*

Similar results may be obtained with a more rudimentary model in the context of small perturbations. Frémond's model may be related to the Khachaturyan-Roytburd-Shalatov theory; Kohn [Kohn, 1991] showed that this last theory is the geometrically linear analogue of the one developed by Ball and James in the context of finite deformations [Ball *et al.*,1992]. The free energy density function is then a function of the small strain tensor ε. The frequency of the oscillations of the layered microstructure depends on the mesh. Due to the choice of the residual stress tensor (introduced below) $\alpha\tau$ parallel to I, the orientation of the microstructure is arbitrary [Pagano, 1998].

3.2 Mesoscopic scale

3.2.1Quasiconvexification and scale transition

If the material is assumed to be non dissipative, a first proposal for the mesoscopic strain energy density function is the quasiconvexification of the microscopic density function. The justification for this rests on the mechanical interpretation of two mathematical properties of the minimizing sequences of the total energy that were invoked in section 3.1.1, [Dacarogna, 82]. First their (weak) limits describe the average, say *mesoscopic*, state of a grain due to the convergence of the gradient averages in every subdomain. Next, these limits do not minimize the true total energy, but a total energy whose density is the quasiconvexification QW of the true density W:

$$QW(F) = \operatorname{Inf}\left\{\frac{1}{|Y|}\int_Y W(F + \nabla u(y))dy, u = 0 \text{ on } \partial Y\right\},$$

$QW(F)$ is the infimum of the average of the strain energy for deformations whose average gradient is F. Thus the mesoscopic behavior of the monocrystal is not described by the true (microscopic) energy density function W but by an *apparent* density QW. The main difficulty with this approach is the computation of QW.
Another approach is to propose a strain energy density W through phenomenological considerations and to check if it is quasiconvex (i. e. $W = QW$!).

3.2.2 An academic two-phase phenomenological modeling

Many phenomenological models of two-phase shape memory material incorporate a so-called interaction term $W^I = h\,x(1-x)$ in the free energy; x is a phase proportion and h a non-negative function of the temperature. Thus, if the material is assumed to be non dissipative, the strain energy is not a convex function of the linearized strain ε. This lack of convexity generates controversies which may lead to a rejection of the previous structure of W^I; we showed [Licht, 2000] that *if h is not too large* the previous structure provides a *good* model.
We confine our attention to the case of two linearly thermoelastic phases with different stress-free strains a_i but with the same elastic stiffness A ; the *energy density function of each phase* is

$$W^i(\varepsilon,T) = \frac{1}{2}\,A(\varepsilon - a_i(T)) : (\varepsilon - a_i(T)) + w_i(T),$$

and a common proposal for a family of free energies and strain energies of the mixture is

$$W_h(\varepsilon, x, T) = \frac{1}{2}\,A(\varepsilon - a_x) : (\varepsilon - a_x) + w_x + h\,x(1-x),$$

$$\overline{W}_h(\varepsilon, T) = \mathrm{Min}\{W_h(\varepsilon, x, T), 0 \le x \le 1\},$$

a_x (resp. w_x) is the convex combination in proportions x and (1-x) of a_i (resp. w_i).
Let $H = Min\left\{\frac{1}{2}A(a-z) : (a-z),\ z = k \otimes_s v,\ k, v \in R^3\right\},\ a = a_2 - a_1$.
Through Fourier analysis as in [Kohn, 91], it can be shown that

$\overline{W}_h(\varepsilon, T)$ is convex if and only if $h \le 0$,

$\overline{W}_h(\varepsilon, T)$ is not convex but quasiconvex if and only if $0 < h \le H$,

$\overline{W}_h(\varepsilon, T)$ is not quasiconvex if and only if $H < h$.

Of course the second case does not occur if H=0, that is to say, if the stress-free strains are compatible. Thus the good range for h ($\underline{W}_h$ quasiconvex or convex) depends strongly on the compatibility of the a_i. On the other hand, incorporating a term like W^i in a more rational definition of the macroscopic energy of a mixture like

$$\operatorname{Inf}\left\{\begin{array}{l}\frac{1}{|Y|}\int_Y \left\{\chi(y)W^1(e(u(y)))+(1-\chi(y))W^2(e(u(y)))\right\}dy, \\ \chi \text{ characteristic function, } \int_Y \chi(y)dy = x|Y| \text{ and } u(y)=\varepsilon y \text{ on } \partial Y;\, e=\nabla_s u\end{array}\right\},$$

yields a free energy equal to W_{H+h}. Thus quasiconvexity of $\underline{W}_{H+h}$ implies $h \leq 0$ which agrees with a proposal by [James *et al.*, 1993] for mismatch energy. Moreover, if $0 < h \leq H$, the function $\underline{W}_h(\cdot,T) - \Sigma:\cdot$ has a double-well structure for suitable Σ, so that the total mechanical energy of a body subjected to a uniform surface loading may have two strong local minimizers. Thus in some displacement-controlled tests, a negative slope of the stress-strain response curve of a specimen can be observed, while in some load-controlled test a kind of hysteresis occurs.
We also proved that approximations of the previous local minimizers can be obtained by classical finite element methods.

3.2.3 D. C. algorithm: a convenient numerical tool

The previous approach may be extended to a three phase mixture, as is usual for some models of SMAs with an austenitic phase (proportion x_0) and two variants of martensite (proportions x_- and x_+). We assume that the stress-free strain of the austenite is zero ($a_0 = 0$) and the stress-free strains of the martensites are opposite ($a_- = -a$ and $a_+ = a$). Similarly the term $w_-(T)$ is equal to $w_+(T)$ and we define $c(T) = w_+(T) - w_0(T)$. To relate this formulation with others ([Frémond, 1987, Raniecki *et al.*, 1992]), it is useful to introduce a residual stress tensor τ and to express the stress-free strain with respect to it, to the elastic tensor A and a scalar function $\alpha(T)$ of the temperature: $a = \alpha A^{-1}\tau$. We have then 3 interaction terms between 2 phases, h_{0-}, h_{0+} and h_{-+}. But it seems natural enough to suppose that $h_{0-} = h_{0+} = h_0$. By choosing $h_+ = 4h_0 = 2G$ [Leclercq *et al.*, 1996, Pagano *et al.*, 1999], the previous analysis about a two-phase mixture is preserved and the convexity and quasiconvexity properties may be recovered with respect to the single parameter G. Consequently, a family of free energies of the mixture may be expressed as follows:

$$W_G(\varepsilon,x,T) = \frac{1}{2}A(\varepsilon - a_x):(\varepsilon - a_x) + w_x + h(x_+ - x_-)^2,$$

where

$$a_x = a_x(T) = (x_+ - x_-)\alpha A^{-1}\tau,\quad h = h(T) = \frac{1}{2}\left(G(T) - \alpha(T)^2\tau : A^{-1}\tau\right).$$

Similarly to the 2-phase mixture, we can then distinguish 2 important values G_c and G_q which are determined according to our assumptions. For a bidimensional modeling, the isotropic elastic tensor (λ and μ the Lamé coefficients) and $\tau = diag(\tau_1,\tau_2)$ we have,

$$G_c = \frac{-\alpha^2}{4\mu(\lambda+\mu)}\left[(\lambda+\mu)\left(\tau_1^2+\tau_2^2\right)-2\lambda\tau_1\tau_2\right] \text{ and}$$

$$G_q = \begin{cases} G_c & \text{if } (\lambda+2\mu)\tau_1 - \lambda\tau_2 < 0 \text{ and } \lambda\tau_1 - (\lambda+2\mu)\tau_2 < 0 \\ \dfrac{-\alpha^2}{\lambda+\mu}\max\left(\tau_1^2,\tau_2^2\right) & \text{otherwise} \end{cases}$$

by convexification and quasiconvexification of the bulk energy proposed by Frémond, which corresponds to G equal to zero [Frémond, 1987]. By neglecting dissipative effects, we define the bulk energy density as the marginal function of the free energy for all admissible volume proportions ($x=(x_+,x_-)$ belongs to the simplex C, $C=\left\{0\leq x^{\pm}, x^{+}+x^{-}\leq 1\right\}$: $MW_G(\varepsilon,T)=\min\{W_G(\varepsilon,x,T);x\in C\}$.

This bulk energy density is then a three-well or a two-well potential with five or three regimes described below. For an efficient numerical treatment it is useful to split it into the difference between two convex functions [Pagano *et al.*, 1998]; the first is quadratic, represents the bulk energy density of the austenitic phase and does not depend on G: $MW_G(\varepsilon,T)=W_a(\varepsilon,T)-W_b(\varepsilon)$ with $W_a(\varepsilon,T)=W_0(\varepsilon,T)=\frac{1}{2}\varepsilon:A\varepsilon+w_0(T)$.

If $c>\frac{G}{2}$, the second convex function presents five regimes:

$$W_b(\varepsilon)=\begin{cases} -\alpha\tau:\varepsilon-c & \textit{if } \alpha\tau:\varepsilon\leq -c-\frac{G}{2} \\ \frac{1}{2G}\left(\alpha\tau:\varepsilon+c-\frac{G}{2}\right)^2 & \textit{if } -c-\frac{G}{2}\leq\alpha\tau:\varepsilon\leq -c+\frac{G}{2} \\ 0 & \textit{if } \alpha\left|\tau:\varepsilon\right|\leq c-\frac{G}{2} \\ \frac{1}{2G}\left(\alpha\tau:\varepsilon-c+\frac{G}{2}\right)^2 & \textit{if } c-\frac{G}{2}\leq\alpha\tau:\varepsilon\leq c+\frac{G}{2} \\ \alpha\tau:\varepsilon-c & \textit{if } \alpha\tau:\varepsilon\geq c+\frac{G}{2} \end{cases}.$$

If $c\leq\frac{G}{2}$, we recover a two-well potential with three regimes:

$$W_b(\varepsilon)=\begin{cases} -\alpha\tau:\varepsilon-c & \textit{if } \alpha\tau:\varepsilon\leq -G \\ \frac{1}{2G}(\alpha\tau:\varepsilon)^2+c+\frac{G}{2} & \textit{if } \alpha\left|\tau:\varepsilon\right|\leq G \\ \alpha\tau:\varepsilon-c & \textit{if } \alpha\tau:\varepsilon\geq G \end{cases}.$$

To reproduce local instabilities inherent to phase transition [James, 1987, Zhong *et al.*, 1996], we search for only a local minimum of the total energy Π on a domain Ω which may still be split into two convex parts, including the bulk energy and the work of the external loading $l(v)$:

$$\Pi(v)=\phi_1(v)-(\phi_2\circ\varepsilon)(v),$$

$$\phi_1(v)=\int_\Omega W_a(\varepsilon)dx - l(v) \textit{ and } \phi_2(\varepsilon)=\int_\Omega W_b(\varepsilon)dx.$$

This decomposition leads us quite naturally to introduce a type II Lagrangian according to the terminology of Auchmuty [Auchmuty, 1989], depending on the Fenchel conjugate function of ϕ_2, and associated with a min-min problem:

$$L_{II}(v,\tau_2) = \phi_1(v) + \phi_2^*(\tau_2) - \int_\Omega \tau_2 : \varepsilon(v)dx.$$

The D.C. algorithm [Stavroulakis *et al.*,1993] consists in constructing a minimizing sequence of L_{II} descending, alternately, in the primal variable and the dual one.

D.C. algorithm: $\left(u^0, \sigma_2^0\right)$ given, with $\left(u^{n-1}, \sigma_2^{n-1}\right)$ known, determine successively u^n, σ_2^n as follows :

Step 1: $L_{II}\left(u^{n-1}, \sigma_2^{n-1}\right) \le L_{II}\left(v, \sigma_2^{n-1}\right) \forall v \in U_{\text{ad}}$,

Step 2: $L_{II}\left(u^n, \sigma_2^n\right) \le L_{II}\left(u^n, \tau_2\right) \forall \tau_2 \in U_{\text{ad}}$.

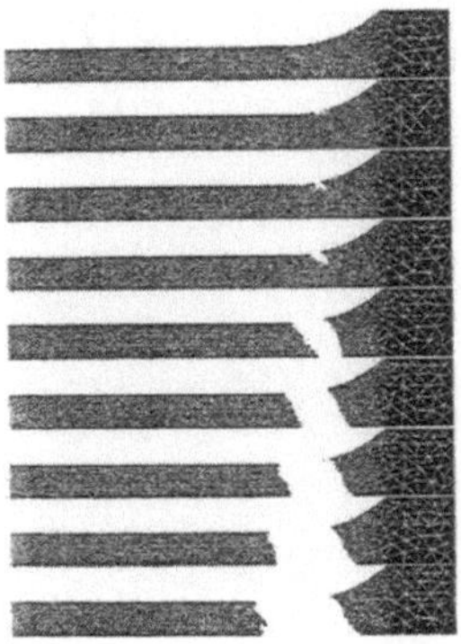

As an example of application (figure 2), numerical simulations [Pagano *et al.*, 1999] on a tensile test showed a propagation of the phase transition initialized at the connection zone of the sample and spreading out over the gauge length in accordance with some recent experiments.

***Figure 2** ; Evolution of phase transition during the loading*

3.2.4 The phase change: a thermomechanical coupling mechanism

In a previous paragraph, we pointed out the properties of the quasiconvex energy QW. This energy was deduced from an analysis at a microscopic scale. Another way to identify the mesoscopic potential is based on experiments on monocrystalline samples. The phenomenological features we saw led us to consider a convex energy at this scale. The confrontation between the micro and meso approaches and the phenomenological one was profitable. In this section, we present the retained mesoscopic free energy $W(T, \varepsilon, x_k)$ and the derived constitutive equations of the monocrystal, in order properly to describe different phenomena such as pseudoelasticity, reorientation effect and thermomechanical couplings accompanying the phase change.

The experiments [Balandraud, 2000] were carried out on a monocrystalline CuZnAl specimen. First, uniaxial load-unload tests were performed in the "low temperature region" (typically 22°C). The apparent permanent strain observed in Figure 3 can be recovered by a heating-cooling process under zero stress. This effect, induced by martensite variant reorientation, leads us to consider, at low temperatures, a non-strictly convex property of the marginal energy :

$$\underline{W}(T,\varepsilon) = \underset{x_k}{Min}\, W(T,\varepsilon,x_k)\,.$$

Note also that, once the reorientation strain is annealed, the stress, the strain and the temperature recover their initial values. Therefore, and from a thermomechanical point of view, we do not refer to plasticity to explain reorientation strain.

Another interesting test concerns pseudoelasticity. If the load-unload is imposed in the "high temperature region" (typically 30°C), we can observe (Figure 4a) a hysteresis loop in the strain-stress diagram that must be studied in the light of the thermodynamic framework presented in part 2. Moreover, temperature variations of the sample can be observed (Figure 4b) around the prescribed value (the temperature of the environmental chamber). These variations, though small, are induced by significant heat sources. Energy balance makes it possible to split the volume heat source w_{ch} into a part due to dissipative effects and another one due to thermomechanical coupling mechanisms. In all the tests performed, the first part was always small compared to the second (<2%).

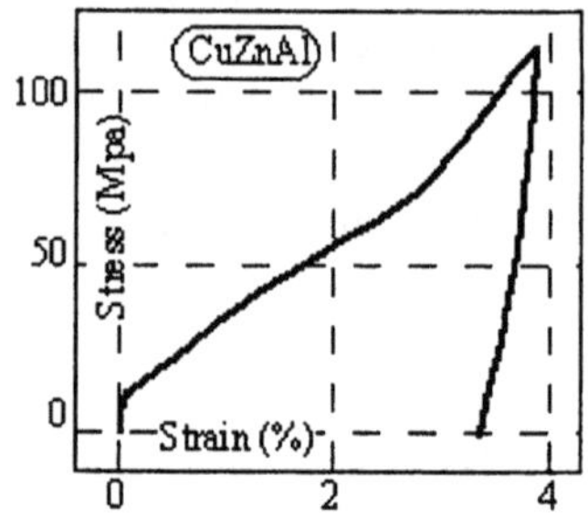

Figure 3 *; Stress-strain curve in a reorientation test*

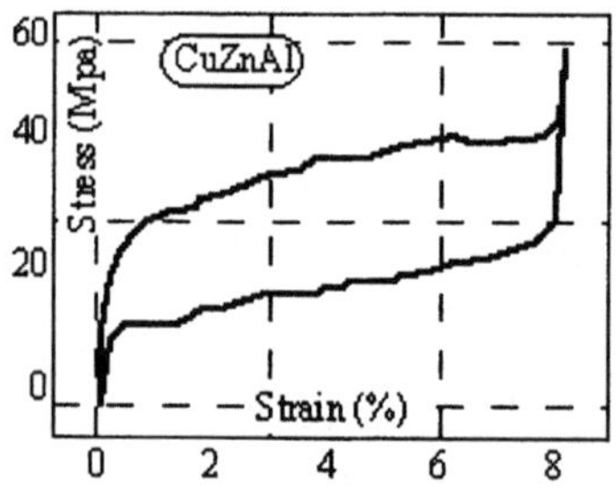

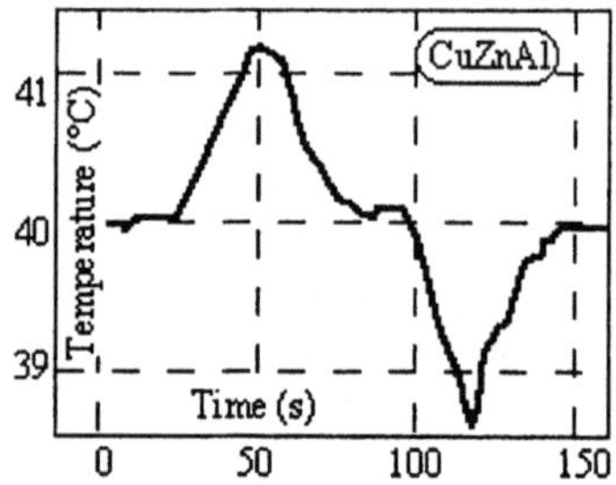

Figure 4a *; Stress-strain curve in a pseudoelastic test*

Figure 4b *; Temperature variation in a pseudoelastic test*

Note that, for all uniaxial tests performed, we did not observe non-monotonous stress-strain curves. In all cases where non-monotonous load-displacement curves occurred, strain localization effects were seen (propagation of transformation band, [Wattrisse,1999, Balandraud, 2000]). All these remarks lead us to consider phase change in SMAs as an anisothermal process, and to use a quasiconvex energy density function. For the sake of simplicity and without any data on multiaxial behaviour, we chose a convex function to suitably describe the behaviour in terms of the kinetics of phase change. The retained free energy W is

$$W(T,\varepsilon,x) = W^{\text{el}}(T,\varepsilon,x) + W^{\text{pc}}(T,\varepsilon,x) + I_C(x).$$

Here I_C is the characteristic function of admissible values of x, and

$$W^{\text{el}}(T,\varepsilon,x) = \tfrac{1}{2}A\left(\varepsilon - \varepsilon^{\text{pc}}(x) - \alpha T\right):\left(\varepsilon - \varepsilon^{\text{pc}}(x) - \alpha T\right) - \tfrac{1}{2}\left(\frac{\rho C}{T_0} + \alpha A \alpha\right)T^2,$$

where the subscript pc stands for phase change, and $\varepsilon^{\text{pc}}(x) = x_1\beta^1 + x_2\beta^2 + \ldots + x_n\beta^n$.

The values β^k characterise the phase change strain associated with the variant k and derive directly from microscopic studies of the crystallographic lattice.

$$W^{\text{pc}} = \frac{L}{T_0}\left(T - T_A\right)(x_1 + x_2 + \ldots + x_n),$$

with L the latent heat and T_0 the reference temperature.

Now, in the formalism of standard materials, a choice of a dissipation potential ϕ leads to complementary laws, and the writing of intrinsic dissipation gives the expression for the transition domain:

$$-A\left[\varepsilon - \alpha\theta - x_k\beta^k\right]:\beta^k + \rho\frac{L}{T_0}\left(T - T_A\right) \in -\rho\partial_{\dot{x}_k}\phi - \rho\partial_{x_k} I_C,\ k = 1,\ldots,n,$$

where $\partial_s f$ stands for the partial sub-differential of f with respect to s.

In the particular case of a zero dissipation potential ϕ, the last expression reduces to

$$-\sigma:\beta^k + \rho\frac{L}{T_0}\left(T - T_A\right) \in -\rho\partial_{x_k} I_C,\ k = 1,\ldots n,$$

and represents the transition domain in a temperature-stress plane (T_A is the zero stress transition temperature). Figure 5 represents the numerical results obtained in a pseudoelastic test by using this model. The hysteresis loop is correctly predicted, and the temperature variations are consistent with experiment.

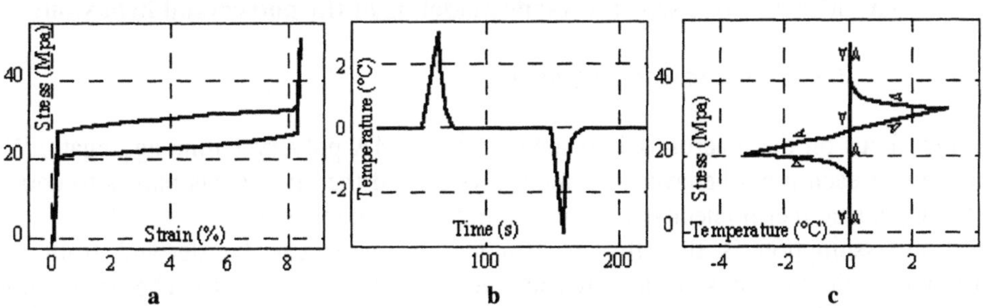

***Figure 5; a)** hysteresis due to thermomechanical couplings, **b)** temperature variations due to "stress-induced" phase change, **c)** thermodynamic path in the phase diagram corresponding to the load-unload anisothermal cycle.*

Even if thermal effects due to dissipation are difficult to measure, it is legitimate to suppose that a slight irreversibility accompanies the phase change. To test the influence of slight dissipation, we chose

$\phi(\dot{x}_1,\dot{x}_2,...,\dot{x}_n)=\frac{1}{2}K_v\left(\dot{x}_1^2+\dot{x}_2^2...+\dot{x}_n^2\right)$.

The existence of terms in $\dot{x}_k$ in the expression for the transition diagram expresses a dependence on the sense of transformation (M→A or A→M). Figure 6 shows that this slight dissipation has significant effects on the size of the hysteresis loop area, while the temperature evolutions with and without dissipation are hardly distinguishable.

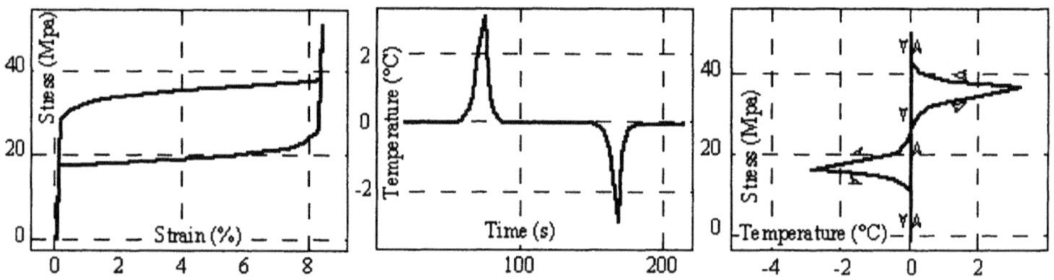

***Figure 6**; **a)** hysteresis due to thermomechanical couplings and to a slight intrinsic dissipation, **b)** temperature variations due to a "stress-induced" dissipative phase change, **c)** corresponding path in the phase diagram.*

These constitutive equations and the heat equation have been included in a finite element program to simulate the fully coupled thermomechanical behavior of monocrystalline SMAs [Peyroux, 1998, Balandraud, 2000]. It allows a correct prediction of pseudoelasticity, reorientation effect and recovery strain, it predicts phenomena such as asymmetry in tension-compression tests, time effects due to heat diffusion, inducing "relaxation".

3.3 Macroscopic scale; towards modeling of the polycrystal behaviour

3.3.1. Macroscopic variables

We now focus on the macroscopic behaviour of a SMA polycrystalline aggregate. We suppose that each grain behaves as a monocrystal, and the aim of this part is to derive the homogenized thermomechanical characteristics of the material.

The first assumption is that the macroscopic behavior can be derived from the thermomechanical response of a representative volume element (RVE). The observation of a polycrystalline sample by means of electronic and optical microscopy, allowed us to gather data on the crystallographic texture, the shape and the statistical representation of the grains in the RVE. (Figure 7)

In the following, we will consider that all the grains differ only by their crystallographic orientation.

It is established [Suquet, 1984] that for microscopic constitutive equations containing internal variables, "the homogenized law does not reduce to a single equation" on the macroscopic domain. "The knowledge of the macroscopic law requires as data the (mesoscopic) state variables". An alternative and pragmatic attitude is to link the macroscopic thermomechanical variables together, by prescribing thermomechanical loading and numerically solving the problems on the RVE. The RVE is considered as a virtual sample, and the finite element code as a virtual thermomechanical testing device.

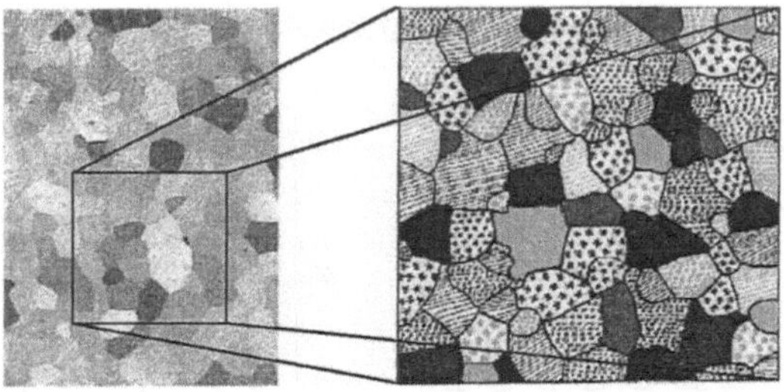

***Figure** 7; RVE of a polycrystalline sample of CuZnAl SMA obtained by image analysis; Different colors stand for different crystallographic orientations*

The set of variables used at the mesoscopic level is (T, ε, x_k), and additional variables such as the stress tensor σ or the heat influx vector q, can be deduced from the initial set owing to the transition diagram or the heat equation.
On the macroscopic level, we consider the classic and natural macroscopic strain and stress tensors E and Σ defined as average values of ε and σ:

$$E = \langle \varepsilon \rangle \text{ and } \Sigma = \langle \sigma \rangle .$$

At first we assume a quasi-homogeneous mesoscopic temperature field, the value of which is identified with the macroscopic temperature. Recall that the quasi-homogeneity of the temperature field does not imply the homogeneity of its gradient.
At this point E^{pc}, part of macroscopic strain due to phase change, is given by

$$E^{pc} = E - A^{-1}\Sigma - \alpha(T - T_0).$$

It remains to define X: macroscopic equivalence to the volume proportion of phase change. Deriving $\overline{L}$, macroscopic latent heat, from $\overline{\rho L} = \langle \rho L \rangle$, we can propose as a definition of X: $\overline{\rho L} X = \langle \rho L x \rangle$. This macroscopic proportion has to be regarded as an energy indicator of the advancement of phase change. The value of X is 0 if the polycrystal is completely austenitic, and 1 if completely martensitic. On the other hand a value between 0 and 1 can be reached under several microscopic configurations.

3.3.2 Some numerical results on the macroscopic behavior of a polycrystalline SMA
A convenient set of macroscopic variables being established, we present in this part some results of finite element simulations on the RVE.
First, we performed a so-called pseudoelasticity test consisting of a load-unload path at a room temperature upper than T_A and the different results are plotted in Figure 8. The imposed strain is increased from point *a* to point *d*, and maintained at a certain level between points *b* and *c*. In the particular stage *bc*, we can observe the role played by the thermomechanical couplings. The material keeps on transforming (X increases), the stress relaxes while the temperature returns to the imposed value. Now, concerning the entire test, the evolution of the different variables is consistent with experimental results obtained on polycrystalline SMA. This macroscopic behaviour is of course different to the monocrystalline one and corresponds to the particular ordering of grains chosen in the RVE.

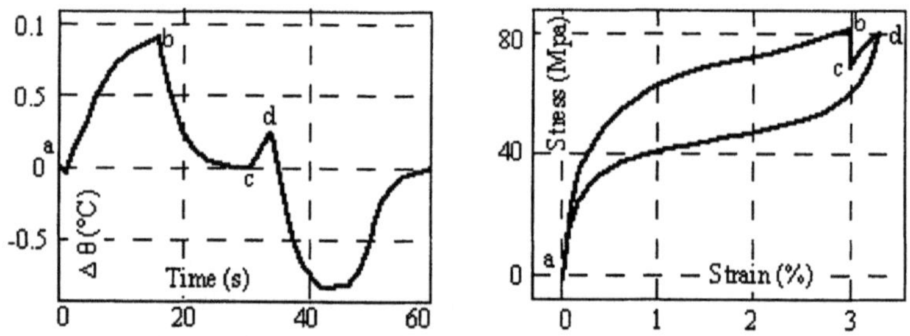

Figure 8; *Pseudoelastic "numerical" test*

From Figures 7 we can deduce some interesting points in the macroscopic transition domain. If we perform this analysis at several room temperatures, we obtain isovalues of transformation in the stress-temperature plane (Figure 9). An interesting result is that, according to the slope of the isovalues, it appears to be difficult to reach a complete transformation under reasonable stresses. In fact, some grains in the RVE have a non-compatible crystallographic orientation with respect to the imposed stress, while others transform easier. Note that taking into account a non zero dissipation leads to "moving lines" with regard to the direction of transformation.

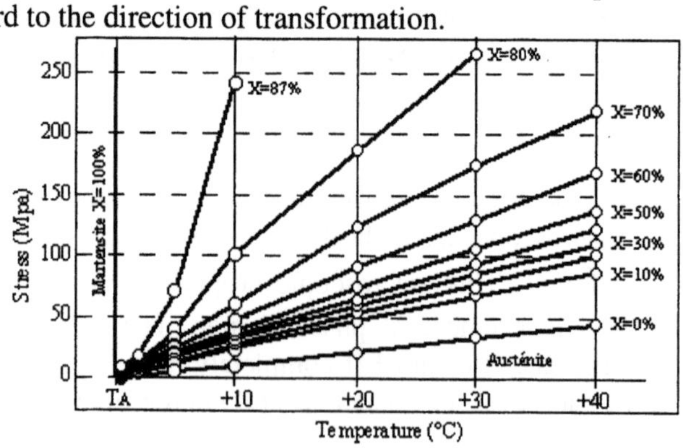

Figure 9 ; *Phase diagram for uniaxial tension test performed ;* $T \geq T_A$ *and* $d_I=0$.

The possibilities of this approach have been widely developed in [Balandraud, 2000]. For instance, low-temperature tests allowed determining the value of the macroscopic recoverable strain due to variants reorientation at zero-stress. This numerical tool could also be used to investigate multiaxial behavior of polycrystalline SMAs and to study the influence of various dissipation potentials. Finally, this approach giving access to mesoscopic behaviour, the evolution of phase change in each grain might be soon correlated with optical observations to check the consistency of this multiscale modeling.

4. CONCLUDING COMMENTS

We have tried to present the different points of view developed at the LMGC concerning the understanding of the SMA behaviour. This variety of approaches comes from the different analysis methods and from the chosen scales of description. Even though certain theoretical concepts were borrowed from other works in the literature, all our approaches led to original numerical tools. From an experimental point of view, the LMGC has developed for more than ten years infrared image processing techniques allowing us to perform calorimetric balances. For of SMAs, these tools made it possible to observe the strong thermomechanical coupling that the martensite transformation represents. At meso and macroscopic scales, the experiments highlighted the quasi-static and slightly dissipative character of the kinetics of this solid-solid phase transition.

REFERENCES

[Abeyaratne *et al.*, 1993] Abeyaratne, R, Knowles, J. K., A continuum model of thermoelastic solid capable of undergoing phase transitions, *J. Mech. Phys. Solids*, Vol.41, pp. 541-571.

[Auchmuty, 1989] Auchmuty, G.; Duality algorithms for non convex variational principles; In: *Numerical Functional Analysis and Optimization*, Vol. 10, pp 211-264.

[Balandraud, 2000] Balandraud, X., Changement de phase et changements d'echelle dans les alliages à mémoire de forme, *Thesis of University Montpellier II*.

[Ball *et al.*, 1987] Ball, J.M., James, R., Fine phase mixtures as minimizers of energy, *Arch. Ration. Analysis*, 100, pp. 15-52.

[Ball *et al.*, 1992] Ball, J.M., James, R., Proposed experimental tests of a theory of fine microstructure and the two-well problem, *Phil. Trans. R. Soc. London*, A, 338, pp; 389-450.

[Bhattacharya *et al.*, 1997], Battacharya, K., Kohn, R., Elastic energy minimization and the recoverable strains of polycrystalline shape-memory materials, *Arch. Rat. Mech. and Ana.*, pp.99-181.

[Chrysochoos *et al*, 1993], Chrysochoos, A., Pham, H., Maisonneuve, O., An experimental analysis of a thermomechanical behaviour of a CuZnAl shape memory alloy, *C.R. Acad. Sci, Paris*, t.316, Série II, pp.1031-1036.

[Chrysochoos *et al*, 1995], Chrysochoos, A., Löbel, M., Maisonneuve, O., Thermomechanical coupling of pseudoelastic behaviour of CuZnAl and NiTi alloys, *C.R. Acad. Sci, Paris*, t.320, Série II*b*, pp.217-223.

[Chrysochoos *et al.*, 2000], Chrysochoos, A., Louche, H., An infrared image processing to analyse calorific effects accompanying strain localisation, to appear in : *Int. J. Engng. Sci.*.

[Collins, 1993], Collins, C., Computation of twinning, microstructure and phase transition, *IMA*, vol54.

[Dacorogna, 1982], Dacorogna, B., Quasiconvexity and relaxation of nonconvex problems in the calculus of variation. *Journal of functional analysis,* Vol. 46, pp.102-118.

[Ericksen, 1984], Ericksen, J.L., The Cauchy-Born hypotheses for crystals. *Phase Transformations and Material Instabilities in Solids,* Ed. M. Gurtin, Academic Press, Vol. 46, pp.61-78.

[Frémond, 1987], Frémond, M.; Adhérence des solides; *Journal de Mécanique Théorique et Appliquée,* Vol 6, pp 383-407.

[Frémond, 1987], Frémond, M.; Matériaux à mémoire de forme; *C. R. Acad. Sci. Paris*, ", Vol 304, Série 2, pp 239-244.

[Germain, 1973], Germain, P, Cours de Mécanique des Milieux Continus, Masson et Cie Ed, p.417.

[Germain *et al.*, 1983], Germain, P, Nguyen, Q. S., Suquet, P., Continuum thermodynamics, *J. of Applied Mechanics*, vol. 50, pp. 1010-1020.

[Glansdorff *et al.*, 1971], Glansdorff, P., Prigogine, I., *Structure, Stabilités, Fluctuations*, Masson et Cie Ed., p.288.

[James, 1987], James, R.D.; The stability and metastability of quartz, *Metastability and Incompletely Posed Problems*, Springer-Verlag, IMA Vol 3, pp 147-176 .

[James *et al.*, 1989], James, R., Kinderlehrer, D., Theory of diffusionless phase transformations, *Lecture Notes in Physics*, 344, pp. 51-84.

[James *et al.*, 1993], Biaxial loading experiments on CuAlNi single crystals., AMD-Vol 181, *Experiments in Smart Materials and Structures*, ASME.

[Kohn, 1991], Kohn,R.V., The relaxation of a double-well energy, *Continuum Thermodyn.*, 3, pp. 193.

[Kondepudi *et al.*, 1998], Kondepudi, D, Prigogine, I., *Modern Thermodynamics*, Wiley & Sons Ed., p.486.

[Leclercq *et al.*, 1996], Leclercq, S.; Lexcellent, C.; A general macroscopic description of the thermodynamical behavior of shape memory alloys; *J. Mech. Phys. Solids*, Vol. 44, pp.953-980.

[Licht, 1997], Licht, C., Numerical experiments about non-convex bulk energy functions, proceedings of *Engineering Mechanics Today 97*, Hanoi Vietnam, 1997.

[Licht, 2000], Licht, C., Interaction energy in the macroscopic modelling of phase transition in solids, *Euromech 2000*, Metz.

[Müller *et al.*, 1991], Müller, I, Xu, H, On pseudoelastic hysteresis, *Acta Met. Mater.*, 39, 263,

[Pagano *et al.*, 1998], Pagano, S.; Alart, P.; Maisonneuve O.; Solid-solid phase transition modeling. Local and global minimizations of non-convex and relaxed potentials. Isothermal case for shape memory alloys; *Int. J. of Eng. Sc. 36,* pp. 1143-172.

[Pagano *et al.*, 1999], Pagano, S.; Alart, P.; Solid-solid phase transition modeling: relaxation procedures, configurational energies and thermomechanical behaviors; *Int. J. of Eng. Sc.* to appear.

[Papon *et al.*, 1999], Papon, P., Leblond, J., Meijer, Paul H.E., Physique des Transitions de Phases, Dunod Ed., p.390.

[Patoor *et al.*, 1994], Patoor, E., Berveiller, M, Technologie des AMF, Hermes ed., p. 286.

[Peyroux *et al.*, 1998], Peyroux, R., Chrysochoos, A., Licht, C., Löbel, M., Thermo-mechanical couplings and pseudoelasticity of shape memory alloys, *I. J. Eng. Sci.*, Vol. 36, N°4, pp. 489-509.

[Raniecki *et al.*, 1992], Raniecki, B.; Lexcellent, C.; Tanaka, K.; Thermodynamic models of pseudoelastic behavior of shape memory allows; *Arch. Mech.* Vol; 44, pp. 261-284.

[Stavroulakis *et al.*, 1993], Stavroulakis, G.E.; Panagiotopoulos,P.D.; Convex multilevel decomposition algorithms for non-monotone problems; *Int. J. for Num. Meth. in Eng.*, Vol. 36, pp. 1945-1966.

[Suquet, 1984], Suquet, P.M.; Approach by homogenization of some linear and nonlinear problems in solids mechanics; Ed. J.P. Boelher, *Colloques Internationaux du CNRS*, n°319, pp.77-117.

[Wattrisse, 1999], Wattrisse, B., Etude cinématique de la localisation dans des aciers par intercorrélation d'images, *Thesis*, University of Montpellier II.

[Zhong *et al.*, 1996], Zhong, S.G.; Batra, R.C.; Modeling of macroscopic response of phase transformation materials under quasi-static loading ; *J. Elasticity 44*, pp. 145-160.

Regularization of flutter ill-posedness in fluid-saturated porous media

Benjamin Loret[(1)]***, F.M.F. Simões***[(2)] ***and J.A.C. Martins***[(2)]
[(1)] *Laboratoire Sols, Solides, Structures BP 53 X 38041 Grenoble Cédex - France*
Benjamin.Loret@hmg.inpg.fr
[(2)] *Instituto Superior Técnico, DECivil and ICIST, 1049-001 Lisboa - Portugal*

Abstract: Elastic-plastic fluid-saturated porous media with non-associative flow rules may undergo strain localization and flutter. The simultaneous analysis of the nature and evolution of acceleration and harmonic waves has revealed a clear distinction between flutter instability and flutter ill-posedness. We show here that, as for solids, regularization methods may evidence infimum wavelengths below which any of these critical events is excluded.
Keywords: elasto-plasticity, porous media, flutter, regularization

1. INTRODUCTION

We consider fluid-saturated porous media in the framework of the theory of mixtures. The nature of the wave-speeds and of the propagation modes of acceleration waves in these media were analyzed in Loret and Harireche [1991], and their growth or decay were studied in Loret et al. [1997]. The nature of the speeds of harmonic waves propagating in the same media was studied in Simões et al. [1999]. The speeds of the acceleration waves are equal to the limit of the speeds of the harmonic waves for infinitely small wavelengths or for infinitely small viscous damping associated to Darcy's law. In addition, situations exist such that the harmonic waves have a growing time behavior that is controlled by a coefficient that equals, in the limit of an infinitely small wavelength, the finite growth coefficient of the corresponding acceleration waves, provided these propagate with real speeds: these situations correspond to linear instability. On the other hand, situations exist such that acceleration waves are not real, and, simultaneously, harmonic waves have an exponential growth in time that is unboundedly magnified as the wavelengths are decreased to zero: these are situations of linear ill-posedness, in the sense of the definitions proposed by Schaeffer [1990] and Benallal [1992]. The cases of complex wave-speeds are known as flutter (instability or ill-posedness) and the non-associative character of the plastic behavior of the solid skeleton is crucial for their occurrence. The onset of ill-posedness due to strain localization corresponds to the occurrence of stationary acceleration and harmonic waves.

To cure ill-posedness due to strain localization in solids, several regularization methods have been attempted in the past. They essentially consist in introducing an

G.A. Maugin et al. (eds.), Continuum Thermomechanics, 265-276.
© 2000 *Kluwer Academic Publishers.*

internal length-scale in the appropriate constitutive equations. We show here that these methods can also be used to cure strain localization and flutter ill-posedness in fluid-saturated porous media. Notice that the material length-scale to be introduced should address the source of the troubles, namely the behavior of the drained solid. Notice also that a length-scale already exists in the field equations due to Darcy's diffusion. However it does not affect the acceleration wave-speeds, and consequently it does not affect the onset of strain localization or flutter ill-posedness. Of course, coupled effects between material and diffusion length-scales will exist, the latter may even have a destabilizing effect, but, on its own, it will have no direct regularization effect on ill-posedness.

2. FIELD AND CONSTITUTIVE EQUATIONS

The porous media we consider are made up of two constituents, a solid and a fluid, namely water, which are viewed as two independent overlapping continua, see e.g. Bowen [1976]. Phases (solid (s) and fluid (w)) represent the constituents when viewed as part of the mixture. Each constituent has a mass M_α and a volume V_α, $\alpha{=}s, w$, which make up the total mass $M{=}M_s{+}M_w$ and the total volume $V{=}V_s{+}V_w$ of the mixture, which is also the volume of the solid phase. The intrinsic mass density is defined as $\rho_\alpha{=}M_\alpha/V_\alpha$, whereas the apparent mass density is defined by $\rho^\alpha{=}M_\alpha/V$; hence $\rho^\alpha{=}n^\alpha\,\rho_\alpha$ where $n^\alpha{=}V_\alpha/V$ is the volume fraction of phase α. Since we consider fluid-saturated porous media, the volume fractions satisfy $n^s{+}n^w{=}1$. The equations of balance of momentum of the phases involve two (symmetric) *partial stress tensors* $\boldsymbol{\sigma}^s$ and $\boldsymbol{\sigma}^w$, which make up the total stress tensor of the mixture $\boldsymbol{\sigma}{=}\boldsymbol{\sigma}^s + \boldsymbol{\sigma}^w$. The partial stress tensor of the fluid phase is defined in terms of its partial p^w or intrinsic p_w pressure as $\boldsymbol{\sigma}^w{=}{-}p^w\,\mathbf{I}{=}{-}n^w p_w\,\mathbf{I}$, with $\mathbf{I}$ identity tensor. Each phase α is endowed with its own (infinitesimal) apparent strain tensor $\boldsymbol{\epsilon}^\alpha$ defined as the symmetrized gradient of its macroscopic displacement field. The constituents are generally compressible but we restrict the present analysis to the case where both constituents are incompressible. Then the rates of volume changes must satisfy the linearized constraint in terms of velocities $\mathbf{v}_\alpha$ or strain-rates $\dot{\boldsymbol{\epsilon}}^\alpha$,

$$n^s\,\mathrm{div}\,\mathbf{v}_s + n^w\,\mathrm{div}\,\mathbf{v}_w = n^s\,\mathbf{I}:\dot{\boldsymbol{\epsilon}}^s + n^w\,\mathbf{I}:\dot{\boldsymbol{\epsilon}}^w = 0\,. \tag{1}$$

For each phase, $\alpha{=}s$, w, of the porous medium, the balance of momentum

$$\mathrm{div}\,\boldsymbol{\sigma}^\alpha + \hat{\mathbf{p}}_\alpha + \rho^\alpha\,(\mathbf{b}_\alpha - \mathbf{a}_\alpha) = \mathbf{0} \qquad \text{(no sum over } \alpha) \tag{2}$$

involves, in addition to the usual terms present in single phase solids, namely the divergence of the stress tensor, the body force per unit mass $\mathbf{b}_\alpha$ and the acceleration $\mathbf{a}_\alpha$, the apparent mass density ρ^α and the momentum supply $\hat{\mathbf{p}}_\alpha$ to the phase α by the rest of the mixture. Momentum supplies are subject to the closure condition

$\hat{\mathbf{p}}_s + \hat{\mathbf{p}}_w = \mathbf{0}$. The momentum supplies describe diffusion of the fluid through the solid skeleton, via an isotropic Darcy's law, $\hat{\mathbf{p}}_s = -\hat{\mathbf{p}}_w = -\xi\,(\mathbf{v}_s - \mathbf{v}_w)$, that introduces a single constant material parameter $\xi > 0$, proportional to the inverse of the permeability. Since we envisage a linearized analysis around a natural equilibrium state, the mass density ρ^α in eqn (2) is fixed to its reference value and the acceleration $\mathbf{a}_\alpha$ is approximated by the time derivative of the velocity $\dot{\mathbf{v}}_\alpha$.

Due to the incompressibility condition (1), the fluid pressure p_w is indeterminate and the elastic constitutive equations specify solely Terzaghi's effective stress $\bar{\boldsymbol{\sigma}} = \boldsymbol{\sigma} + p_w\,\mathbf{I}$ in terms of the strain of the solid phase $\boldsymbol{\epsilon}^s$,

$$\bar{\boldsymbol{\sigma}} = \mathbb{E}^{\text{DS}} : \boldsymbol{\epsilon}^s\,. \tag{3}$$

For isotropic elasticity, the positive definite tensor moduli of the *drained solid* $\mathbb{E}^{\text{DS}}$ are defined by the two Lamé constants λ^{DS} and μ, with $\lambda^{\text{DS}} + 2\,\mu/3 > 0$ and $\mu > 0$.

Plastic behavior emanates from the solid phase. However, both solid and fluid *phases* will develop irreversible strains. The plastic properties are defined in terms of the effective stress $\bar{\boldsymbol{\sigma}}$. The yield function, with unit ouward normal $\mathbf{Q}$, is defined in the space of effective stresses $\bar{\boldsymbol{\sigma}}$, $f{=}f(\bar{\boldsymbol{\sigma}}, \mathcal{H})$, where $\mathcal{H}$ is a set of hardening variables. The strain rates of the two phases are additively decomposed into an elastic part and a plastic part, namely $\dot{\boldsymbol{\epsilon}}^\alpha{=}\dot{\boldsymbol{\epsilon}}^\alpha_{\text{el}} + \dot{\boldsymbol{\epsilon}}^\alpha_{\text{pl}}$, $\alpha{=}s, w$. The plastic strain rate of the solid skeleton is defined by the flow rule $\dot{\boldsymbol{\epsilon}}^s_{\text{pl}} = \dot{\Lambda}\,\mathbf{P}$, where $\dot{\Lambda} \geq 0$ is the plastic multiplier and $\mathbf{P}$ a unit symmetric tensor indicating the direction of flow. In order to satisfy the incompressibility constraint for both elastic unloading and plastic loading, the plastic strain rate in the fluid phase is defined by the *"plastic coupling equation"* $\mathbf{I} : \dot{\boldsymbol{\epsilon}}^w_{\text{pl}} = \boldsymbol{\beta} : \dot{\boldsymbol{\epsilon}}^s_{\text{pl}}$ where $\boldsymbol{\beta}{=}-(n^s/n^w)\,\mathbf{I}$. The evolution of the hardening variables $\mathcal{H}$ is introduced through the hardening modulus h. The plastic consistency condition $\dot{f}(\bar{\boldsymbol{\sigma}}, \mathcal{H}){=}0$ provides the plastic multiplier $\dot{\Lambda}$,

$$\dot{\Lambda} = \frac{1}{H^{\text{DS}}}\,\mathbf{Q} : \mathbb{E}^{\text{DS}} : \dot{\boldsymbol{\epsilon}}^s\,, \qquad H^{\text{DS}} = h + \mathbf{Q} : \mathbb{E}^{\text{DS}} : \mathbf{P} > 0 \tag{4}$$

where the modulus H^{DS} is supposed to be strictly positive. The resulting elastic-plastic rate equations emerge as

$$\dot{\bar{\boldsymbol{\sigma}}} = \mathbb{E}^{\text{DS}}_{\text{ep}} : \dot{\boldsymbol{\epsilon}}^s\,, \qquad \mathbb{E}^{\text{DS}}_{\text{ep}} = \mathbb{E}^{\text{DS}} - \frac{1}{H^{\text{DS}}}\,\mathbb{E}^{\text{DS}} : \mathbf{P} \otimes \mathbf{Q} : \mathbb{E}^{\text{DS}}\,. \tag{5}$$

Hence, both the elastic and plastic constitutive moduli associated to the effective stress are equal to their drained counterparts defined by the drainage condition $p_w = 0$ or $\dot{p}_w = 0$. When the plastic potential and the yield function are of the Drucker-Prager type, each of the unit normals $\mathbf{P}$ and $\mathbf{Q}$ is defined by a single parameter, dilatancy angle χ for $\mathbf{P}$ and friction angle ψ for $\mathbf{Q}$, and $\mathbf{P} = \cos\chi\,\hat{\mathbf{S}} + \sin\chi\,\mathbf{I}/\sqrt{3}$, $\mathbf{Q}{=}\cos\psi\,\hat{\mathbf{S}} + \sin\psi\,\mathbf{I}/\sqrt{3}$, $\hat{\mathbf{S}}$ being the normalized deviatoric stress.

The subsequent analysis of harmonic waves will make use of the elastic and elastic-plastic acoustic tensors of the drained solid, $\mathbf{B}^{\text{DS}}$ and $\mathbf{B}^{\text{DS}}_{\text{ep}}$ respectively, namely

$$\mathbf{B}^{\text{DS}} = \mathbf{n} \cdot \frac{\mathbb{E}^{\text{DS}}}{\rho^s} \cdot \mathbf{n}\,, \quad \mathbf{B}^{\text{DS}}_{\text{ep}} = \mathbf{n} \cdot \frac{\mathbb{E}^{\text{DS}}_{\text{ep}}}{\rho^s} \cdot \mathbf{n} = \mathbf{B}^{\text{DS}} - \frac{1}{H^{\text{DS}}}\,\frac{\mathbf{p} \otimes \mathbf{q}}{\rho^s}\,, \tag{6}$$

where $\mathbf{p}$ and $\mathbf{q}$ are vectors, $\mathbf{p}=\mathbf{p}(\mathbf{n})=(\mathbb{E}^{DS}:\mathbf{P})\cdot\mathbf{n}$ and $\mathbf{q}=\mathbf{q}(\mathbf{n})=\mathbf{n}\cdot(\mathbf{Q}:\mathbb{E}^{DS})$.

3. HARMONIC WAVE ANALYSIS AND FLUTTER ILL-POSEDNESS

A linearized analysis of the stability of the flow in elastic-plastic fluid-saturated porous media with incompressible constituents is performed. Compressibility of the constituents does not affect the major qualitative features of the results presented here, Simões [1997]. The dynamic growth or decay of a small plane harmonic perturbation is analyzed. Of particular interest is the limit behavior of the wave-speeds and of the growth coefficient as the limit of a zero wavelength, corresponding to acceleration waves, is approached.

The exclusive consideration of the plastic loading regime in the present linearized stability analysis precludes an assessment of the full consequences of the detected flutter instabilities, since the growing oscillatory nature of the flutter solutions leads to situations of local elastic unloading that are quite distinct from the linearized loading regime considered in this stability analysis. Nevertheless, it makes sense to compare the results of the present harmonic wave analysis with those of acceleration waves, because only the linearized constitutive equations for the plastic loading regime are used in both analyses.

Assuming that the body forces $\mathbf{b}_\alpha, \alpha=s,w$, are constant in time, we seek solutions in terms of velocity fields
- that satisfy the constitutive equations (5), the incompressibility constraint (1) and the rate form of the equations of linear momentum balance for each phase (2),
- and that represent plane harmonic waves of assigned wavenumber k propagating with the (possibly complex) speed of propagation c along the direction $\mathbf{n}$, namely:

$$\mathbf{v}_\alpha(\mathbf{x},t)=\hat{\mathbf{v}}_\alpha \exp\Big(\mathrm{i}\,k\,(\mathbf{n}\cdot\mathbf{x}-c\,t)\Big)=\hat{\mathbf{v}}_\alpha \exp(k\,\mathcal{I}m\,c\;t)\,\exp\Big(\mathrm{i}\,k\,(\mathbf{n}\cdot\mathbf{x}-\mathcal{R}e\,c\;t)\Big)\,, \quad (7)$$

where the $\hat{\mathbf{v}}_\alpha$'s, $\alpha=s,w$, are eigenvectors to be defined. The wavenumber k is linked to the wavelength L by the usual relation $k=2\pi/L$. Notice that Darcy's law introduces a characteristic length-scale L_{dif} in the field equations, namely

$$\frac{1}{L_{\mathrm{dif}}}=\frac{\xi}{c_s^e}\Big(\frac{1}{\rho^s}+\frac{1}{\rho^w}\Big)\,. \quad (8)$$

The indeterminate rate of the partial pressure in the fluid p^w is also assumed to be a periodic quantity:

$$\dot{p}^w=\hat{p}^w\,n^w\,k\,\exp\Big(\mathrm{i}\,k\,(\mathbf{n}\cdot\mathbf{x}-c\,t)\Big)\,. \quad (9)$$

Introduction of the rate constitutive equations (5) in the rate form of the balance of momentum of the solid phase provides one set of three equations for the eigenmode $(\hat{\mathbf{v}}_s, \hat{\mathbf{v}}_w, \hat{p}^w)$. The rate form of the balance of momentum in the fluid phase provides another set of three equations while the incompressibility constraint (1) yields the seventh relation. The resulting algebraic system can be cast in the form

$$\begin{bmatrix} \rho^s\,\mathbf{B}_{\rm ep}^{\rm DS} - (\mathrm{i}\,c\,\xi k^{-1} + \rho^s\,c^2)\mathbf{I} & \mathrm{i}\,c\,\xi k^{-1}\mathbf{I} & \mathrm{i}\,n^s\,\mathbf{n} \\ \mathrm{i}\,c\,\xi k^{-1}\mathbf{I} & -(\mathrm{i}\,c\,\xi k^{-1} + \rho^w\,c^2)\,\mathbf{I} & \mathrm{i}\,n^w\,\mathbf{n} \\ \mathrm{i}\,n^s\,\mathbf{n}^{\rm T} & \mathrm{i}\,n^w\,\mathbf{n}^{\rm T} & 0 \end{bmatrix} \begin{bmatrix} \hat{\mathbf{v}}_s \\ \hat{\mathbf{v}}_w \\ \hat{p}^w \end{bmatrix} = \begin{bmatrix} \mathbf{0} \\ \mathbf{0} \\ 0 \end{bmatrix}. \tag{10}$$

The characteristic equation for the normalized wave-speed $X{=}c/c_s^e$ can be factorized as a product of polynomials in X, namely $X^3\;F(\mathrm{i}\,X)\,G(\mathrm{i}\,X) = 0$. $G(\mathrm{i}\,X)$ contains only elastic information while $F(\mathrm{i}\,X)$ contains elastic-pastic information and it has in general one purely imaginary root X, and two pairs of roots X symmetric with respect to the imaginary axis, i.e. if $X = \mathcal{R}e\,X + \mathcal{I}m\,X$ is a root, so is $-\mathcal{R}e\,X + \mathcal{I}m\,X$. Both F and G depend on the nondimensional wavelength α,

$$\alpha = \frac{1}{2\,\pi}\frac{L}{L_{\rm dif}}. \tag{11}$$

One can show the following

Proposition 1: *As α tends to zero, i.e. for infinitely small wavelengths or for infinitely small viscous damping associated to Darcy's law, the harmonic wave-speeds tend to the acceleration wave-speeds.*

Assuming that during a deformation process the modulus $H^{\rm DS}$ (or equivalently the plastic modulus h) decreases continuously starting from $H^{\rm DS}{=}\infty$ (elasticity), we look for situations in which divergence or flutter-type growth occurs along at least one direction $\mathbf{n}$. As we already know, Loret and Harireche [1991], the viscous effects due to Darcy's law play no role on the determination of the onset of stationary waves i.e. $X = 0$. Since the solutions X are in general complex, transitions to divergence ($\mathcal{I}m\,X > 0, \mathcal{R}e\,X = 0$) can occur either when $\mathcal{R}e\,X = 0$ and $\mathcal{I}m\,X = 0$ (a stationary wave) or when $\mathcal{I}m\,X > 0$ and $\mathcal{R}e\,X \to 0$, see Figure 1. However in the last case divergence occurs necessarily after the occurrence of flutter ($\mathcal{I}m\,X > 0, \mathcal{R}e\,X \neq 0$). We therefore do not investigate these events in detail and concentrate on flutter.

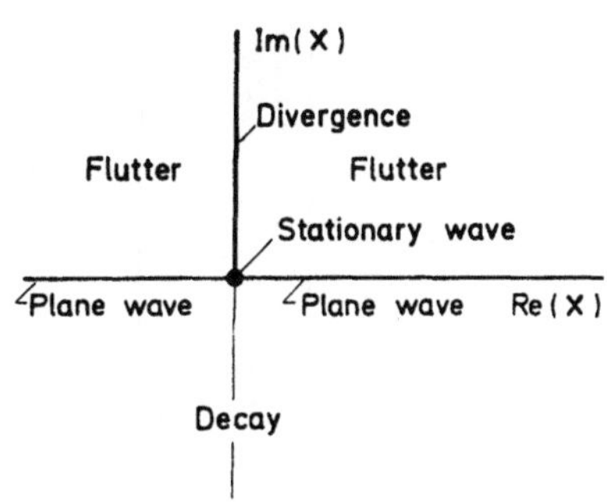

Figure 1. *Sketch of the five possibilities for the nature of the normalized wave-speed X. $\mathcal{I}m\,X$ negative, zero or positive implies motions with an amplitude which is, respectively, exponentially decaying, constant or exponentially growing in time.*

Notice that, for elastic mixtures ($H^{\rm DS}=\infty$) or for elastic-plastic mixtures with associative plasticity ($\mathbf{P}=\mathbf{Q}$), the matrix in (10) is symmetric. Therefore the speeds X are real or complex with negative imaginary parts, so that flutter is excluded. For non-associative plasticity ($\mathbf{P}\neq\mathbf{Q}$), one of the roots X may become complex with positive imaginary part, leading to an harmonic solution with flutter-type growth. As already observed, only the plastic wave-speeds X that are roots of $F(\mathrm{i}\,X)=0$ may become complex leading to flutter. When the acceleration wave-speeds are real, then, according to Proposition 1, $\lim_{\alpha\to 0}\mathcal{I}m\,X(\alpha)=0$. The limit as $\alpha\to 0$ of the coefficient that controls the growing time behavior of the harmonic waves, namely $k\,\mathcal{I}m\,c=c_s^e/L_{\rm dif}\,\mathcal{I}m\,X/\alpha$, can be estimated by l'Hôpital's rule and the continuity of $d\mathcal{I}m\,X(\alpha)/d\alpha$ at $\alpha=0$. After algebraic manipulations, this limit is found to be equal to the coefficient that controls the growth or decay of the acceleration waves obtained in Loret et al. [1997]. We can thus state

Proposition 2: *In the regions of the parameter space where the acceleration waves have real speeds, the corresponding harmonic waves are found to grow or decay at the same rate when their wavelengths are decreased to zero.*

On the other hand, if the normalized acceleration wave-speed X is not real, then the limit as $\alpha\to 0$ of $\mathcal{I}m\,X(\alpha)$ is equal to $\mathcal{I}m\,X(\alpha=0)>0$ and

$$\lim_{\alpha\to 0} k\;\mathcal{I}m\,c(\alpha)=\frac{c_s^e}{L_{\rm dif}}\lim_{\alpha\to 0}\mathcal{I}m\,\frac{X(\alpha)}{\alpha}=\infty\,. \tag{12}$$

So, complementary to the previous Proposition, we have

Proposition 3: *Inside the regions of the parameter space where there are some speeds of acceleration waves that are not real, the exponentially growing time behavior of the harmonic solutions is unboundedly magnified for vanishing small values of the arbitrary wavelengths, and the problem becomes (linearly) ill-posed.*

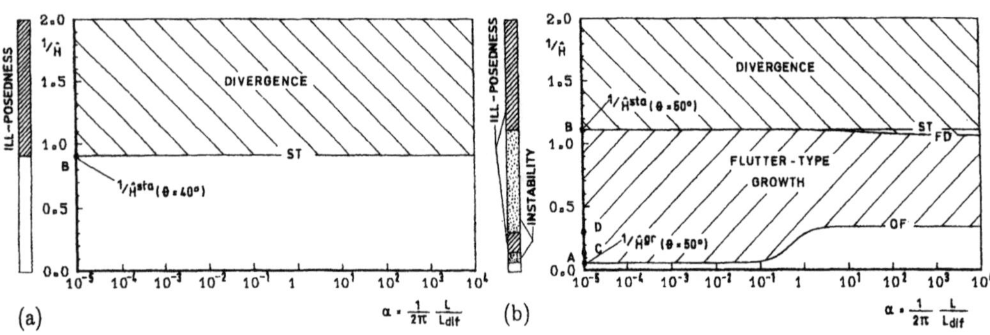

Figure 2. Regions of divergence and flutter growth in the plane ($1/\hat{H}$, α). Material parameters: $\chi=0°$; $\psi=30°$; $\lambda^{DS}/\mu=15$, (a) $\theta=40°$; (b) $\theta=50°$; ST: Stationary discontinuity, OF: Onset of flutter, FD: Flutter-divergence transition.

In Figure 2, we represent the regions of divergence and flutter growth in the parameter plane $(1/\hat{H}, \alpha)$ ($\hat{H}=H^{DS}/2\mu$ is the normalized modulus), for two values of λ^{DS}/μ, one excluding flutter for acceleration waves, the other not, and for two directions of propagation $\mathbf{n}$ belonging to the plane defined by the eigenvectors of $\hat{\mathbf{S}}$ that correspond to the maximum and minimum eigenvalues, $\mathbf{E}_1$ and $\mathbf{E}_3$ respectively, and we denote θ=angle$(\mathbf{E}_1, \mathbf{n})$. It can be seen that it is not possible to find, for given normalized plastic modulus and diffusion length, an infimum wavelength which would preclude linear ill-posedness. It can also be seen that the limit as $\alpha \to 0$ of the divergence boundary in the harmonic wave analysis coincides with the onset of stationary waves in the acceleration wave analysis (point B). In contrast, the region of flutter is discontinuous as $\alpha \to 0$: the limit of the flutter region in the harmonic wave analysis (line AB) is larger than the flutter region in the acceleration wave analysis (line CD). This means that in the harmonic wave analysis for the non-associative case, the viscous damping effects due to Darcy's law may anticipate, or even trigger, the occurrence of flutter relatively to the acceleration wave analysis (the no-damping case).

Although the characteristic length introduced by Darcy's law does not prevent small wavelength unstable modes, the exponentially growing time behavior of the unstable modes is unboundedly magnified for vanishing small wavelengths only when values of H^{DS} are used such that non-real speeds of propagation exist for the corresponding acceleration waves. A clear distinction has thus been made between (linear) ill-posedness and (linear) instability, precisely as defined in the Introduction. The regions AC and DB in Fig. 2-(b) are representative of linear instability, while the regions CD and B to ∞ are representative of linear ill-posedness.

4. REGULARIZATION BY INTERNAL LENGTH

In the analysis of harmonic waves, we stressed the (linear) ill-posedness character of the equations that govern the behavior of an elastic-plastic fluid saturated porous medium for some ranges of the material parameters. On the other hand, it has been long recognized that ill-posedness due to stationarity wave/strain localization can be fixed by using some enriched constitutive descriptions of the continuum that incorporate internal (or characteristic) lengths of the material. A related study for a case of flutter ill-posedness but in the context of contact problems involving the (non-associative) friction law of Coulomb was recently presented by Simões and Martins [1998]. In the present context of flutter ill-posedness of incompressible fluid-saturated porous media, we discuss now the effect of two classical enrichments of the material behavior: the gradient plasticity theory, e.g. Mühlhaus and Aifantis [1991], and the non-local plasticity theory, e.g. Bažant and Lin [1988].

4.1 Gradient plasticity

We use here the simple modified form of the plastic behavior proposed by Mühlhaus and Aifantis [1991] that consists of assuming that the yield function f at a point $\mathbf{x}$ depends not only on the cumulated plastic strain κ, $\dot{\kappa} = \| \dot{\boldsymbol{\epsilon}}^s_{\rm pl} \| = \dot{\Lambda}$, but also on its Laplacian $\Delta\kappa$: $f = f(\bar{\boldsymbol{\sigma}}, \kappa, \Delta\kappa)$, all three arguments of f being estimated at $\mathbf{x}$. The strain decomposition into elastic and plastic components as well as the flow rule are preserved: $\dot{\boldsymbol{\epsilon}}^s = \dot{\boldsymbol{\epsilon}}^s_{\rm el} + \dot{\boldsymbol{\epsilon}}^s_{\rm pl}$ and $\dot{\boldsymbol{\epsilon}}^s_{\rm pl} = \dot{\Lambda}\,\mathbf{P}$. The plastic consistency condition

$$\dot{f} = \frac{\partial f}{\partial \boldsymbol{\sigma}} : \dot{\boldsymbol{\sigma}} + \frac{\partial f}{\partial \kappa} : \dot{\kappa} + \frac{\partial f}{\partial \Delta\kappa} : \Delta\dot{\kappa} = 0 \,, \tag{13}$$

provides the plastic multiplier $\dot{\Lambda}$,

$$\dot{\Lambda} = \frac{1}{H^{\rm DS}} (\mathbf{Q} : \mathbb{E}^{\rm DS} : \dot{\boldsymbol{\epsilon}}^s + b\,\Delta\dot{\kappa}) \,, \tag{14}$$

where $H^{\rm DS}$ keeps its classic form, eqn (4), and

$$\mathbf{Q} = \frac{\partial f}{\partial \boldsymbol{\sigma}} \,, \quad h = -\frac{\partial f}{\partial \kappa} \,, \quad b = \frac{\partial f}{\partial \Delta\kappa} \geq 0 \,. \tag{15}$$

The rate form of the elastic-plastic constitutive law (for the plastic loading regime) reads:

$$\dot{\boldsymbol{\sigma}} = \mathbb{E}^{\rm DS}_{\rm ep} : \dot{\boldsymbol{\epsilon}}^s - \frac{b\,\Delta\dot{\kappa}}{H^{\rm DS}} \mathbb{E}^{\rm DS} : \mathbf{P} \tag{16}$$

with $\mathbb{E}^{\rm DS}_{\rm ep}$ exactly as in (5). These are the equations that we wish to solve, together with the equations (2) of linear momentum balance and the incompressibility condition (1). We seek harmonic solutions for them exactly of the form indicated in (10). Note that the only calculations that are now different from those performed in Section 3 are those involving the Laplacian of the cumulated plastic strain, present in the term that makes the constitutive equation (16) different from (5). For monochromatic velocity $\mathbf{v}_s$ and plastic strain-rate $\dot{\kappa}$,

$$\begin{bmatrix} \mathbf{v}_s \\ \dot{\kappa} \end{bmatrix} = \begin{bmatrix} \hat{\mathbf{v}}_s \\ \hat{\kappa} \end{bmatrix} \exp\Big(\mathrm{i}\,k\,(\mathbf{n}\cdot\mathbf{x} - c\,t)\Big) \,, \tag{17}$$

with wavelength L and wavenumber k linked by $k = 2\,\pi/L$, it follows that

$$\dot{\boldsymbol{\epsilon}}^s = \frac{\mathrm{i}\,k}{2} (\mathbf{v}_s \otimes \mathbf{n} + \mathbf{n} \otimes \mathbf{v}_s) \,, \quad \Delta\dot{\kappa} = -k^2\,\dot{\kappa} \,. \tag{18}$$

Insertion of these relations in (14) yields

$$\dot{\kappa} = \frac{\mathrm{i}\,k}{H^{\rm DS}_{\rm grad}}\,\mathbf{q}\cdot\mathbf{v}_s \,, \quad \Delta\dot{\kappa} = -\frac{\mathrm{i}\,k^3}{H^{\rm DS}_{\rm grad}}\,\mathbf{q}\cdot\mathbf{v}_s \,, \tag{19}$$

with $\mathbf{q}$ defined in Section 2 and $H^{\rm DS}_{\rm grad}$ is a simple modification of $H^{\rm DS}$:

$$H^{\rm DS}_{\rm grad} = H^{\rm DS} + b\,k^2 \,. \tag{20}$$

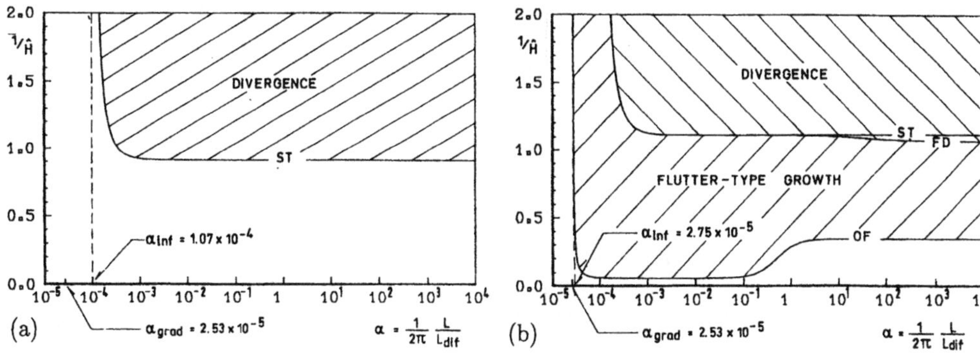

Figure 3. *Same as Fig. 2 but with a gradient regularization showing a minimum wavelength below which (a) neither divergence, (b) nor flutter modes appear.* $L_{\text{grad}}/L_{\text{dif}}=10^{-3}/2\pi$ *(the plots in Fig. 2 correspond to* $L_{\text{grad}}/L_{\text{dif}}=0$*).*

If we now insert the above constitutive equations in the harmonic wave analysis of Section 3, we easily observe that only the balance of momentum for the solid phase is modified due to eqn (16) with $\Delta\dot{\kappa}$ given by (19). The algebraic system of equations (10) providing the acceleration wave-speeds preserves its form, with the only distinction that the acoustic tensor $\mathbf{B}^{\text{DS}}_{\text{ep}}$, eqn (6), is replaced by $\mathbf{B}^{\text{DS}}_{\text{ep, grad}}$,

$$\mathbf{B}^{\text{DS}}_{ep,\,\text{grad}} = \mathbf{B}^{\text{DS}} - \frac{1}{H^{\text{DS}}_{\text{grad}}}\,\frac{\mathbf{p}\otimes\mathbf{q}}{\rho^s}\,. \tag{21}$$

Thus the sole modification introduced in the plasticity model turns out to be equivalent, for the propagation of harmonic waves, to replace the hardening modulus H^{DS} by the modified value $H^{\text{DS}}_{\text{grad}}$.

The scalar b is assumed positive, eqn (15), it has dimension stress$\times$length^2 and it introduces an *internal length-scale* L_{grad},

$$L_{\text{grad}} = \sqrt{\frac{b}{\mu}}\,, \tag{22}$$

but recall that there is another length-scale induced by Darcy's diffusion, namely L_{dif}, eqn (8).

The dependence of the occurrence of flutter and divergence on the wavelengths is illustrated in Figure 3 for this gradient plasticity version of the material behavior discussed in Section 3. The existence of an infimum wavelength for the occurrence of flutter and divergence modes is made clear in those figures. As the wavelength

L, and consequently α, decreases, the modified normalized modulus $\hat{H}_{\text{grad}}$,

$$\hat{H}_{\text{grad}} = \hat{H} + \frac{1}{2\,\alpha^2}\frac{L^2_{\text{grad}}}{L^2_{\text{dif}}}\,. \tag{23}$$

becomes larger and larger, which explains the disappearance of flutter and divergence modes. In fact, at a given wavelength, the critical moduli for any of these events for the plasticity model, $\hat{H}^{\text{crit}}(L_{\text{grad}}{=}0)$, and for the gradient plasticity model, $\hat{H}^{\text{crit}}(L_{\text{grad}})$ differ by a simple shift,

$$\hat{H}^{\text{crit}}(L_{\text{grad}}) = \hat{H}^{\text{crit}}(L_{\text{grad}}{=}0) - \frac{1}{2\,\alpha^2}\frac{L^2_{\text{grad}}}{L^2_{\text{dif}}}\,. \tag{24}$$

The assumption of a positive modulus $\hat{H}$ thus implies the existence of an infimum nondimensional wavelength $\alpha_{\text{inf}}= (2/\hat{H}^{\text{crit}}(L_{\text{grad}}{=}0))^{1/2}\,L_{\text{grad}}/L_{\text{dif}}$. Notice that the infimum wavelength $L_{\text{inf}}= 2\,\pi\,(2/\hat{H}^{\text{crit}}(L_{\text{grad}}{=}0))^{1/2}\,L_{\text{grad}}$ is independent of L_{dif}.

4.2 Non-local plasticity

We provide here a non-local modification of the yield function with respect to the classic elastic-plastic model. It turns out to be not as satisfactory as the gradient modification. In fact, as shown by another modification of the plastic strain-rate in Loret et al. [1999], the regularizing effects are very sensitive to the quantities that are averaged.

The modification introduced here in the plastic behavior consists in assuming that the yield function f at each point of the body depends of the cumulated plastic strain $\kappa(\mathbf{x})$ at that point and on a weighted average of the cumulated plastic strains in a neighborhood $V(\mathbf{x})$ of that point:

$$f = f\,(\,\bar{\boldsymbol{\sigma}}(\mathbf{x}),\,\kappa(\mathbf{x}),\,\tilde{\kappa}(\mathbf{x}) - \kappa(\mathbf{x})\,)\;. \tag{25}$$

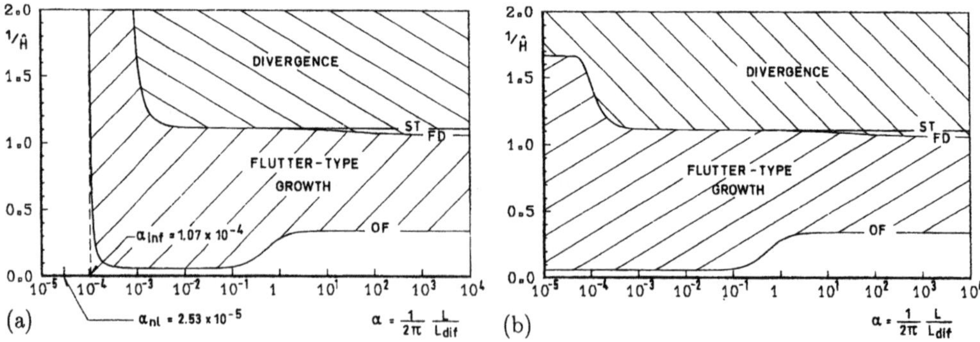

Figure 4. *Same as Fig. 2-(b) but with a non-local regularization which is (a) effective for $b/2\,\mu = 25$ and (b) not effective for $b/2\,\mu = 0.3$; $L_{\text{nl}}/L_{\text{dif}}{=}10^{-3}/2\,\pi$.*

with

$$\tilde{\kappa}(\mathbf{x}) = \frac{1}{V(\mathbf{x})} \int_{V(\mathbf{x})} w(\mathbf{x}-\mathbf{s})\, \kappa(\mathbf{s})\, d\,V(\mathbf{s})\,. \tag{26}$$

For the present calculations, we shall use as weighting function the normalized Gaussian function

$$w(\mathbf{y}) = \frac{1}{(2\,\pi)^{3/2}} \exp\Big(-\frac{\|\,\mathbf{y}\,\|^2}{2\,L_{\mathrm{nl}}^2}\Big)\,, \tag{27}$$

where L_{nl} is the characteristic length of this modified plasticity model. Again, it is a consequence of the plastic consistency condition that

$$\dot{\Lambda} = \frac{1}{H^{\mathrm{DS}}}\Big(\mathbf{Q} : \mathbb{E}^{\mathrm{DS}} : \dot{\boldsymbol{\epsilon}}^{s} + b\,(\dot{\tilde{\kappa}} - \dot{\kappa})\Big)\,, \tag{28}$$

where $b = \partial f/\partial\tilde{\kappa} \geq 0$. The rate form of the elastic-plastic constitutive law (for the plastic loading regime) reads:

$$\dot{\boldsymbol{\sigma}} = \mathbb{E}_{\mathrm{ep}}^{\mathrm{DS}} : \dot{\boldsymbol{\epsilon}}^{s} - \frac{b\,(\dot{\tilde{\kappa}} - \dot{\kappa})}{H^{\mathrm{DS}}}\, \mathbb{E}^{\mathrm{DS}} : \mathbf{P}\,. \tag{29}$$

The consideration of harmonic solutions for $\dot{\kappa}$, eqn (18), leads to

$$\dot{\tilde{\kappa}}(\mathbf{x}) = \dot{\kappa}(\mathbf{x}) \exp\Big(-\frac{k^2\,L_{\mathrm{nl}}^2}{2}\Big)\,, \tag{30}$$

Then for the harmonic problem, the elastic-plastic rate constitutive equations can be cast in the usual format $\dot{\boldsymbol{\sigma}} = \mathbb{E}_{\mathrm{ep,nl}}^{\mathrm{DS}} : \dot{\boldsymbol{\epsilon}}^{s}$ with $\mathbb{E}_{\mathrm{ep,nl}}^{\mathrm{DS}} = \mathbb{E}^{\mathrm{DS}} - 1/H_{\mathrm{nl}}^{\mathrm{DS}}\, \mathbb{E}^{\mathrm{DS}} : \mathbf{P} \otimes \mathbf{Q} : \mathbb{E}^{\mathrm{DS}}$ and the sole modification with respect to the local formulation is due to the shift in modulus,

$$\hat{H}_{\mathrm{nl}} = \hat{H} + \frac{b}{2\,\mu}\Big(1 - \exp\Big(-\frac{k^2\,L_{\mathrm{nl}}^2}{2}\Big)\Big)\,. \tag{31}$$

The resulting modification of the governing algebraic system of equations (10) affects the acoustic tensor of the drained solid only, as in (21), but with now H^{DS} replaced by $H_{\mathrm{nl}}^{\mathrm{DS}}$. The moduli defining any critical event are related, similarly to (24), by,

$$\hat{H}^{\mathrm{crit}}(L_{\mathrm{nl}}) = \hat{H}^{\mathrm{crit}}(L_{\mathrm{nl}}{=}0) - \frac{b}{2\,\mu}\Big(1 - \exp\Big(-\frac{k^2\,L_{\mathrm{nl}}^2}{2}\Big)\Big)\,. \tag{32}$$

Unlike for the gradient plasticity model, the modification due to small wavelengths or large wavenumbers is finite and coincides with a shift of normalized plastic modulus equal to $b/2\,\mu$. The existence of an infimum wavelength below which neither flutter nor divergence would occur requires $\hat{H} + b/2\,\mu$ to be greater than any of the critical moduli $\hat{H}^{\mathrm{crit}}(L_{\mathrm{nl}} = 0)$, which may result in a large value of $b/2\,\mu$, see Figure 4. The same reasoning applies to individual critical events. For example, in their one-dimensional analysis of strain localization, Vermeer and Brinkgreve [1994] stress that the small wavelength regularization of strain localization requires, in our notation, $h + b$ to be strictly positive. Moreover, the non-local problem is not exactly similar to the local one as the shift in apparent modulus increases when the wavelength decreases, so that one can tailor the material parameters b and $L_{\rm nl}$ to imply prescribed critical wavelengths, or conversely.

REFERENCES

Bažant, Z.P. and Lin, F.B. (1988). Non-local yield limit degradation. *Int. J. Num. Meth. Engineering*, **26**, 1805-1823.

Benallal, A. (1992). Ill-posedness and localization in solid structures. *Computational Plasticity*, D.R.J. Owen et al. eds., Pineridge Press, Swansea, U.K., 581-592.

Bowen, R.M (1976). Theory of mixtures. *Continuum Physics*, vol. 3, 1-127, A.C. Eringen ed., Academic Press, New York.

Loret, B. and Harireche, O. (1991). Acceleration waves, flutter instabilities and stationary discontinuities in inelastic porous media. *J. Mech. Phys. Solids*, **39**(5), 569-606.

Loret, B., Simões, F.M.F. and Martins, J.A.C. (1997). Growth and decay of acceleration waves in non-associative elastic-plastic fluid-saturated porous media. *Int. J. Solids Structures*, **34**(13), 1583-1608.

Loret, B., Simões, F.M.F. and Martins, J.A.C. (1999). Flutter instability and ill-posedness in solids and fluid-saturated porous media, CISM course *Material instabilities in elastic and plastic solids*, Udine, September 13-17, to be published by Springer Verlag, H. Petryk ed.

Mühlhaus, H.B. and Aifantis, E. (1991). A variational principle for gradient plasticity. *Int. J. Solids Structures*, **28**, 845-857.

Schaeffer, D. G. (1990). Instability and ill-posedness in the deformation of granular materials. *Int. J. Num. Anal. Meth. Geomechanics*, **14**, 253-278.

Simões, F.M.F. (1997). Instabilidades em problemas não associados da Mecânica dos Sólidos. Ph. D. dissertation, 12/12/1997, Universidade Técnica de Lisboa, Portugal.

Simões, F.M.F. and Martins, J.A.C. (1998). Instability and ill-posedness in some friction problems. *Int. J. Eng. Science*, **36**, 1265-1293.

Simões, F.M.F., Martins, J.A.C. and Loret, B. (1999). Instabilities in elastic-plastic fluid-saturated porous media: harmonic wave versus acceleration wave analyses. *Int. J. Solids Structures*, **36**, 1277-1295.

Vermeer, P.A. and Brinkgreve, R.B.J. (1994). A new effective non-local strain-measure for softening plasticity. *Localisation and Bifurcation Theory for Soils and Rocks*, R. Chambon, J. Desrues and I. Vardoulakis eds., Balkema, Amsterdam, The Netherlands, 89-100.

A 3D-Numerical Thermomechanical Approach for Materials Cutting

Ph. Lorong, O. Pantale**, R. Rakotomalala** and M. Touratier**

**LM2S - UPRES A 8007 - ENSAM - 151. Bd de l'Hôpital - 75013 Paris - France*
*** LGP - ENIT - BP 1629 - 65016 Tarbes - France*
lms@paris.ensam.fr

Abstract : Materials cutting during machining is an extremely complex thermomechanical problem due to the severe physical conditions associated to materials cutting mechanisms : very high elasto-viscoplastic strains and strain rates ; material passes from room temperature to the heated state in milliseconds, with the heat coming from internal dissipation and friction. Based on continuum thermomechanics and including friction and large deformations in dynamics with viscoplasticity, a materials cutting three-dimensional numerical model is briefly described. Weak forms of conservation laws are introduced in an Arbitrary Lagrangian Eulerian configuration allowing for an arbitrary surface of separation of the material, and an automatic and continuous rezoning. The Coulomb friction law is introduced to model the tool-chip and workpiece-tool contacts, and heat generation and heat transfer at these interfaces are taken into account. The flow stress including temperature and strain rate effects is based on the Johnson-Cook law proposed for high strain rate conditions. A three–dimensional example is shown when simulating the oblique cutting. Finally, some needs for the future are detailed to improve constitutive laws as well as friction, and to include more physics, for example chemical diffusion in workpiece and tool.
Keywords : Continuum thermomechanics, viscoplasticity, friction, ALE, metal cutting.

1. MACHINING AS A MULTI-SCALE PROBLEM

Despite significant technological progress in machining, today's machining operations are very difficult to model so as to predict cutting parameters, and are still

G.A. Maugin et al. (eds.), Continuum Thermomechanics, 277-288.
© 2000 *Kluwer Academic Publishers.*

open to scientific challenge. New features on interactions between workpiece and tool materials are welcome both from an experimental and computational point of view, so as to know thermomechanical loading in dynamics accurately for both tool and workpiece. Physical and chemical properties of useful materials for machining are examined to improve performance. All these responses can then permit reaching an expected machined surface and a tool life according to production requirements. Many papers on machining are available in the literature, and a review has for instance been published recently [Ehmann et al., 1997].

Materials cutting is a very complex forming process involving (within the workpiece-tool system) high shear and friction mechanisms in dynamics, liberated as heat. Therefore, thermomechanical coupling occurs at large strains and strain rates in viscoplasticity. In addition, metallurgical and chemical transformations are often happening inside the chips and possibly near the machined surface of the workpiece, as well as dynamic phenomena produced by the cutting process. Chatter is often observed between workpiece and tool, and this has to be avoided in order to reach the expected final surface of the workpiece. Chips may be either continuous or not, or serrated. Their metallurgical analysis will bring out important features on material flow during cutting, cutting capability of a given material, as well as phase changes ; plus chemical diffusion between chip and tool. It is clear that all the above phenomena observed during the cutting process need appropriate reference scales to be properly and consistently analysed by separating physical effects with great length of variation from those with smaller length. We introduced three different scales [Touratier, 1999] to consistently study the cutting operation :

- a microscopic scale at the material grain level. In addition, a nanoscopic scale may be introduced at the material crystal level,
- a mesoscopic scale at the tool tip - workpiece material level (more often we would call this the "tool-tip – chip" level),
- a macroscopic scale either at the workpiece-tool or machine tool structure level.

The microscopic (or possibly nanoscopic) scale will be required : to analyse the material microstructure ; to construct for example a polycrystalline thermomechanical constitutive law to include grain distribution ; to take into account phase changes, chemical diffusion, damage, wear, ... The mesoscopic scale is classic to the study of continuous chip flow, in three - dimensional form from numerical simulations in order to predict chip geometry, stresses and temperatures, as well as cutting forces. Finally, the macroscopic scale will be more appropriate to analyse dynamics of the workpiece-tool system. On the production level, a megascopic scale should be introduced to achieve corresponding studies. The main difficulty remains in some cases how to find a consistent way to pass from one scale to another, due to various nonlinearities and also a big computational effort associated with the very different scale sizes. The physical bases of cutting models are conservation laws including phase change, diffusion ; and of course elastoviscoplasticity, friction, and damage.

2. MATERIALS CUTTING MODELING

The modeling of chip formation is extremely complex due to the severe physical conditions associated with chip formation mechanisms : high viscoplastic strains ($\sim 10^3$) and strain rates ($\sim 10^5 s^{-1}$) ; residual state ; material passes from room temperature to the heated state ($\sim$ 700 °C) in milliseconds, with most of the heat coming from friction between tool and chip, and tool and workpiece. Considerable effort towards chip formation modeling is therefore still needed in order to bring out all thermo-metallurgico-mechanical information concerning the cutting process and to optimise it within the field of high - speed machining in turning, milling, drilling and grinding. In addition, the modeling master can help to modify features of both cutting tools and workpiece material in order to improve production.

In the past, Eulerian, Lagrangian and Arbitrary Lagrangian-Eulerian methods have been adopted, see for example [Caroll et al., 1998], [Sekhon et al., 1993], [Rakotomalala et al., 1993]. Hereafter, we will consider only predictive models which allow for an arbitrary surface of separation between workpiece and chip. Models based on a predetermined line of separation (a lot of these exist in the literature) at the tool tip in the workpiece material are not useful as they are not capable of predicting surface roughness and chip morphology, nor are they able to predict the extremely large strains encountered near the tool tip inside the workpiece material. The Arbitrary Lagrangian-Eulerian (ALE) formulation adopted in this section is an extension of both classical Lagrangian and Eulerian formulations where grid points may have an arbitrary motion. In such a description, material points are represented by a set of Lagrangian coordinates X_i, spatial points with a set of Eulerian coordinates x_i, and reference points (grid nodes) with a set of arbitrary coordinates a_i. All physical quantities are computed at points x_i occupied by grid nodes at time t. All conservation laws must be expressed taking into account the meshing evolution during the calculation.

Considering a space and time dependent quantity g, one must express all the conservation laws using the material $\overset{\bullet}{g}$, spacial g' and mixed $\overset{\circ}{g}$ time derivatives. Conservation laws are usually written using material time derivatives in an Eulerian formulation [Germain, 1986]. Therefore, we have to introduce the relationship below between material and mixed time derivatives

$$\overset{\bullet}{g} = \overset{\circ}{g} + g_{,i} c_i \tag{1}$$

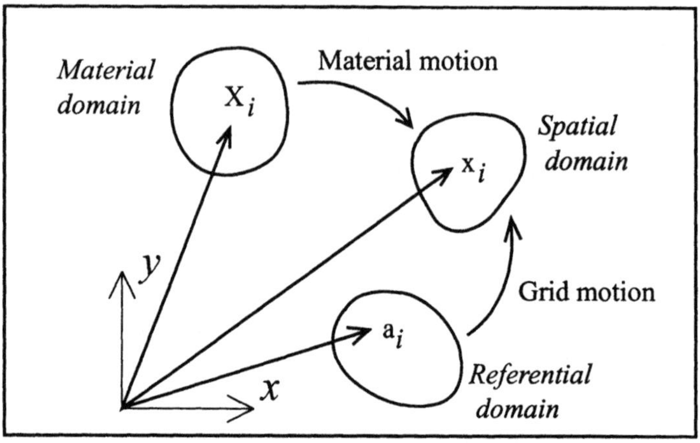

Figure 1 ; Domains and coordinates for the ALE formulation

where $c_i = v_i - \hat{v}_i$ is a convective term representing the relative velocity between the material v_i and grid $\hat{v}_i$ velocities and $g_{,i}$ represents the space derivative of g with respect to x_i. Thus, the ALE method combines both the advantages of Eulerian and Lagrangian representations in a single description which can be considered as an automatic and continuous rezoning method. The model is ALE until the steady state conditions are reached. This permits especially free and contact surfaces to be updated. Once these conditions are obtained, then this model is equivalent to an Eulerian one.

2.1 Discretised weak forms

Let $\Omega(t)$ be the domain to be considered in modeling at the current configuration and $\Gamma(t)$ its boundary. We denote by : $\vec{v}$ the material velocity, $\hat{\vec{v}}$ the mesh (or grid) velocity, ρ the mass density, $\vec{f}$ the external given body forces, $\vec{F}$ the external surface forces, $\boldsymbol{\sigma}$ the Cauchy stress tensor, $\mathbf{d}$ the strain rate tensor, e the specific internal energy, w the given external body heat generation, Q the external given heat and $\vec{q}$ the heat flux. The unknowns are $\vec{v}$, ρ, e and constitutive laws have to be found for $\boldsymbol{\sigma}$, e and $\vec{q}$. Supposing that $e = e(T,t)$, the final unknowns will be $\vec{v}$, ρ, and the temperature T. The formulation of such an intrinsic problem requires defining the following functional spaces :

$$\Re = \{\, \rho \,;\, \rho(x_i,t) \in H^1(\Omega(t)) \,;\, \text{impermeable wall}\} \tag{2}$$

$$\vartheta = \{\, \vec{v} \,/\, \vec{v}(x_i,t) = (v_1,v_2,v_3) \,;\, v_i \in H^1(\Omega(t)) \,;\, v_i = \overline{v}_i \text{ on } \Gamma_V(t)\} \tag{3}$$

$$\epsilon = \{ e \ ; \ e(T,t) \in H^1(\Omega(t)) \ ; \ T = \overline{T} \text{ on } \Gamma_T(t)\} \tag{4}$$

where H^1 is the Sobolev space of order one, and $\Gamma_v(t)$ and $\Gamma_T(t)$ are the boundary parts, respectively, where material velocity and temperature are prescribed. To achieve numerical results, we use a discretisation of the boundary value problem. We then note ρ^h, $\overrightarrow{v}^h$ and e^h as approximated fields on ρ, $\overrightarrow{v}$ and e respectively. The superscript h refers to the association of spaces $\Re^h$, ϑ^h, ϵ^h containing ρ^h, $\overrightarrow{v}^h$, e^h with a mesh $\cup\Omega_e^h(t)$ (or discretisation) of the continuous domain $\Omega(t)$ which is classically parametrised by a characteristic length scale h, and $\Omega_e^h(t)$ refers to a finite element or a finite volume, depending on the kind of discretization chosen. We also need to introduce weighting function sets as follows :

$$\overset{*}{\Re}{}^h = \{ \overset{*}{\rho}{}^h \ ; \ \overset{*}{\rho}{}^h(x_i) \in H^1(\Omega(t))\} \tag{5}$$

$$\overset{*}{\vartheta}{}^h = \{ \overset{*}{\overrightarrow{v}}{}^h \ / \ \overset{*}{\overrightarrow{v}}{}^h(x_i) = (\overset{*}{v}{}^h_1, \overset{*}{v}{}^h_2, \overset{*}{v}{}^h_3) \ ; \ \overset{*}{v}{}^h_i \in H^1(\Omega(t)) \ ; \ \overset{*}{v}{}^h_i = 0 \text{ on } \Gamma_V(t)\} \tag{6}$$

$$\overset{*}{\epsilon}{}^h = \{ \overset{*}{e}{}^h \ ; \ \overset{*}{e}{}^h(\overset{*}{T}{}^h) = \overset{*}{T}{}^h \in H^1(\Omega(t)) \ ; \ \overset{*}{T}{}^h = 0 \text{ on } \Gamma_T(t)\} \tag{7}$$

Using both the relationship (1) between material and mixed time derivatives, and its corresponding following expression

$$\overset{\bullet}{K}(t) = \overset{\circ}{K}(t) + \int_{\Omega(t)} (gc_i)_{,i} \, d\Omega(t) \tag{8}$$

available for a volume integral of a function g, the dicrete weak forms associated respectively with the mass, momentum and energy conservation laws can be expressed from classical Eulerian ones as :

i - find $\rho_e^h \in \Re^h$:

$$\int_{\cup\Omega_e^h(t)} \overset{*}{\rho}{}^h_e \left(\frac{\partial \rho_e^h}{\partial t} + \overrightarrow{c}^h.\overrightarrow{Grad_x}\rho_e^h + \rho_e^h div_x \overrightarrow{v}^h\right) d(\cup\Omega_e^h) = 0, \quad \forall \overset{*}{\rho}{}^h_e \in \overset{*}{\Re}{}^h \tag{9}$$

ii - find $\overrightarrow{v}^h \in \vartheta^h$:

$$\int_{\cup\Omega_e^h(t)} \rho_e^h \left(\frac{\partial \overrightarrow{v}^h}{\partial t} + \mathbf{Grad}_x \overrightarrow{v}^h.\overrightarrow{c}^h\right) \overset{*}{\overrightarrow{v}}{}^h \, d(\cup\Omega_e^h) = -\int_{\cup\Omega_e^h(t)} tr(\boldsymbol{\sigma}^h \, \overset{*}{\mathbf{d}}{}^h) \, d(\cup\Omega_e^h)$$

$$+\int_{\cup\Omega_e^h(t)} \vec{f}\,.\,\overset{*}{\vec{v}}^{\,h}\, d(\cup\Omega_e^h) + \int_{\cup\Gamma_{e_F}^h(t)} \vec{F}\,.\,\overset{*}{\vec{v}}^{\,h}\, d(\cup\Gamma_e^h), \quad \forall\, \overset{*}{\vec{v}}^{\,h} \in \overset{*}{\vartheta}^{\,h} \qquad (10)$$

iii - find $T^h \in \epsilon^h$:

$$\int_{\cup\Omega_e^h(t)} \rho_e^h\Big(\frac{\partial e^h}{\partial t} + \overrightarrow{Grad}_x e^h.\vec{c}^{\,h}\Big)\, \overset{*}{T}^{\,h}\, d(\cup\Omega_e^h) \;=\; \int_{\cup\Omega_e^h(t)} tr(\boldsymbol{\sigma}^h \mathbf{d}^h)\, \overset{*}{T}^{\,h}\, d(\cup\Omega_e^h)$$

$$+\int_{\cup\Omega_e^h(t)} \vec{q}^{\,h}.\overrightarrow{Grad}_x \overset{*}{T}^h\, d(\cup\Omega_e^h) + \int_{\cup\Omega_e^h(t)} \rho_e^h\, w\, \overset{*}{T}^h\, d(\cup\Omega_e^h) \qquad (11)$$

$$+\int_{\cup\Gamma_{e_q}^h(t)} Q\, \overset{*}{T}^h d(\cup\Gamma_e^h), \quad \forall\, \overset{*}{T}^{\,h} \in \overset{*}{\epsilon}^{\,h}$$

These functionals are strongly coupled. Thus, when the Updated Lagrangian configuration is chosen : $\vec{v}^h = \hat{\vec{v}}^h$, while if we choose the Eulerian we must take $\hat{\vec{v}}^h = \vec{0}$. Here, the suggested grid motion control algorithm has been previously proposed in [Donea et al., 1982]. If we adopt finite element approximations of equations (9) to (11), then fields $\vec{v}^h$ and $\overset{*}{\vec{v}}^{\,h}$ will be interpolated by using the same functions. Finite volume approximations will be equivalent to taking $\overset{*}{\vec{v}}^{\,h}$ as a unitary field. Writting the boundary value problem as given above by equations (9) to (11) is very appropriate to an explicit approach that we will retain hereafter. It will be used to solve our cutting problem, being a dynamic one.

2.2 Contact and friction

The explicit integration algorithm used hereafter allows taking the contact between the bodies into account by adding an external force to the contacting nodes. This can be done by the introduction of such a force into the external surface load $\vec{F}$ described above. Normal components of this force vector are set equal to prevent penetration, and the tangential component is set with respect to the Coulomb friction law given by :

$$\left\|\vec{T}_\tau^h\right\| \;<\; \mu\, \left\|\vec{T}_n^h\right\|, \quad \text{for stick ;} \qquad (12)$$

$$\vec{T}_\tau^h \;=\; -\,\mu\, \left\|\vec{T}_n^h\right\| \frac{\vec{V}_\tau}{\left\|\vec{V}_\tau\right\|}, \quad \text{for slip ;} \qquad (13)$$

where μ is the coefficient of friction we would prefer to assume as depending on the nature of the contacting bodies, $\vec{T}_\tau^h$ and $\vec{T}_n^h$ are respectively the tangential and normal components of the stress vector along the interface, and $\vec{V}_\tau$ is as defined below. Efforts to find such a coefficient from experiment associated to machining are given in [Joyot et al., 1998]. The contact algorithm must also incorporate thermal capacities. Heat generation and heat transfers at the interfaces between bodies have to be taken into account.

The heat generation in the slipping contact surfaces is given by

$$dQ^h_{friction} = \left\|\overrightarrow{T}^h_\tau\right\| . \left\|\overrightarrow{V}^h_\tau\right\| \; dt \tag{14}$$

where $\overrightarrow{V}^h_\tau$ is the tangential slip relative velocity between contacting bodies. According to the explicit scheme, calculated heat flux is then re-introduced as an external thermal flux for each contacting node. Generated heat flux is shared among bodies in contact in a ratio depending on their thermal and geometric features, but this point of view is not sufficient since sliding motion occurs at the interface. In addition, we may introduce a thermal resistance in order to represent the thermal discontinuity at the interfaces, but the reality is as above with the shared coefficient. Because of motion between contacting bodies, sharing thermal coefficient and thermal resistance are very difficult to identify separately from the experiment. We will come back later to this essential difficulty.

2.3 Constitutive laws

For the heat flux $\overrightarrow{q}^h$, the Fourier conduction law is generally appropriate :

$$\overrightarrow{q}^h = - \underline{\underline{k}}(T^h) \, \overrightarrow{Grad_x} T^h \tag{15}$$

where $\underline{\underline{k}}$ is the conduction matrix available for an anisotropic material. It may depend on temperature. Otherwise, the specific internal energy e^h is usually linked to the temperature T^h by the following relationship (state law) :

$$de^h = C_v(T^h) \, dT^h \tag{16}$$

where C_v is the specific heat which may depend on temperature. Today, the material constitutive law used in cutting modeling takes into account elasto-visco-plasticity, temperature and strain rates within a standard plasticity formulation. In large strains, the well known strain rate decomposition is allowed if elastic strains are small (< 2%) as we will suppose below. The material constitutive law is then written as :

$$\mathbf{s}^h = \mathbf{C}^h : \mathbf{d}^h \qquad \text{and} \qquad \boldsymbol{\sigma}^{Jh} = \frac{dp^h}{dt}\mathbf{I} + \mathbf{s}^{Jh} \tag{17}$$

where $\mathbf{s}^h$ is the deviatoric part of the Cauchy stress tensor $\boldsymbol{\sigma}^h$, $\mathbf{s}^{Jh}$ is the Jaumann derivative of the $\mathbf{s}^h$ stress tensor which depends on the spin tensor and on a convective term in ALE. Otherwise, p^h is the hydrostatic pressure which is determined from a polynomial expression. **I** is the second order identity tensor and $\mathbf{C}^h$ is the constitutive tensor defined as in [Soua et al., 1999]. Then, introducing the yield function as

$$f^h = \overline{\sigma}^h - R^h \tag{18}$$

where $\overline{\sigma}^h$ is the Von-Mises equivalent stress and R^h the flow stress, according to the author's knowledge, two different choices have been made in the past for R^h so as to simulate numerically chip formation within the above thermo-mechanical requirements.

i- Johnson-Cook [Johnson et al., 1983] flow stress in [Joyot et al., 1998], [Rakotomalala et al., 1993], [Pantalé et al., 1998] :

$$R^h = [A + B(\overline{\varepsilon}^h)^{\frac{1}{n}}][1 + CLog(\dot{\overline{\varepsilon}}^h/\dot{\overline{\varepsilon}}_0)][1 - (T^h - T_0)^m/(T_{melt} - T_0)^m] \tag{19}$$

where $\overline{\varepsilon}^h$ is the equivalent plastic strain, $\dot{\overline{\varepsilon}}^h$ is the equivalent plastic strain rate, T^h is the current temperature, T_0 is a reference temperature ; and A, B, C, T_{melt}, are material characteristics, while $\dot{\overline{\varepsilon}}_0 = 10^3 s^{-1}$ in order to normalise equation (19). Note also that n is the hardening exponent, and m is the softening exponent.

ii- Lemonds-Needleman [Lemonds et al., 1986] hardening rule in [Marusich et al., 1995] :

$$R^h = [1 - \alpha(T^h - T_0)][1 + (\varepsilon^{ph}/\varepsilon_0^p)]^{\frac{1}{n}} \sigma_0 \tag{20}$$

where α is a softening coefficient, σ_0 the yield stress at T_0, ε_0^p a reference plastic strain, and ε^{ph} the accumulated plastic strain.

3. MATERIALS CUTTING SIMULATIONS. AN EXAMPLE

In this section, results from an ALE model are represented to simulate three-dimensional steady metal cutting allowing to take into account lateral expansion of the chip and to represent oblique cutting avoiding any assumption on the chip separation at the tool tip. The thermomechanical equations above discretised by finite elements (momentum)/finite volumes (mass and energy) are kept strongly coupled. The Johnson-Cook hardening rule is used, contact conditions with friction are included between chip and tool, and also workpiece and tool. Material properties are given Reference [Pantalé et al., 1998] for a 42 CD-4 steel workpiece material, and a tungsten carbide for the assumed rigid tool (SECO TPGN-160302 P10). The quite coarse mesh created involves 5200 three-dimensional eight node finite elements (same mesh for the finite volume discretization). Cutting conditions are given in Reference [Pantalé et al., 1998] as well as thermomechanical boundary conditions. The time integration of the above discretised equations is accomplished from an explicit central difference scheme of the third order. The time increment Δt is subject to the Courant stability criterion defined by $\Delta t \leq \Delta t_{crit.}$ where $\Delta t_{crit.}$ is usually a function of the sound and convective speeds. The ALE method also requires the use of an appropriate grid (mesh) motion algorithm controlled by both space and time criteria, coming here from [Donea et al., 1982]. Knowledge of the grid speed allow us to compute the convective speed. Thus, integrating successively equations (9) to (11) gives the unknown values of $\overrightarrow{v}^h(t+\Delta t)$, $\rho^h(t+\Delta t)$ and $e^h(t+\Delta t)$. Finally, integrating the material constitutive law given by equation (17) allow one to determine the Cauchy stress tensor $\boldsymbol{\sigma}^h(t+\Delta t)$ using an explicit elastic prediction and a radial return algorithm. Of course, we also needed to take into account the Fourier conduction law, the state law for internal energy and contact with friction as given above. We assumed k, c, and m as constant. Figures *2* and *3* give results on three-dimensional chip flow within oblique cutting conditions (angle between the edge of the tool and the cutting speed direction equal to 75°), [Pantalé et al., 1998].

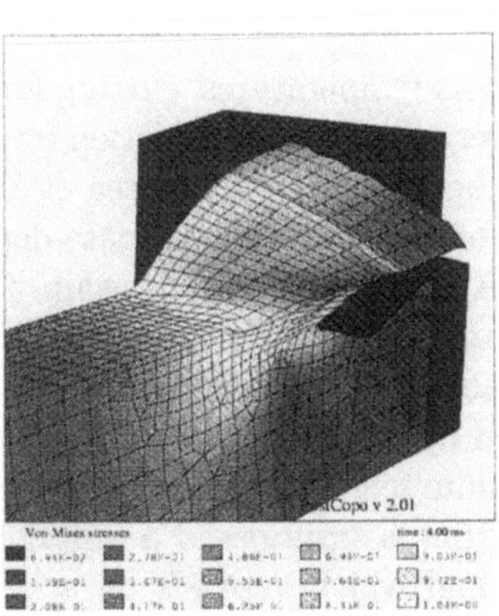

Figure 2 ; Von Mises stress contour plot

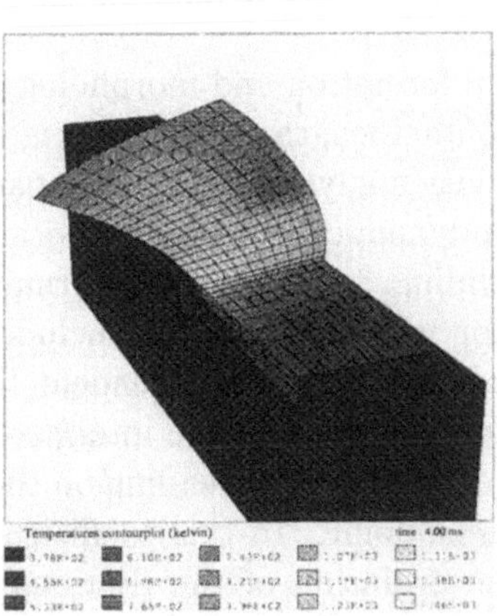

Figure 3 ; Temperature distribution in the chip and workpiece

Comparing oblique and orthogonal chip flow based on a three-dimensional approach, temperature distributions are found quite similar, but cutting forces and advancing forces have been reduced by the oblicity while lateral forces increase in a wide range. A very good agreement on forces are found for orthogonal chip flow when comparing with available experimental results [Pantalé et al., 1998].

4. NEEDS FOR THE FUTURE

Chip formation and chip morphology during machining is mainly dependent on very intense shear inside the chip in the so-called primary shear zone where shear banding occurs. Crystal plasticity employing the fundamental principle of sliding in crystalline slip systems can provide a better approach to model the phenomena inside the primary shear zone and to capture shear band localisation. In addition, this makes it easier to include induced plastic anisotropy which appears when material is being machined. Finally, this seems to be the only way to reach an accurate description of the material microstructure evolution and to control it during machining to improve product performance. The combination of the crystal plasticity theory, techniques for measurement of texture, quantitative texture analysis and the finite element method enables the study of gradients in deformation which follows from the interplay of the material state and the boundary conditions. It is even possible in this way to introduce inhomogeneous material if the number of grains is not large or as soon as computational means will be adequate. Finally, a crystal plasticity constitutive law provides a way to consistently pass from the microscopic scale to the mesoscopic defined in the above Section 1.

The chip formation and morphology, as well as temperatures, cutting forces and tool-chip contact length are found numerically very dependent on friction from a friction sensitivity analysis. Recently, a paper has been published [Oancea et al., 1997] on a thermodynamically consistent formulation of thermomechanical rate-dependant frictional sliding, within large deformations. A constitutive relationship has been proposed for frictional traction including thermal softening of frictional response, rate dependence, pressure dependent for contact thermal resistance across the interface, and heat sinks on the interfaces created by trapped wear debris. Then, the coefficient of friction is depending on the maximum temperature on the two surfaces at the current point. To this coefficient of friction is surimposed a function which controls the evolution of the frictional stress and which depends on a new state variable to be determined from a supplementary equation derived from thermodynamics. This state variable allow us to govern transitions of response between a viscous dependence of frictional traction on sliding velocity and a dependence of the coefficient of friction upon the sliding velocity. Finally, another quantity is surimposed to the coefficient of friction : it is a parameter which might be associated with the viscoplastic shearing effects of the asperities tips.

Several microstructural experiments have been done for 32CDV 13 martensitic iron machined either by ceramic or cermet tools. Machining conditions may lead to a white thin layer in the chip on the chip-tool side, not seen on many other tested steels, among them stainless steel. This thin white layer of thickness near $5\mu m$, is attributed to phase transformations and contains austenite (γ-phase), as shown from XR diffraction analysis. From a secondary ion mass spectrometry analysis, chemical diffusion of titanium and nitrogen elements has been observed, both from tool to chip, and tool to workpiece. To include chemical diffusion capability analysis into the boundary value problem shown above one conservation equation has to be added for the mass concentration. The main differences between mass conservation and mass concentration laws are now that we need to express the diffusion flux of the diffusing phase coming out the diffusing boundary, as a function of the concentration. A typical constitutive law for the diffusion flux may be expressed as a Fick law. The diffusion between chip and tool changes the mechanical strength of the tool, resulting in wear. Keeping all the above discretised weak forms and adding a damage law to produce crack initiation, then crack propagation should be an efficient way to study tool wear. Wear mechanisms during cutting may come from mechanical effects : adhesion, abrasion, sliding/delamination, erosion, fretting, ... and from chemical effects : solution, diffusion, oxidation, corrosion, ... Wear is a very important and difficult subject that we have to pursue in order to master cutting, especially concerning tool wear which is an economic as well as a workpiece surface quality challenge. Finally, metallurgical effects have also to be included into the formulation of the boundary value problem to take into account phase transformations. These will involve supplementary coupling when deriving the corresponding discrete boundary value problem in order to simulate chip formation. Corresponding features of the metallurgical model will include proportions of metallurgical constituents and dilatations of various phases as functions of temperature. Then, as above for friction and diffusion, new state variables have to be introduced according to thermodynamic requirements.

REFERENCES

[Caroll et al., 1998] Caroll, J. T.; Strenkowski, J. S.; "Finite element models of orthogonal cutting with application to single point diamond turning". *International Journal of Mechanical Sciences*, Vol. 30, N°12, pp. 899–920.

[Donea et al., 1982] Donea, J.; Giuliani, S.; and Halleux, J. P.; "An Arbitrary Lagrangian-Eulerian finite element method for transient dynamic fluid-structure interactions". *Computer Methods in Applied Mechanics and Engineering*, Vol. 33, pp. 689–723.

[Ehmann et al., 1997] Ehmann, K. F.; Kapoor, S. G.; DeVor, R. E.; Lazoglu, I.; "Machining process modelling". *ASME-Journal of Manufacturing Science and Engineering*, Vol. 119, pp. 655–663.

[Germain, 1986] Germain, P.; *Mécanique*. Ellipses-Paris, 1986.

[Johnson et al., 1983] Johnson, R.; Cook, W. K.; "A constitutive model and data for metals subjected to large strains, high strain rates and high temperatures". In *Proceedings 7th International Symposium on Ballistics*, The Hague, pp. 541–547.

[Joyot et al., 1998] Joyot, P.; Rakotomalala, R.; Pantalé, O.; Touratier, M.; and Hakem, N.; "A numerical simulation of steady state metal cutting". *Journal of Mechanical Engineering Science*, Vol. 212, pp. 331–341.

[Lemonds et al., 1986] Lemonds, J.; Needleman, A.; "Finite element analysis of shear localization in rate and temperature dependent solids". *Mechanics of Materials*, Vol. 5, pp. 339–361.

[Marusich et al., 1995] Marusich, T. D.; Ortiz, M.; "Modelling and simulation of high-speed machining". *International Journal for Numerical Methods in Engineering*, Vol. 38, pp. 3675–3694.

[Oancea et al., 1997] Oancea, V. G.; Laursen, A.; "A finite element formulation of thermomechanical rate-dependent friction sliding". *International Journal for Numerical Methods in Engineering*, Vol. 40, pp. 4275–4311.

[Pantalé et al., 1998] Pantalé, O.; Rakotomalala, R.; Touratier, M.; "An ALE three-dimensional model of orthogonal and oblique metal cutting processes". *International Journal of Forming Processes*, Vol. 1, N°3, pp. 371–389.

[Rakotomalala et al., 1993] Rakotomalala, R.; Joyot, P.; Touratier, M.; "Arbitrary Lagrangian-Eulerian thermomechanical finite element model of material cutting". *Communications in Numerical Methods in Engineering*, Vol. 9, N°3, pp. 975–987.

[Sekhon et al., 1993] Sekhon, G. S.; Chenot, J. L.; "Numerical simulation of continuous chip formation during non-steady orthogonal cutting". *Engineering Computations*, Vol. 10, pp. 31–48.

[Soua et al., 1999] Soua, A.; Touratier, M.; Polac, L.; "Computations of an engine to analyse cylinder distorsion". *Engineering Computations*, Vol. 16, N°1, pp. 9–25.

[Touraticr, 1999] Touratier, M.; "Computational models of chip formation and chip flow in machining in a multi-scale approach. Present status and future needs". In *Proceedings of the International Workshop on Modeling of Machining Operations*, Nantes, pp. 1–29.

From Clausius-Duhem and Drucker-Ilyushin inequalities to standard materials

Jean-Jacques MARIGO
LPMTM, Institut Galilée, Université Paris 13,
Avenue JB Clément, 93430 Villetaneuse (France)
marigo@lpmtm.univ-paris13.fr

Abstract : We study the power of restriction of Clausius-Duhem and Drucker-Ilyushin inequalities on the constitutive relations of several classes of materials, such as elastic, elastoplastic, brittle damaging and viscous ones. The goal is to see whether these two physical principles can justify the very useful but formal notion of standard materials.
Keywords : Thermodynamics, Constitutive relations, Stability

1. INTRODUCTION

Following P. Germain's lead, the french mechanics community embraced the concept of (Generalised) Standard Material (GSM) : Are my constitutive relations those of a standard material? How can I change or present them in order to agree with the rule? The reasons of this justified success clearly appear in the remarkable survey [Germain *et al.*, 1983] : *(i) Its implementation is systematic — (ii) It can be generalised to any kind of thermo-mechanical behaviour — (iii) It does not disagree with well established thermodynamical principles and even permits to automatically satisfy the second law — (iv) It agrees with the most usual constitutive relations — (v) It generally leads to "nice" initial-boundary value problems with "good" mathematical properties —*. In other words, choose a set of state variables, define two (convex) thermodynamical potentials and you have obtained an efficient constitutive law. This *plug in and play* instrument was a godsend for the engineer in that it enabled him, with the help of faster and faster computers, to develop more and more elaborate material models and to use them to design or to check the reliability of engineering structures.

But the success of an approach should not prevent an investigation into the soundness of its foundations. In the present case, the weakness of the GSM rule comes from its too formal character. This can lead to excess. Moreover it proceeds from assumptions which are not easily checked from experimental tests : How is one to find the relevant set of state variables? Or the relevant forms of the free energy and dissipation potential? It would be completely satisfactory and could even be promoted to the rank of a physical law, if we knew how the approach is related to some more fundamental and well established physical quantities or principles. What are the candidates? It is now clear that the usual universal

G.A. Maugin et al. (eds.), Continuum Thermomechanics, 289-300.
© 2000 *Kluwer Academic Publishers.*

principles, like material frame indifference or the second law of thermodynamics, give valuable but very limited restrictions on constitutive relations for complex inelastic materials. In any case they alone cannot fill in the gap up to GSM. More interesting is Drucker and Ilyushin's postulate on the positivity on the work that the external agency has to supply to perform a cyclic loading. Indeed, when applied by its authors to elastoplastic materials (the setting in which it was conceived), it implies the convexity of the yield surface and the normality rule for the plastic strain rate, *i.e.*, precisely the main properties required to be a GSM. Equipped with this cheering result, we must now apply it to various classes of materials and examine whether it excludes any material which is not a GSM. This is the main goal of the present paper, in which we limit our attention to the most usual types of behaviour, elastic, elastoplastic, brittle damaging and viscoelastic materials.

2. CLAUSIUS-DUHEM AND DRUCKER-ILIUSHIN INEQUALITIES

In classical treatises on thermomechanics of continuous media, the second law of thermodynamics leads to the Clausius-Duhem inequality (CD-I), that is, to an inequality that every material element will have to satisfy along every process that it will undergo. For a simple material (in the sense of Noll) in isothermal processes, CD-I reduces to

$$\mathcal{D} = \sigma \cdot \dot{\varepsilon} - \dot{\psi} \geq 0, \tag{1}$$

where $\mathcal{D}$ denotes the volume density of dissipated power, ψ the volume density of free energy, σ and ε the stress and strain tensors.

Drucker [Drucker, 1951] introduced his *stability* postulate in terms of stress cycle and of strain work of *added* forces. Specifically, denoting by σ_0 and σ the initial and current stresses of the material element, the postulate requires that $\int_{\mathcal{C}}(\sigma - \sigma_0) \cdot d\varepsilon \geq 0$ in any stress cycle $\mathcal{C}$, that is in any stress path starting and ending at σ_0. Ilyushin [Ilyushin, 1961] formulated the stability postulate in terms of strain cycle and of total strain work by requiring that

$$\int_{\mathcal{C}} \sigma \cdot d\varepsilon \geq 0 \quad \textit{in any } \text{strain } \textit{cycle } \mathcal{C}. \tag{2}$$

We will adopt Ilyushin's point of view and consider (2) as the work condition (essentially because it can be more easily extended to general thermodynamical systems). However the use of the postulate in its original form would lead to unsatisfactory results for some classes of materials (like viscoelastic one, see Section 4.). We change it slightly by requiring the positivity of the strain work during a strain cycle, *i.e.*, Condition (2), *only* when the material element starts from an *equilibrium state*. In our isothermal context, a state (characterised by the values of the strain and of the internal variables) corresponds to an equilibrium state if it evolves only when the strain does. In the sequel, this revised version of the stability postulate will be referred to as Drucker-Ilyushin Inequality (DI-I).

3. HOW THEY RESTRICT MATERIALS WITH ELASTIC RANGE

This section is devoted to materials that behave elastically in a certain strain range, like elastic, elastoplastic or brittle damaging materials. Since their behaviour is rate-independent, every state is an equilibrium state and, in the statement of DI-I, (2) must be satisfied whatever the initial state may be.

3.1 Elastic materials

First consider elastic materials, that is non dissipative materials in which the stress tensor only depends on the strain tensor. We thus consider the class of materials such that

$$\sigma = f(\varepsilon), \qquad \mathcal{D} = 0. \tag{3}$$

The question then becomes : *Is any function f admissible?* Let us first study the consequences of DI-I. For elastic materials, the strain work can be seen as the circulation of the field f along the strain path. Consequently, if the work along a cycle $\mathcal{C}$ equals W, then it will be equal to $-W$ along the reversal cycle $\mathcal{C}_r$. As in both cases the work must be positive, necessarily $W = 0$. In other words, DI-I requires that the strain work vanishes in any cycle. But, a classical result for differential forms says that this is possible if and only if the stress function f derives from a potential. Hence, DI-I requires that there exists an elastic potential φ such that

$$\sigma = D\varphi(\varepsilon), \tag{4}$$

where the symbol D stands for the derivative. Equation (4) characterises hyperelastic materials. If we insert (4) into (1), we immediately deduce that $\psi = \varphi(\varepsilon) + const.$, that is that, in the present isothermal context, the free energy is equal (up to a constant) to the elastic energy. Conversely, if (4) holds, then the strain work vanishes along any cycle; we have proved the following

Proposition 1 *An elastic material satisfies DI-I* if and only if *it is hyperelastic. Moreover, by virtue of CD-I, its free energy then corresponds to its elastic energy.*

This first fundamental result merits several remarks.

▷ *The legitimacy of DI-I.* Let us first show why the postulate of the positivity of the strain work is legitimate for elastic materials. Assume that there exists a strain cycle $\mathcal{C}$ such that the strain work W is strictly negative. In that case the "external agency" who prescribes the strain loading will receive the energy amount $-W$ at the end of the cycle. By repeating the cycle n times, he will receive the energy amount $-nW$. By increasing n at will, he could then recover as much energy as he wants. This material will then constitute an unfailing energy source. If it were possible, then the energy question would be solved for the entire universe. Clearly, such is not the case. ◁

▷ *The relation between DI-I and usual stability criteria.* Any hyperelastic material satisfies DI-I. We have obtained no restriction on the elastic potential φ. In particular no usual dynamic stability condition, like convexity in linear elasticity or rank-one convexity in the case of finite strain. Conversely, we could define some dynamic stability properties for elastic (but not hyperelastic) materials. Consequently, if DI-I is related to some material stability — as Drucker presented it —, it corresponds to an additional notion of stability. ◁

▷ *The role of CD-I.* The hypothesis $\mathcal{D} = 0$ and CD-I are only used to calculate the free energy and to identify it with the elastic energy. ◁

To emphasise the contribution of DI-I, let us now analyse the consequence of CD-I alone by removing DI-I. Since this question is mishandled in the literature, we return to it by insisting on the importance of the adopted definition for elastic materials.

In their pioneering paper [Coleman *et al.*, 1963] the authors started from a different definition of elastic materials by assuming that in such materials the stress *and* the free energy are functions of the strain, that is $\sigma = f(\varepsilon)$ and $\psi = \varphi(\varepsilon)$. With this definition, (1) becomes $\mathcal{D} = (f(\varepsilon) - D\varphi(\varepsilon)) \cdot \dot{\varepsilon} \geq 0$. Since the inequality must be satisfied for every ε and $\dot{\varepsilon}$ — this corresponds to the implicit assumption that one can prescribe any strain path to the material element —, we immediately deduce that $f = D\varphi$ and $\mathcal{D} = 0$, conditions which are exactly those of (non dissipative) hyperelastic materials. Thereby, a negligent reader could conclude that CD-I, like DI-I, rules out the elastic but not hyperelastic materials. In fact, this is not correct because we started from a restricted definition. Here the crucial (but somewhat hidden) assumption lies in the hypothesis that the free energy depends only on the current strain.

Let us remove this assumption and adopt (3) as the correct definition for elastic materials. Then CD-I becomes an equality and permits us to compute the variation of the free energy along the process : $\dot{\psi} = f(\varepsilon)\cdot\dot{\varepsilon}$. In particular, this relation shows that, for an elastic but not hyperelastic material, the free energy depends on the entire strain history. In any case, such a material is not in contradiction to the second law of thermodynamics. Even, the hypothesis usually adopted, that the free energy is a function of state, is meaningless as long as the set of state variables is not well defined. In the present case, it suffices to set (ε, ψ) as the pair of state variables to abide by this rule. We can then summarise this analysis of the consequence of CD-I by

Proposition 2 *CD-I excludes no elastic material, but simply says that* only *the hyperelastic ones possess a free energy which depends only on the current strain.*

By comparison of Propositions 1 and 2, we immediately see the contribution of DI-I.

3.2 Elastoplastic materials

For the most part, the results presented in this subsection are those first published in [Drucker, 1951]. We consider elastoplastic materials and assume that the stress-strain relation is

$$\sigma = A(\varepsilon - \varepsilon^p), \tag{5}$$

where A is the stiffness tensor (assumed invertible, but not necessarily symmetric) and ε^p the plastic strain which plays the role of a dissipative internal state variable. For its evolution law, we will essentially adopt the usual assumptions of a rate-independent law based on the notion of yield surface. We depart from the custom only by the fact that, as the strain is the controlled variable, we have to consider the elastic domain in strain space rather than in stress space. It is of little importance, because we can deduce one from the other by virtue of (5) : if $\mathcal{E}$ denotes the current elastic domain in the strain space, its image in the stress space is simply $\mathcal{E}^* = \{\sigma \mid A^{-1}\sigma + \varepsilon^p \in \mathcal{E}\}$.

Let us examine the consequences of DI-I. By considering elastic cycles, *i.e.*, strain cycles along which the plastic strain does not evolve, we easily deduce from Proposition 1 that the elasticity tensor A must be symmetric.

Let us now consider inelastic processes. Let ε_0, ε_0^p and $\mathcal{E}_0$ be the initial (total) strain, plastic strain and elastic range of the material element, with $\varepsilon_0 \in \mathcal{E}_0$. Submit that element to an elastic strain path, *i.e.*, a path lying in $\mathcal{E}_0$, in a such manner that the final strain ε_1 lies on the yield surface $\partial\mathcal{E}_0$. Then, apply an infinitesimal strain increment $\delta\varepsilon$ inducing an infinitesimal plastic strain increment $\delta\varepsilon^p$ (with possibly a small change of $\mathcal{E}$). Finally, end the strain cycle by elastically returning to ε_0. Let us compute the strain work needed in the three parts of this cycle. The first (elastic) part easily yields $W_1 = \int_{\varepsilon_0}^{\varepsilon_1} A(\varepsilon - \varepsilon_0^p) \cdot d\varepsilon = \frac{1}{2}A\varepsilon_1 \cdot \varepsilon_1 - \frac{1}{2}A\varepsilon_0 \cdot \varepsilon_0 - A\varepsilon_0^p \cdot (\varepsilon_1 - \varepsilon_0)$. In a similar manner, the third (elastic) part yields $W_3 = \frac{1}{2}A\varepsilon_0 \cdot \varepsilon_0 - \frac{1}{2}A(\varepsilon_1 + \delta\varepsilon) \cdot (\varepsilon_1 + \delta\varepsilon) - A(\varepsilon_0^p + \delta\varepsilon^p) \cdot (\varepsilon_0 - \varepsilon_1 - \delta\varepsilon)$. The second (infinitesimal inelastic) part is evaluated at first order; we obtain $W_2 = A(\varepsilon_1 - \varepsilon_0^p) \cdot \delta\varepsilon + \cdots$. The total strain work performed along the cycle, to first order, is $W = A(\varepsilon_1 - \varepsilon_0) \cdot \delta\varepsilon^p + \cdots$. DI-I requires that $W \geq 0$. Thereby, by passing to the limit when $\delta\varepsilon$ goes to 0, we finally obtain the inequality $(\sigma_1 - \sigma_0) \cdot \dot{\varepsilon}^p \geq 0$ in which $\sigma_1 = A(\varepsilon_1 - \varepsilon_0^p)$ and $\sigma_0 = A(\varepsilon_0 - \varepsilon_0^p)$ represent the stresses at the end and at the beginning of the first elastic path respectively, while $\dot{\varepsilon}^p$ denotes the plastic strain rate at the beginning of the inelastic path. Since ε_0 can be chosen arbitrarily in $\mathcal{E}_0$, while ε_1 can be chosen arbitrarily on $\partial\mathcal{E}_0$ and since ε^p evolves only when ε lies on $\partial\mathcal{E}_0$, the previous inequality can be extended to any real process and becomes

$$(\sigma - \sigma^*) \cdot \dot{\varepsilon}^p \geq 0, \qquad \forall \sigma^* \in \mathcal{E}^* \tag{6}$$

where σ, $\dot{\varepsilon}^p$ and $\mathcal{E}^*$ now denote the current stress, plastic strain rate and elastic stress domain respectively. Inequality (6) is nothing but the famous Hill's maximum dissipation principle. We have then proved that *an elastoplastic element satisfies DI-I* only if *it satisfies Hill's maximum dissipation principle.* In return, it is well known, see [Marigo, 1989] for a complete proof, that (6) is satisfied *if and only if* the elastic domain $\mathcal{E}^*$ (and then $\mathcal{E}$) is convex, and the plastic strain rate follows the normality rule. We can then summarise our analysis in the following

Proposition 3 *An elastoplastic material satisfies DI-I* only if *its elasticity tensor is symmetric, its yield surface is convex and the plastic strain rate is normal to the yield surface.*

Inequality (6) was obtained by using a particular infinitesimal inelastic cycle. The question is to know whether we can obtain further restrictions on the constitutive relations from DI-I by considering more general strain cycles. Clearly possible additional restrictions would only concern the evolution of the elastic domain. So the question becomes : Are some hardening effects excluded by DI-I? We will not try to answer this question in its greatest generality, but will merely prove that ideal plasticity agrees with DI-I. Specifically, assume that $\mathcal{E}^*$ is a fixed, convex set, that the normality rule holds, and that the elasticity tensor is symmetric, positive. Let us consider an arbitrary strain cycle C and denote by σ_0 and σ_1 the initial and final stress tensors, both in $\mathcal{E}^*$. Then computing the strain work yields

$$\begin{aligned} W \equiv \int_C \sigma \cdot d\varepsilon &= \int_{t_0}^{t_1} \sigma(t) \cdot (\dot{\varepsilon}(t) - \dot{\varepsilon}^p(t))\, dt + \int_{t_0}^{t_1} \sigma(t) \cdot \dot{\varepsilon}^p(t)\, dt \\ &= \int_{t_0}^{t_1} A^{-1}\sigma(t) \cdot \dot{\sigma}(t)\, dt + \int_{t_0}^{t_1} \sigma(t) \cdot \dot{\varepsilon}^p(t)\, dt \\ &= \tfrac{1}{2} A^{-1}(\sigma_1 - \sigma_0) \cdot (\sigma_1 - \sigma_0) + \int_{t_0}^{t_1} (\sigma(t) - \sigma_0)) \cdot \dot{\varepsilon}^p(t)\, dt. \end{aligned} \tag{7}$$

Since the two terms on the right hand side of (7) are positive, by virtue of the assumed positivity of A and by virtue of (6), $W \geq 0$ and the result follows. We have then proved the following

Proposition 4 *Assume that the elasticity tensor of a given ideal elastoplastic material is symmetric positive; that material satisfies DI-I* if and only if *the yield surface is convex and the plastic strain evolution satisfies the normality rule.*

At this step, nothing has been said about CD-I. Convexity of the yield surface and normality rule are deduced from DI-I alone. Let us assume, as in the previous subsection, that dissipation occurs only during inelastic processes. Then, by considering elastic processes, we deduce that the free energy is $\psi = \frac{1}{2}A(\varepsilon - \varepsilon^p) \cdot (\varepsilon - \varepsilon^p) + \varphi$, where φ depends *a priori* on time and may only vary when ε^p does. We cannot continue the analysis without an assumption about the dissipation during the inelastic phases. For example, if we assume that for ideal elastoplastic materials the free energy equals the elastic energy, then φ is a constant and the dissipation rate $\mathcal{D}$ reduces to $\sigma \cdot \dot{\varepsilon}^p$. So, if we assume that DI-I (and hence Hill's inequality) holds, then the positivity of $\mathcal{D}$ is ensured provided $\mathcal{E}^*$ contains the null stress.

3.3 Brittle damaging materials

A part of the results of this subsection was first published in [Marigo, 1989]. We consider brittle damaging materials, that is materials with elasticity which evolves with time through specific internal state variables, the evolution of which in turn follow a rate independent law based on the notion of yield surface, like that for elastoplastic materials. We limit our analysis to materials for which the damage evolution can be described by a scalar parameter, say α. Specifically, we assume that the stress-strain relation is

$$\sigma = A(\alpha)\varepsilon \quad with \quad A'(\alpha) < 0 \quad and \quad 0 \in \overset{\circ}{\mathcal{E}}, \tag{8}$$

where $A(\alpha)$ is the current value of the stiffness tensor corresponding to the current damage state α. The assumed inequality on the derivative of $\alpha \mapsto A(\alpha)$ means that the stiffness decreases when damage increases; this can be checked by micromechanical analyses when α is related to microcrack density. As in the previous subsection, the current elastic domain $\mathcal{E}$, which governs the evolution of α, is a connected closed domain in the strain space and we assume that it always contains the natural unstrained state.

Let us examine the consequences of DI-I. First, by considering elastic cycles, *i.e.*, strain cycles along which the plastic strain does not evolve, we easily deduce from Proposition 1 that the elasticity tensor $A(\alpha)$ must be symmetric. Let us now consider inelastic processes by proceeding as in the previous subsection. Let ε_0, α_0 and $\mathcal{E}_0$ be the initial (total) strain, damage state and elastic range of the material element, with $\varepsilon_0 \in \mathcal{E}_0$. Submit this material to an elastic strain path, *i.e.*, a path lying in $\mathcal{E}_0$, in such a manner that the final strain ε_1 lies on the yield surface $\partial\mathcal{E}_0$. Then apply an infinitesimal strain increment $\delta\varepsilon$ in a manner such that $\varepsilon_1 + \delta\varepsilon$ leaves $\mathcal{E}_0$ and induces an infinitesimal damage increment $\delta\alpha$ — the sign of which is not *a priori* known — and a small change of $\mathcal{E}$. Finally, end the strain cycle by elastically returning to ε_0. Let us compute the strain work needed in the three parts of this cycle. The first (elastic) part easily yields $W_1 = \int_{\varepsilon_0}^{\varepsilon_1} A(\alpha_0)\varepsilon \cdot d\varepsilon = \frac{1}{2}A(\alpha_0)\varepsilon_1 \cdot \varepsilon_1 - \frac{1}{2}A(\alpha_0)\varepsilon_0 \cdot \varepsilon_0$. In a similar manner, the third (elastic) part produces $W_3 = \frac{1}{2}A(\alpha_0 + \delta\alpha)\varepsilon_0 \cdot \varepsilon_0 - \frac{1}{2}A(\alpha_0 + \delta\alpha)(\varepsilon_1 + \delta\varepsilon) \cdot (\varepsilon_1 + \delta\varepsilon)$. The second (infinitesimal inelastic) part is evaluated at first order and one simply finds $W_2 = A(\alpha_0)\varepsilon_1 \cdot \delta\varepsilon + \cdots$. The total strain work performed along the cycle reads, at first order, as $W = \left(\frac{1}{2}A'(\alpha_0)\varepsilon_0 \cdot \varepsilon_0 - \frac{1}{2}A'(\alpha_0)\varepsilon_1 \cdot \varepsilon_1\right)\delta\alpha + \cdots$. DI-I requires that $W \geq 0$. Therefore, by considering a small enough strain increment, we obtain the inequality

$$\left(\tfrac{1}{2}A'(\alpha_0)\varepsilon_0 \cdot \varepsilon_0 - \tfrac{1}{2}A'(\alpha_0)\varepsilon_1 \cdot \varepsilon_1\right)\delta\alpha \geq 0 \tag{9}$$

which holds for all $\varepsilon_0 \in \mathcal{E}_0$ and all $\varepsilon_1 \in \partial\mathcal{E}_0$. From (9) we deduce the following

Proposition 5 *A brittle damaging material in the sense of (8) satisfies DI-I* only if *its damage parameter is an increasing function of time and if its yield surface is given by a critical energy release rate criterion, i.e.,*

$$\dot{\alpha} \geq 0, \qquad \exists\kappa > 0 \,:\, \mathcal{E} = \{\varepsilon \mid -\tfrac{1}{2}A'(\alpha)\varepsilon \cdot \varepsilon \leq \kappa\} \tag{10}$$

Proof. ▷ *(i) : α increases.* Let us consider an arbitrary strain path. Let t be a time when damage evolves, *i.e.*, $\dot{\alpha}(t+) \neq 0$. It suffices to put $\varepsilon_0 = 0$, $\varepsilon_1 = \varepsilon(t)$, $\mathcal{E} = \mathcal{E}(t)$ and $\delta\alpha = \dot{\alpha}(t+)$ in (9) to obtain, by virtue of (8), that $\dot{\alpha}(t+) > 0$. Since at the other times $\dot{\alpha} = 0$, the result follows. □

(ii) : Determination of $\mathcal{E}$. Let ε_1 be a particular strain lying on the boundary of the current elastic range $\mathcal{E}$ and set $\kappa = -\frac{1}{2}A'(\alpha)\varepsilon_1 \cdot \varepsilon_1 > 0$. Since α can only increase, we deduce from (9) that $\mathcal{E} \subset \hat{\mathcal{E}} \equiv \{\varepsilon \mid -\frac{1}{2}A'(\alpha)\varepsilon \cdot \varepsilon \leq \kappa\}$. Let us now consider another point ε_0 on $\partial\mathcal{E}$. By inverting the role of ε_0 and ε_1, we deduce from (9) that $-\frac{1}{2}A'(\alpha)\varepsilon_0 \cdot \varepsilon_0 = -\frac{1}{2}A'(\alpha)\varepsilon_1 \cdot \varepsilon_1 = \kappa$. Therefore we also get $\partial\mathcal{E} \subset \partial\hat{\mathcal{E}}$. But both inclusions are possible if and only if $\mathcal{E} = \hat{\mathcal{E}}$, which yields the desired result. □ ◁

The remarkable result is that DI-I forces the form of the elastic range by relating it to the released elastic energy due to damage growth. Only the size of $\mathcal{E}$ remains to be determined, which is the role of the hardening parameter κ. Can we obtain from DI-I restrictions on the evolution of κ with time? Part of the answer is given just below. Let us assume that (10) holds true and let C be an arbitrary strain cycle that the material can undergo. If we remark that, at each time, either $\dot{\alpha} = 0$ or $-\frac{1}{2}A'(\alpha)\varepsilon \cdot \varepsilon = \kappa$, the computation of the strain work gives

$$\begin{aligned} W \equiv \int_C \sigma \cdot d\varepsilon &= \int_{t_0}^{t_1} A(\alpha(t))\varepsilon(t) \cdot \dot{\varepsilon}(t)\, dt \\ &= \left[\tfrac{1}{2}A(\alpha)\varepsilon \cdot \varepsilon\right]_{t_0}^{t_1} - \int_{t_0}^{t_1} \tfrac{1}{2}A'(\alpha(t))\varepsilon(t) \cdot \varepsilon(t)\, \dot{\alpha}(t)\, dt \\ &= \int_{t_0}^{t_1} \left(\tfrac{1}{2}A'(\alpha(t))\varepsilon_0 \cdot \varepsilon_0 + \kappa(t)\right) \dot{\alpha}(t)\, dt. \end{aligned} \tag{11}$$

Therefore, if $\mathcal{E}$ is an increasing function of time, then $\varepsilon_0 \in \mathcal{E}(t), \forall t \geq 0$, and the right hand side of (11) is positive, *i.e.*, DI-I is verified. Thus, DI-I could fail only when $\mathcal{E}$ decreases on some time interval. But then certain strain cycles are not realisable and the analysis is more complicated.

Let us now examine the restrictions due to CD-I. By considering first elastic (and hence non dissipative) processes, we deduce that $\psi = \frac{1}{2}A(\alpha)\varepsilon \cdot \varepsilon + \phi$ where ϕ can evolve only when α does. If we consider a strain cycle starting and ending at $\varepsilon = 0$, since the stresses also vanish at the ends of the cycle, we can suppose that, for such a cycle, the strain work corresponds exactly to the dissipated energy. Under such an hypothesis, let us consider an arbitrary inelastic strain path beginning at the state $(\varepsilon_0, \alpha_0, \mathcal{E}_0)$ and ending at the state $(\varepsilon_1, \alpha_1, \mathcal{E}_1)$. We can always construct a cycle starting and ending at $\varepsilon = 0$ and containing this inelastic path (it suffices to precede it by an elastic path going from 0 to ε_0 and to follow it by an other elastic one going from ε_1 to 0). During this cycle, ϕ can evolve only during the inelastic stage, but, by virtue of our assumption, its final value must be equal to its initial one, hence it remains constant. Thus the free energy coincides with the elastic energy. Inserting this result into (1) leads to $\mathcal{D} = -\frac{1}{2}A'(\alpha)\varepsilon \cdot \varepsilon\, \dot{\alpha} \geq 0$, from which we can deduce only that damage must grow. CD-I gives no information on the form of the yield surface.

4. HOW THEY RESTRICT VISCOELASTIC MATERIALS

The results presented below have never been published. The interested reader should however consult [Day, 1976] for the study of the consequences of a slightly different work axiom, and, [Coleman *et al.*, 1967] for the implications of CD-I.

4.1 The linear viscoelastic models of Maxwell

We first show why it is important to require (2) only when the material starts from an equilibrium state. To this effect, we consider the (family of) well-known Maxwell model(s) for linear viscoelastic material :

$$\sigma = A(\varepsilon - \varepsilon^v), \qquad \dot{\varepsilon}^v = B\sigma, \tag{12}$$

where ε^v is the viscous strain, A the stiffness tensor and B the viscosity tensor. For physical reasons, it is usually admitted that A and B must be symmetric and positive. In particular, owing to the positivity of B, when the strain remains constant the stress (asymptotically) relaxes to 0, while, if B was negative, it exponentially increases with time.

Since this model is well-established, no reliable general principle can rule it out.

Therefore, let us first check that it is in agreement with DI-I. Following the definition, an equilibrium state is a state $(\varepsilon, \varepsilon^v = \varepsilon)$ at which the stress vanishes. Let us assume that the material starts from such a state, submit it to a strain cycle and compute the strain work. We obtain

$$\begin{aligned} \int_C \sigma \cdot d\varepsilon &= \int_{t_0}^{t_1} \sigma(t) \cdot (\dot{\varepsilon}(t) - \dot{\varepsilon}^v(t))dt + \int_{t_0}^{t_1} \sigma(t) \cdot \dot{\varepsilon}^v(t)dt \\ &= \tfrac{1}{2}A^{-1}\sigma(t_1) \cdot \sigma(t_1) + \int_{t_0}^{t_1} B\sigma(t) \cdot \sigma(t)dt \geq 0. \end{aligned}$$

Hence DI-I and Maxwell's models are compatible.

Let us now show that, if the material starts from a non equilibrium state, then there exist strain cycles along which the work is negative. Let $(\varepsilon, \varepsilon^v)$ be the initial state with $\varepsilon \neq \varepsilon^v$. Let us deform the material very quickly in such a manner that its strain evolves from ε to ε^v while the viscous strain (nearly) remains at its initial value ε^v. Thus, the material arrives at ε^v in an equilibrium state. Then, let us return to the initial strain ε very slowly, the stress remaining (nearly) at 0. The strain work performed during the cycle is (nearly) equal to $-\frac{1}{2}A(\varepsilon - \varepsilon^v) \cdot (\varepsilon - \varepsilon^v) < 0$.

4.2 Extended viscoelastic materials of Maxwell's type

We consider a class of viscoelastic materials more general than Maxwell's model. The stress versus strain relation is now given by

$$\sigma = A\varepsilon - \sigma^v, \qquad \dot{\sigma}^v = -\frac{1}{\tau}(\sigma^v - f(\varepsilon)), \tag{13}$$

where, for computational convenience, we prefer to consider the viscous stress σ^v rather than the viscous strain $\varepsilon^v \equiv A^{-1}\sigma^v$ as the internal state variable. Moreover, we assume that the time parameter τ is positive and that the function f is a smooth one-to-one map onto the space of symmetric matrices. The hypothesis on f is not essential, but permits us to simplify the proofs. On the other hand, the assumed positivity of τ plays a fundamental role by allowing the relaxation of stresses. An interesting issue would be to see whether this inequality is required by DI-I or CD-I.

▷ *The extension of Maxwell's linear model.* If we put $B = A^{-1}/\tau$ in (12), then the corresponding Maxwell model belongs to the studied class with $f(\varepsilon) = A\varepsilon$ (a linear model). Now, with this extension, when the strain is held to a given value ε, the stresses do not generally relax to 0 but to $A\varepsilon - f(\varepsilon)$ and therefore the "delayed" elastic behaviour is non linear. ◁

Let us state the main result of this subsection

Proposition 6 *An extended viscoelastic material (in the sense of (13)) satisfies DI-I* if and only if *A is symmetric and f is the gradient of a convex function.*

Proof. ▷ *(i) : Symmetry of A.* It suffices to consider "instantaneous" strain cycles, *i.e.*, limit cases of strain paths travelled at high speed, because the viscous part of the stress has no time to evolve and remains at its initial value $f(\varepsilon_0)$. The stress versus strain relation becomes $\sigma = A\varepsilon - f(\varepsilon_0)$. We conclude by virtue of Proposition 1. □

(ii) : $f = D\varphi$. Here we consider "infinitely" slow strain cycles. Thus the stresses have enough time to relax and the stress versus strain relation becomes $\sigma = A\varepsilon - f(\varepsilon)$. The material behaves elastically and we conclude by virtue of Proposition 1. □

(iii) : Convexity of φ. Let ε_0 be the initial strain and $f(\varepsilon_0)$ the associated relaxed viscous stress, $(\varepsilon_0, f(\varepsilon_0))$ being the initial equilibrium state. We consider the following cycle : first, the material is "instantaneously" deformed up to the strain ε_1; then we maintain the strain a long time at this state so that the viscous stress (nearly) relaxes to $f(\varepsilon_1)$; finally we get the material to return "instantaneously" to its initial strain state ε_0, which completes the cycle. Since the first and the third steps are (nearly) elastic and since the strain work vanishes during the second one (because $\dot{\varepsilon} = 0$), we easily compute the strain work performed along the cycle and check, by virtue of DI-I, that

$$(D\varphi(\varepsilon_1) - D\varphi(\varepsilon_0)) \cdot (\varepsilon_1 - \varepsilon_0) \geq 0. \tag{14}$$

But, since ε_0 and ε_1 are arbitrary, (14) is nothing but the monotonicity of $D\varphi$ which is strictly equivalent to the convexity of φ. □

(iv) : Converse statement. Let us now assume that A is symmetric and $f = D\varphi$ (with f one-to-one). We take an arbitrary equilibrium state $(\varepsilon_0, D\varphi(\varepsilon_0))$ as the initial state of the material element and submit that element to an arbitrary strain cycle $\mathcal{C}$. We denote by $\varepsilon(t)$, $\sigma^v(t)$ and $\sigma(t)$ the strain, viscous stress and total stress at time t, and set $e(t) = f^{-1}(\sigma^v(t))$, $e_1 = e(t_1)$. Then the computation of the strain work yields

$$\begin{aligned}
\int_{\mathcal{C}} \sigma \cdot d\varepsilon &= -\int_{t_0}^{t_1} \sigma^v(t) \cdot \dot{\varepsilon}(t)\, dt \\
&= \int_{t_0}^{t_1} \sigma^v(t) \cdot \left(\dot{e}(t) - \dot{\varepsilon}(t)\right)\, dt - \int_{t_0}^{t_1} D\varphi(e(t)) \cdot \dot{e}(t)\, dt \\
&= -\int_{t_0}^{t_1} \dot{\sigma}^v(t) \cdot (e(t) - \varepsilon(t))\, dt + [\sigma^v \cdot (e - \varepsilon) - \varphi(e)]_{t_0}^{t_1} \\
&= \frac{1}{\tau}\int_{t_0}^{t_1} (D\varphi(e(t)) - D\varphi(\varepsilon(t)) \cdot (e(t) - \varepsilon(t))\, dt \\
&\quad + D\varphi(e_1) \cdot (e_1 - \varepsilon_0) - \varphi(e_1) + \varphi(\varepsilon_0)
\end{aligned}$$

and, since both terms in the last right hand side are positive by virtue of the convexity of φ, the strain work is positive. □ ◁

Let us now check CD-I and construct the free energy. For the sake of simplicity, we assume *a priori* that ψ only depends on ε and σ^v — though this can be deduced from a less brutal assumption on the dissipation. By considering an instantaneous strain path (along which the material does not dissipate any energy), we conclude that the free energy takes the form $\psi = \frac{1}{2}A\varepsilon \cdot \varepsilon - \sigma^v \cdot \varepsilon + \phi(\sigma^v)$. Now consider a relaxation process, the material element starting from a viscous stress state σ^v and the strain being fixed at ε. By virtue of CD-I, the free energy decreases with time until the viscous stress has relaxed to $D\varphi(\varepsilon)$. Since the initial state can be chosen arbitrarily, we deduce that the function $\sigma^v \mapsto -\sigma^v \cdot \varepsilon + \phi(\sigma^v)$ reaches its minimum at $D\varphi(\varepsilon)$. Therefore, from the assumption that f is one-to-one, we obtain as an optimal condition that $f^{-1} = D\phi$ and hence that ϕ is (up to an arbitrary constant) the Legendre transform of φ, *i.e.*,

$$\psi = \tfrac{1}{2}A\varepsilon \cdot \varepsilon - \sigma^v \cdot \varepsilon + \varphi^*(\sigma^v), \qquad \varphi^*(\sigma^v) = \sup_\varepsilon\{\sigma^v \cdot \varepsilon - \varphi(\varepsilon)\}. \tag{15}$$

If we insert (15) into (1), we obtain the expression for the dissipation rate

$$\mathcal{D} = \frac{1}{\tau}\left(D\varphi^*(\sigma^v) - D\varphi^*(D\varphi(\varepsilon))\right)\cdot(\sigma^v - D\varphi(\varepsilon)) \geq 0, \tag{16}$$

its positivity being ensured by the convexity of φ^*.

5. THEIR LINK TO STANDARD MATERIALS

In this concluding section we consider the results established in the two previous sections concerning the power of restriction CD-I and DI-I so as to evaluate how close they take us to a GSM. To that end, let us recall the main ingredients needed in the construction of a GSM, see [Halphen *et al.*, 1975] or [Germain *et al.*, 1983] : *(i) Choice of the state variables, i.e., in the present context choice of the pair* (ε, α)*,* α *representing the possible (dissipative) internal variables — (ii) Choice of the (possibly convex) function free energy* $\psi(\varepsilon, \alpha)$ *as a thermodynamical potential giving the stress tensor* $\sigma = \partial_\varepsilon \psi(\varepsilon, \alpha)$ *and the "thermodynamical internal forces"* $a = -\partial_\alpha \psi(\varepsilon, \alpha)$ *— (iii) Choice of a potential of dissipation, convex function* ϕ *of* a*, giving the evolution law of the internal variables :* $\dot{\alpha} = D\phi(a)$*, where the symbol* D *denotes the usual derivative when* ϕ *is differentiable, a subgradient otherwise —* Revisiting the two previous sections leads us to question the notion of a convex dissipation potential governing the evolution of the internal state variables. Indeed, that the free energy acts as a potential for the stress is always checked (but maybe not the convexity of ψ). Let us examine each class of materials.

• *Elastoplastic materials.* We obtained the convexity of the yield surface and the normality rule for $\dot{\varepsilon}^p$. Thus, for ideal plastic materials, upon assuming that the free energy coincides with the elastic energy, we end up with a GSM. Indeed, in such a case, $a = \sigma$ and if we set ϕ as the indicator function of the closed convex set $\mathcal{E}^*$ ($\phi(\sigma) = 0$ when $\sigma \in \mathcal{E}^*$, $+\infty$ otherwise), then the normality rule forces $\dot{\varepsilon}^p$ to be a subgradient of ϕ at σ. More problematic is the case (for ideal plastic materials again) when $\psi = \frac{1}{2}A(\varepsilon - \varepsilon^p)\cdot(\varepsilon - \varepsilon^p) + \varphi(\varepsilon^p)$, a situation that neither DI-I nor CD-I can exclude. In such a case $a = \sigma - D\varphi(\varepsilon^p)$ and the potential ϕ also depends on ε^p. It seems however that in both approaches the most important properties are the convexity of $\mathcal{E}^*$ and the normality of $\dot{\varepsilon}^p$.

• *Brittle damaging materials.* In fact, this case is nearly identical to the previous one. Indeed, now $a = -\frac{1}{2}A(\alpha)\varepsilon\cdot\varepsilon$, *i.e.*, a is the energy release rate. Therefore if we consider the interval $K = [-\kappa, \kappa]$ and set ϕ as the indicator function of K, then the evolution law deduced from CD-I and DI-I also reads as : $a \in K$ (really, $0 \leq a \leq \kappa$), $\dot{\alpha} \in \partial\phi(a)$, which corresponds to a GSM. However, since κ can vary, ϕ is κ-dependent. Furthermore the two principles leave the evolution of K free of any constraint.

• *Viscoelastic materials.* For the linear Maxwell models, since $\alpha = \varepsilon^v$ and $\psi = \frac{1}{2}A(\varepsilon - \varepsilon^v)\cdot(\varepsilon - \varepsilon^v)$, we get $a = \sigma$ and by introducing $\phi(a) = \frac{1}{2}Ba\cdot a$ the evolution law of ε^v reads as that of a GSM. The non linear cases are more involved. Indeed, in such a case (now with $\alpha = \sigma^v$), (15) yields $a = \varepsilon - D\varphi^*(\sigma^v) = \varepsilon - e$, where e is the strain with associated relaxed viscous stress σ^v. On the other hand, the evolution law is

$$\dot{\sigma}^v = \frac{1}{\tau}(D\varphi(\varepsilon) - D\varphi(e)).$$

This is not the desired form (except when φ is quadratic, *i.e.,* except in the linear case). Thus, the non linear viscoelastic materials allowed by CD-I and DI-I are not strictly GSM. In order to abide by the rule, we could add σ^v as a variable in the potential. In any case, the convexity of φ is ensured.

From this very partial analysis, it seems that DI-I is a good complementary principle, which, coupled with CD-I, will imply important properties for real materials even if it cannot completely fill in the gap up to GSM. Its application to more general thermodynamical systems will be an interesting issue, but we will first need to generalise the formulation (by adding for example thermal terms in the definition of the "strain work"). A nice challenge.

REFERENCES

[Coleman et al., 1967] Coleman B.D.; Gurtin M.E.; Thermodynamics with internal state variables; *J. Chem. Phys.,* **47**, *597–613.*

[Coleman et al., 1963] Coleman B.D.; Noll W.; The thermodynamics of elastic materials with conduction and viscosity; *Arch. Rat. Mech. Anal.,* **13**, *167–178.*

[Day, 1976] Day W.A.; Entropy and hidden variables in continuum thermodynamics; *Arch. Rat. Mech. Anal.,* **62**, *367–389.*

[Drucker, 1951] Drucker D.C.; A more fundamental approach to plastic stress-strain relations; In: *Proc. First US Congress of Applied Mechanics, ed. ASME (New York), 487–491.*

[Germain et al., 1983] Germain P.; Nguyen Q.S.; Suquet P.; Continuum thermodynamics; *J. Appl. Mech., Transactions of the ASME,* **50**, *1010–1020.*

[Halphen et al., 1975] Halphen B.; Nguyen Q.S.; Sur les matériaux standards généralisés; *J. Mécanique,* **14**, *39–63.*

[Ilyushin, 1961] Ilyushin A.A.; On the postulate of stability; *J. Appl. Math. Mech.,* **25**, *746–752.*

[Marigo, 1989] Marigo J.-J.; Constitutive relations in plasticity, damage and fracture mechanics based on a work property; *Nuclear Eng. Design,* **114**, *249-272.*

Constitutive relations involving internal variables based on a micromechanical analysis

*J.C. Michel *, U. Galvanetto ** and P. Suquet * (+)*

* *L.M.A./ C.N.R.S. 31 Chemin Joseph Aiguier.*
13402. Marseille. Cedex 20. France.
** *Department of Aeronautics. Imperial College.*
Prince Consort Road. London. SW7 2BY. UK.
(+) *Corresponding author. Email: suquet@lma.cnrs-mrs.fr*

Abstract: The exact description of the overall behavior of composites with nonlinear dissipative phases requires an infinity of internal variables. Approximate models involving only a finite number of those can be obtained by considering a decomposition of the microscopic anelastic strain field on a finite set of shape functions. The *Transformation Field Analysis* of [Dvorak, 1992] is obtained as a special case of the proposed theory by considering shape functions which are uniform within a given subdomain. The interest of considering nonuniform shape functions is shown.

Keywords: Constitutive relations, internal variables, nonlinear composites, micromechanics, Transformation Field Analysis.

1. INTRODUCTION

Among all Paul Germain's scientific achievements, his contribution to the formulation of constitutive relations in the coherent framework of Continuum Thermodynamics has a prominent place. His book *Mécanique des Milieux Continus* [Germain, 1973], contemporary with the seminal work of [Halphen et al, 1975] *Generalized Standard Materials* (GSM), opened a way which was further pursued, developed and illustrated by many authors inside and outside France (see [Germain et al, 1983, Lemaître et al, 1988, Lubliner, 1990, Maugin, 1992] for a review).

This theory relies on two fundamental concepts, the notion of internal variables, often denoted $\boldsymbol{\alpha}$, and the notion of thermodynamic potentials, the free energy ρw and the dissipation potential φ, endowed with specific mathematical properties.

The internal variables $\boldsymbol{\alpha}$ at time t are supposed to contain all the relevant information about the material history for times $\tau \leq t$. The choice of these variables depends obviously on the constitutive relations under consideration. Some of these variables have a purely macroscopic interpretation. Other variables are sometimes *interpreted* as microstructural variables but most often the equations governing their evolution are not really derived from micromechanical considerations.

G.A. Maugin et al. (eds.), Continuum Thermomechanics, 301-312.

© 2000 *Kluwer Academic Publishers.*

The aim of this paper is to give an example of an approximate micromechanical scheme in which internal variables at the macroscopic level have a well understood micromechanical interpretation. More specifically, we consider a nonhomogeneous material with elementary constituents which are standard is the simplest possible form. The only internal variable is the anelastic strain. It is well known that the overall behavior of such a composite has a GSM structure but with infinitely many internal variables which are the fields of local internal variables. This result will be recalled in section 2 following the presentation of [Suquet, 1985]. Similar ideas were already present in different forms in the works of [Rice, 1970] and [Mandel, 1972] among others.

In practice it is necessary to reduce the number of internal variables by considering that the local fields of anelastic strains depend only on a finite number of *shape functions* describing the variations of the local plastic fields. This reduction is done in section 3. It is remarked that the proposed method reduces to the *Transformation Field Analysis* of [Dvorak et al, 1994], when the plastic strain fields are assumed to be uniform within each individual phase. This corresponds to the case where the shape functions are the characteristic functions of the phases. The proposed theory is more general in that the shape functions are not required to be uniform within a given domain. The method is illustrated in section 4 and its merits are assessed by comparison with the TFA.

2. STABILITY OF THE STANDARD STRUCTURE BY CHANGE OF SCALE

Consider a *representative volume element* (r.v.e) V of a nonhomogeneous material composed of N different phases or subdomains. This r.v.e is subjected to an average loading characterized by a given path in the space of overall strain or stress. Attention is limited to isothermal evolutions and infinitesimal deformations. The overall Cauchy stress $\overline{\boldsymbol{\sigma}}$ and the overall infinitesimal deformation $\overline{\boldsymbol{\varepsilon}}$ are the averages of their local counterparts $\boldsymbol{\sigma}$ and $\boldsymbol{\varepsilon}$

$$\overline{\boldsymbol{\sigma}} = \langle \boldsymbol{\sigma} \rangle, \quad \overline{\boldsymbol{\varepsilon}} = \langle \boldsymbol{\varepsilon} \rangle, \quad \text{where } \langle f \rangle = \frac{1}{|V|} \int_V f(\boldsymbol{x})\, d\boldsymbol{x}. \tag{1}$$

The local stress and strain fields are determined through the resolution of the *local* evolution problem posed for the r.v.e. and consisting of equilibrium equations, boundary conditions and constitutive relations. The boundary conditions are assumed to be such that Hill's micro-macro localization condition is satisfied: for any compatible strain field $\boldsymbol{\varepsilon}$ and any stress field $\boldsymbol{\sigma}$ in equilibrium, both meeting the boundary conditions imposed on the boundary of the r.v.e., the following equality holds

$$\langle \boldsymbol{\sigma} : \boldsymbol{\varepsilon} \rangle = \langle \boldsymbol{\sigma} \rangle : \langle \boldsymbol{\varepsilon} \rangle = \overline{\boldsymbol{\sigma}} : \overline{\boldsymbol{\varepsilon}}. \tag{2}$$

Examples of boundary conditions meeting (2) include uniform strains, uniform stresses, periodicity conditions (see [Suquet, 1987] for more details). Periodicity boundary conditions will be assumed in the following.

2.1 Standard constitutive relations

The behavior of the individual constituent at point $\boldsymbol{x}$ is defined by a *standard* model, *i.e.* by two thermodynamic potentials. The free energy ρw defines (through the state laws) the forces available in the system at rest to drive the dissipative mechanisms, and the dissipation potential φ relates the rate of the dissipative mechanisms (complementary laws) with the associated forces. For simplicity we shall consider that the only dissipative mechanism comes through an anelastic strain (plastic or viscoplastic strain) $\boldsymbol{\varepsilon}^{\mathsf{an}}$ and that the free energy is a quadratic function of the elastic strain

$$\boldsymbol{\varepsilon} = \boldsymbol{\varepsilon}^{\mathsf{e}} + \boldsymbol{\varepsilon}^{\mathsf{an}}, \quad \rho w(\boldsymbol{\varepsilon}, \boldsymbol{\varepsilon}^{\mathsf{an}}) = \frac{1}{2}(\boldsymbol{\varepsilon} - \boldsymbol{\varepsilon}^{\mathsf{an}}) : \boldsymbol{L} : (\boldsymbol{\varepsilon} - \boldsymbol{\varepsilon}^{\mathsf{an}}). \tag{3}$$

Then the state laws and the complementary laws read respectively:

$$\text{State laws:} \qquad \boldsymbol{\sigma}(\boldsymbol{x}) = \rho \frac{\partial w}{\partial \boldsymbol{\varepsilon}}(\boldsymbol{x}, \boldsymbol{\varepsilon}(\boldsymbol{x}), \boldsymbol{\varepsilon}^{\mathsf{an}}(\boldsymbol{x})) = \boldsymbol{L}(\boldsymbol{x}) : (\boldsymbol{\varepsilon}(\boldsymbol{x}) - \boldsymbol{\varepsilon}^{\mathsf{an}}(\boldsymbol{x})), \tag{4}$$

$$\text{Complementary laws:} \qquad \boldsymbol{\sigma} = \frac{\partial \varphi}{\partial \dot{\boldsymbol{\varepsilon}}^{\mathsf{an}}}(\boldsymbol{x}, \dot{\boldsymbol{\varepsilon}}^{\mathsf{an}}(\boldsymbol{x})). \tag{5}$$

The potentials ρw and φ may depend on $\boldsymbol{x}$ (nonhomogeneity of the volume element) and are assumed to be convex with respect to their other arguments.

2.2 Generalized standard structure of the overall constitutive relations

The free energy is an additive quantity or, in other words, the overall free energy of the r.v.e. V is the average of the local free energy. As shown in [Suquet, 1987], the state variables at the macroscopic level consist of the average strain $\overline{\boldsymbol{\varepsilon}}$ and of an infinite number of internal variables which are the values of the anelastic strain field at every microscopic point $\boldsymbol{x} \in V$:

$$\boldsymbol{\alpha} = \{\boldsymbol{\varepsilon}^{\mathsf{an}}(\boldsymbol{x})\}_{\boldsymbol{x} \in V}. \tag{6}$$

The overall free energy reads

$$\widetilde{\rho}\widetilde{w}\left(\overline{\boldsymbol{\varepsilon}}, \{\boldsymbol{\varepsilon}^{\mathsf{an}}(\boldsymbol{x})\}_{\boldsymbol{x} \in V}\right) = \langle \rho w(\boldsymbol{\varepsilon}, \boldsymbol{\varepsilon}^{\mathsf{an}}) \rangle, \quad \text{with} \quad \widetilde{\rho} = \langle \rho \rangle. \tag{7}$$

The forces associated with the state variables are

$$\widetilde{\rho}\frac{\partial \widetilde{w}}{\partial \overline{\boldsymbol{\varepsilon}}}\left(\overline{\boldsymbol{\varepsilon}}, \{\boldsymbol{\varepsilon}^{\mathsf{an}}(\boldsymbol{x})\}_{\boldsymbol{x} \in V}\right), \quad \mathcal{A} = \{\mathcal{A}_{\boldsymbol{x}}\}_{\boldsymbol{x} \in V}, \quad \mathcal{A}_{\boldsymbol{x}} = -\widetilde{\rho}\frac{\partial \widetilde{w}}{\partial \boldsymbol{\varepsilon}^{\mathsf{an}}(\boldsymbol{x})}\left(\overline{\boldsymbol{\varepsilon}}, \{\boldsymbol{\varepsilon}^{\mathsf{an}}(\boldsymbol{x})\}_{\boldsymbol{x} \in V}\right). \tag{8}$$

The first force can be computed using Hill's lemma

$$\widetilde{\rho}\frac{\partial \widetilde{w}}{\partial \overline{\boldsymbol{\varepsilon}}}\left(\overline{\boldsymbol{\varepsilon}}, \{\boldsymbol{\varepsilon}^{\mathsf{an}}(\boldsymbol{x})\}_{\boldsymbol{x} \in V}\right) = \left\langle \rho\frac{\partial w}{\partial \boldsymbol{\varepsilon}}(\boldsymbol{\varepsilon}, \boldsymbol{\varepsilon}^{\mathsf{an}}) : \frac{\partial \boldsymbol{\varepsilon}}{\partial \overline{\boldsymbol{\varepsilon}}} \right\rangle = \left\langle \boldsymbol{\sigma} : \frac{\partial \boldsymbol{\varepsilon}}{\partial \overline{\boldsymbol{\varepsilon}}} \right\rangle = \langle \boldsymbol{\sigma} \rangle : \left\langle \frac{\partial \boldsymbol{\varepsilon}}{\partial \overline{\boldsymbol{\varepsilon}}} \right\rangle = \overline{\boldsymbol{\sigma}},$$

since $\left\langle \frac{\partial \varepsilon}{\partial \overline{\varepsilon}} \right\rangle = \boldsymbol{I}$. Similarly, the set of forces $\{\mathcal{A}_{\boldsymbol{x}}\}_{\boldsymbol{x}\in V}$ coincides with the local stress field $\{\boldsymbol{\sigma}(\boldsymbol{x})\}_{\boldsymbol{x}\in V}$. The effective dissipation potential reads

$$\widetilde{\varphi}(\{\dot{\boldsymbol{\varepsilon}}^{\text{an}}(\boldsymbol{x})\}_{\boldsymbol{x}\in V}) = \langle \varphi(\dot{\boldsymbol{\varepsilon}}^{\text{an}}) \rangle. \tag{9}$$

With the choices (6) (7) (9), the effective constitutive relations of the composite have a generalized standard structure:

$$\text{State variables:} \qquad \overline{\boldsymbol{\varepsilon}}, \quad \boldsymbol{\alpha} = \{\boldsymbol{\varepsilon}^{\text{an}}(\boldsymbol{x})\}_{\boldsymbol{x}\in V}, \tag{10}$$

$$\text{State laws:} \qquad \overline{\boldsymbol{\sigma}} = \widetilde{\rho}\frac{\partial \widetilde{w}}{\partial \overline{\boldsymbol{\varepsilon}}}(\overline{\boldsymbol{\varepsilon}}, \boldsymbol{\alpha}), \quad \mathcal{A} = -\widetilde{\rho}\frac{\partial \widetilde{w}}{\partial \boldsymbol{\alpha}}(\overline{\boldsymbol{\varepsilon}}, \boldsymbol{\alpha}), \tag{11}$$

$$\text{Complementary laws:} \qquad \mathcal{A} = \frac{\partial \widetilde{\varphi}}{\partial \dot{\boldsymbol{\alpha}}}(\dot{\boldsymbol{\alpha}}). \tag{12}$$

2.3 Green operator $\boldsymbol{\Gamma}$

At rest (no evolution of the system) the stress and strain field in the r.v.e. solve the following linear elastic problem, with appropriate boundary conditions

$$\boldsymbol{\sigma}(\boldsymbol{x}) = \boldsymbol{L}(\boldsymbol{x}) : (\boldsymbol{\varepsilon}(\boldsymbol{x}) - \boldsymbol{\varepsilon}^{\text{an}}(\boldsymbol{x})), \quad \text{div}\,(\boldsymbol{\sigma}(\boldsymbol{x})) = 0, \quad \langle \boldsymbol{\varepsilon} \rangle = \overline{\boldsymbol{\varepsilon}}. \tag{13}$$

The strain field $\boldsymbol{\varepsilon}(\boldsymbol{x})$ solution of this problem is a nonlocal function of the anelastic strain field and can be expressed as:

$$\boldsymbol{\varepsilon}(\boldsymbol{x}) = \boldsymbol{A}(\boldsymbol{x}) : \overline{\boldsymbol{\varepsilon}} + \frac{1}{|V|}\int_V \boldsymbol{D}(\boldsymbol{x}, \boldsymbol{x}') : \boldsymbol{\varepsilon}^{\text{an}}(\boldsymbol{x}')\, d\boldsymbol{x}' = \boldsymbol{A}(\boldsymbol{x}) : \overline{\boldsymbol{\varepsilon}} + \boldsymbol{D} * \boldsymbol{\varepsilon}^{\text{an}}(\boldsymbol{x}), \tag{14}$$

where

$$\boldsymbol{D} * \boldsymbol{\varepsilon}^{\text{an}}(\boldsymbol{x}) \stackrel{def}{=} \frac{1}{|V|}\int_V \boldsymbol{D}(\boldsymbol{x}, \boldsymbol{x}') : \boldsymbol{\varepsilon}^{\text{an}}(\boldsymbol{x}')\, d\boldsymbol{x}'. \tag{15}$$

In this relation $\boldsymbol{A}(\boldsymbol{x})$ denotes the elastic strain-localization tensor, and the nonlocal operator $\boldsymbol{D}(\boldsymbol{x}, \boldsymbol{x}') = \boldsymbol{\Gamma}(\boldsymbol{x}, \boldsymbol{x}') : \boldsymbol{L}(\boldsymbol{x}')$ gives the strain at point $\boldsymbol{x}$ created by an eigenstrain $\boldsymbol{\varepsilon}^{\text{an}}(\boldsymbol{x}')$ at point $\boldsymbol{x}'$. $\boldsymbol{\Gamma}(\boldsymbol{x}, \boldsymbol{x}')$ denotes the nonlocal elastic Green operator which can be expressed in terms of the derivatives of the Green function for the Navier equations with elastic coefficients $\boldsymbol{L}$. More specifically, given a field of eigenstress $\boldsymbol{\tau}(\boldsymbol{x})$, the solution of the elasticity problem:

$$\boldsymbol{\sigma}(\boldsymbol{x}) = \boldsymbol{L}(\boldsymbol{x}) : \boldsymbol{\varepsilon}(\boldsymbol{x}) + \boldsymbol{\tau}(\boldsymbol{x}), \quad \text{div}\,(\boldsymbol{\sigma}(\boldsymbol{x})) = 0, \quad \langle \boldsymbol{\varepsilon} \rangle = \boldsymbol{0}, \tag{16}$$

can be expressed as:

$$\boldsymbol{\varepsilon}(\boldsymbol{x}) = -\langle \boldsymbol{\Gamma} * \boldsymbol{\tau} \rangle(\boldsymbol{x}). \tag{17}$$

$\boldsymbol{\Gamma}$ has simple properties which will be useful in the sequel

$$\int_V \boldsymbol{\Gamma}(\boldsymbol{x}, \boldsymbol{x}')\, d\boldsymbol{x}' = \mathbf{0}, \quad \int_V \boldsymbol{\Gamma}(\boldsymbol{x}, \boldsymbol{x}')\, d\boldsymbol{x} = \mathbf{0}, \quad \boldsymbol{\Gamma}(\boldsymbol{x}, \boldsymbol{x}') = {}^{\mathsf{T}}\boldsymbol{\Gamma}(\boldsymbol{x}', \boldsymbol{x}). \tag{18}$$

3. REDUCTION OF THE NUMBER OF INTERNAL VARIABLES

3.1 State variables

In order to reduce the number of internal variables, we assume throughout the following that the field of anelastic strains can be expressed as a function of a finite number of *shape functions* θ_α:

$$\boldsymbol{\varepsilon}^{\mathsf{an}}(\boldsymbol{x}) = \sum_{\alpha=1}^{A} \boldsymbol{\varepsilon}_\alpha^{\mathsf{an}}\, \theta_\alpha(\boldsymbol{x}). \tag{19}$$

The *reduced* state variables of the model are the overall strain and the component of the anelastic strain field on the shape functions:

$$\overline{\boldsymbol{\varepsilon}} \quad \text{and} \quad \boldsymbol{\alpha} = \{\boldsymbol{\varepsilon}_\alpha^{\mathsf{an}}\}_{\alpha=1,..,A}. \tag{20}$$

A typical example of such shape functions is provided by the characteristic functions of phases or subdomains within the same phase (we shall consider the subdomains as separate phases even though a single mechanical phase can be divided into several subdomains):

$$\chi_r(\boldsymbol{x}) = 1 \text{ if } x \in V_r, \quad \chi_r(\boldsymbol{x}) = 0 \text{ otherwise}. \tag{21}$$

This specific choice leads to the *Transformation Field Analysis* of [Dvorak, 1992] where the plastic strain field is assumed to be uniform over each individual subdomain V_r. However it may be interesting in certain circumstances to consider shape functions which are richer than the characteristic functions to account for spatial nonuniformity of the (anelastic) strain field within one phase or subdomain. The resulting theory will be called the *Nonuniform Transformation Field Analysis* (NTFA).

It is assumed that the shape functions have their support entirely contained in a single phase. In more mathematical terms it is assumed that

$$\theta_\alpha(\boldsymbol{x})\chi_r(\boldsymbol{x}) = 0 \text{ or } \theta_\alpha(\boldsymbol{x}) \quad \forall r = 1, ..., N. \tag{22}$$

Therefore one can define $\boldsymbol{L}_\alpha$ and φ_α as the stiffness tensor and the dissipation potential of the phase r in which the shape function θ_α has its support.

3.2 State laws

With the decomposition (19), the effective free-energy (7) and the thermodynamic forces associated with the state variables $(\overline{\varepsilon}, \{\varepsilon_\alpha^{\mathrm{an}}\}_{\alpha=1,\ldots,A})$ read as

$$\left.\begin{aligned}
\widetilde{\rho}\widetilde{w}\left(\overline{\varepsilon}, \{\varepsilon_\alpha^{\mathrm{an}}\}_{\alpha=1,\ldots,A}\right) &= \frac{1}{2}\left\langle (\varepsilon - \sum_{\alpha=1}^{A} \varepsilon_\alpha^{\mathrm{an}}\theta_\alpha) : \boldsymbol{L} : (\varepsilon - \sum_{\beta=1}^{A} \varepsilon_\beta^{\mathrm{an}}\theta_\beta) \right\rangle, \\
\overline{\sigma} = \widetilde{\rho}\frac{\partial \widetilde{w}}{\partial \overline{\varepsilon}}\left(\overline{\varepsilon}, \{\varepsilon_\alpha^{\mathrm{an}}\}_{\alpha=1,\ldots,A}\right) &= \left\langle \boldsymbol{L} : (\varepsilon - \sum_{\alpha=1}^{A} \varepsilon_\alpha^{\mathrm{an}}\theta_\alpha) \right\rangle, \\
\mathcal{A}_\alpha = -\widetilde{\rho}\frac{\partial \widetilde{w}}{\partial \varepsilon_\alpha^{\mathrm{an}}}\left(\overline{\varepsilon}, \{\varepsilon_\alpha^{\mathrm{an}}\}_{\alpha=1,\ldots,A}\right) &= \langle \boldsymbol{L} : (\varepsilon - \varepsilon^{\mathrm{an}})\theta_\alpha \rangle = \langle \sigma\theta_\alpha \rangle.
\end{aligned}\right\} \tag{23}$$

The state laws can be more easily expressed in terms of the following generalized stresses and strains :

$$\boldsymbol{a}_\alpha = \frac{\langle \sigma\theta_\alpha \rangle}{\langle \theta_\alpha \rangle}, \quad \boldsymbol{e}_\alpha = \frac{\langle \varepsilon\theta_\alpha \rangle}{\langle \theta_\alpha \rangle}, \quad \boldsymbol{e}_\alpha^{\mathrm{an}} = \frac{\langle \varepsilon^{\mathrm{an}}\theta_\alpha \rangle}{\langle \theta_\alpha \rangle}. \tag{24}$$

Note that

$$\mathcal{A}_\alpha = \boldsymbol{a}_\alpha \langle \theta_\alpha \rangle, \quad \varepsilon_\alpha^{\mathrm{an}} = \sum_{\beta=1}^{A} (\boldsymbol{g})_{\alpha\beta}^{-1} \langle \theta_\beta \rangle \boldsymbol{e}_\beta^{\mathrm{an}}, \quad \text{where} \quad g_{\alpha\beta} = \langle \theta_\alpha\theta_\beta \rangle. \tag{25}$$

Multiplying the local state law (4) by θ_α and averaging over V yields

$$\boldsymbol{a}_\alpha = \boldsymbol{L}_\alpha : (\boldsymbol{e}_\alpha - \boldsymbol{e}_\alpha^{\mathrm{an}}), \quad \alpha = 1, \ldots, A. \tag{26}$$

Under the approximation (19), (14) becomes

$$\varepsilon(\boldsymbol{x}) = \boldsymbol{A}(\boldsymbol{x}) : \overline{\varepsilon} + \sum_{\beta=1}^{A} (\boldsymbol{D} * \theta_\beta)(\boldsymbol{x}) : \varepsilon_\beta^{\mathrm{an}}. \tag{27}$$

Upon multiplication of equation (27) by θ_α and averaging over V, one obtains

$$\boldsymbol{e}_\alpha = \boldsymbol{A}_\alpha : \overline{\varepsilon} + \sum_{\beta=1}^{A} \boldsymbol{D}_{\alpha\beta} : \varepsilon_\beta^{\mathrm{an}}, \tag{28}$$

where

$$\boldsymbol{A}_\alpha = \frac{\langle \boldsymbol{A}\theta_\alpha \rangle}{\langle \theta_\alpha \rangle}, \quad \boldsymbol{D}_{\alpha\beta} = \frac{\langle \theta_\alpha \boldsymbol{D} * \theta_\beta \rangle}{\langle \theta_\alpha \rangle}. \tag{29}$$

Similarly, incorporating (27) into (23b) yields

$$\overline{\boldsymbol{\sigma}} = \langle \boldsymbol{L} : \boldsymbol{A} \rangle : \overline{\boldsymbol{\varepsilon}} + \sum_{\alpha=1}^{A} \left(\langle \boldsymbol{L} : \boldsymbol{D} * \theta_\alpha \rangle - \boldsymbol{L}_\alpha \langle \theta_\alpha \rangle \right) : \boldsymbol{\varepsilon}_\alpha^{\mathsf{an}}. \tag{30}$$

The state laws, expressing the forces $\overline{\boldsymbol{\sigma}}$ and $\mathcal{A}_\alpha$ in terms of the state variables $\overline{\boldsymbol{\varepsilon}}$ and $\boldsymbol{\varepsilon}_\alpha^{\mathsf{an}}$ consist of (30), (26) with (28) and (25).

3.3 Complementary laws

Under the approximation (19), the dissipation potential (9) can be expressed in terms of the rates of the internal variables $\{\boldsymbol{\varepsilon}_\alpha^{\mathsf{an}}\}_{\alpha=1,\ldots,A}$:

$$\widetilde{\varphi}\left(\dot{\boldsymbol{\alpha}}\right) = \langle \varphi\left(\dot{\boldsymbol{\varepsilon}}^{\mathsf{an}}\right) \rangle, \quad \text{where } \dot{\boldsymbol{\varepsilon}}^{\mathsf{an}}(\boldsymbol{x}) = \sum_{\alpha=1}^{A} \dot{\boldsymbol{\varepsilon}}_\alpha^{\mathsf{an}} \theta_\alpha(\boldsymbol{x}). \tag{31}$$

Then

$$\frac{\partial \widetilde{\varphi}}{\partial \dot{\boldsymbol{\varepsilon}}_\alpha^{\mathsf{an}}} \left(\{ \dot{\boldsymbol{\varepsilon}}_\alpha^{\mathsf{an}} \}_{\alpha=1,\ldots,A} \right) = \left\langle \frac{\partial \varphi}{\partial \dot{\boldsymbol{\varepsilon}}^{\mathsf{an}}} (\dot{\boldsymbol{\varepsilon}}^{\mathsf{an}}) \theta_\alpha \right\rangle = \langle \boldsymbol{\sigma} \theta_\alpha \rangle = \mathcal{A}_\alpha. \tag{32}$$

Therefore the constitutive relations of the composite (in reduced form) have a generalized standard structure defined by the state variables (20), the free energy (23) and the dissipation potential (31).

There is however a difficulty in applying the complementary law (32) in *exact form.* This would require the knowledge of $\langle \varphi(\dot{\boldsymbol{\varepsilon}}^{\mathsf{an}}) \rangle$ which cannot be expressed simply in terms of the $\dot{\boldsymbol{\varepsilon}}_\alpha^{\mathsf{an}}$'s. Another approximation has to be introduced. Note that thanks to the convexity of φ_α one has:

$$\frac{\langle \theta_\alpha \varphi_\alpha(\dot{\boldsymbol{\varepsilon}}^{\mathsf{an}}) \rangle}{\langle \theta_\alpha \rangle} \geq \varphi_\alpha \left(\frac{\langle \dot{\boldsymbol{\varepsilon}}^{\mathsf{an}} \theta_\alpha \rangle}{\langle \theta_\alpha \rangle} \right) \tag{33}$$

The right-hand-side of (33) can be considered as an approximation of its left-hand-side. Then

$$\langle \theta_\alpha \boldsymbol{\sigma} \rangle = \left\langle \theta_\alpha \frac{\partial \varphi_\alpha}{\partial \dot{\boldsymbol{\varepsilon}}^{\mathsf{an}}} (\dot{\boldsymbol{\varepsilon}}^{\mathsf{an}}) \right\rangle \simeq \left\langle \theta_\alpha \frac{\partial \varphi_\alpha}{\partial \dot{\boldsymbol{\varepsilon}}^{\mathsf{an}}} \left(\frac{\langle \dot{\boldsymbol{\varepsilon}}^{\mathsf{an}} \theta_\alpha \rangle}{\langle \theta_\alpha \rangle} \right) \right\rangle = \langle \theta_\alpha \rangle \frac{\partial \varphi_\alpha}{\partial \dot{\boldsymbol{\varepsilon}}^{\mathsf{an}}} \left(\frac{\langle \dot{\boldsymbol{\varepsilon}}^{\mathsf{an}} \theta_\alpha \rangle}{\langle \theta_\alpha \rangle} \right).$$

Therefore the complementary equations (32) can be expressed (in approximate form) as:

$$\boldsymbol{a}_\alpha = \frac{\partial \varphi_\alpha}{\partial \dot{\boldsymbol{\varepsilon}}^{\mathsf{an}}} \left(\dot{\boldsymbol{e}}_\alpha^{\mathsf{an}} \right). \tag{34}$$

3.4 Link with the Transformation Field Analysis

The link with the Transformation Field Analysis of [Dvorak, 1992] can be made by choosing the shape functions to coincide with the characteristic functions (21) of the phases or subdomains:

$$\theta_\alpha(\boldsymbol{x}) = \chi_r(\boldsymbol{x}), \quad \langle\theta_\alpha\rangle = c_r, \quad \langle\theta_\alpha\theta_\beta\rangle = c_r\delta_{rs}, \quad r, s = 1, ..., N,$$

where c_r is the volume fraction of phase r and δ_{rs} is the Kronecker symbol. The anelastic strain is assumed to be uniform within each subdomain V_r. The generalized stress and strains $\boldsymbol{a}_\alpha$ and $\boldsymbol{e}_\alpha$ reduce to the average stress and strain over the subdomain V_r.

The state laws (26) and the complementary laws (34) read:

$$\overline{\boldsymbol{\sigma}} = \sum_{r=1}^{N} c_r\boldsymbol{\sigma}_r, \quad \boldsymbol{\sigma}_r = \boldsymbol{L}_r : \left(\boldsymbol{\varepsilon}_r - \boldsymbol{\varepsilon}_r^{\text{an}}\right), \quad \boldsymbol{\sigma}_r = \frac{\partial\varphi_r}{\partial\dot{\boldsymbol{\varepsilon}}^{\text{an}}}\left(\dot{\boldsymbol{\varepsilon}}_r^{\text{an}}\right), \quad r = 1, ..., N, \tag{35}$$

where the average strains $\boldsymbol{\varepsilon}_r$ in the different subdomains are given by:

$$\boldsymbol{\varepsilon}_r = \boldsymbol{A}_r : \overline{\boldsymbol{\varepsilon}} + \sum_{s=1}^{N} \boldsymbol{D}_{rs} : \boldsymbol{\varepsilon}_s^{\text{an}}, \quad r = 1, ..., N. \tag{36}$$

The fourth-order tensors $\boldsymbol{A}_\alpha$ and $\boldsymbol{D}_{\alpha\beta}$ are the average strain localization tensors $\boldsymbol{A}_r$ and the influence tensors $\boldsymbol{D}_{rs}$ which are the basic ingredients of the TFA ([Dvorak, 1992]):

$$\boldsymbol{A}_r = \frac{1}{c_r}\frac{1}{|V|}\int_V \boldsymbol{A}(\boldsymbol{x})\chi_r(\boldsymbol{x})\, d\boldsymbol{x},$$

and

$$\boldsymbol{D}_{rs} = \frac{1}{c_r}\frac{1}{|V|}\frac{1}{|V|}\int_V\int_V \chi_r(\boldsymbol{x})\boldsymbol{\Gamma}(\boldsymbol{x}, \boldsymbol{x}') : \boldsymbol{L}(\boldsymbol{x}')\chi_s(\boldsymbol{x}')\, d\boldsymbol{x}'d\boldsymbol{x}.$$

A few well known algebraic properties of these tensors are useful in their computations

$$\sum_{r=1}^{N} \boldsymbol{D}_{sr} = \boldsymbol{I} - \boldsymbol{A}_s, \quad \sum_{r=1}^{N} \boldsymbol{D}_{sr}\boldsymbol{L}_r^{-1} = 0, \quad c_s\boldsymbol{L}_s\boldsymbol{D}_{sr} = c_r{}^{\mathsf{T}}\boldsymbol{D}_{rs}\boldsymbol{L}_r.$$

4. EXAMPLES

The relative merits of the above models are assessed by means of the following two-dimensional example. The r.v.e. consists of a square unit cell containing a square inclusion (phase 1) located at its center (with volume fraction $c_1 = 0.25$). The unit cell is subjected to periodic boundary conditions. The inclusion is linear elastic. The surrounding matrix is elastic perfectly plastic (von Mises criterion with yield stress $\sigma_0 = 100$ MPa). The inclusion and the matrix have the same elastic moduli $E = 100$ GPa, $\nu = 0.25$.

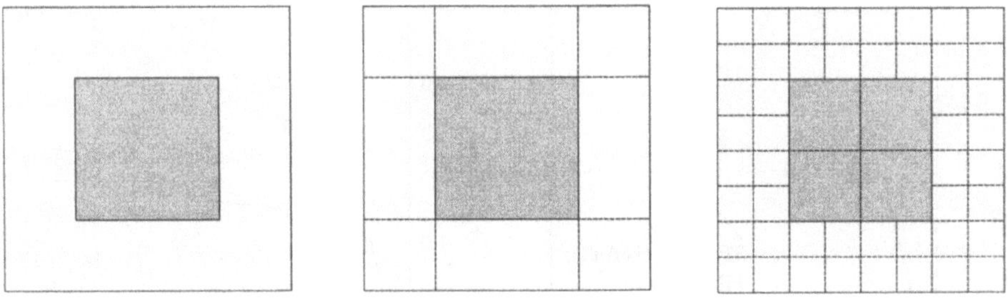

(a) (b) (c)

Figure 1: *Unit cell. Subdivisions used for the implementation of the TFA. (a): 2 subdomains, matrix and inclusion. (b): 8 subdomains in the matrix. (c): 48 subdomains in the matrix, 4 subdomains in the inclusion.*

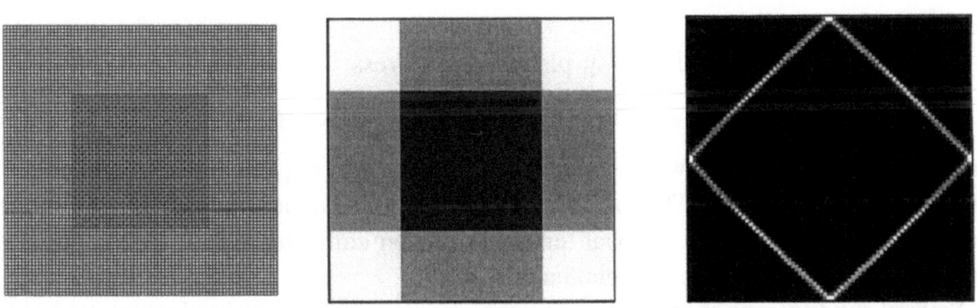

(a) (b) (c)

Figure 2: *(a) Mesh used in the FEM calculations. (b) Shape function 1 (flow mode in simple shear). (c) Shape function 2 (flow mode in uniaxial tension).*

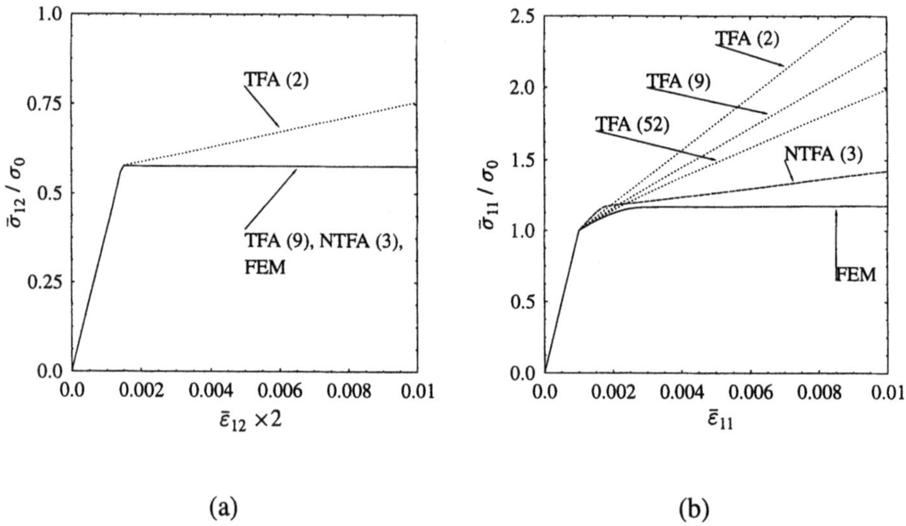

Figure 3: *Effective stress strain curves. (a) Pure shear. (b) Uniaxial tension. The number in parentheses denotes the number of subdomains or shape functions used in the implementation of the TFA or NTFA.*

The unit cell is subjected to an in-plane overall stress

$$\overline{\boldsymbol{\sigma}} = \overline{\sigma}_{11}\boldsymbol{e}_1 \otimes \boldsymbol{e}_1 + \overline{\sigma}_{22}\boldsymbol{e}_2 \otimes \boldsymbol{e}_2 + 2\overline{\sigma}_{12}\boldsymbol{e}_1 \otimes_s \boldsymbol{e}_2. \tag{37}$$

Attention has been paid to two specific cases, simple shear ($\overline{\sigma}_{11} = \overline{\sigma}_{22} = 0$) and uniaxial tension ($\overline{\sigma}_{12} = \overline{\sigma}_{22} = 0$). The exact responses (up to numerical errors) of the unit cell under the simple shear and uniaxial tension has been computed by the FEM. The TFA and the NTFA have also been implemented.

Several subdivisions of the unit cell have been considered in the implementation of the TFA. The cruder subdivision considers the plastic strain to be uniform in the matrix (Figure 1a). Finer subdivisions were also investigated in which the matrix was divided into 8 and 48 subdomains respectively, as shown in Figure 1b and 1c. The elastic properties of the inclusion and of the matrix being identical, the elastic strain localization is trivial $\boldsymbol{A}(\boldsymbol{x}) = \boldsymbol{I}$. The influence matrices $\boldsymbol{D}_{rs}$ are computed numerically by the Finite Element Method using a regular mesh of 80×80 quadrilateral 8 nodes elements shown in Figure 2a. A uniform eigenstrain ε^{an} is prescribed to the subdomain V_s and the average strain in subdomain V_r caused by this perturbation is computed. The corresponding relation yields the influence tensor $\boldsymbol{D}_{rs}$.

Only three shape functions were used in the NTFA (more shape functions could have been used but it is worth noting that satisfactory results can be obtained with relatively few shape functions). The first shape function θ_1 is the characteristic function of the inclusion. The second shape function θ_2 is the flow mode in pure shear. The third shape function θ_3 is the flow mode under uniaxial tension. These modes and the corresponding influence matrices $\boldsymbol{D}_{\alpha\beta}$ were computed numerically by the FEM using the mesh shown in Figure 2a. Snapshots are shown in Figure 2b (pure shear) and 2c (uniaxial tension). Whiter zones denote higher strains. Remarkably enough, the pattern flow mode under pure shear shows zones with uniform strains. An exact solution for this problem can be constructed in closed form. The strain field, solution of the elasto-plastic evolution problem, is a pure slip solution which can be expressed in terms of $\varepsilon^{\mathsf{e}} = \dfrac{\sigma_0}{2\sqrt{3}\mu}$ as

$$\varepsilon_{12} = \varepsilon^{\mathsf{e}} \text{ in } A, \quad \varepsilon_{12} = \frac{\overline{\varepsilon}_{12} - \varepsilon^{\mathsf{e}}}{1 - \sqrt{c_1}} + \varepsilon^{\mathsf{e}} \text{ in } B, \quad \varepsilon_{12} = \frac{\overline{\varepsilon}_{12} - \varepsilon^{\mathsf{e}}}{2(1 - \sqrt{c_1})} + \varepsilon^{\mathsf{e}} \text{ in } C, \tag{38}$$

where A denotes the inclusion, B denotes the subdomains located at the top right, top left, bottom right, bottom left in Figure 1b, and C denotes the remaining subdomains in Figure 1b.

The responses of the unit cell to the imposed loadings as predicted by the TFA and NTFA models are shown in Figure 3. The predictions of the TFA with the cruder discretization *i.e.* with one subdomain for the inclusion and one subdomain for the matrix are unrealistically stiff (this is well known [Suquet, 1997], [Chaboche et al, 1999]). When the discretization is refined (the number of subdomains is increased), the predictions of the TFA become more realistic. In simple shear, the exact solution is recovered by the discretization with 8 subdomains as follows straightforwardly from the exact solution (38). But in general the convergence towards the exact solution can be slow as the number of subdomains is increased. This is shown in Figure 3b where it is seen that, even with 48 subdomains in the matrix, the prediction of the TFA is not very accurate. The prediction of the NTFA with only 3 modes is as accurate as the prediction of the TFA with many more subdomains. This is due to the fact that the solution is nonuniform and that this nonuniformity is built-in into the shape functions.

In conclusion, we have shown that the reduction of the number of internal variables achieved by the TFA can be improved. We have considered a decomposition of the microscopic anelastic field on shape functions which are nonuniform and which capture the expected nonuniformity of the exact fields.

ACKNOWLEDGEMENTS

Ugo Galvanetto acknowledges the financial support of the research project ex-40% 1996 *Problemi di meccanica dei materiali compositi e innovativi*. Part of this study was completed while Pierre Suquet was on residence at the Isaac Newton Institute for Mathematical Sciences in Cambridge. Financial support of the Institute is gratefully acknowledged.

[Chaboche et al, 1999] J.L. Chaboche, S. Kruch, J.F. Maire and T. Pottier. Towards a micromechanics based inelastic and damage modeling of composites. *Submitted.*

[Dvorak, 1992] G.J. Dvorak. Transformation field analysis of inelastic composite materials. *Proc. R. Soc. Lond. A*, **437**:311–327, 1992.

[Dvorak et al, 1994] G.J. Dvorak, Y.A. Bahei-El-Din and A.M. Wafa. The modeling of inelastic composite materials with the transformation field analysis. *Modelling Simul. Mater. Sci. Eng*, **2**:571–586, 1994.

[Germain, 1973] P. Germain. *Cours de Mécanique des Milieux Continus*, volume I. Masson, Paris, 1973.

[Germain et al, 1983] P. Germain, Q.S. Nguyen and P. Suquet. Continuum Thermodynamics. *J. Appl. Mech.*, **50**:1010–1020, 1983.

[Halphen et al, 1975] B. Halphen and Q.S. Nguyen. Sur les matériaux standard généralisés. *J. Mécanique*, **14**:39–63, 1975.

[Lemaître et al, 1988] J. Lemaître and J.L. Chaboche. *Mécanique des Matériaux Solides*. Dunod, Paris, 1988.

[Lubliner, 1990] J. Lubliner. *Plasticity Theory*. Macmillan, New-York, 1990.

[Mandel, 1972] J. Mandel. *Plasticité classique et Viscoplasticité*, volume **97** of *CISM Lecture Notes*. Springer-Verlag, Wien, 1972.

[Maugin, 1992] G.A. Maugin. *The Thermomechanics of Plasticity and Fracture*. Cambridge University Press, Cambridge, 1992.

[Rice, 1970] J.R. Rice. On the structure of stress-strain relations for time-dependent plastic deformation in metals. *J. Appl. Mech.*, **37**:728–737, 1970.

[Suquet, 1985] P. Suquet. Local and global aspects in the mathematical theory of plasticity. In A. Sawczuk and G. Bianchi, editors, *Plasticity Today: Modelling, Methods and Applications*, pages 279–310. Elsevier, London, 1985.

[Suquet, 1987] P. Suquet. Elements of homogenization for inelastic solid mechanics. In E. Sanchez-Palencia and A. Zaoui, editors, *Homogenization Techniques for Composite Media*, volume **272** of *Lecture Notes in Physics*, pages 193–278. Springer Verlag, Wien, 1987.

[Suquet, 1997] P. Suquet. Effective properties of nonlinear composites. In P. Suquet, editor, *Continuum Micromechanics*, volume **377** of *CISM Lecture Notes*, pages 197–264. Springer Verlag, Wien, 1997.

Adiabatic shear banding as an example of viscoplastic flow instability

Alain Molinari
ISGMP-LPMM, Université de Metz, Ecole Nationale d'Ingénieurs
Ile du Saulcy, 57045 Metz-cedex, France.
Email :molinari@lpmm.univ-metz.fr

Abstract : Viscoplastic flow instability is an important mode of ductile failure of structural elements. When impact loading conditions are considered, thermal effects are significant and can lead to adiabatic shear banding. At very high rates of straining, inertia effects have an increasing influence and can produce multiple shear banding. A modeling of adiabatic shearing is presented here with special emphasis on the experiments made with the torsional Kolsky bars set-up. Analogies with other types of plastic instabilities will be occasionally discussed.
Keywords : viscoplastic instabilities, thermal effects, adiabatic shear banding, Kolsky bars.

1. INTRODUCTION

The localization of plastic deformation is a phenomenon consisting in the development of large deformations in narrow zones, and is the result of plastic flow instability. Such a localization process can be a mode of ductile failure of materials sustaining large plastic deformations. A typical example is provided by the wall breakage duc to localized necking during the deep drawing of metallic sheets. [Considère, 1885] was the first to characterize the onset of necking in a bar under simple tension. The fundamental solution consisting in the uniform deformation of the bar, becomes unstable when the maximum of the tensile force is reached. Then a non-uniform mode of deformation develops, leading to the neck formation. The Considère criterion determines the critical tensile strain at the maximum force as being $\varepsilon_c = n$, for a rigid plastic material whose hardening law is of the form : $\sigma = k\varepsilon^n$; σ is the tensile component of the Cauchy stress, and it is supposed that the strain hardening exponent n is positive. From this result, it appears that the onset of necking is delayed (and therefore the plastic flow is more stable) in a material with strong strain hardening. Other material parameters can have important effects on strain localization. Superplastic materials are quite ductile, with elongation strains in simple tension tests as large as 1000% and more. This remarkably stable deformation process can be interpreted in terms of the viscoplastic material response. More precisely the strain rate sensitivity, characterized by the parameter $m = \partial \ln\sigma / \partial \ln\dot{\varepsilon}$ (with $\dot{\varepsilon}$ being the strain rate), can be

G.A. Maugin et al. (eds.), Continuum Thermomechanics, 313-330.
© 2000 *Kluwer Academic Publishers.*

shown to stabilize the plastic flow. The large ductility of superplastic materials is attributed to their large strain rate sensitivity ($m \approx 0.3$).

Plastic flow localization is also very sensitive to the loading conditions. When large extension rates are considered, new phenomena are observed. Consider the example of copper shaped charge jets used for armor penetration. The following values, typical of the state of the jet at a given time of its history (diameter = few millimeters, length = 0.3 m , tip velocity = 7 km/s, tail velocity = 4 km/s) indicate that elongation rates of straining are of the order of $\dot{\varepsilon} = 10^4 s^{-1}$. At these large values of $\dot{\varepsilon}$, the jet breaks up in a multitude of fragments, contrary to what happens at low strain rates where a single neck is observed. Multiple necking can be attributed to inertia effects which become significant in the necking zones at large values of the overall strain rate. The jet fragmentation can be interpreted as the consequence of the competition between inertia effects which are stabilizing on long wave-length perturbations, and stress triaxiality effects within necks which tend to stabilize small wave-length perturbations. The outcome of this competition is the selection of the critical size of fragments that are likely to be formed.

At high rate of straining, thermal effects play a crucial role in the development of plastic flow instabilities. Examination of a metallic plate impacted by a spherical projectile at the velocity of $3000 ms^{-1}$, reveals the existence of a family of narrow shear bands with a thickness of the order of $10 \mu m$, and a characteristic separation distance of the order of $1 mm$, see the fig.1.1 p.2 of [Bai and Dodd, 1992]. These bands, called Adiabatic Shear Bands (ASB), emanate from the impacted surface and propagate through the plate thickness. Very large localized shear strains can develop within the bands (depending on the test and on the material considered, values ranging from 1 to 10 and more, can be measured) accompanied by a significant temperature rise limited to the band area (several hundred of degrees).

ASB are the result of a thermomechanical instability, which can be schematically presented as follows. The instability process is initiated at a local stress concentration due to a microscopic defect (geometrical or material defects) or related to boundary conditions (e.g. stress concentration at the edges of a projectile entering into the target). Localized plastic deformations are generated at stress concentrations, leading to a local temperature increase - due to plastic work dissipation- which in turns produces a local decay of the flow stress in a thermal softening material. This material weakening again favors the development of localized deformations, producing a local growth of temperature, and clearly an autocatalytic process is generated. Thus adiabatic shear banding is the result of a thermomechanical instability.

Two important conditions have to be satisfied in order to produce ASB :

- the material must be thermal softening.
- strain rates must be high, because the local temperature rise is not possible in slow processes where heat diffusion has enough time to make the temperature field uniform. Heath transfer is reduced in high strain rate processes, but is not negligible near a band where temperature gradients are large. The formation of an ASB is not really an

adiabatic process although this terminology is commonly used to qualify this type of shear banding.

Adiabatic shear banding is frequently at the origin of the catastrophic failure of structures under dynamic loading. For instance in a plugging experiment, the energy used for the piercing of a thin metallic plate by a projectile is much smaller than the energy necessary to go through the plate, with the projectile used as a punch progressing at a low speed. During plugging, adiabatic shear banding emanates from the edges of the projectile and propagates through the plate thickness. The material within the bands suffers from a dramatic drop in the flow stress, and offers only a weak resistance to the progression of the projectile. Almost no plastic deformations are observed outside the ASB in a plugging experiment. On the contrary, in quasistatic loading, the advance of the punch is accompanied by plastic deformations spread over a large area, leading to the formation of a dome until the plate is finally broken. For that reason the quasistatic process is much more energy-consuming than the impact process. These observations have a general character, and have important consequences for the security of impacted structures.

Adiabatic shear banding is a phenomenon commonly observed in metals, polymers and geological materials (e.g. ice). It is related to reaction-diffusion problems encountered in combustion theory or in chemical kinetics (blow up of an explosive is triggered by thermomechanical instabilities such as ASB). ASB are observed in high speed metal forming processes such as forging, magneto-forming. In high speed machining, ASB can lead to chip segmentation.

Adiabatic shear banding is a very rich physical problem, which will be discussed in this paper, and used as a guideline to illustrate the process of viscoplastic flow instability. In particular, the role of material parameters and of loading conditions on the onset, development and propagation of ASB will be analyzed. Links and analogies with other types of plastic instabilities will be occasionally discussed.

2. TORSIONAL KOLSKY BARS EXPERIMENTS.

Throughout this paper, the high speed torsion of a thin tube in a Kolsky bars set-up, [Marchand and Duffy,1988] will be considered as the reference test. In these experiments adiabatic shear bands are formed under well controlled conditions and precise measurements.

The evolution of the shear stress τ in terms of the shear strain γ is reported in fig.1 for a HY100 steel sustaining a nominal shear strain rate of $\dot{\gamma} = 1600 s^{-1}$ [Marchand and Duffy, 1988]. Three stages in the deformation process can be defined. Except for a local peak due to dynamical effects, most of the shear stress evolution during stage I consists in a growth of the shear stress related to strain hardening. Plastic flow is almost homogeneous. However the heat generated by the dissipation of part of the mechanical work, produces a thermal softening which finally overcomes strain hardening when the maximum value of the stress is attained, at the deformation $\gamma_m \approx 0.3$. This maximum stress marks the onset of stage II, which is characterized by a slow decrease in the stress due to the influence of thermal softening. In addition, a weak heterogeneity in the plastic flow is observed. Finally stage III manifests itself by a sudden drop in the bearing capacity resulting from the development of an adiabatic shear band. Ultimately this band propagates along the tube circumference. A crack can be generated within the band, leading to the fracture of the specimen. These different stages marking the onset and evolution of adiabatic shear banding will be analyzed in the following.

Here again, it is essential to note the differences between dynamic and quasistatic loading as they appear in fig.2. A HY100 steel has been sheared at different values of nominal strain rate, $\dot{\gamma} = 1500 s^{-1}$ and $\dot{\gamma} = 5x10^{-4} s^{-1}$. In dynamic loading, the fracture of the specimen is obtained by ASB at the strain $\gamma \approx 0.5$, while in quasistatic loading, fracture occurs at $\gamma \approx 1.7$ with no ASB. The energy absorbed during the test- which is mostly the area delimited by the τ vs γ diagram- is much smaller in the dynamic test, in agreement with the foregoing discussion of the plugging test.

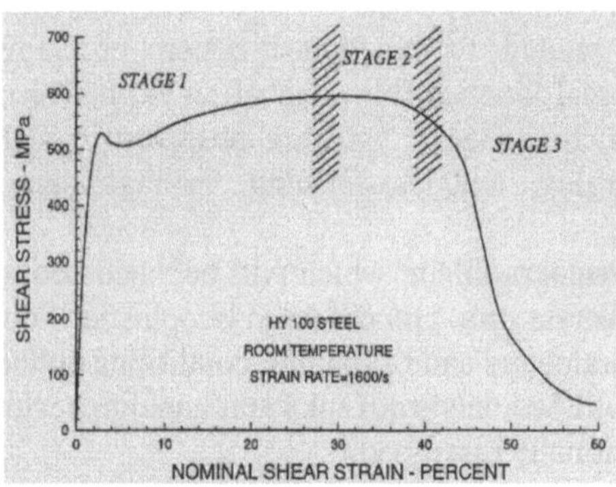

Figure1 ; Stress-strain response in the high-speed torsion of a tube, [Marchand and Duffy, 1988]

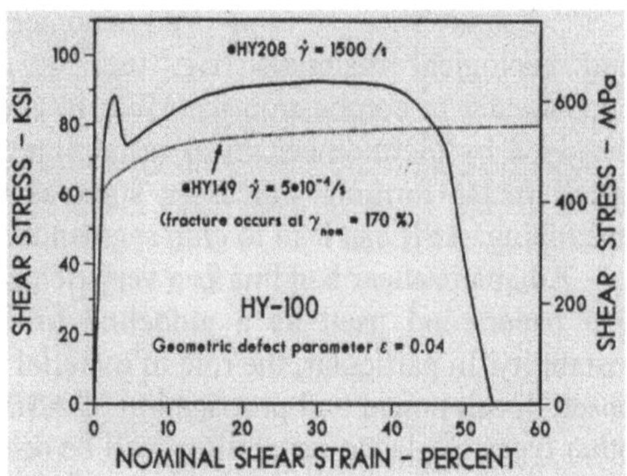

Figure2 ; Dynamic and quasi-static response [Chi and Duffy, 1989]

3. MODELLING OF VISCOPLASTIC SHEAR AT HIGH STRAIN RATES.

To model adiabatic shear banding in Marchand and Duffy's experiments, it is convenient to consider that the tube is cut along a direction parallel to the rotation axis, and that the surface of the tube is developed into a plane. The problem is modeled as that of a layer infinitely extended in the shear direction x and in the out of plane direction z, with finite thickness 2h in the direction y, see fig.3. In the Kolsky bar experiments, it can be considered that the boundary conditions consist of constant velocities $\pm V$ parallel to the x direction, applied at the upper and lower surfaces :

$$v = \pm V \qquad \text{at} \qquad y = \pm h \tag{1}$$

In most of the experiments the flow is laminar within the shear zone. Note however that vortices have been observed in very thin bands obtained during the dynamic compaction of two-phase materials at nominal strain rates of $10^4 s^{-1}$, [Nesterenko et al., 1994,1995b]. In a general context, two dimensional modes of instability have been analyzed by [Molinari and Leroy,1991] and [Leroy and Molinari, 1993]. First restricting the analysis to laminar flow, the problem is formulated in a one dimensional framework, the variables depending solely upon the coordinate y and the time t. The velocity of a particle is parallel to the shear direction x and is denoted by v. As high strain rates are considered, adiabatic conditions can be assumed at the boundaries :

$$\partial\theta/\partial y = 0 \quad \text{at} \quad y = \pm h \tag{2}$$

where θ is the temperature.

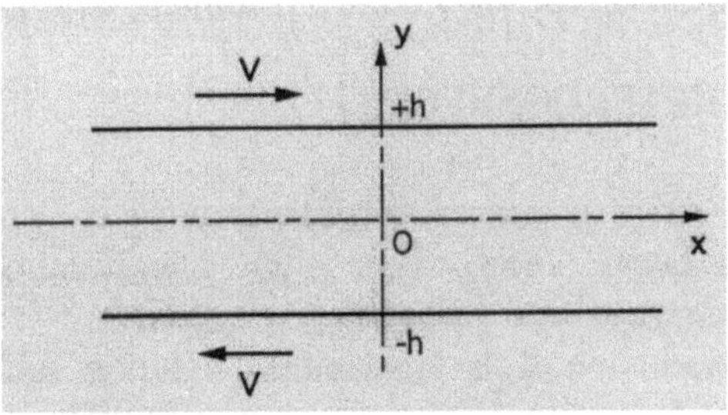

Figure 3 : Geometry of the sheared layer

The problem is modeled by five basic equations. The balance of momentum and the conservation of energy can be expressed as :

$$\partial\tau/\partial y = \rho\,\partial v/\partial t \tag{3}$$

$$\rho c\,\partial\theta/\partial t - k\,\partial^2\theta/\partial y^2 = \beta\tau\dot{\gamma}^p \tag{4}$$

where ρ is the mass density, c the heat capacity, k the heat conductivity, τ the shear stress and $\dot{\gamma}^p$ the plastic shear rate. The source term in the right side of the heat equation represents the part β (Taylor-Quinney coefficient) of the rate of plastic work transformed into heat. The compatibility equation has the form :

$$\partial v/\partial y = \dot{\gamma}^p + \dot{\gamma}^e \tag{5}$$

where $\dot{\gamma}^e$ is the elastic shear rate. The viscoplastic flow law can be expressed in the general form

$$\dot{\gamma}^p = \psi(\tau, \gamma^p, \theta) \qquad \text{or} \qquad \tau = \phi(\dot{\gamma}^p, \gamma^p, \theta) \tag{6}$$

where the strain γ^p is defined by $\gamma^p(t) = \int_0^t \left|\dot{\gamma}^p(t_1)\right| dt_1$.

The dependence of the flow stress upon the plastic strain, strain rate and temperature, is necessary to account for the strain hardening, strain rate sensitivity and thermal softening effects observed in the experiments. A more physical approach would assume that the flow stress depends on the whole deformation history ; however for analytical convenience the simpler law (6) will be used here. Finally Hooke's law

$$\dot{\gamma}^e = \dot{\tau}/\mu \tag{7}$$

with μ the shear modulus, completes the set of equations for the five unknowns, $\tau,\theta,v,\gamma^p,\gamma^e$. In most of the applications considered, the elastic deformation can be neglected (an exception is the analysis of shear band propagation). Therefore in the following we shall have $\gamma = \gamma^p$ and $\gamma^e = 0$.

A homogeneous and time dependent fundamental solution exists because there is no heat flux at the boundaries. This solution corresponds to homogeneous laminar flow with uniform increasing temperature, and is denoted by

$$\underline{S}^0(y,t) = (v^0(y),\gamma^0(t),\tau^0(t),\theta^0(t))$$

with $v^0(y) = V\,y/h$. When entering into stage II, the fundamental solution becomes unstable. The onset of instability is described in the next section with use of a perturbation method, [Clifton, 1978], [Bai, 1982], [Molinari, 1985].

4. THE ONSET OF HETEROGENEOUS FLOW.

Just after the maximum stress is passed, a weak flow heterogeneity is observed to develop, [Marchand and Duffy, 1988]. This is the manifestation of a viscoplastic instability which can be analyzed with a perturbation approach.

Consider a small perturbation of the fundamental solution, initiated at time t_0 and evolving with time t : $\delta\underline{S}(y,t) = (\delta v(y,t),\delta\gamma(y,t),\delta\tau(y,t),\delta\theta(y,t))$

By Fourier decomposition, $\delta\underline{S}$ can be seen as the superposition of elementary spatial modes of the form :

$$\delta\underline{S}(y,t) = \delta\hat{\underline{S}}(t)\exp(i\xi y) \tag{8}$$

characterized by the wave number ξ. Substitution of the perturbed solution $\underline{S}(y,t) = \underline{S}^0(y,t) + \delta\underline{S}(y,t)$ into the governing equations (3)-(6) and linearization provide a linear system of first order equations for the amplitude of the perturbation :

$$\underline{\underline{B}}.\frac{d}{dt}\delta\hat{\underline{S}}(t) = \underline{\underline{A}}(\xi,t).\delta\hat{\underline{S}}(t) \tag{9}$$

where $\underline{\underline{A}}$ and $\underline{\underline{B}}$ are matrices, $\underline{\underline{A}}$ depends on the wave number ξ and on time since the fundamental solution evolves with time. In the particular problem considered here, the matrix $\underline{\underline{B}}$ is a constant, but without more complications it could depend on the same arguments as $\underline{\underline{A}}$. The system (9) has to be solved for a given value of the initial amplitude of the perturbation :

$$\delta\hat{\underline{S}}(t_0) = \delta\hat{\underline{S}}_0 . \tag{10}$$

Because of the time dependence of $\underline{\underline{A}}$, the system (9) is non-autonomous, and therefore difficult to solve by classical means. Assume for a moment that the coefficients of $\underline{\underline{A}}$ are frozen to their value at time t_0. Then the solution of (9) with $\underline{\underline{A}}(\xi,t) \to \underline{\underline{A}}(\xi,t_0)$ has the general form :

$$\delta\underline{\tilde{S}}(t) = \sum_i b_i \exp\big(\eta_i(t_0)(t-t_0)\big)\underline{S}_i^*(t_0) \tag{11}$$

where $\eta_i(t_0)$ and $\underline{S}_i^*(t_0)$ are solutions of the eigenvalue problem :

$$\eta_i(t_0)\underline{B}.\underline{S}_i^*(t_0) = \underline{A}(\xi,t_0).\underline{S}_i^*(t_0) \tag{12}$$

η_i are the complex roots of the characteristic polynomial $\det\big(\eta\underline{B} - \underline{A}\big) = 0$, which has the form of a cubic equation :

$$a_3\eta^3 + a_2\eta^2 + a_1\eta + a_0 = 0 \tag{13}$$

The coefficients a_i depend on the material parameters, on the wave number ξ of the perturbation, and on the fundamental solution. The coefficients b_i in (11) are determined so as to satisfy the initial condition (10)

$$\sum b_i \underline{S}_i^*(t_0) = \delta\underline{\hat{S}}_0$$

Since $\delta\underline{\tilde{S}}(t)$ satisfies equation (9) at the time t_0 together with the initial condition $\delta\underline{\tilde{S}}(t_0) = \delta\underline{\hat{S}}_0$, it is related to the solution $\delta\underline{\hat{S}}(t)$ of the non-autonomous system (9) by

$$\delta\underline{\tilde{S}}(t_0) = \delta\underline{\hat{S}}(t_0) \quad \text{and} \quad \underline{B}.\frac{d}{dt}\delta\underline{\tilde{S}}(t_0) = \underline{B}.\frac{d}{dt}\delta\underline{\hat{S}}(t_0)$$

In that sense, $\delta\underline{\tilde{S}}(t)$ can be viewed as an approximate solution of (9) and (10) close to the initial time of the perturbation t_0. The eigenvalues $\eta_i(t_0)$ of the associated autonomous problem, provide exact information on the rate of growth of the perturbation $\delta\underline{\hat{S}}(t)$ at the initial time t_0. A negative real part of all the roots $\eta_i(t_0)$ for every wave-number ξ implies the initial decay of any perturbation ; this situation will be referred to in the following as linear stability at time t_0. A positive real part of one root $\eta_i(t_0)$ for a certain value of ξ is sufficient for the existence of a perturbation having an instantaneous growth; this will be referred to as linear instability at time t_0.

It should be noted that this absolute perturbation analysis with frozen coefficients refers only to the instantaneous rate of growth of the perturbation at time t_0 at which the perturbation is introduced. Therefore limitations of the analysis appear as an immediate consequence of the time-dependence of the fundamental solution : an instantaneous growth of the perturbation is not a rigorous proof of a long-term instability since the rate of growth evolves with time , can become negative at a later time and thus can lead to an attenuation of the perturbation. However, in many circumstances of practical interest, the instantaneous stability analysis provides useful indications.

When at a given time t^* of a deformation process, a linear absolute instability is detected, it is indicative of the onset of a possible critical phenomenon. Then one has to pursue the same analysis and check the evolution of the rate of growth of the perturbations from the calculation of the roots $\eta_i(t_0)$ for $t_0 > t^*$. This analysis furnishes in general coherent information concerning the long range trends, although it is limited to small perturbations and cannot give all the information needed for the description of the final localization process.

Another circumstance of practical interest is that in which the instantaneous rate of growth of the perturbation is large compared to the rate of evolution of the fundamental solution ; then it is justified to consider the coefficients of the differential system as frozen, and the analysis provides pertinent qualitative information.

In any case, a better evaluation of the intensity of the instability process is obtained by normalizing the η_i's with the nominal strain rate, as for example in fig.4.

Now the initial development of the viscoplastic instability in the stage II is analysed with a linearized perturbation method.

Onset of instability

The nominal strain corresponding to the maximum in the stress strain curve (fig.1) is denoted by γ_m. Calculations for different materials (CRS1018, HY100 steels) show in all cases that for nominal strains smaller than γ_m, all the real parts of the η_i's are negative, indicating stability of the fundamental solution, as observed in the experiments. On the contrary, as soon as the strain is larger than γ_m, a root with positive real part is detected, indicating the emergence of a critical phenomena. However the rate of growth of the perturbation is rather small just after the shear strain γ_m is passed, indicating a weak instability as observed by [Marchand and Duffy, 1988] at the beginning of stage II.

A CRS1018 steel is now considered whose mechanical properties are provided by [Clifton et al, 1984] ; the constitutive function ϕ is :

$$\tau = \phi = \mu_0(\gamma + \gamma_i)^n \dot{\gamma}^m \theta^\nu \tag{14}$$

The strain hardening, the strain rate sensitivity and the thermal softening coefficients have respectively the values : $n = 0.015, m = 0.019, \nu = -0.38$. $\gamma_i = 0.01$ is a prestrain ; $\mu_0 = 3579x10^6 SI$. Calculations show that $\gamma_m \cong 0.12$ [Molinari, 1988], in accordance with experimental results. For a nominal strain of 0.15 and a nominal strain rate of $10^3 s^{-1}$, the dominant eigenvalue, which appears to be real, is represented in terms of the wave number of the perturbation ξ in the fig.4. The eigenvalue obtained while disregarding inertia and conduction is also represented (upper horizontal line). Effects of inertia and conduction are stabilizing respectively on large wavelength (small ξ) and small wavelength (large ξ) perturbations, [Molinari, 1985]. This leads to the selection of a critical wavenumber ξ_c for which the rate of growth of the perturbation is maximum. Only a discrete set of wavenumbers is compatible with the boundary conditions : $\left(\xi_n = n\frac{\pi}{h}, n = 1,2...\right)$, h being half the thickness of the sheared layer. Fig.4 shows that the perturbation associated with the wavenumber ξ_1 (longest wavelength compatible with the boundary conditions) is the one more likely to develop, since other modes have a lower rate of growth due to the stabilizing effects of heat conduction. That information is in accordance with the wavelength pattern during stage II of the Marchand and Duffy experiments, and with the fact that a single band is observed to develop. Note that for the mode ξ_1, the inertia effects are negligible, and the conduction effects are small but not negligible (compare with the upper horizontal line corresponding to the quasistatic and adiabatic process).

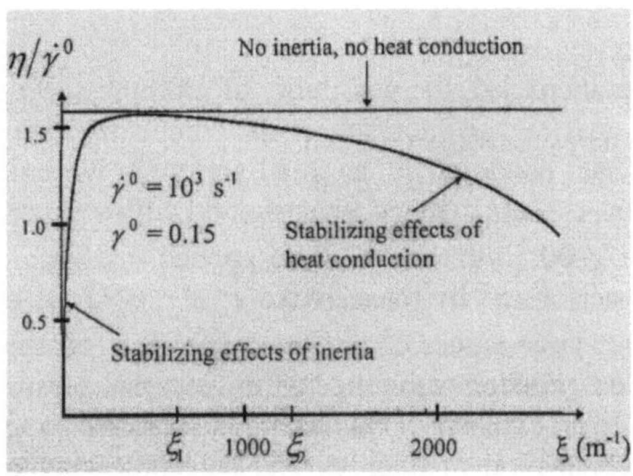

Figure 4 : Normalized rate of growth of the instability in terms of the wave number.

For larger values of the nominal strain rate, the picture of fig.4 is changed by shifting the maximum of the curve to the right; thus the dominant mode of instability has a smaller wavelength Then it appears that for high rates of straining, multiple shear bands can be generated, [Molinari, 1988]. This point will be discussed further in the following.

Influence of strain rate sensitivity

Another result directly obtained from the linearized stability analysis concerns the role of the strain rate sensitivity parameter m. The analysis is made easier by considering the particular case of no inertia and no heat conduction. Note that both can be reasonably neglected at the nominal strain rate $10^3 s^{-1}$. In that instance the leading eigenvalue takes the form :

$$\eta = \frac{\partial \psi}{\partial \gamma} + \frac{\beta \tau}{\rho c} \frac{\partial \psi}{\partial \theta} \tag{15}$$

and reduces further, when considering the power law (14) :

$$\eta = 1/m\left(-n\dot{\gamma}/(\gamma + \gamma_i) - \nu\dot{\theta}/\theta\right) \tag{16}$$

The strain hardening ($n > 0$) and the thermal softening ($\nu < 0$) terms have respectively stabilizing and destabilizing effects. η vanishes at the stress maximum, for the value of the nominal strain :

$$\gamma_m = \left[\frac{n(1-\nu)(1-a\gamma_i^{\,n+1})}{-a(\nu+n)}\right]^{\frac{1}{n+1}} - \gamma_i \quad \text{with} \quad a = \frac{(1-\nu)\mu_0\beta\dot{\gamma}^m}{(n+1)\theta^{-\nu+1}\rho c} \tag{17}$$

The eigenvalue (16) is proportional to $1/m$. Thus a material with a very small strain rate sensitivity m should exhibit a rapid development of the instability in stage II of the test. An increase of the strain rate sensitivity has a stabilizing effect ; this is a general feature in any localization problem in viscoplastic materials (see the discussion on superplastic materials in the introduction). Actually, a non-linear analysis reveals that a small strain rate exponent of the order m=0.01 is sufficient to create the latency period that characterizes stage II.

Multiple shear banding

When the strain rate is increased, the possibility of multiple shear banding is revealed by the linear stability analysis, as the result of the competition between the stabilizing effects of inertia and heat conductivity, at small and large wavenumbers respectively. While a single band was created in the Marchand and Duffy's experiments, where strain rates in the range of $1200 s^{-1}$ to $1500 s^{-1}$ were applied, multiple shear banding was observed in the tests performed by [Nesterenko et al., 1995] at nominal rates of the order of $10^4 s^{-1}$. These large values of the strain rate were obtained during the radial collapse of thick-walled cylinders submitted to the external pressure generated by the controlled detonation of an explosive. This technique was used to investigate the shear band initiation and propagation in titanium and austenitic stainless steel under pure shear deformation. Shear bands were reported to emanate from the inner boundary of the cylinder, with spiral trajectories and regular spacing. The spacing at the initiation of shear banding was found to be of the order of 1mm for the materials investigated.

Two different theoretical approaches have been proposed for characterizing the shear band spacing. [Grady and Kipp, 1987] have obtained the shear band spacing by accounting for momentum diffusion due to unloading within bands. [Wright and Ockendon, 1996] and [Molinari, 1997] have used a perturbation analysis to characterize the dominant mode of instability. It was argued that the wavelength associated with that dominant mode corresponds to the most probable minimum spacing of shear bands. The analysis of [Wright and Ockendon, 1996] is restricted to perfectly plastic materials, that of [Molinari, 1997] includes the effects of strain hardening. In simple shear loading, the shear band spacing L_s was characterized in terms of the rate of loading and of the material parameters as follows, [Molinari, 1997] :

$$L_s = \min_{\gamma} \left\{ 2\pi (mkc)^{\frac{1}{4}} \left(\beta^2 \dot{\gamma}^0 \tau^0 \left(\frac{\partial \psi^0}{\partial \theta} \right)^2 \right)^{-\frac{1}{4}} \left(1 + \frac{3}{4} \rho c \frac{\partial \psi^0}{\partial \gamma} \Big/ \beta \tau^0 \frac{\partial \psi^0}{\partial \theta} \right)^{-1} \right\} \tag{18}$$

L_s is calculated by minimization over the strain γ. For a non hardening material $\partial\psi/\partial\gamma = 0$, and the minimization can be shown to be realized for $\gamma = 0$, i.e. at the beginning of the process :

$$L_s = 2\pi \left(mkc \Big/ \beta^2 \dot{\gamma}^0 \tau^0 \left(\frac{\partial \psi^0}{\partial \theta} \right)^2 \right)^{\frac{1}{4}} \tag{19}$$

If in addition the material behavior is described by the power law (14) (with $n = 0$), then

$$L_s = 2\pi \left(\frac{m^3 k(\rho c)\theta^{0^2}}{\beta^2 \dot{\gamma}^{0^3} \tau^0 \nu^2 \rho} \right)^{\frac{1}{4}} \tag{20}$$

It has to be noted that in these results, $\dot{\gamma}^0$, τ^0, θ^0 represent the fundamental homogeneous solution of the problem (3)-(6) with boundary conditions (1) and (2). ψ^0 is the value of ψ calculated with use of the fundamental solution.

The effects of material parameters and of the loading conditions are clearly illustrated by these results. When the nominal strain rate $\dot{\gamma}^0$ is increased, the shear band spacing is reduced, in agreement with experimental results. In addition, it appears that the parameters which were shown to have a stabilizing effect on plastic flow localization, such as the strain rate sensitivity m, the thermal conductivity k and the heat capacity ρc, contribute here to increase the shear-band spacing. It is also the case for inertia which is represented by the mass density ρ in the denominator of the relationship (20). The opposite effect is observed for the thermal softening parameter ν, known to be a destabilizing factor.

Results for a strain hardening material are presented in the fig.5, where the stabilizing role of n appears clearly.

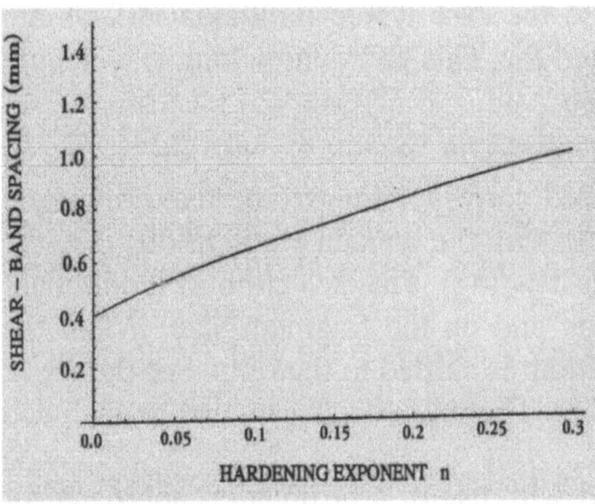

Figure 5 : Dependence of the shear band spacing on strain hardening.

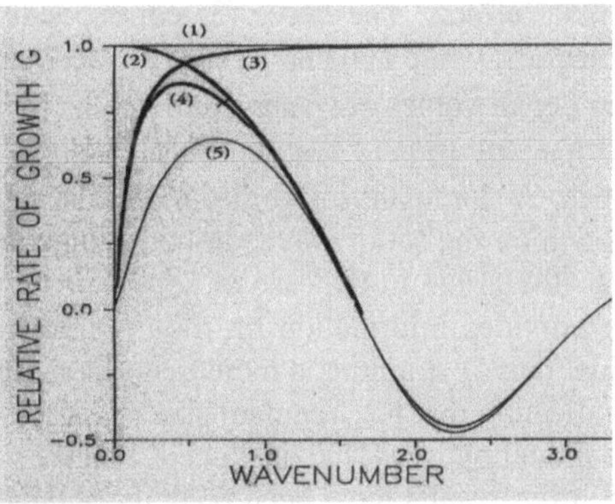

Figure 6 : Relative rate of growth of the perturbation in terms of the wave number. A stretching sheet is considered.

Comparisons with the experimental results of [Meyers and Nesterenko, 1995] show a good qualitative and quantitative agreement. For instance for a titanium alloy sustaining a nominal strain rate of $\dot{\gamma}^0 = 10^4 s^{-1}$, the shear band spacing was measured as $L_{exp} = 1mm$. The results of [Grady and Kipp, 1987] and of [Wright and Ockendon, 1996] are : $L_{GK} = 1.8mm$ $L_{WO} = 0.3mm$. With inclusion of hardening [Molinari, 1997], the perturbation analysis gives : $L_{MO} = 0.75mm$. To appreciate these results one has to remember the lack of exact characterization of the material behavior at the extreme conditions of deformations and temperature considered here.

Fragmentation
An analogy can be traced between the analysis of multiple shear banding and the fragmentation of a shaped charge jet. Using perturbation techniques it was found that the fragment size was controlled by the competition between the stabilizing effects of inertia, which prevail on long wavelength modes, and the stabilizing effects due to stress triaxiality within the necking zones, which act on the short wavelength modes, [Fressengeas and Molinari, 1994]. This is illustrated in fig.6 for a rapidly stretching sheet. A non-hardening material was studied. The relative rate of growth of the perturbation is shown in terms of the longitudinal wave number, see [Fressengeas and Molinari, 1994] for details concerning the data. In fig.6 , the horizontal line (1) corresponds to the results where inertia and multidimensional effects are both neglected (a simple 1-D analysis is considered in which the axial stress is assumed uniform in a cross-section). With these hypothesis, all the instability modes have the same growth rate, controlled by the factor $1/m$, where m is the strain rate sensitivity (note the analogy with adiabatic shearing where the same factor appeared in formula (16)). Multiaxial effects are accounted for in the curve (2) of fig.6, but inertia effects are switched off. This result shows how short wavelength modes (large values of ξ) are stabilized, [Hutchinson et al., 1978], while long wavelength modes are unaffected by multiaxial effects. The curve (3) corresponds to the case where multiaxial effects are switched off, while dynamic effects are accounted for. This shows how long wavelength modes (small values of ξ) are stabilized by inertia.

When all effects are working together, the result shown by the curve (4) is obtained. A dominant instability mode is selected corresponding to the maximum rate of growth on the set of wavenumbers compatible with the boundary conditions (see the similar discussion in the case of adiabatic shear banding). This selection of a dominant mode provides information on multiple necking and on the fragment size. When the nominal rate of straining is increased, the maximum is shifted to the right, see the curve (5), indicating that the fragment size is decreasing. Thermal effects can also be included in the modeling, see [Dudzinski et al., 1992].

Similar results have been recently obtained with use of a bifurcation approach for rate independent plastic materials, [Shenoy and Freund, 1999].

5. THE ONSET OF LOCALIZATION.

During the stage II of the torsional test, the rate of the flow instability gradually increases, until the stage III is entered, with a sudden stress drop marking the onset of an intense strain localization. The development of strain localization cannot be analyzed with the preceding linearized perturbation method, since nonlinear effects become important at that stage of the process. With some simplified hypothesis, a nonlinear analysis of the localization process can be conducted, and full explicit analytical results can be obtained, [Molinari and Clifton, 1983], [Molinari, 1985], [Molinari and Clifton, 1987], [Dinzart et al., 1993]. A simple example is first presented to illustrate some difficulties in the analysis of thermoviscoplastic instabilities

5.1 A simple nonlinear analysis

The material is non-linearly viscous and thermally sensitive, with a constitutive law in shear of the form :

$$\tau = \mu(\theta)(\gamma + \gamma_i)^n \dot{\gamma}^m \tag{21}$$

where γ_i is a prestrain. The same geometry as in fig.3 is considered, except that now the width $l(y)$ might be variable, as in fig.7. Inertia and conduction effects are neglected

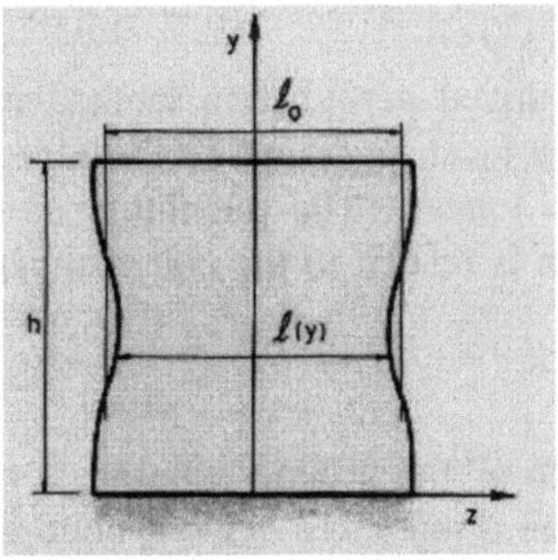

Figure 7 : Variation in the width of the specimen.

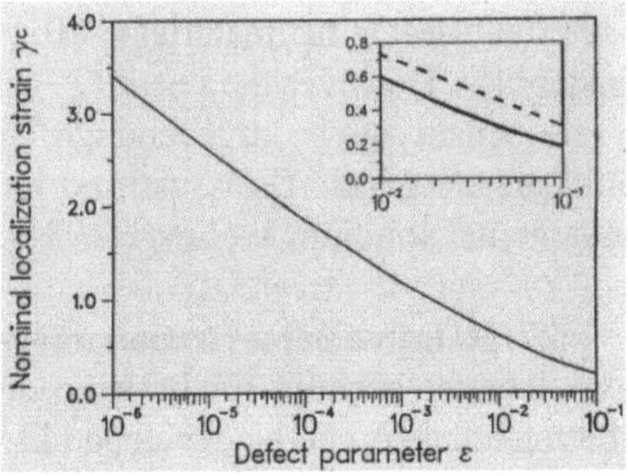

Figure 8 : Localization strain in terms of the normalized amplitude of the geometrical defect.

in this simplified analysis. These hypothesis are justified in a range of moderate strain rates $\dot{\gamma} \approx 1500 s^{-1}$. All quantities are assumed to depend solely on the position y and time t. Equilibrium between two layers A and B of width l_A and l_B, can be written as

$$\tau_A l_A = \tau_B l_B \tag{22}$$

The energy equation has the form :

$$\rho c\, \partial\theta / \partial t = \beta \tau \dot{\gamma} \tag{23}$$

The case of no-strain hardening is considered in this subsection *(n=0)*.
Elimination of $\dot{\gamma}$ between the constitutive law and the energy equation, leads to

$$\rho c\, \partial\theta / \partial t = \beta \tau \left(\tau / \mu(\theta) \right)^{1/m} \tag{24}$$

By division term by term of the equations (24) considered at two different sections A and B, use of the equilibrium condition (22) and integration, the following relationship is obtained between the temperatures θ_A and θ_B in the two sections :

$$l_A^{\frac{m+1}{m}} \int_{\theta_A^0}^{\theta_A} \mu(\xi)^{\frac{1}{m}} d\xi = l_B^{\frac{m+1}{m}} \int_{\theta_B^0}^{\theta_B} \mu(\xi)^{\frac{1}{m}} d\xi \tag{25}$$

with θ_A^0 and θ_B^0 being the initial values of the temperature.

We shall say that a L_∞ localization of the temperature occurs at B, if the temperature tends to infinity at B, while it remains finite at all other sections A, [Molinari and Clifton,1983], [Molinari, 1985] :

$$\theta_B \to \infty; \quad \theta_A \to \theta_{A_c} < \infty \quad \forall A \neq B \tag{26}$$

Clearly this condition is realized if there exists a temperature or a geometrical defect and a scalar $\alpha > 0$ such that

$$\int_\alpha^\infty \mu(\xi)^{\frac{1}{m}} d\xi < \infty \tag{27}$$

For instance if the following power law is considered :

$$\mu(\theta) = \mu_0 \theta^\nu \tag{28}$$

the conditions for localization are

$$L_\infty \ localization \Leftrightarrow \nu + m < 0 \tag{29}$$

For a thermal softening material $\nu < 0$; the linearized perturbation method predicts flow instability. However the result (29) shows that localization can be forbidden if the strain rate sensitivity is large enough, so that $\nu + m > 0$. The possibility of having instability of the plastic flow, with no localization is related to the non-stationarity of the fundamental solution, Molinari (1985).

5.2 Relative perturbation analysis

To have a better account of the time dependence of the ground solution, a relative perturbation analysis can be developed in which one is interested in the evolution of the relative temperature defect $\frac{\delta\theta}{\theta^0}$ (with θ^0 the temperature given by the fundamental solution) contrary to the classical linearized approach where the absolute defect $\delta\theta$ is considered, see [Molinari and Clifton, 1983], [Molinari, 1985] and [Fressengeas and Molinari, 1987]. The relative perturbation approach (relative defect theory) gives better information concerning the occurrence of localized flow than the absolute perturbation approach (absolute defect theory). Note that for a stationary fundamental flow, both theories give identical results.

For a strain-hardening material, the absolute and relative perturbation analyses provide the following conditions for instability (growth of the absolute defect $\delta\theta$) and relative instability (growth of the relative defect $\delta\theta / \theta^0$), respectively:

$$\nu + n < 0 \Leftrightarrow \textit{absolute instability} \tag{30}$$

$$\nu + n + m < 0 \Leftrightarrow \textit{relative instability} \tag{31}$$

For a non-hardening material, it appears that the criterion of relative instability (31) is identical to the localization criterion (29).

5.3 Localization strain

An important result for engineering purpose is the value of the nominal localization strain at which the bearing capacity of the structure suffers a sudden drop. The localization strain in the Marchand and Duffy's experiments corresponds to the beginning of stage III ; it will be denoted by γ_c.

Considering the model of the preceding paragraph, one obtains a quantitative evaluation of the localization strain γ_c by analytical means, [Molinari and Clifton, 1987]. Neglecting inertia and heat conductivity is a valid hypothesis for the analysis of the onset of localization (end of stage II, beginning of stage III), as shown by finite element computations. However at later times in the stage III, the stress decreases rapidly and inertia forces may be accounted for. Heat conduction effects can also have an important role in building the shear band morphology, as will be discussed later.

We shall consider separately the cases of strain hardening materials and materials with no strain hardening. The following calculations of the localization strain are based on a defect analysis.

No strain hardening

Providing that the localization condition (27) is satisfied, the relationship (25) implies that localization will occur at the section B where the following function attains a minimum value : $M \rightarrow l_M^{\frac{1+m}{m}} \int_{\theta_M^0}^{\infty} \mu(\xi)^{\frac{1}{m}} d\xi$

At localization, the temperature θ_A^c at any point A is given by the implicit relationship :

$$l_A^{\frac{m+1}{m}} \int_{\theta_A^0}^{\theta_A^c} \mu(\xi)^{\frac{1}{m}} d\xi = l_B^{\frac{m+1}{m}} \int_{\theta_B^0}^{\infty} \mu(\xi)^{\frac{1}{m}} d\xi \qquad (32)$$

Using the energy equation, we calculate the strain at any point A at the moment of localization. The case of an exponential dependence of the flow stress upon temperature is analyzed here:

$$\tau = \mu_0 \exp(-\alpha\theta)\dot{\gamma}^m \qquad (33)$$

The results depend on the type of loading considered. For a stress-boundary condition (constant shear stress applied at the boundaries), the critical strain at A when localization occurs at B is given by

$$\gamma_A^c = -\frac{\rho c m}{\alpha\beta\tau} \ln\left[1 - \left(\frac{l_B}{l_A}\right)^{\frac{1+m}{m}} \exp\left(-\alpha(\theta_B^0 - \theta_A^0)\right)\right] \qquad (34)$$

When constant velocities $\pm V$ are applied at the boundaries, one has

$$\gamma_A^c = \frac{\rho c}{\alpha\beta\tau_A^0}\left[\left(1 - \left(\frac{l_B}{l_A}\right)^{\frac{1}{m}} \exp(-\alpha(1-m))(\theta_B^0 - \theta_A^0)/m)\right) - 1\right] \qquad (35)$$

where $\tau_A^0 = \mu_0 \dot{\gamma}^{0m} \exp(-\alpha\theta_A^0)$; $\dot{\gamma}^0 = \frac{V}{h}$ is the nominal strain rate. The stabilizing effect of the strain rate sensitivity ($m > 0$) and the destabilizing effect of thermal softening ($\alpha > 0$) are easily analysed from these expressions.

The defect parameter is defined as

$$\chi = 1 - \frac{l_B}{l_A}\exp\left(-\alpha(\theta_B^0 - \theta_A^0)\right) \tag{36}$$

For small defects $\chi << 1$, the localization strain can be approximated, for a velocity boundary condition, by

$$\gamma_A^c = \frac{\rho c}{\alpha\beta\tau_A^0}\left(-m\ln\chi + const\right) \tag{37}$$

Thus, the critical strain at localization shows a weak logarithmic dependence upon the defect parameter χ.

Materials with strain hardening

Using similar lines, we can calculate the localization strain in a strain hardening material [Molinari and Clifton, 1987]. For instance, considering the constitutive law (21) and a constant velocity applied on the boundaries, we obtain the following relationship between the strains in sections A and B :

$$l_A^{\frac{1}{m}} \int_0^{\gamma_A} \mu(\theta_A(\varsigma))^{\frac{1}{m}}\left(\varsigma + \gamma_i^A\right)^{\frac{n}{m}} d\varsigma = l_B^{\frac{1}{m}} \int_0^{\gamma_B} \mu(\theta_B(\varsigma))^{\frac{1}{m}}\left(\varsigma + \gamma_i^B\right)^{\frac{n}{m}} d\varsigma \tag{38}$$

As before, L_∞-localization occurs if and only if the integral on the right side of equation (38) remains bounded as $\gamma_B \to \infty$. For $\mu(\theta) = \mu_0\theta^\nu$, the condition for L_∞-localization becomes

$$\nu + n + m(1-\nu) < 0 \tag{39}$$

Note that $\nu < 0$ for a thermal softening material.

Using the relationship (38) with $\gamma_B \to \infty$, we can calculate the critical strain γ_A^c at localization, and the nominal localization strain γ_c is obtained by spatial averaging, [Molinari and Clifton, 1987], [Dinzart et al., 1994]. It was shown that γ_c has a logarithmic dependence with respect to the initial imperfection in the specimen geometry, fig.8. In the upper part of fig.8, the theoretical predictions(continuous line) for a CRS-1018 steel [Molinari and Clifton, 1987], are compared with the experimental data (dashed line) [Duffy et al., 1991]. Although the trends are similar, the theoretical curve is shifted below the experimental curve. The calculations were made by assuming a geometrical defect with a sinusoidal shape. With a different defect-shape, the predictions can be improved [Dinzart et al., 1994].

The stabilizing influences of the strain rate sensitivity *m* and of the strain hardening exponent *n* appear clearly in the condition (39). In addition, *m* has a strong influence on the localization strain γ_c. A small augmentation of the strain rate sensitivity considerably increases the duration of stage II, i.e. the incubation time to localization.

6. CONCLUDING REMARKS.

The development of adiabatic shear banding in a tubular specimen under high speed torsion has been analyzed by using linear perturbation methods together with nonlinear approaches. These methods were successful in describing by analytical means the weak flow instability after the maximum stress is passed (stage II of the process) and the early stage of localization (beginning of stage III). Additional analyses are needed to describe further stages in the localization process. The shear band builds up an internal structure. A band-width can be defined and calculated by accounting for the competition between destabilizing factors (e.g. thermal softening, non-linear effects) and stabilizing factors (e.g. heat conductivity, strain rate sensitivity) which tend respectively to decrease and to increase the shear band width, [Wright and Ockendon, 1992] and [Dinzart and Molinari, 1998].

At the end of the process the band is observed to propagate along the circumference of the tubular specimen. Few works have been devoted to the analysis of shear band propagation, see [Mercier and Molinari, 1998] for a literature survey together with a new analysis of the stationary propagation of an adiabatic shear band.

REFERENCES

[Bai, Y., 1982] Thermo-plastic instability in simple shear, *J. Mech. Phys.Solids*,30, p.195

[Bai, Y. ; Dodd, B., 1992] Adiabatic shear localization, Pergamon Press, Oxford.

[Duffy, J. ; Chi, Y.C., 1989] On the measurement of local strain and temperature during the formation of adiabatic shear bands in steels, *Brown University report* N°4, DAAL 03-88-K-0015, pp.1-56.

[Clifton, R.J., 1978] Report to the NRC committee on material responses to ultrasonic loading rates.

[Clifton et al., 1984] Clifton, R.J. ; Duffy, J. ; Hartley, K.A. ; Shawki, T.G. ;On critical conditions for shear band formation in high-strain rates; *Scripta Mater.*, 18, p.443.

[Considère, A., 1885] *Ann. Des Ponts et Chaussées*,9, p. 574.

[Dinzart et al., 1994] Dinzart, F. ; Fressengeas, C. ; Molinari, A.; The catastrophic development of shear localization in thermo-viscoplastic materials, *Int. Conf. on High Strain Rates*, Oxford, *J. de Phys.*, C8, pp. 435-440.

[Dinzart, F. ; Molinari A., 1998] Structure of adiabatic shear bands in thermo-viscoplastic materials, *Eur. J. Mech. A/Solids*, 17, pp. 923-938.

[Dudzinski et al., 1992] Dudzinski, D. ; El Majdoubi, M. ; Molinari, A. ; Plastic flow localization at high strain rates; in *Shock Waves and High Strain-Rate Phenomena in Materials*, Eds. M.A. Meyers, L.E. Murr and K.P. Staudhammer, M. Dekker, N.Y., pp. 193-201.

[Fressengeas, C.; Molinari A., 1987] Instability and localization of plastic flow in shear at high strain rates, *J. of Mech. and Phys. of Solids*, 35, 2, pp.185-211.

[Fressengeas, C.; Molinari A., 1994] Fragmentation of rapidly stretching sheets, *Eur. J. Mech. Solids*, 13, pp. 251-268.

[Grady, D.E. and Kipp, M.E., 1987] The growth of unstable thermoplastic shear with application to steady-wave shock compression in solids, *J. Mech. Phys. Solids*, 35, p.95.

[Hutchinson et al., 1978] Hutchinson, J.W. ; Neale, K. ; Needleman A. ; Sheet necking I-Validity of plane stress assumptions on the long-wavelength approximation, in *Mechanics of Sheet Metal Forming*, Koistinen D.P. and Wang N.M. Eds, 1, pp. 111-126, Plenum, N.Y.

[Leroy, Y. M.; Molinari, A., 1993] Spatial patterns and size effects in shear zones: a hyperelastic model with higher-order gradients, *J. Mech. Phys. Solids*, 41, pp. 631-663.

[Marchand, A. ; Duffy, J.,1988] An experimental study of the formation process of adiabatic shear bands in a structural steel, *J. Mech. Phys. Solids*, 36, p. 251.

[Mercier, S. ; Molinari, A., 1998] Steady-state shear band propagation under dynamic conditions, *J. Mech. Phys. Solids*, 46, pp.1463-1495.

[Molinari, A ; Clifton, R.J., 1983] Localisation de la déformation viscoplastique en cisaillement simple, résultats exacts en théorie non linéaire, *Comptes Rendus Académie des Sciences*, II, t. 296, pp.1-4.

[Molinari, A., 1985] Instabilité thermo-visco-plastique en cisaillement simple, *J.Méc. Théor. Appl.*, 4, p.659-684.

[Molinari, A. ; Clifton R.J., 1987] Analytical characterization of shear localization, *J. of Appl. Mech.*, 54, pp. 806-912.

[Molinari, A., 1988], Shear band analysis, *Solid St. Phenomena*, 3-4, pp. 447-468.

[Molinari, A. ; Leroy, Y.M., 1991], Structures in shear zones, *Comptes Rendus Académie des Sciences*, 313, Série II, pp. 7-13.

[Molinari, A., 1997], Collective behavior and spacing of adiabatic shear bands, *J. Mech. Phys. Solids*, 45, pp. 1551-1575.

[Nesterenko, V.F. et al., 1994], Nesterenko, V.F. ; Meyers, M.A. ; Chen, H.C. ; LaSavia, J.C.; Controlled high-rate localized shear in porous reactive media, *Appl. Phys. Lett.*, 65, pp. 3069-3071.

[Nesterenko, V.F. et al., 1995a], Nesterenko, V.F ; Meyers, M.A. ;Wright, T.W. ; Collective behavior of shear bands, in *Metallurgical and Materials Applications of Shock-Wave and High-Strain-Rate Phenomena*, Ed. L.E. Murr, K.P. Staudhammer and M.A . Meyers, Elsevier Science B.V., pp. 397-404.

[Nesterenko, V.F. et al., 1995b], Nesterenko, V.F. ; Meyers, M.A. ; Chen, H.C. ; LaSavia, J.C. ; The structure of controlled shear bands in dynamically deformed reactive mixtures, *Metall. Mater. Trans. A*, 26A, pp.2511-2519.

[Shenoy, V.B. ; Freund L.B., 1999], Necking and bifurcation during high strain rate extension, *J. Mech. Phys. Solids*, 47, pp. 2209-2233.

[Wright, T.W. ; Ockendon, H., 1992], A model for fully-formed shear bands scaling law for the effect of inertia on the formation of adiabatic shear bands, *J. Mech. Phys. Solids*, 40, pp. 1217-1226.

[Wright, T.W. ; Ockendon, H., 1996], A scaling law for the effect of inertia on the formation of adiabatic shear bands, *Int. J. Plasticity*, 12, pp. 927-934.

On the evaluation of damping in a structure with viscoelastic components

Patrick MULLER
Laboratoire de Modélisation et Mécanique des Structures
UPRESA C. N. R. S. n°8007
8 rue du Capitaine Scott,, 75 015 PARIS France
muller@ccr.jussieu.fr

Abstract : This paper deals with an extension of the loss factor introduced by Ungar in terms of energy concepts to evaluate, in view of design, damping in structures where high damping is introduced by means of viscoelastic components.

Keywords : Damping – Loss factors – Modal strain energy - Viscoelasticity

1. INTRODUCTION

The development of industrial softwares in order to improve passive damping capacities of structures is still a topical question. The state of the art concerning damping may be found in the pioneering book by Nashif & al. [Nashif, Jones & Henderson , 1985] or in the more recent one by Sun & Lu [Sun & Lu, 1995].
The structures considered here are N-degree-of freedom (N-d.o.f.) elastic structures subjected to harmonic excitations (whose circular frequency will be denoted by ω), in which damping is introduced by means of inclusions of linear viscoelastic parts whose constitutive relations may be represented by so-called rheological models.
Such structures are not necessarily "close to a conservative system" and damping cannot in this case be considered as an uncertain adjustment parameter, but has to be the subject of an accurate evaluation. Damping in a structure depends not only on the intrinsic material characteristics of its components, but also on its geometry, on its various connections, and, unfortunately, on the applied displacements and/or external forces. For design purpose, it would be better if a numerical investigation could define the individual contribution of each of these factors.
It seems reasonable to think that any evaluation of damping must rest on an evaluation of the dissipation in the structure. Thus, the dimensionless loss factor $\eta(\omega)$ introduced in the early sixties by Ungar [Ungar & Kerwin, 1962], namely

$$\eta(\omega)=\frac{1}{2\pi}\frac{D_{cycle}}{W_0} \quad , \qquad [1]$$

where D_{cycle} is the dissipation per cycle and W_0 an "amplitude" of the instantaneous elastic energy stored in the structure taken as a reference, will be the central tool to

G.A. Maugin et al. (eds.), Continuum Thermomechanics, 331-342.
© 2000 *Kluwer Academic Publishers.*

evaluate damping in this paper. This positive and homogeneous-of-order-zero functional of the displacement field could be compared to some Rayleigh-quotient for the couple {dissipation ; elastic energy} instead of the couple {elastic energy ; kinetic energy}.
One starts from the 1-d.o.f. oscillator with classical damping, which will be the subject of the SECTION 2, to introduce some basic constants and state the notations, but also to emphasize the essential unability of 1-d.o.f. structures to account for "directional properties" of the displacement field which play an essential role in such a functional as [1].
SECTION 3 is devoted to N-d.o.f. structures with classical viscous damping, which may be considered as the first step towards viscoelastic damping (associated with Kelvin-Voigt rheological model). When the damping is proportional, the structure may be considered as the superposition of N independent modal oscillators, and [1] takes the form

$$\eta(\omega)=\sum_{p=1}^{N}\eta_p(\omega)\,w_p(\omega) \quad , \qquad [2]$$

where $\eta_p(\omega)$ is the loss factor [1] associated with the p-th modal oscillator and $w_p(\omega)$ is the so called *modal strain energy* (that is the fraction of the total elastic energy stored by this oscillator at the frequency ω). The simplicity of this expression, where the contribution of each one of the modal oscillators may clearly be identified, makes it a remarkable potential numerical tool for suggesting what should be changed in such a structure in order to improve its damping capacities. According to the author's knowledge, the use of [1] is not usual when the damping is not proportional. Some questions raised by the use of [1] in this case are considered.
SECTION 4 deals with the case of damping by use of linear viscoelastic materials. By using complex viscoelastic moduli, one shows that the typical system of governing equations for the complex amplitude $\underline{Q}(\omega)$ of the steady-state response of the structure has, after discretisation by any finite element method, the form

$$\left\{\left(\underline{\underline{K}}'(\omega)+i\,\underline{\underline{K}}''(\omega)\right)-\underline{\underline{M}}\,\omega^2\right\}\underline{Q}(\omega)=\underline{f} \quad , \qquad [3]$$

where, as a consequence of the presence of complex viscoelastic moduli, the stiffness matrix $\left(\underline{\underline{K}}'(\omega)+i\,\underline{\underline{K}}''(\omega)\right)$ is now complex, with real and imaginary parts depending on ω. It is shown that, in this case, the loss factor [1] is given by the formula :

$$\eta(\omega)=\frac{\overline{\underline{Q}}^T(\omega)\underline{\underline{K}}''(\omega)\,\underline{Q}(\omega)}{\overline{\underline{Q}}^T(\omega)\underline{\underline{K}}'(\omega)\,\underline{Q}(\omega)} \quad . \qquad [4]$$

SECTION 5 is a brief conclusion for this paper proposing ideas for further research and mentioning other emerging methods to evaluate damping in structures with viscoelastic components.

2. BASIC CONSTANTS FROM THE 1 D.O.F. ELEMENTARY SYSTEM

The position $x_0(t)$ at the time t of a classical 1-d.o.f. damped oscillator consisting of a mass m_0 attached to a Kelvin-Voigt spring {elastic spring [stiffness k_0] in parallel with a dash-pot [cœfficient of viscosity a_0]} is governed, when the mass is excited by a harmonic force $f_0\cos\omega t$, by the well-known elementary equation :

$$m_0\ddot{x}_0 + a_0\dot{x}_0 + k_0 x_0 = f_0\cos(\omega t) \quad , \tag{5}$$

From [5], one introduces the usual basic constants, namely the circular frequency ω_0 of the conservative associated system ; and the three equivalent constants $\tilde{\eta}_0$ [modal loss factor] or δ_0 [rate of critical damping] or τ_0 [decreasing time of natural oscillations], respectively defined by

$$\omega_0 = \sqrt{\frac{k_0}{m_0}} \;\; ; \;\; \frac{a_0}{m_0} = \tilde{\eta}_0\omega_0 = 2\delta_0\omega_0 = \frac{2}{\tau_0} \quad . \tag{6}$$

The steady-state response (vibrating at the circular frequency ω of the excitation) is

$$x_0(t) = X_0(\omega)\cos\{\omega t + \Phi_0(\omega)\} \quad , \tag{7}$$

where the amplitude $X_0(\omega)$ is given by

$$X_0(\omega) \;=\; \frac{f_0}{k_0}\,\frac{1}{\sqrt{\left[1-\left[\frac{\omega}{\omega_0}\right]^2\right]^2 + \tilde{\eta}_0^2\left[\frac{\omega}{\omega_0}\right]^2}} \quad . \tag{8}$$

Thus, $D_{cycle} = \oint a_0\dot{x}_0^{\,2}(t)\,dt = \pi a_0\omega X_0^{\,2}(\omega)$ and the amplitude of W(t) is

$$W_0 = \tfrac{1}{2}k_0\,X_0^2(\omega) \quad , \tag{9}$$

so that the loss factor [1] of this 1-d.o.f. oscillator, defined for any ω, is given by

$$\eta_0(\omega) = \frac{1}{2\pi}\frac{D_{cycle}}{W_0} = \frac{a_0}{k_0}\,\omega \quad , \tag{10}$$

so that, taking [6] into account, it can be written

$$\eta_0(\omega) = \tilde{\eta}_0\,\frac{\omega}{\omega_0} \quad . \tag{11}$$

A distinction has to be made between two concepts which will play an essential role in the following analysis : the (rheological) loss factor $\eta_0(\omega)$ of the 1-d.o.f. oscillator (a function of the external frequency ω which depends on the material constants a_0 and k_0 only) and the (structural) modal loss factor $\tilde{\eta}_0$ (a constant which, according to [6], depends on a_0, k_0, *and* m_0). Formula [11] clarifies the relation between them :

$$\eta_0(\omega_0) = \tilde{\eta}_0 \quad , \tag{12}$$

and this matching justifies the 2π-factor in the definition [1]. But it is worth noting that the amplitude $X_0(\omega)$ given by [8] has been eliminated from the ratio [10] : in the 1-d.o.f. oscillator, this ratio is independent of the external applied force.

3 BASIC CONCEPTS FOR N D.O.F. SYSTEMS WITH VISCOUS DAMPING

3.1 The classical system of equations

Let the position at the instant t of the N-d.o.f. structure be defined by the column-vector

$$\begin{bmatrix} q_1(t) \\ q_2(t) \\ \vdots \\ q_N(t) \end{bmatrix} , \qquad [13]$$

which will be noted $\underline{q}(t)$. Below, an (NxN) matrix will be noted $\underline{\underline{M}}$, and $[\bullet]^T$ will note the transposition of the quantity $[\bullet]$ (so, $\underline{q}^T$ is the line-vector $[q_1(t), q_2(t), \ldots, q_N(t)]$).
The classical system we consider is

$$\underline{\underline{M}}\,\underline{\ddot{q}} + \underline{\underline{A}}\,\underline{\dot{q}} + \underline{\underline{K}}\,\underline{q} = \underline{f}\cos\omega t \quad , \qquad [14]$$

where $\underline{f}$ is constant, and where the three matrices $\underline{\underline{M}}$ [mass], $\underline{\underline{A}}$ [damping] and $\underline{\underline{K}}$ [stiffness] are assumed to be real, constant and positive-definite.
For this N-d.o.f. structure, one has now, for D_{cycle} and W_0 respectively :

$$D_{cycle} = \oint \underline{\dot{q}}^T \,\underline{\underline{A}}\,\underline{\dot{q}}\; dt , \qquad [15]$$

$$W_0 = \text{"Amplitude"} \left\{ \tfrac{1}{2} \underline{q}^T \,\underline{\underline{K}}\,\underline{q} \right\} \; . \qquad [16]$$

The N modes $\underline{Q}_i$ of the so called *associated conservative system* [$\underline{\underline{A}} = \underline{\underline{0}}$], which are associated with its eigenfrequencies $\omega_1 \leq \omega_2 \leq \ldots \leq \omega_N$, are real (their components are in phase) and satisfy the following well-known orthogonality relations :

$$\underline{Q}_i^T \underline{\underline{M}}\,\underline{Q}_j = \begin{cases} 0 \text{ if } i \neq j \\ m_i > 0 \text{ if } i = j \end{cases} , \quad \underline{Q}_i^T \underline{\underline{K}}\,\underline{Q}_j = \begin{cases} 0 \text{ if } i \neq j \\ k_i > 0 \text{ if } i = j \end{cases} . \qquad [17]$$

3.2 Proportional viscous damping

Looking for a steady-state response (vibrating at the external pulsation ω) in the form

$$\underline{q}(t) = \sum_{p=1}^{N} r_p(t)\,\underline{Q}_p \quad , \qquad [18]$$

and using the modal basis of the associated conservative system, one is led, after "modal projection", to the system of generally coupled differential equations :

$$m_i\,\ddot{r}_i + \sum_{p=1}^{N} \left(\underline{Q}_i^T \,\underline{\underline{A}}\,\underline{Q}_p \right) \dot{r}_p + k_i\, r_i = f_i \cos\omega t \quad , \quad i = 1,2,\ldots,N \; , \qquad [19]$$

in which the modal projections of the amplitude of the excitation has been introduced :

$$f_i = \underline{Q}_i^T \underline{f} \quad . \qquad [20]$$

When the damping matrix $\underline{\underline{A}}$ satisfies the so-called hypothesis of proportionality :

$$\underline{Q}_i^T \underline{\underline{A}}\,\underline{Q}_j = \begin{cases} 0 \text{ if } i \neq j \\ a_i \text{ if } i = j \end{cases} , \qquad [21]$$

the system [19] splits into a system of uncoupled equations for N independent 1-d.o.f. oscillators :

$$m_i\,\ddot{r}_i + a_i\,\dot{r}_i + k_i\, r_i = f_i \cos\omega t \quad , \quad i = 1,2,\ldots,N. \qquad [22]$$

Since Rayleigh [Rayleigh, 1945], various authors such as Caughey & al. ([Caughey, 1960], [Caughey & O'Kelly, 1965]), and, more recently, Liang & Lee [Liang & Lee,

1991] have given more and more refined algebraic conditions for the matrix $\underline{\underline{A}}$ to satisfy the hypothesis [21]. Alternately, Shahruz [Shahruz, 1990] contributed to the estimation of the error made by cancelling roughly the p≠i-terms in [19] so as to obtain an approximated system satisfying [21].
Comparing [22] to [5], one introduces the same basic notations than in [7], [8] and [6] with index $_0$ replaced by index $_p$, so that [18] takes the explicit form :

$$\underline{q}(t)=\sum_{p=1}^{N}\left[X_p(\omega)\cos\left[\omega t+\Phi_p(\omega)\right]\underline{Q}_p\right] . \qquad [23]$$

By substituting [23] into [15] and into [16], and taking [21] into account, one is led to :

$$D_{cycle}=\pi\omega\sum_{p=1}^{N}a_p X_p{}^2(\omega) \ ; \qquad [24]$$

and, with an obvious choice for the "Amplitude", to :

$$W_0=\tfrac{1}{2}\sum_{p=1}^{N}\left\{X_p(\omega)\underline{Q}_p{}^T\right\}\underline{\underline{K}}\sum_{q=1}^{N}\left[X_q(\omega)\underline{Q}_q\right] . \qquad [25]$$

that is, taking $[17]_2$ into account :

$$W_0=\tfrac{1}{2}\sum_{p=1}^{N}k_p X_p{}^2(\omega) . \qquad [26]$$

Then, according to [1], one has :

$$\eta(\omega)=\omega\frac{\sum_{p=1}^{N}a_p X_p{}^2(\omega)}{\sum_{p=1}^{N}k_p X_p{}^2(\omega)} . \qquad [27]$$

By substituting the a_p's by their expressions from [6], one obtains :

$$\eta(\omega)=\omega\frac{\sum_{p=1}^{N}\frac{\tilde{\eta}_p k_p}{\omega_p}X_p{}^2(\omega)}{\sum_{p=1}^{N}k_p X_p{}^2(\omega)} , \qquad [28]$$

and this expression appears to be the so-called modal strain energy formula [2] where $\eta_p(\omega)$ is the material loss factor [11] associated with the p-th modal oscillator and $w_p(\omega)$ is the modal strain energy, respectively given by :

$$\eta_p(\omega)=\tilde{\eta}_p\frac{\omega}{\omega_p} \quad \text{and} \quad w_p(\omega)=\frac{k_p X_p^2(\omega)}{\sum_{r=1}^{N}k_r X_r^2(\omega)} , \qquad [29]$$

REMARK 1 : It is worth noticing that $\eta(0)=0$, and that, when $\omega=\omega_i$, one has :

$$\eta(\omega_i)=\sum_{p=1}^{N}\eta_p(\omega_i)\, w_p(\omega_i) , \qquad [30]$$

so that, except in the unusual case of a strictly unimodal excitation [$f_p=0$ when $p \neq i \Rightarrow \eta(\omega)=\eta_i(\omega)$] where one has *exactly* $\eta(\omega_i)=\tilde{\eta}_i$ as in [12], $\eta(\omega)$ may not be considered as an interpolation of the modal loss factors $\tilde{\eta}_i$.

REMARK 2 : When the $\tilde{\eta}_p$'s are (very) small, so that the contribution of $X_p(\omega)$ to $w_p(\omega)$ is confined to a (very) narrow peak located at $\omega^{(a)}{}_p \# \omega$, one has $\eta(\omega_i) \# \tilde{\eta}_i$. But, even in this case, the values of $\eta(\omega)$ between ω_p and ω_{p+1} may appreciably deviate from a smooth interpolation of the $\tilde{\eta}_i$'s : in other words, the knowledge of the $\tilde{\eta}_i$'s only gives an approximate information on damping (see *Figure 1*).

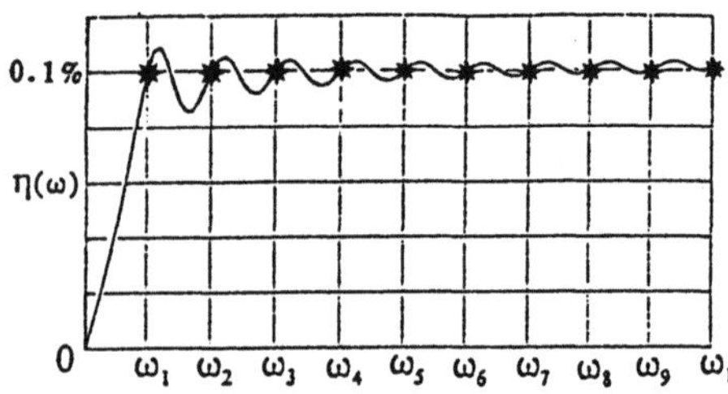

***Figure 1a** : Little damping*
10 d.o.f. - $\omega_n=n\omega_1 - f_n \equiv f_0 \forall n$
$m_n \equiv m_0 \forall n$ - (✱) $\tilde{\eta}_n \equiv 0{,}1\% \; \forall n$

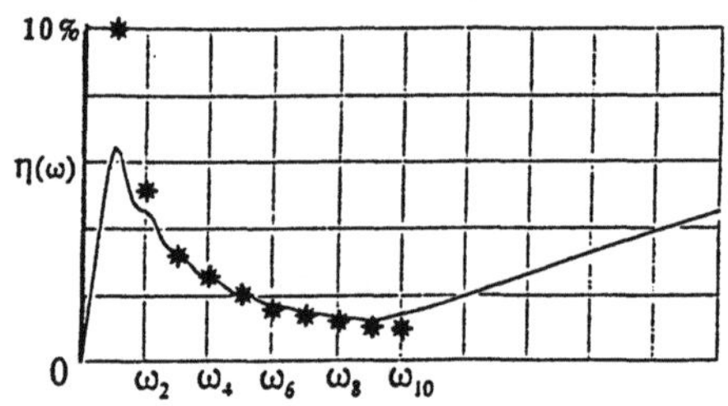

***Figure 1b** : High damping*
10 d.o.f. - $\omega_n=n\omega_1 - f_n \equiv f_0 \forall n$
$m_n \equiv m_0 \forall n$ - (✱) $\tilde{\eta}_n \equiv 10\% / n$

REMARK 3 : The possible strong variability of the $\tilde{\eta}_i$'s from one mode to the next reinforces Remark 1 : the elementary 2-d.o.f. symmetrical structure shown in *Figure 2*, where the damping is proportional, has a first (symmetrical) mode (where the central Kelvin-Voigt block moves as a rigid body) which is obviously nondissipative, while its second (antisymmetrical) one is fully dissipative.

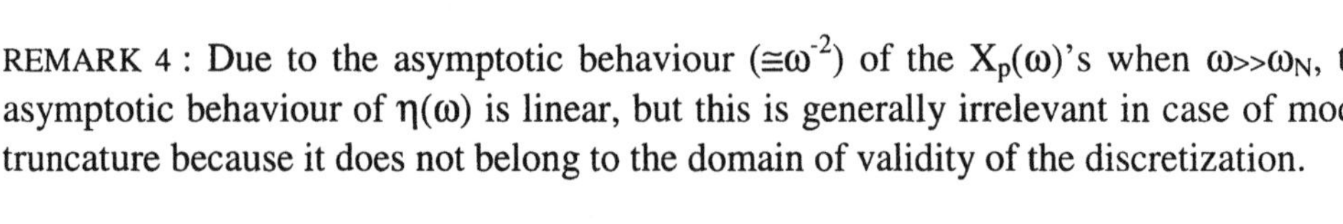

***Figure 2** : 2-d.o.f. symmetrical elementary structure*

REMARK 4 : Due to the asymptotic behaviour ($\equiv \omega^{-2}$) of the $X_p(\omega)$'s when $\omega >> \omega_N$, the asymptotic behaviour of $\eta(\omega)$ is linear, but this is generally irrelevant in case of modal truncature because it does not belong to the domain of validity of the discretization.

REMARK 5 : In the structures (such as those for which, as a particular case of Rayleigh-condition, the matrix $\underline{A}$ is directly proportional to the matrix $\underline{K}$) where all the loss factors $\eta_p(\omega)$'s are equal to the same, the loss factor has, due to the obvious property $\sum_{p=1}^{N} w_p = 1$, the same expression as the 1-d.o.f. loss factor [11].

An example of application of [2] to the detailed analysis in the form of histograms of the contribution of each mode and of each material to the damping in a composite shell structure for design purpose may be found in Macé [Macé, 1991].
Below, it will be convenient to use the *complex notation* according to which [14] and [23] should respectively be written as :

$$\underline{\underline{M}}\,\ddot{\underline{q}}+\underline{\underline{A}}\,\dot{\underline{q}}+\underline{\underline{K}}\,\underline{q}=\underline{f}\,e^{i\omega t} \quad , \tag{31}$$

$$\underline{q}(t)=\sum_{p=1}^{N}\left\{X_p(\omega)\,e^{i[\omega t+\Phi_p(\omega)]}\,\underline{Q}_p\right\}=\underline{Q}(\omega)e^{i\omega t} \quad , \tag{32}$$

where one has introduced the complex amplitude $\underline{Q}(\omega)$, which, in the case of proportional damping, is then given by :

$$\underline{Q}(\omega)=\sum_{p=1}^{N}\left\{X_p(\omega)e^{i\Phi_p(\omega)}\,\underline{Q}_p\right\} . \tag{33}$$

One may notice that, with this notation, if $\overline{\underline{q}}$ denotes the complex conjugate of $\underline{q}$, the non-linear expressions D_{cycle} and W_0 have now respectively to be written as follows :

$$D_{cycle}=\oint \tfrac{1}{2}\dot{\overline{\underline{q}}}^{T}\,\underline{\underline{A}}\,\dot{\underline{q}}\;dt \quad , \qquad W_0=\tfrac{1}{2}\overline{\underline{q}}^{T}\,\underline{\underline{K}}\,\underline{q} \;. \tag{34}$$

Actually, substituting first [32] in $[34]_1$ one has generally with this complex notation :

$$D_{cycle}=\pi\omega\overline{\underline{Q}}(\omega)^{T}\,\underline{\underline{A}}\,\underline{Q}(\omega) \quad , \tag{35}$$

which renders [24], if, in addition, one takes into account the proportionality of damping via [33] and [21].
In the same way, substituting [32] in $[34]_2$ one has generally for this complex notation :

$$W_0=\tfrac{1}{2}\left\lfloor\overline{\underline{Q}}^{T}(\omega)\underline{\underline{K}}\,\underline{Q}(\omega)\right\rfloor \quad , \tag{36}$$

which renders [26] if, in addition, one takes the proportionality of damping into account via [33] and $[17]_2$

3.3 Non-proportional viscous damping

A *first method* should consist in using, here again, the modes of the associated conservative structure and in looking again for a solution in the form [18]. If one sets :

$$\underline{Q}_i^{\,T}\,\underline{\underline{A}}\,\underline{Q}_j=a_{ij} \quad , \tag{37}$$

where the matrix of cœfficients a_{ij} is no longer diagonal, the system [19] takes the form :

$$m_i\,\ddot{r}_i+\sum_{p=1}^{N}a_{ip}\,\dot{r}_p+k_i\,r_i=f_i\cos\omega t \quad , \quad i=1,2,\ldots,N. \tag{38}$$

Hence, from [15] and [16], the expression of the loss factor $\eta(\omega)$ defined by[1] becomes :

$$\eta(\omega)=\frac{\displaystyle\sum_{p=1}^{N}\sum_{q=1}^{N}a_{pq}\oint\dot{r}_p\dot{r}_q\,dt}{\pi\ \text{"Amplitude"}\left\{\displaystyle\sum_{q=1}^{N}k_q r_q^{\,2}(t)\right\}} \quad , \tag{39}$$

which renders [27] when the damping is proportional. Of course, the difficulty of this method lies in the calculation of $r_p(t)$'s which are no longer given by $X_p(\omega)\cos[\omega t+\Phi_p(\omega)]$, but are solutions of the coupled differential system [38].

A *variant of this first method* should consist in using, the (complex) modes $\underline{Q}^*{}_i$ of the dissipative structure itself. As one can find in the pioneering paper by Foss [Foss, 1958], or in the more recent one by Veletsos & Ventura concerning earthquake mechanics [Veletsos & Ventura, 1986], the search of solutions for free vibrations ($\underline{f}=\underline{0}$) of the structure with non-proportional damping in the form $\underline{q}(t)=\underline{Q}^*e^{st}$ shows that s must be a root of the 2N-degree-polynomial with real cœfficients :

$$\operatorname{Det}\left\{\underline{\underline{M}}\,s^2+\underline{\underline{A}}\,s+\underline{\underline{K}}\right\}=0 \quad , \tag{40}$$

the 2N roots ($s_1, s_2,\ldots, s_{2N}$) of which are N pairs of complex conjugates with negative real parts [as before, one assumes "underdamping"], which may be written in the form :

$$-\frac{1}{\tau_k}\pm i\,\omega_k{}^{(a)} \quad , \tag{41}$$

to which N pairs of complex conjugates eigenmodes ($\underline{Q}^*{}_1, \underline{Q}^*{}_2,\ldots, \underline{Q}^*{}_{2N}$) are associated. The orthogonality relations [17] for the conservative structure are then to be replaced by :

$$\begin{cases}(s_i+s_j)\,\overline{\underline{Q}}^*{}_j{}^T\underline{\underline{M}}\,\underline{Q}^*{}_i+\overline{\underline{Q}}^*{}_j{}^T\underline{\underline{A}}\,\underline{Q}^*{}_i=0 \ , \\ \overline{\underline{Q}}^*{}_j{}^T\underline{\underline{K}}\,\underline{Q}^*{}_i-s_is_j\,\overline{\underline{Q}}^*{}_j{}^T\underline{\underline{M}}\,\underline{Q}^*{}_i=0 \ ,\end{cases} \tag{42}$$

where $\overline{\underline{Q}}^*{}_i$ denotes the conjugate of $\underline{Q}^*{}_i$. Although this method is not currently used because of the difficulties of numerical manipulation of these complex modes, it could be interesting to explore that idea.

A *second method* should consist in using the complex notation by searching for [31] a steady-state response of the system under the form

$$\underline{q}(t)=\underline{Q}(\omega)e^{i\omega t} \quad , \tag{43}$$

where $\underline{Q}(\omega)$ denotes now the complex amplitude of the forced vibration (without reference to any modal basis as in [33]). Substituting directly [43] in [31] one is led to the following algebraic system for $\underline{Q}(\omega)$:

$$\left\{\left(\underline{\underline{K}}+i\,\underline{\underline{A}}\omega\right)-\underline{\underline{M}}\,\omega^2\right\}\underline{Q}(\omega)=\underline{f} \quad . \tag{44}$$

With this complex notation, the loss factor formula [1] has now to be calculated from expressions [35] for D_{cycle} and [36] for W_0, and one gets :

$$\eta(\omega)=\frac{1}{2\pi}\frac{\pi\omega\overline{\underline{Q}}(\omega)^T\underline{\underline{A}}\,\underline{Q}(\omega)}{\frac{1}{2}\left\{\overline{\underline{Q}}^T(\omega)\underline{\underline{K}}\,\underline{Q}(\omega)\right\}}=\omega\,\frac{\overline{\underline{Q}}^T(\omega)\underline{\underline{A}}\,\underline{Q}(\omega)}{\overline{\underline{Q}}^T(\omega)\underline{\underline{K}}\,\underline{Q}(\omega)} \quad . \tag{45}$$

4 SYSTEMS WITH VISCOELASTIC DAMPING

The pioneering paper concerning this approach to the evaluation of damping in viscoelastic structures is by Johnson & Kienholtz [Johnson & Kienholtz, 1982].

4.1 The structures taken into account

One is looking for the steady-state response of a three-dimensional structure S with elastic parts E and linear viscoelastic parts V (such that S = E + V), excited by distributed harmonic forces $\{\underline{F}e^{i\omega t}\}$, the amplitude $\underline{F}$ of which may be complex (see *Figure 3*).

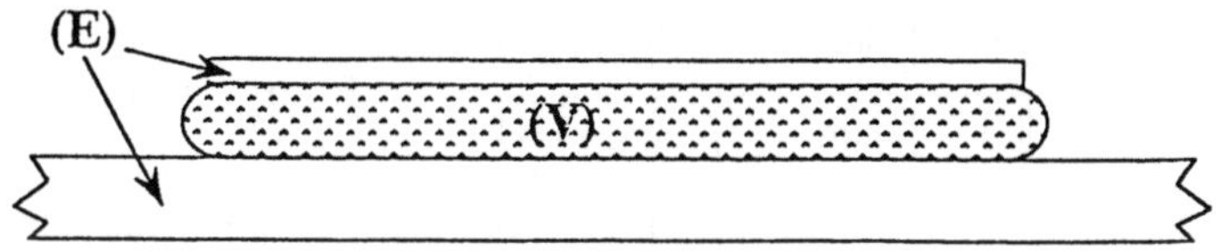

Figure 3 *: Example of structure taken into account : sandwich plate*

Denoting the displacement field by $\vec{u}(x,t)$, the (small) deformation field by $\overline{\overline{\varepsilon}}(x,t)$, and the stress field by $\overline{\overline{\sigma}}(x,t)$, the steady-state three-dimensional response is in the form :

$$\vec{u}(x,t) = \vec{u}(x)\,e^{i\omega t} \;;\; \overline{\overline{\varepsilon}}(x,t) = \overline{\overline{\varepsilon}}\{\vec{u}(x)\}e^{i\omega t} \;;\; \overline{\overline{\sigma}}(x,t) = \overline{\overline{\sigma}}(x)\,e^{i\omega t} \quad . \tag{46}$$

In the elastic parts E, the stress-strain relations are in the form :

$$\sigma_{ij}(x) = a^{\circ}{}_{ijkl}(x)\varepsilon_{kl}(\vec{u}) \quad , \tag{47}$$

where the matrix $a^{\circ}{}_{ijkl}$ is independent of ω, and, in the viscoelastic parts V :

$$\sigma_{ij}(x) = a^{*}{}_{ijkl}(x,\omega)\varepsilon_{kl}(\vec{u}) \quad , \tag{48}$$

where the complex stiffness matrix $a^{*}{}_{ijkl}(x,\omega)$ depends on ω.

One will find for instance in P. Germain [P. Germain, 1964] a close and complete demonstration of [48] starting from the classical stress-strain linear viscoelastic relations which may be written

$$\sigma_{ij}(x,t) = \int_{-\infty}^{t} a_{ijkl}(x,t-t')\frac{D\varepsilon_{kl}(x,t')}{Dt'}dt \quad , \tag{49}$$

and concluding that the steady-state expression of $\sigma_{ij}(x,t)$ due to a harmonic deformation $\varepsilon_{ij}(x,t)=\varepsilon_0 e^{i\omega t}$ is in the form [48], where, if $\hat{f}(p)$ denotes the Laplace-Carson transform of a (regular enough) function f(t), given by :

$$\hat{f}(p) = p\int_{-\infty}^{\infty} f(t)e^{-pt}\,dt \quad , \tag{50}$$

the complex stiffness matrix $a^{*}{}_{ijkl}(x,\omega)$ is given by

$$a^{*}{}_{ijkl}(x,\omega) = \left[\hat{a}_{ijkl}(x,p)\right]_{p=i\omega} \quad . \tag{51}$$

4.2 On the complex moduli

Introductions to rheological models (elastic springs and dash-pots in series or in parallel connections) and to complex viscoelastic moduli can be found in Bland [Bland, 1960] or in Flügge [Flügge, 1967].

For instance, in the case of an isotropic viscoelastic material, two elastic complex moduli are required at each point of the viscoelastic domain V in the expression of the $a^{*}{}_{ijkl}(\omega)$'s, such as the complex Young modulus $E^{*}(\omega)$ and the complex Poisson ratio $\nu^{*}(\omega)$ (which may often be considered as constant).

Any viscoelastic relaxation modulus, for instance E(t), may be represented by a so called relaxation spectrum, that is :

$$E(t)=E_\infty+(E_0-E_\infty)\sum_{k=1}^{K}\lambda_k e^{-\frac{t}{\tau_k}} \quad , \quad \text{with} \quad E_0>E_\infty \quad \text{and} \quad \sum_{k=1}^{K}\lambda_k=1 \quad , \qquad [52]$$

so that, after reduction, it may always be written (*cf* Bland [Bland, 1960]) in the form :

$$E^*(\omega)=E_0\frac{\left(i\omega+\frac{1}{\theta_1}\right)\left(i\omega+\frac{1}{\theta_2}\right)\ldots\left(i\omega+\frac{1}{\theta_K}\right)}{\left(i\omega+\frac{1}{\tau_k}\right)\left(i\omega+\frac{1}{\tau_2}\right)\ldots\left(i\omega+\frac{1}{\tau_K}\right)}=\ldots=E'(\omega)+iE''(\omega) \quad , \qquad [53]$$

where the relaxation times τ_i and the so-called creep times θ_i are arranged as follows :

$$0\le\tau_1\le\theta_1\le\tau_2\le\theta_2\le\cdots\cdots\le\tau_N\le\theta_N \quad . \qquad [54]$$

The usual rheological loss factor $\eta_E(\omega)$, associated with $E^*(\omega)$, given by :

$$\eta_E(\omega)=\frac{E''(\omega)}{E'(\omega)} \quad , \qquad [55]$$

is nothing else than the natural extension to a viscoelastic material of the rheological loss factor [1] (for a Kelvin-Voigt material, for which one has $E^*(\omega)=E_\infty+i\alpha\omega$, its expression agrees with [10]).
The dependancies of $E'(\omega)$, $E''(\omega)$ and $\eta_E(\omega)$ (whose equations are easy to obtain from [53] and [55]) versus ω have the same general aspect as those deduced from the three-parameters solid model of Zener obtained from [52] where K=1 (see *Figure 4*) :

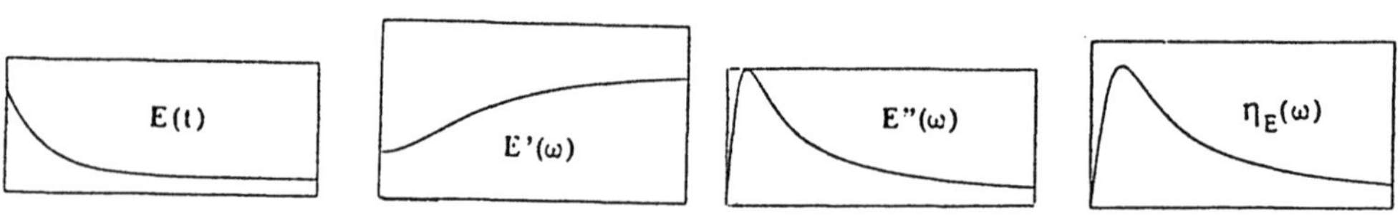

Figure 4 *: Complex modulus of a three-parameter model : $E'(\omega)$, $E''(\omega)$, $\eta_E(\omega)$*

4.3 Finite-element discretization

The discretized solution is in the form :

$$\vec{u}(x)=\sum_{p=1}^{N} q_p\,\vec{e}_p(x) \quad , \qquad [56]$$

where the q_p's are the (complex) nodal displacements and the $\vec{e}_p(x)$'s are the (real) interpolation fields.
The elements of the stiffness matrix $\underline{\underline{K}}$ are given by :

$$K_{pq}=\int_S a_{ijkl}\,\varepsilon_{ij}(\vec{e}_p)\varepsilon_{kl}(\vec{e}_q)\,dx \quad , \qquad [57]$$

so that, due to S = E + V, $\underline{\underline{K}}$ splits up into $\underline{\underline{K}}=\underline{\underline{K}}^e+\underline{\underline{K}}^v$, with :

$$\underline{\underline{M}}\,\ddot{\underline{q}}+\underline{\underline{A}}\,\dot{\underline{q}}+\underline{\underline{K}}\,\underline{q}=\underline{f}\,e^{i\omega t} \quad , \qquad [58]$$

where the matrix K^e_{pq} is independent of ω, and :

$$K^v_{pq}=\int_V a^*_{ijkl}\,\varepsilon_{ij}(\vec{e}_p)\varepsilon_{kl}(\vec{e}_q)\,dx=K^{v\,\prime}_{pq}(\omega)+i\,K^{V\,\prime\prime}_{pq}(\omega) \quad , \qquad [59]$$

so that $\underline{\underline{K}}=\underline{\underline{K}}'(\omega)+i\,\underline{\underline{K}}''(\omega)$, with $\underline{\underline{K}}'(\omega)=\underline{\underline{K}}^e+\underline{\underline{K}}^{v\,\prime}(\omega)$ and $\underline{\underline{K}}''(\omega)=\underline{\underline{K}}^{v\,\prime\prime}(\omega)$.
Thus, the discretized problem is in the general form :

$$\left\{\left(\underline{\underline{K}}'(\omega)+i\,\underline{\underline{K}}''(\omega)\right)-\underline{\underline{M}}\,\omega^2\right\}\underline{Q}(\omega)=\underline{f}\quad, \qquad [60]$$

which includes the Kelvin-Voigt case :

$$\underline{\underline{K}}'(\omega)=\underline{\underline{K}}\ ,\ \underline{\underline{K}}''(\omega)=\omega\underline{\underline{A}}\quad . \qquad [61]$$

According to [35] and [36], the loss factor [1] (depending on the resolution of [60] in terms of $\underline{Q}(\omega)$) is now given by formula [4], which turns into [45] in the Kelvin-Voigt case [61].

The "frequency scanning method", which consists in solving [60] for a narrow sequence of values of ω, is only possible for moderate numbers N of d.o.f. and intervals of frequency : at each step, one has to invert the complex matricial system [60] and this may rapidly lead to very heavy calculations.

In other methods, the interval of frequencies is cut into sub-intervals where the matrices $\underline{\underline{K}}'(\omega)$ and $\underline{\underline{K}}''(\omega)$ may be approximated by simplifed expressions, often constant or linear with respect to ω.

Examples of application of [4] to the analysis of damping capacities of a sandwich plate structure for design purpose may be found in Blanchard [Blanchard, 1995].

5 CONCLUSION

For further research, a close mathematical study of the properties of the basic system [60], whose resolution is necessary to calculate η(ω), should be welcome : the complex stiffness matrix $\underline{\underline{K}}^*(\omega)$ in [60] depends on ω through the complex moduli, and this dependence is not arbitrary (*cf* [53]). What role does the associated conservative system :

$$\left\{\underline{\underline{K}}'(\omega)-\underline{\underline{M}}\,\omega^2\right\}\underline{Q}(\omega)=\underline{0}\quad , \qquad [62]$$

with its associated so-called "non-linear eigenvalues problem" play ? What about the "modes" of such a system ? In relation to this problem, it is worth mentioning that Fergusson & Pilkey [Fergusson & Pilkey, 1992] have studied systems in the form :

$$\left\{\underline{\underline{K}}(\omega)-\underline{\underline{M}}(\omega)\,\omega^2\right\}\underline{Q}(\omega)=\underline{f}\quad . \qquad [63]$$

It is also worth noting that the so-called "structural damping" approximation, which leads to the solution of simplified problems in the form :

$$\left\{\left(\underline{\underline{K}}'_0+i\,\underline{\underline{K}}''_0\right)-\underline{\underline{M}}\,\omega^2\right\}\underline{Q}(\omega)=\underline{f}\quad , \qquad [64]$$

where the complex stiffness matrix $\underline{\underline{K}}^*$ is constant, may be used only in narrow bands of frequencies : in a famous paper, Crandall [Crandall, 1970] has shown that such a model for damping is not causal, so that using it in large bands of frequencies is not correct.

At last, it is worth mentioning alternative emergent methods to evaluate damping in view of design, such as the use of fractional derivatives in the rheological models (*cf* Rossikhin & Shitikova, [Rossikhin & Shitikova, 1997]) ; or the technique of modal reduction proposed by Balmès ([Plouin & Balmès, 1998]).

6 REFERENCES

[Blanchard,1995], Blanchard L. ; "Modélisation d'une tôle sandwich acier-polymère-acier et optimisation de ses propriétés amortissantes", Thèse de Doctorat de l'Université P. et M. Curie, Paris

[Bland,1960], Bland D. R. ; "The theory of linear viscoelasticity", Pergamon Press

[Caughey & O'Kelly, 1965], Caughey T. K. ; O'Kelly M. E. J. ; "Classical normal modes in damped linear dynamic systems", J. Appl. Mech., Trans. A.S.M.E., pp. 583-588

[Caughey, 1960], Caughey T. K. ; "Classical normal modes in damped linear dynamic systems", J. Appl. Mech., Trans. A.S.M.E., pp. 269-271

[Crandall, 1970], Crandall S. H. ; "The role of damping in vibration theory", J. Sound & Vibration, **11**, 1, pp. 3-18

[Fergusson & Pilkey, 1992], Fergusson N. J. ; Pilkey W. D. ; "Frequency-dependent element mass matrices", J. Appl. Mech., Trans. A.S.M.E., **59**, pp. 136-139

[Flügge, 1967], Flügge W. ; "Viscoelasticity", Blaisdell Pub. Co

[Foss, 1958], Foss K. A. ; "Co-ordinates which uncouple the equations of motion of damped linear dynamic systems", J. Appl. Mech., Trans. A.S.M.E., pp. 361-364

[Liang & Lee, 1991], Liang Z. ; Lee G. C. ; "Representation of damping matrix", J. Engng Mech., **117**, 5, 1005-1020

[Macé, 1991], Macé M. ; "Modélisation de structures amorties par film viscoélastique", Thèse de Doctorat de l'Université P. et M. Curie, Paris

[Nashif, Jones & Henderson, 1985], Nashif A. D. ; Jones D. I. G. ; Henderson J. P. ; "Vibration damping", John Wiley & Sons

[P. Germain, 1964], Germain P. ; "Viscoélasticité linéaire", cours de D.E.A. de Mécanique des Solides (notes rédigées par les étudiants) Institut Henri Poincaré, Paris

[Plouin & Balmès, 1998], Plouin A.-S. ; Balmes E. ; "Pseudo-modal representations of large models with viscoelastic behaviour", Proceedings of the International Modal Analysis Conference, pp. 1440-1446, Santa-Barbara, U.S.A.

[Rayleigh, 1945], Rayleigh, Lord ; "The theory of sound", Vol.1, Dover

[Rossikhin & Shitikova, 1997], Rossikhin Y. A. ; Shitikova M. V. ; "Applications of fractional calculus to dynamic problems of linear and nonlinear hereditary mechanics of solids", Appl. Mech. Rev., **50**, 1, pp. 15-67

[Shahruz, 1990], Shahruz S. M. ; "Approximate decoupling of the equations of motion of damped linear systems", J. Sound & Vibration, **136**, 1, pp. 51-64

[Sun & Lu, 1995], Sun C. T. ; Lu Y. P. ; "Vibration damping of structural elements", Prentice Hall PTR

[Ungar & Kerwin, 1962], Ungar E. E. ; Kerwin E. M. Jr ; "Loss factors of viscoelastic systems in terms of energy concepts", J. Acoust. Soc. Amer., **34**, 7, pp. 954-957

[Veletsos & Ventura, 1986], Veletsos A. S. ; Ventura C. E. ; "Modal analysis of non-classically damped linear systems", Earthquake Engng & Struct. Dyn., **14**, pp. 217-243

Standard dissipative systems and stability analysis

Quoc-Son Nguyen
Laboratoire de Mécanique des Solides, CNRS, Umr-7649
Ecole Polytechnique, 91128 - Palaiseau, France
Son@lms.polytechnique.fr

Abstract : Stability and bifurcation analyses are discussed here for time-independent standard dissipative systems of materials and structures. This discussion is illustrated by some simple applications in plasticity and fracture mechanics.
Keywords : generalized standard model, standard dissipative system, static or dynamic stability, bifurcation, stability criterion.

1. INTRODUCTION

It is well known that the concept of energy and dissipation potentials, cf. for example [Germain, 1973], [Halphen et al., 1975], [Lemaitre et al., 1985], [Maugin, 1992], leads to a general framework in the study of dissipative effects in materials and structures. This framework is considered again in order to derive an operational and general formulation of stability and bifurcation criteria in the stability analysis of equilibrium of a time-independent standard dissipative system.

It is recalled that in finite and isothermal deformation, a generalized standard material admits as state variables $q = (\nabla u, \alpha)$ and as energy density per unit volume $\mathcal{W}(\nabla u, \alpha)$. The dissipation is a product of force and flux: $d_{in} = (b - b_R) : \nabla \dot{u} + A \cdot \dot{\alpha}$, where b_R and A denote the associated forces $b_R = \mathcal{W}_{,\nabla u}$, $A = -\mathcal{W}_{,\alpha}$. Complementary laws must be introduced to relate force and flux. These laws are written in terms of a dissipation potential as $A = D_{,\dot{\alpha}}\ (\dot{\alpha}, \alpha)$. It has been assumed that dissipation potential $D = D(\dot{\alpha}, \alpha)$ is convex with respect to flux $\dot{\alpha}$ and may depend on the present state through the present value of α. The dual function of D, obtained by Legendre-Fenchel's transform as $D^*(A, \alpha) = \max_{\dot{\alpha}}\ A \cdot \dot{\alpha} - D(\dot{\alpha}, \alpha)$, permits an equivalent expression of complementary laws $\dot{\alpha} = D^*{}_{,A}\,(A, \alpha)$.

When the strain path $\nabla u(t)$ is given, the associated internal parameter can be obtained from its initial value by solving the system of equations

$$A = -\mathcal{W}_{,\alpha}, \quad \dot{\alpha} = D^*{}_{,A}\,(A, \alpha), \quad \alpha(0) = \alpha^0 \tag{1}$$

which can also be written as a differential equation, called Biot's equation cf. [Biot, 1965]

$$\mathcal{W}_{,\alpha} + D_{,\dot{\alpha}} = 0, \qquad \alpha(0) = \alpha^0. \tag{2}$$

G.A. Maugin et al. (eds.), Continuum Thermomechanics, 343-354.
© 2000 *Kluwer Academic Publishers.*

The assumption of dissipation potential can be expressed with any pair of associated force and flux. Indeed, the dissipation is

$$d = A \cdot \dot{\alpha} = G \cdot \gamma, \quad \gamma = S(\dot{\alpha}), \quad A = S^T(G)$$

where S denotes any state-dependent linear operator, γ is the transported flux resulting from this operator and G is the associated force. If there exists a dissipation potential $D(\gamma)$, depending on the present state such that $G = D_{,\gamma}$ then the function $\mathbf{D}(\dot{\alpha}) = D(\gamma) = D(S(\dot{\alpha}))$ is a dissipation potential in the sense that $A = \mathbf{D}_{,\dot{\alpha}}$. The convexity of $D(\gamma)$ is also equivalent to the convexity of $\mathbf{D}(\dot{\alpha})$ since convexity is conserved in a linear transformation. The dissipation potential is *a priori* state-dependent, the dependence on the present state α has been here omitted for the sake of clarity.

In particular, the notion of generalized standard materials is stable with respect to a change of variables. Indeed, a change of variables $\beta = \beta(\alpha)$ leads to new force $B = -\mathcal{W}_{,\beta} = A \cdot \alpha_{,\beta}$ while $\dot{\beta} = \beta_{,\alpha} \cdot \dot{\alpha}$. It is clear that $d = A \cdot \dot{\alpha} = B \cdot \dot{\beta}$ and all the ingredients of the model (energy potential, force, dissipation potential, convexity) remain valid.

For example, the model of plasticity with relaxed configuration, discussed by Lee and by Mandel, cf. [Mandel, 1971], is a generalized standard model defined by state variables (F, α, T) with $\alpha = P$. The energy density is $\mathcal{W}(F, P)$ in isothermal transformation. In this case, the rate $\dot{\alpha} = \dot{P}$ is associated with force $A = -\mathcal{W}_{,P}$ while the rate $\gamma = \dot{P}P^{-1}$ is associated with the stress $G = \Psi$. The existence of a convex function (dual dissipation potential) $D^*(\Psi)$ such that $\dot{P}P^{-1} = D^*_{,\Psi}$, is strictly equivalent to the existence of a (state-dependent) convex function $\mathbf{D}^*(A)$ such that $\dot{P} = \mathbf{D}^*_{,A}$. The introduction of suitable expressions of force and flux is only a matter of choice, principally motivated by physical considerations.

The generalized standard model can be extended to a mechanical system of solids if Biot's equation is the governing equation of the system in a quasi-static transformation

$$\mathbf{E}_{,q} + \mathbf{D}_{,\dot{q}} = 0, \qquad q(0) = q^0. \tag{3}$$

If $q = (u, \alpha)$ where u denotes displacement components and α the internal parameters, dynamic transformation of the system can be introduced with governing equation

$$J + \mathbf{E}_{,u} + \mathbf{D}_{,\dot{u}} = 0, \qquad \mathbf{E}_{,\alpha} + \mathbf{D}_{,\dot{\alpha}} = 0, \qquad q(0) = q^0, \quad \dot{u} = v^0 \tag{4}$$

where the generalized inertia force J depends linearly on $\ddot{u}$.

By definition, such a system is denoted as a **dissipative standard system**. *Dissipative standard systems are governed in quasi-static transformation by Biot's differential equation (3), and in dynamic transformation by the second order differential equation (4).*

If u is a reversible variable, then $\mathbf{D} = \mathbf{D}(\dot{\alpha}, q)$. In this case, the governing equations in quasi-static transformation can also be written as

$$\mathbf{E}_{,u} = 0, \quad \mathbf{E}_{,\alpha} + \mathbf{D}_{,\dot{\alpha}} = 0, \quad \alpha(0) = \alpha^0, \tag{5}$$

or, in an equivalent way as the system of equations

$$\mathbf{E}_{,u} = 0, \quad A = -\mathbf{E}_{,\alpha}, \quad \dot{\alpha} = \mathbf{D}^*_{,A}, \quad \alpha(0) = \alpha^0. \tag{6}$$

Equation $\mathbf{E}_{,u}(u, \alpha, \lambda) = 0$ gives an implicit representation $u = u(\alpha, \lambda)$ when the quadratic form $\mathbf{E}_{,uu}(u, \alpha, \lambda)[\delta u, \delta u]$ is positive-definite. In this case, let $\mathbf{E}^r$ be the reduced energy potential

$$\mathbf{E}^r(\alpha, \lambda) = \mathbf{E}(u(\alpha, \lambda), \alpha, \lambda). \tag{7}$$

The quasi-static evolution of the system is also described by the reduced equations of evolution

$$A = -\mathbf{E}^r{}_{,\alpha}, \quad \dot{\alpha} = \mathbf{D}^*{}_{,A}, \quad \alpha(0) = \alpha^0. \tag{8}$$

The abstract equations (3) or (4) can be broadly understood by following the nature of variables u and α. For discrete systems, these variables are vectors. But for continuous systems, they may be vector functions defined on a curve, a surface or a domain. In each case, it is sufficient to define the meaning of the differentiation operations and the associated duality of force and flux.

If energy and dissipation potentials $\mathbf{E}$, $\mathbf{D}$ are regular functions, a system of first-order differential equations is obtained for quasi-static transformation. For example, the study of the quasi-static evolution of a visco-elastic structure obeying a generalized standard model of visco-elasticity and subjected to implied forces and displacements can be given in this framework. Elastic visco-plastic or elastic-plastic materials however are associated with non-smooth dissipation potentials. In this case, the concept of sub-differential of a convex function can be introduced, cf. [Moreau, 1971], to generalize the operation of differentiation and to write the governing equations of materials and structures in the same framework.

2. TIME-INDEPENDENT STANDARD DISSIPATIVE SYSTEMS

2.1 Evolution equation

A non-viscous or time-independent behaviour arises when the dissipation potential is positively homogeneous of degree 1 with respect to the flux:

$$\mathbf{D}(m\, \dot{\alpha}, \alpha) = m\, \mathbf{D}(\dot{\alpha}, \alpha) \quad \forall\ m > 0. \tag{9}$$

Such a function is not differentiable at point $\dot{\alpha} = 0$, but is sub-differentiable. The set $\mathbf{C}$ of sub-gradients at this point

$$\mathbf{C} = \partial_{\dot{\alpha}}\mathbf{D}(0, \alpha) = \{\ A^* \ \mid\ A^* \cdot \delta\alpha \leq \mathbf{D}(\delta\alpha, \alpha)\ \ \forall\, \delta\alpha\ \} \tag{10}$$

is a convex domain of admissible forces. The dual dissipation potential is in this case the indicator function of the convex domain of admissible forces. Force-flux relation $\dot{\alpha} = \mathbf{D}^*{}_{,A}$ can be written under the form of the normality law

$$\dot{\alpha} = \mathbf{N}_{\mathbf{C}}(A) \tag{11}$$

which states that rate $\dot{\alpha}$ must be an external normal to the admissible domain at the present state of force A. It is well known that this evolution law can also be equivalent to the maximum dissipation principle which is classical in plasticity under the name of the principle of maximum plastic work

$$\mathbf{D}(\dot{\alpha}, \alpha) = \max_{A^* \in C}\ A^* \cdot \dot{\alpha}. \tag{12}$$

Governing equations (1) can be written as

$$\mathbf{E}_{,u} = 0, \quad A = -\mathbf{E}_{,\alpha}, \quad \dot{\alpha} = \mathbf{N}_{,\mathbf{C}}(A), \quad \alpha(0) = \alpha^0 \tag{13}$$

or in an equivalent form

$$\mathbf{E}_{,u} \cdot (\delta u - \dot{u}) + \mathbf{E}_{,\alpha} \cdot (\delta\alpha - \dot{\alpha}) + \mathbf{D}(\delta\alpha, \alpha) - \mathbf{D}(\dot{\alpha}, \alpha) \geq 0 \quad \forall \quad (\delta u, \delta\alpha). \tag{14}$$

2.2 Rate problem

Let $\gamma(t)$ and $G(t)$ be two histories of flux and force associated by the normality law. Then, flux $\gamma(t)$ and force rate $\dot{G}(t)$ are related by the following proposition

Proposition 1 *Let $\gamma(t)$ and $G(t)$ denote a flux and force associated by the normality law with a convex* $\mathbf{C}$ *of non void interior, depending on a given function $y(t)$ for $t \in [0,T]$. If $\dot{y}(t), \gamma(t)$ and $\dot{G}(t)$ are piecewise continuous, then the following expressions hold for right-hand-side (r.h.s.) derivatives*

$$-\dot{G} \cdot \gamma + \dot{y} \cdot \mathbf{D}_{,y}\ (\gamma, y) = 0, \tag{15}$$

$$-\dot{G} \cdot \gamma^* + \dot{y} \cdot \mathbf{D}_{,y}\ (\gamma^*, y) \geq 0 \quad \forall \ \gamma^* \ \textit{admissible}. \tag{16}$$

By definition, a rate γ^* is admissible if $\gamma^* \in \mathbf{N}_{\mathbf{C}(y)}$.

The proof of this proposition follows simply from the maximum dissipation principle (12) and can be found in [Nguyen, 2000]. As a consequence of the proposition, it should be noted that, if the assumption of state-independence is satisfied, i.e. if the dissipation potential does not depend on the present state, then the r.h.s. rates satisfy $\dot{G} \cdot \gamma = 0$. This orthogonality property is classical in perfect plasticity and gives $\dot{\sigma} : \dot{\varepsilon}^p = 0$.

Relations (15), (16), written for $y(t) = q(t)$, together with equations $A = -\mathbf{E}_{,\alpha}$, $\mathbf{E}_{,u} = 0$, lead to the following description of the rate problem which consists in obtaining the rate response of the system $\dot{q}$ as a function of rate data $\dot{\lambda}$ when the present state is assumed to be known:

Proposition 2 *The rate response $\dot{q}$ is a solution of the variational inequality*

$$(\mathbf{E}_{,qq} \cdot \dot{q} + \mathbf{E}_{,q\lambda} \cdot \dot{\lambda}\ ) \cdot (\delta q - \dot{q}) + \dot{q} \cdot (\mathbf{D}_{,q}(\delta q, q) - \mathbf{D}_{,q}(\dot{q}, q)) \geq 0 \tag{17}$$

$$\forall \ \delta q = (\delta u, \delta\alpha) \ \textit{admissible}.$$

If D depends on the present state, this variational inequality is symmetric if

$$\dot{q} \cdot \mathbf{D}_{,q}(q^*, q) = q^* \cdot \mathbf{D}_{,q}(\dot{q}, q) \quad \forall \text{ admissible rates } \dot{q}, \ q^*. \tag{18}$$

In this case, the obtained variational inequality is also quadratic. The uniqueness of the rate response $\dot{q}$ is ensured if the following positivity is ensured for all admissible rates $\delta q_1 \neq \delta q_2$

$$(\delta q_1 - \delta q_2) \cdot \mathbf{E}_{,qq} \cdot (\delta q_1 - \delta q_2) + (\delta q_1 - \delta q_2) \cdot (\mathbf{D}_{,q}(\delta q_1, q) - \mathbf{D}_{,q}(\delta q_2, q)) > 0. \tag{19}$$

The following proposition holds

Proposition 3 *Under the assumptions of symmetry (18) and of positivity (19), the rate solution $\dot{q}$ minimizes among all admissible rates the rate functional*

$$\mathbf{H}(q^*) = \frac{1}{2}\ (\ q^* \cdot \mathbf{E}_{,qq} \cdot q^* + q^* \cdot \mathbf{D}_{,q}\,(q^*, q)\ ) + \dot{\lambda}\ \mathbf{E}_{,q\lambda} \cdot q^*. \tag{20}$$

Indeed, for all admissible rate q^*

$$\mathbf{H}(q^*) - \mathbf{H}(\dot{q}) = \frac{1}{2}\{(q^* - \dot{q}) \cdot \mathbf{E}_{,qq} \cdot (q^* - \dot{q}) + (q^* - \dot{q}) \cdot (\mathbf{D}_{,q}\,(q^*, q) - \mathbf{D}_{,q}\,(\dot{q}, q))\} +$$

$$(\mathbf{E}_{,qq} \cdot \dot{q} + \mathbf{E}_{,q\lambda} \cdot \dot{\lambda}\ ) \cdot (q^* - \dot{q}) + \dot{q} \cdot (\mathbf{D}_{,q}\,(q^*, q) - \mathbf{D}_{,q}\,(\dot{q}, q)) \geq 0$$

after (17) and (19). This functional is an extension of Hill's functional to standard dissipative systems. It is difficult however to eliminate the internal parameters as a function of displacement components in order to obtain a rate functional of the displacement rates. The existence of a solution $\dot{q}$ is ensured for all $\dot{\lambda}$ under the condition

$$\mathbf{E}_{,qq}\,[q^*, q^*] + q^* \cdot \mathbf{D}_{,q}\,(q^*, q) > 0 \ \ \forall\, q^* \ \text{ admissible } \neq 0. \tag{21}$$

3. STABILITY AND BIFURCATION ANALYSIS

3.1 Stability criterion

Condition (21) can be interpreted as a criterion of stability of the current equilibrium in a certain sense, cf. [Petryk, 1985], in relation to the notion of static or directional or dynamic stability. Indeed, in a perturbation of the system out of equilibrium, the energy injected by the external world in a time interval $[0, \tau]$ is

$$\mathbf{W}_{per}(\tau) = \mathbf{W}^s(\tau) + C_\tau, \quad \mathbf{W}^s(\tau) = [\mathbf{E}(q, \lambda)]_0^\tau + \int_0^\tau \mathbf{D}(\dot{q}, q)\ dt$$

where $C_t \geq 0$ denotes the kinetic energy of the system. Thus if τ is sufficiently small, the expansion

$$\mathbf{W}^s = W_2^s\ \frac{\tau^2}{2} + \ o(\tau^2), \qquad W_2^s = q_1 \cdot \mathbf{E}_{,qq} \cdot q_1 + q_1 \cdot \mathbf{D}_{,q}\,(q_1, q)$$

can be associated with the expansion $q_\tau = q + q_1\ \tau + q_2\ \frac{\tau^2}{2} + ...$ It follows from condition (19) that it is necessary to inject energy initially into the system in order to remove it from the considered equilibrium, i.e. static stability as well as directional stability, are obtained. It is expected that the symmetry condition (18) is necessary to interpret this criterion as a sufficient condition of dynamic stability of the considered equilibrium.

In some particular cases, the dissipated energy may be a function of the present state if there exists a function $\mathbf{W}^d(q)$ such that

$$\int_0^t \mathbf{D}(\dot{q}, q)\ dt = \mathbf{W}^d(q(t)) - \mathbf{W}^d(q(0)). \tag{22}$$

These systems, denoted in [Ehrlacher,1985] as *simple dissipative systems*, are frequently found in fracture and damage mechanics. Such a system is almost a conservative system in the sense that a total potential energy $\mathbf{E}^a$ exists:

$$\mathbf{E}^a(q,\lambda) = \mathbf{E}(q,\lambda) + \mathbf{W}^d(q),$$

but the rate $\dot{q}$ must be admissible. For simple dissipative systems, condition (21) can be written as

$$\delta^2\mathbf{E}^a = \mathbf{E}^a{}_{,qq}\,[\delta q, \delta q] > 0 \quad \forall \quad \delta q \ \text{ admissible.} \tag{23}$$

The following proposition holds, cf. [Nguyen, 2000]

Proposition 4 *Condition (21) is a static or directional criterion of stability. This criterion ensures also the dynamic stability of the considered equilibrium if the assumption of state-independent potential is satisfied or if the system is a simple dissipative system.*

3.2 Non-bifurcation criterion

The non-uniqueness of the rate response indicates a critical point and eventually a bifurcation point. This idea leads to Hill's criterion of non-bifurcation. This criterion is available in the study of angular bifurcation as well as of tangent bifurcation.

Proposition 5 *Condition (19) is a non-bifurcation criterion in the sense of Hill.*

From the definition of admissible rates, the linear space $V(A)$ generated by the external normals to the convex of admissible forces $\mathbf{C}$ at the present value A can be introduced. Since $\delta\alpha_1 - \delta\alpha_2 \in V(A)$ if the rates $\delta\alpha_1, \delta\alpha_2$ are admissible, the non-bifurcation condition can also be written as

$$\mathbf{E}_{,qq}\,[\delta q, \delta q] + \delta q \cdot \mathbf{D}_{,q}\,(\delta q, q) > 0 \quad \forall \ \delta q = (\delta u, \delta\alpha), \ \ \delta\alpha \ \in \ V(A). \tag{24}$$

4. ILLUSTRATION IN PLASTICITY

The case of a generalized standard elastic-plastic material admitting as elastic domain a non-smooth convex, defined by several inequalities:

$$f^i(A,\alpha) \leq 0, \qquad i = 1, N, \tag{25}$$

is considered in order to illustrate the symmetry condition (18). Such a model is necessary in the study of monocrystals for example, where plastic strains arise from different gliding mechanisms obeying Schmid's law. Functions f^i are classically denoted as plastic potentials. The normality law states that

$$\dot{\alpha} = \mu^i f^i{}_{,A} \quad \text{with} \quad \mu^i \geq 0, \ \ \mu^i f^i = 0, \ \ f^i \leq 0. \tag{26}$$

The rate equations can be written after the computation of the plastic multipliers μ_i in terms of $\nabla\dot{u}$. For this, it can be noted that

$$\mu^i \geq 0, \quad (m^i - \mu^i)\, \dot{f}^i \geq 0 \ \ \forall \ m^i \geq 0.$$

It follows that μ must satisfy

$$(m^i - \mu^i)\,(C_{ij}\mu^j + B_i : \nabla\dot{u}) \geq 0, \quad \forall\, m \in M, \quad \text{with} \tag{27}$$

$$M = \{\ m = (m^1, ...m^N)\ \mid\ m^i \geq 0,\ \ m^i\, f^i = 0\ \ (\text{no sum})\ \}$$

$$C_{ij} = f^i{}_{,A} \cdot \mathcal{W}_{,\alpha\alpha} \cdot f^j{}_{,A} - f^i{}_{,A}\, f^j{}_{,\alpha} \quad , \quad B_j = f^i{}_{,A} \cdot \mathcal{W}_{,\alpha\nabla u}\,. \tag{28}$$

This variational inequality gives μ^i in terms of $\nabla\dot{u}$ in a unique manner if the matrix C is positive-definite. However the explicit expressions of these relations cannot be derived when $N > 1$. Matrix C is not necessarily symmetric; its symmetry is ensured only if the interaction matrix $\mathcal{H}$ defined by

$$\mathcal{H}_{ij} = f^i{}_{,A}\, f^j{}_{,\alpha} \tag{29}$$

is symmetric. This matrix satisfies the relation

$$\delta\alpha \cdot D_{,\alpha}(\dot{\alpha}, \alpha) = -\ m^j \mathcal{H}_{ji}\, \mu^i\ \ \forall\ \ \delta\alpha = m^j f^j{}_{,A}\,,\ \ \dot{\alpha} = \mu^j f^j{}_{,A}\,. \tag{30}$$

Indeed, the expression of the dissipation potential

$$D(\dot{\alpha}, \alpha) = A \cdot\ \dot{\alpha} -\ \mu^i f^i(A, \alpha),$$

gives after differentiation in the direction $\delta\alpha = m^j f^j{}_{,A}$

$$\delta\alpha \cdot D_{,\alpha}(\dot{\alpha}, \alpha) = \delta A \cdot\ (\dot{\alpha} - \mu^i f^i{}_{,A}) - \mu^i f^i{}_{,\alpha} \cdot \delta\alpha - \delta\mu^i f^i.$$

Thus

$$\delta\alpha \cdot D_{,\alpha}(\dot{\alpha}, \alpha) = -\ \mu^i f^i{}_{,\alpha} \cdot \delta\alpha = -\ m^j H_{ji}\, \mu^i.$$

The symmetry of the interaction matrix is exactly the symmetry (18). If I denotes the set of index of active mechanisms, i.e. $f^i = 0,\quad \dot{f}^i = 0$ for $i \in I$, the plastic modulus $\mathcal{L}^p_I$ associated with these active mechanisms can be computed. Indeed since

$$C_{ij}\mu^j + B_i : \nabla\dot{u} = 0 \quad i, j\ \in\ I$$

it follows that $\dot{b} = \mathcal{L}^p : \nabla\dot{u}$ with

$$L^p_I = \mathcal{W}_{,\nabla u \nabla u} - B^T_I\ C^{-1}_I\ B_I$$

where C_I and B_I are sub-matrices of C, B and related to active index I. It is concluded that the plastic modulus admits the major symmetry if and only if the interaction matrix $\mathcal{H}$ is symmetric.

5. CRACK PROPAGATION, STABILITY AND BIFURCATION

The analyses of crack nucleation, crack propagation and crack stability are the objective of fracture mechanics. In brittle fracture, the stability of a Griffith crack has been considered in many discussions. Its generalization to study the propagation of a system of interacting linear cracks or of a plane crack of arbitrary shape in an elastic solid is relatively straightforward as a particular example of standard dissipative systems.

5.1 System of interacting linear cracks

We consider the equilibrium problem of a solid in two-dimensional deformation, i.e. plane strain or plane stress, admitting in its volume V a system of linear cracks of lengths $\ell_i, i = 1, n$, undergoing small transformation under the action of implied forces and implied displacements defined by a load parameter λ. To simplify, it is assumed that surface forces $r^d(\lambda)$ are applied on the portion S_r of the boundary and on the complemenntary part S_u, displacements $u^d(\lambda)$ are implied. If the solid is elastic, the response of the system is reversible when there is no crack propagation and irreversible when the crack lengths change. Variables ℓ_i thus describe the irreversible behaviour of the system and represent the internal variables α. The set of admissible displacements of the solid depends on the present state of cracks, and can be written as

$$U(\ell, \lambda) = \{u \mid u = u^d \text{ on } S_u,\ [u_n] \geq 0 \text{ on } \Gamma_i\} \tag{31}$$

where Γ_i denotes the crack surfaces. If $W(\epsilon)$ is the elastic energy density, the total potential energy of the system is

$$\mathbf{E}(u, \ell, \lambda) = \int_{V_\ell} W(\epsilon(u))\, dV - \int_{S_r} r^d \cdot u\, dS. \tag{32}$$

The displacement at equilibrium must satisfy the virtual work equation which can be written in the form of a variational inequality in order to take into account the possibility of unilateral contact, assumed to be frictionless, on crack surfaces:

$$\mathbf{E}_{,u}(u, \ell, \lambda) \cdot \delta u \geq 0 \quad \forall\, \delta u = v - u,\ v \in U(\ell, \lambda).$$

If the energy is strictly convex, the equilibrium displacement must minimise the total potential energy of the system, and permits the introduction of the energy at equilibrium $\mathbf{E}^r(\ell, \lambda)$:

$$\mathbf{E}^r(\ell, \lambda) = \mathbf{E}(u(\ell, \lambda), \ell, \lambda) = \min_{u^* \in U(\ell,\lambda)} \mathbf{E}(u^*, \ell, \lambda). \tag{33}$$

The associated generalized force $G_i = -\mathbf{E}^r{}_{,\ell_i}$ is by definition the energy release rate associated with crack length ℓ_i. Griffith's law states that

$$\begin{cases} \text{If } G_i < G_c \text{ then } \dot{\ell}_i = 0 & \text{(no propagation)} \\ \text{If } G_i = G_c \text{ then } \dot{\ell}_i \geq 0 & \text{(possible propagation).} \end{cases}$$

The critical surface energy G_c is often considered as a constant of the material. In order to interpret the resistance effects due to the presence of plastic strains, it has been also assumed in certain appplications that the value G_c can depend on the effective length of propagation $\Delta\ell_i = \ell_i - \ell_i^0$; function $G_c = G_c(\Delta\ell_i)$ describes the resistance curve. Such a criterion introduces a domain of admissible forces:

$$\mathbf{C} = \{G \mid G_i - G_c(\Delta\ell_i) \leq 0 \quad i = 1, n\}$$

and a dissipation potential $\mathbf{D}(\dot{\ell},\ell) = G_c(\Delta\ell_i)\dot{\ell}_i$ when $\dot{\ell}_i \geq 0$. The energy dissipated by crack propagation $\mathbf{W}^d(\ell) = \sum_i \int_{\ell_i^0}^{\ell_i} G_c(\Delta\ell_i)\, d\ell_i$ depends only on the present state of cracks. A system of simple dissipation is thus obtained with total energy $\mathbf{E}^{ra}(\ell,\lambda) = \mathbf{E}^r(\ell,\lambda) + \mathbf{W}^d(\ell)$.

In order to obtain the description of the rate problem at a given state, let I be the set of indices i such that the propagation limit is reached: $G_i = G_c(\Delta\ell_i)$. Admissible rate $\delta\ell$ must satisfy $\delta\ell_i \geq 0$ for all $i \in I$. The previous discussion leads to the following statements:

- Propagation rate $\dot{\ell}$ is a solution of the quadratic and symmetric variational inequality

$$(\mathcal{A}_{ij}\cdot\dot{\ell}_j + \mathbf{E}^r{}_{,\lambda i}\cdot\dot{\lambda})\ (\delta\ell_i - \dot{\ell}_i) \geq 0 \quad \forall\ \delta\ell \ \text{ admissible.}$$

where $\mathcal{A}$ denotes the symmetric matrix

$$\mathcal{A}_{ij} = \mathbf{E}^r{}_{,ij} + \frac{dG_c}{d\ell}(\ell_i - \ell_i^o)\ \delta_{ij} \quad \text{with} \quad \mathbf{E}^r{}_{,ij} = \frac{\partial^2\mathbf{E}^r}{\partial\ell_i\partial\ell_j}, \quad i,j \in I. \tag{34}$$

This problem can also be written as a linear complementarity problem (cf. [Cottle et al., 1992]) which consists of finding X such that

$$Y = \mathcal{A}X - F, \quad Y_i \geq 0, \ \ X_i \geq 0, \ \ X_iY_i = 0 \quad \text{(no sum)} \tag{35}$$

where X, F are respectively vectors of components $\dot{\ell}_i$ and $\dot{\lambda}\mathbf{E}^r{}_{,\lambda i}$.

- The present equilibrium state is stable with respect to crack propagation and to displacement in the dynamic sense if matrix $\mathcal{A}$ satisfies the co-positivity condition:

$$\delta\ell\cdot\mathcal{A}\cdot\delta\ell > 0 \quad \forall\ \delta\ell \neq 0, \ \ \delta\ell \geq 0. \tag{36}$$

This condition can also be written as

$$(\delta G_c - \delta G)\cdot\delta\ell > 0 \quad \forall\ \delta\ell \ \neq 0, \ \ \delta\ell \geq 0. \tag{37}$$

- The present equilibrium state is not a bifurcation state if the matrix $\mathcal{A}$ is positive-definite. The stability criterion is less restrictive than the non-bifurcation criterion since positive-definiteness is more restrictive than co-positivity.

The computation of matrix $\mathcal{A}$ or of the second derivatives of energy has been discussed in [Nguyen et al., 1990], [Suo et al., 1992]. The difference between co-positivity and positive-definiteness has been illustrated in many simple analytical examples, cf. [Nguyen, 2000].

5.2 Stability and configurational stability of plane cracks

This problem is of interest in various applications, for example in the debonding of interface cracks, of surface coating by thin films, cf. [Berest, 1989], [Hutchinson et al., 1992] [Jensen, 1995], in the delamination of multi-layer composites, cf. [Cochelin, 1994], [Destuynder, 1987], [Pradeilles-Duval,1992], [Storakers,1988]. Moreover, the mechanical modelling of brittle damage also leads to the study of the propagation of damage zones in an elastic solid, a problem of the same mathematical nature, cf. for example [Dems et al., 1985], [Stolz, 1987].

For a plane crack, the crack surface is a plane domain Ω of contour Λ; this domain represents the irreversible variable α. The displacement at equilibrium $u(\Omega, \lambda)$ leads again to the energy of the system at equilibrium $\mathbf{E}^r(\Omega, \lambda)$. The generalized force associated with the irreversible variable is defined from the partial derivative of energy at equilibrium with respect to Ω. In order to compute this partial derivative, a rate of variation of the boundary of domain Ω can be described by the rate of normal extension. This normal rate, which is a scalar function defined on the present contour Λ, is denoted as $\dot{\Omega}(s)$. Since the crack surface can only increase, it follows that

$$\dot{\Omega} \text{ admissible } \Leftrightarrow \dot{\Omega}(s) \geq 0 \quad \forall \ s \in \Lambda. \tag{38}$$

Let $\mathbf{E}^r{}_{,\Omega} \cdot \partial\Omega$ be the directional derivative of energy with respect to domain Ω in the direction $\partial\Omega$. This directional derivative is a linear form for plane cracks, and can be expressed as

$$\mathbf{E}^r{}_{,\Omega} \cdot \partial\Omega = -\mathbf{G} \cdot \partial\Omega = -\int_\Lambda G(s)\ \partial\Omega(s)\ ds \tag{39}$$

where $G(s)$ is a function defined on the present contour Λ. By definition, the local value $G(s)$ is denoted as the local energy release rate, and the associated generalized force to the crack extension or the crack driving force $\mathbf{G} = -\mathbf{E}^r{}_{,\Omega}$ is function $\mathbf{G}$.

The computation of the crack driving force $\mathbf{G}$ has been discussed for different particular cases. For example, for a plane crack in a three-dimensional solid, $G(s)$ is still given by the limit value of the J–integral. The dissipation of the whole system due to crack extension is

$$D_{in} = \mathbf{G} \cdot \dot{\Omega} = \int_\Lambda G(s)\dot{\Omega}(s)\ ds. \tag{40}$$

In particular, if G_c is a constant, the dissipated energy by crack propagation is proportional to the cracked surface. The system of solid with crack is then an irreversible system of simple dissipation, of total energy $\mathbf{E}^{ra}(\Omega, \lambda) = \mathbf{E}^r(\Omega, \lambda) + \mathbf{W}^d(\Omega)$, and leads to the variational inequality

$$\int_{\Lambda_c} (\dot{G} - \dot{G}_c)\ (\partial\Omega - \dot{\Omega})\ ds \leq 0 \quad \forall\ \partial\Omega \geq 0 \ \text{ on } \ \Lambda_c, \tag{41}$$

where Λ_c is the portion at yield of the contour. The rate problem is thus described by the variational inequality

$$(\partial\Omega - \dot{\Omega}) \cdot (\mathbf{E}^r{}_{,\Omega\Omega} \cdot \dot{\Omega} + \mathbf{E}^r{}_{,\Omega\lambda} \cdot \dot{\lambda}) + \dot{\Omega} \cdot (\mathbf{D}_{,\Omega}(\partial\Omega, \Omega) - \mathbf{D}_{,\Omega}(\dot{\Omega}, \Omega)) \geq 0 \tag{42}$$

for all admissible $\partial\Omega$, i.e. satisfying $\partial\Omega \geq 0$ on Λ_c. The general form of the rate problem is then recovered. In particular, the stability criterion follows:

$$\int_{\Lambda_c} (\partial G_c - \partial G)\ \partial\Omega\ ds > 0 \quad \forall\ \partial\Omega \neq 0 \text{ admissible.} \tag{43}$$

In particular, if G_c is constant, the stability criterion is reduced to the co-positivity condition

$$\partial\Omega \cdot \mathbf{E}^{ra}{}_{,\Omega\Omega} \cdot \partial\Omega > 0 \quad \forall\ \partial\Omega \neq 0 \ \text{ admissible,} \tag{44}$$

while the non-bifurcation criterion requires the positive-definiteness of the same quadratic form.

The principal difficulty stems from the calculation of the rate ∂G following the crack motion. Several analytical discussions have been given recently in the literature for the circular crack problem, cf. [Gao et al., 1987], [Berest, 1987] or of the tunnel crack problem cf. [Leblond et al., 1996], [Jensen et al., 1995]. For example, a circular crack may be stable in displacement control, but stable bifurcation in a star-shaped mode can be observed. The fact that the circular form may be lost is also known as a configurational instability in the literature. For the tunnel crack, a bifurcated mode to a wavy form has been also computed.

6. CONCLUDING REMARKS

For standard systems satisfying the symmetry condition, stability and bifurcation analysis leads to a general expression of stability and non-bifurcation criteria in terms of energy and dissipation potentials. This is an extension of the classical second variation criterion to dissipative systems.

The fact that stability and bifurcation analysis can be discussed in a satisfactory manner in fracture mechanics and in plasticity is due principally to the symmetry of the rate problem which permits a static analysis. For unsymmetric systems, for example in the context of unilateral contact with friction, it is clear that a static analysis presents less interest since stability must be considered in dynamics, in the same spirit as for non-conservative systems, cf. for example [Nguyen, 2000] for a review on the subject.

REFERENCES

[Bazant et al., 1991] Bazant, Z.; Cedolin, L.; *Stability of Structures: Elastic, Plastic, Fracture and Damage Theories*; Oxford University Press.

[Berest, 1989] Bérest, P. *Problèmes de Mécanique Associés au Stockage Souterrain*, Thèse de l'Ecole des Mines.

[Biot, 1965] Biot, M.A. *Mechanics of Incremental Deformation*, Wiley, New York.

[Cochelin et al.] Cochelin, B. ; Potier-Ferry, M.; Interaction entre les flambages locaux et la propagation du délaminage dans les composites. *Calcul des structures et intelligences artificielles*. Pluralis, Paris.

[Cottle et al] Cottle, R.; Pang, J.S.; Stone, R.E.; *The linear complementary problem*; Academic Press.

[Dems, 1985] Dems, K; Mroz, Z.; Stability condition for brittle plastic structures with propagating damage surfaces. *J. Struc. Mech.*, 13:95-122.

[Destuynder et al., 1987] Destuynder, P.; Nevers, T.; Un modèle de calcul des forces de délaminage dans les plaques minces. *J. Mec. Th. Appl.*; 6:179-207.

[Ehrlacher, 1985] Ehrlacher, A.; Contribution à l'étude thermodynamique de la progression de fissure et à la mécanique de l'endommagement brutal. *Thèse d'Etat*, Paris.

[Gao et al., 1987] Gao, H.J.; Rice, J.; Somewhat circular tensile crack. *Int. J. Fracture*; 33:155-174.

[Germain, 1973] Germain, P.; *Cours de Mécanique des Milieux Continus*; Masson, Paris.

[Halphen et al., 1975] Halphen, B; Nguyen, Q.S.; Sur les matériaux standard généralisés. *J. Mécanique*; 14:1-37.

[Hill, 1958] Hill, R.; A general theory of uniqueness and stability in elastic/plastic solid; *J. Mech. Phys. Solids*; 6:236-249.

[Hutchinson et al., 1992] Hutchinson, JW.; Thouless, M.D.; Liniger, E.G.; Growth and configurational stability of circular buckling driven thin film delamination. *Act. Metal. Mat.*; 40:455-466.

[Jensen et al., 1995] Jensen, H.M.; Thouless, M.D.; Buckling instability of stable edge cracks. *J. Appl. Mech.*, 62:620-625.

[Leblond et al., 1996] Leblond, J.B.; Mouchrif, S.E.; Perrin, G.; The tensile tunnel crack with a slightly wavy form. *Int. J. Solids Structures*; 33:1995-2022.

[Lemaitre et al., 1985] Lemaitre, J.; Chaboche, J.L.; *Mécanique des Matériaux Solides.* Dunod, Paris.

[Mandel, 1971] Mandel, J.; *Plasticité Classique et Visco-plasticité.* CISM, Springer-Verlag, Wien.

[Maugin, 1992] Maugin, G.A.; *The thermomechnics of plasticity and fracture*; Cambridge University Press.

[Moreau, 1974] Moreau, J.J.; *On unilateral constraints, friction and plasticity*; CIME, Springer-Verlag, Wien.

[Nguyen, 1994] Nguyen, Q.S.; Bifurcation and stability in dissipative media (plasticity, fracture, friction); *Appl. Mech. Rev.*, 47:1-31.

[Nguyen et al., 1990] Nguyen, Q.S.; Stolz, C.; Debruyne, G.; Energy methods in fracture mechanics: stability, bifurcation and second variation; *Eur. J. Mech.*;9:157-173.

[Nguyen, 2000] Nguyen, Q.S. *Stabilité et Mécanique Non Linéaire*, Hermes, Paris.

[Petryk, 1985] Petryk, H.; On energy criteria of plastic instability. *Plastic instability, Considère mem.*; ENPC-Presses, Paris.

[Pradeilles-Duval, 1992] Pradeilles-Duval, R.M.; Evolution des systèmes avec frontières de discontinuité mobiles. *Thèse*, Ecole Polytechnique; Paris.

[Storakers, 1992] Storakers, B.; Nonlinear aspects of delamination in structural members. *Proceedings ICTAM*, Springer-Verlag, Berlin.

[Suo et al, 1992] Suo, X.Z.; Combescure, A.; Second variation of energy and an associated line-independent integral in fracture mechanics. *Eur. J. Mech.*, 11:609-624.

Enriched damage models for continuum failure analyses

Gilles Pijaudier-Cabot and René de Borst***
**Laboratoire de Génie Civil de Nantes Saint-Nazaire*
Ecole Centrale de Nantes, BP 92101, 44321 Nantes Cedex 03, France
Email: Gilles.Pijaudier-Cabot@ec-nantes.fr
***Koiter Institute, Delft University of Technology*
P.O. Box 5058, 2600 GB Delft, The Netherlands

Abstract: Continuum damage mechanics is a framework for describing the variations of the elastic properties of a material due to microstructural degradations. In this contribution two issues are reviewed: the first one deals with standard damage models, damage induced anisotropy and relations with other existing approaches such as smeared crack models. The second issue is concerned with strain localisation due to softening, which is inherent to continuum damage modelling of quasi-brittle materials. Nonlocal approaches to damage are reviewed. We conclude with possible extensions of these enriched damage models to the case of damage induced anisotropy.
Keywords: Damage, isotropy, anisotropy, strain localisation, nonlocality.

1 INTRODUCTION

Computational modelling and failure analysis of quasi-brittle materials and structures require adequate constitutive models coupled to robust computational schemes. Some of the issues which received most of the attention in this field over the past two decades are progressive cracking and strain localisation due to the inherent softening of the material. With respect to the first item, continuous damage models have become among the most popular ones. Isotropic and anisotropic formulations were proposed. Whether they derive from smeared crack approaches including the rotating crack models [see e.g. Feenstra, 1993], microplane models [Bazant and Prat, 1988], or pure phenomenological constitutive relations [Mazars 1984, Lemaitre 1992], the basic principle remains the same: the compliance of the material can be seen as an internal variable, in the thermodynamic sense, which is indexed on the state of microcracking of the material.

The issue of strain localisation due to strain softening is a bit more intricate since it reflects on the ability of models to capture the inception of failure and its propagation when solving boundary value problems. It became apparent at the end of the 70's that strain softening causes a loss of well posedness of incremental boundary value problems and, more importantly, that it yields prediction of failure without energy dissipation [Bazant, 1976, Rice 1976]. The remedy to this physically unrealistic feature was found

G.A. Maugin et al. (eds.), Continuum Thermomechanics, 355-366.
© 2000 *Kluwer Academic Publishers.*

in the early 90's, surprisingly inspired by earlier theories of materials with microstructure to situations where the material behaviour is not reversible anymore. Among such models, called enriched models for failure analysis, are the nonlocal damage model [Pijaudier-Cabot and Bazant, 1987], gradient plasticity approaches [de Borst and Muhlhaus, 1992], Cosserat models [de Borst, 1991] and more recently gradient damage models [Peerlings et al., 1996]. Their common feature is the incorporation in the constitutive relations of an internal length which controls the failure process and thus precludes any dissipative process to occur in a region of the solid of zero volume.

It is also interesting to observe that, in fluid mechanics, similar issues were faced with transonic flow and boundary layer problems [Joseph et al. 1985]. The most popular techniques for circumventing those problems were the introduction of a characteristic time in the problem, for instance by considering that the fluid is viscous. A small amount of viscosity (damping) provides a regularisation of the governing equations. Similar solutions exist in solid mechanics. Needleman [1988] and Sluys [1992] demonstrated that material rate dependency is a proper regularisation method, i.e. that it restores well posedness of the BVP at the inception of softening. Although the technique is very attractive, meaning that physical motivations for rate dependency might be easier to provide than for nonlocality of constitutive relations, some inconsistencies remain because the response of a quasi brittle material is slightly more complex compared to simple fluids: rate dependency provides for instance an evolution of the material strength for homogeneously strained specimens at different loading rates. At the same time, it provides also a *width* of the fracture process zone and eventually controls the fracture energy of the material. The fracture energy and the evolution of the material strength with the loading rate are two characteristics which should be fitted with a single parameter (the viscosity), which is not always in accordance with experimental data.

In this chapter, we will concentrate on rate independent modelling of damage and on two regularisation techniques: the nonlocal and gradient approaches. In the first part, we will review standard damage models, isotropic and anisotropic. The second part will be devoted to the integral and gradient enrichments of such models.

2 STANDARD DAMAGE MODELS

Let us consider, for the sake of simplicity, the case of tension dominated mechanical actions. The development of microcracks results in a progressive degradation of the material elastic stiffness. In the reversible (elastic) domain, the stress-strain relation reads:

$$\sigma_{ij} = C_{ijkl}^{damaged} \varepsilon_{kl}, \tag{1}$$

where σ_{ij} is the stress component, ε_{kl} is the strain component, and $C_{ijkl}^{damaged}$ is the stiffness coefficient of the damaged material. The simplest approach to material damage is to assume that the material stiffness (for isotropic materials) remains isotropic. The stress strain relations becomes :

$$\varepsilon_{ij} = \frac{3}{2E_0(1-d)}(\sigma_{ij} - \frac{\sigma_{kk}}{3}\delta_{ij}) + \frac{(1-2\nu_0)}{2E_0(1-D)}[\sigma_{kk}\delta_{ij} - \sigma_{ij}], \tag{2}$$

where E_0 and ν_0 are the Young's modulus and Poisson's ratio of the undamaged isotropic material, and δ_{ij} is the kronecker symbol. d and D are two independent damage variables which vary between 0 and 1. It should be pointed out that isotropic damage means indeed two damage state variables. The subsequent assumption $d = D$ yields the stress-strain relationship used by Mazars (1984):

$$\varepsilon_{ij} = \frac{1+\nu_0}{E_0(1-d)}\sigma_{ij} - \frac{\nu_0}{E_0(1-d)}[\sigma_{kk}\delta_{ij}], \tag{3}$$

or

$$\sigma_{ij} = (1-d)C^0_{ijkl}\varepsilon_{kl}, \tag{4}$$

where C^0_{ijkl} is the stiffness of the undamaged material. According to this assumption, the Poisson's ratio is not affected by damage. The elastic (i.e. free) energy per unit mass of material is:

$$\rho\psi = \frac{1}{2}(1-d)\varepsilon_{ij}C^0_{ijkl}\varepsilon_{kl}. \tag{5}$$

This energy is assumed to be the state potential. The damage energy release rate is:

$$Y = -\rho\frac{\partial\psi}{\partial d} = \frac{1}{2}\varepsilon_{ij}C^0_{ijkl}\varepsilon_{kl}, \tag{6}$$

with the rate of dissipated energy:

$$\dot{\phi} = -\frac{\partial\rho\psi}{\partial d}\dot{d}. \tag{7}$$

The second principle of thermodynamics constrains the dissipation rate to be zero or positive. In this constitutive relation (with a single damage variable), it means that damage must either increase (irreversible response) or remain constant (reversible response).

The thermodynamic framework depicted above allows also for the incorporation of damage induced anisotropy. To this end, the free energy needs to be modified in order to account for the directionality of damage. Another technique due to Fichant et al. (1997) consists in an approximation of the effective stress σ^t_{ij}:

$$\sigma^t_{ij} = C^0_{ijkl}\varepsilon^e_{kl} \text{ or } \sigma^t_{ij} = C^0_{ijkl}\left(C^{damaged}\right)^{-1}_{klmn}\sigma_{mn}, \tag{8}$$

where C^0_{ijkl} is the initial stiffness of the (undamaged) material. Let us define the relation between the stress and the effective stress along a finite (or infinite) set of directions defined by unit vectors $\vec{n}$:

$$\sigma = (1 - d(\vec{n}))n_i\sigma^t_{ij}n_j \,,$$
$$\tau = (1 - D(\vec{n}))\sqrt{\sum_{i=1}^{3}\left(\sigma^t_{ij}n_j - \left(n_k\sigma^t_{kl}n_l\right)n_i\right)^2} \,, \tag{9}$$

where σ and τ are the normal and tangential components of the stress vector respectively. Here $d(\vec{n})$ and $D(\vec{n})$ are scalar valued quantities which define the influence of damage on the relation between the effective stress and the stress vectors in direction $\vec{n}$. The overall resolved stress tensor is the solution of the virtual work equation:

$$\text{find } \sigma_{ij} \text{ such that } \forall\varepsilon^*_{ij} : \frac{4\pi}{3}\sigma_{ij}\varepsilon^*_{ij} =$$
$$\int_\Omega \left(\left[(1 - d(\vec{n}))n_k\sigma^t_{kl}n_l n_i + (1 - D(\vec{n}))\left(\sigma^t_{ij}n_j - n_k\sigma^t_{kl}n_l n_i\right)\right] \cdot \varepsilon^*_{ij}n_j\right)d\Omega, \tag{10}$$

where Ω is a sphere of radius 1. This equation can be recast as follow:

$$\text{find } \sigma_{ij} \text{ such that } \forall\varepsilon^*_{ij} : \frac{4\pi}{3}\sigma_{ij}\varepsilon^*_{ij} =$$
$$\int_\Omega \left(\left[(1 - d(\vec{n}))E_N n_k\varepsilon^t_{kl}n_l n_i + (1 - D(\vec{n}))E_T\left(\varepsilon^t_{ij}n_j - n_k\varepsilon^t_{kl}n_l n_i\right)\right] \cdot \varepsilon^*_{ij}n_j\right)d\Omega, \tag{11}$$

where E_N and E_T are initial stiffness moduli which are functions of the Young's modulus and Poisson's ratio of the material. Eq. (11) is very similar to the equation which relates the resolved stress to the microplane stress and strains in the microplane model [Bazant and Prat 1988]. In fact, $d(\vec{n})$ and $D(\vec{n})$ can either be defined for each angular directions $\vec{n}$ independently, or can be interpolated by angular functions. In the first form, one recovers a microplane damage model and the most general form of damage induced anisotropy [Carol et al., 1991]. In the second form, damage induced anisotropy is fixed by the angular functions defining $d(\vec{n})$ and $D(\vec{n})$. It is a restriction of the first form which can be more efficient in computations as it involves less variables: a finite number of directions $\vec{n}$ only are needed in order to define the material stiffness. It should be also noted that the integral form in Eq. (11) can be transformed again in order to arrive to a format which is very similar to the multiple fixed crack model:

$$\sigma_{ij} = \sum_{\vec{n}} \omega^n \left(\left[(1 - d(\vec{n}))E_N n_k\varepsilon_{kl}n_l n_i + (1 - D(\vec{n}))E_T\left(\varepsilon_{ij}n_j - n_k\varepsilon_{kl}n_l n_i\right)\right] \cdot n_j\right), \tag{12}$$

where ω^n is a weighting factor (see for instance the review in [de Borst, 1999]).

The evolution of $d(\vec{n})$ and $D(\vec{n})$ for each direction $\vec{n}$ must be consistent with the second principle of thermodynamics. For isothermal conditions the rate of energy dissipation must be positive or zero. Energy dissipation is slightly more complex than for the scalar damage model (Eq. 7):

$$\dot{\phi} = \int_{\Omega} \left[\dot{d}(\vec{n}) n_k \sigma^t_{kl} n_l n_i + \dot{D}(\vec{n}) \left(\sigma^t_{ij} n_j - n_k \sigma^t_{kl} n_l n_i \right) \right] \varepsilon^e_{ij} n_j d\Omega \geq 0 . \quad (13)$$

The evolution of damage is controlled by the same loading function f:

$$f(\vec{n}) = \vec{n} \varepsilon^e \vec{n} - \varepsilon_d - \chi(\vec{n}), \quad (14)$$

where χ is a hardening - softening variable. If the damage angular functions are interpolated, e.g. with ellipsoidal functions, there are six scalar loading surfaces, one for each principal direction and for each damage variable. The hardening-softening variables are also angular functions which must be interpolated (with similar ellipsoidal functions). In the more general microplane approach, there is one loading function and one hardening-softening variable per microplane.

The loading function and the rate of the hardening-softening variable must comply finally with the Kuhn-Tucker conditions:

$$f \leq 0, \ \dot{\chi} \geq 0, \ f\dot{\chi} = 0. \quad (15)$$

3 NONLOCAL AND GRADIENT DAMAGE

We will first restrict attention to isotropic damage and consider a material which contains growing voids with isotropic characteristics; the nonlocal generalisation of the scalar damage model will be recalled first. Then, we will turn toward the gradient damage model and the influence of the internal length on the bifurcation condition involved in strain localisation analyses.

3.1 Nonlocal damage model

Consider for instance the case in which damage is a function of the positive strains (which means that it is mainly due to micro cracks opening in mode I). The following scalar called equivalent strain is introduced first (see for instance Mazars, 1984):

$$\tilde{\varepsilon} = \sqrt{\sum_{i=1}^{3} \left(\langle \varepsilon_i \rangle_+ \right)^2}, \quad (16)$$

where $\langle . \rangle_+$ is the Macauley bracket and ε_i are the principal strains. In classical damage models, the loading function f reads $f(\tilde{\varepsilon}, \chi) = \tilde{\varepsilon} - \chi$. The principle of nonlocal continuum models with local strains is to replace $\tilde{\varepsilon}$ with its average:

$$\bar{\varepsilon}(x) = \frac{1}{V_r(x)} \int_\Omega \psi(x-s)\tilde{\varepsilon}(s)ds \text{ with } V_r(x) = \int_\Omega \psi(x-s)ds, \tag{17}$$

where Ω is the volume of the structure, $V_r(x)$ is the representative volume at point x, and $\psi(x-s)$ is the weight function, for instance:

$$\psi(x-s) = \exp\left(-\frac{4\|x-s\|^2}{l_c^2}\right). \tag{18}$$

l_c is the internal length of the non local continuum. $\bar{\varepsilon}$ replaces the equivalent strain (Eq. 16) in the evolution of damage. In particular, the loading function becomes $f(\bar{\varepsilon}, \chi) = \bar{\varepsilon} - \chi$. It should be noticed that this model is easy to implement in the context of explicit, total strain models. Its implementation to plasticity and to implicit incremental relations is awkward. The local tangent stiffness operator relating incremental strains to incremental stresses becomes non symmetric, and more importantly its bandwidth can be very large due to nonlocal interactions. This is one of the reasons why gradient damage models have become popular over the past few years.

3.2 Isotropic gradient damage model

A simple method to transform the above nonlocal model to a gradient model is to expand the effective strain into Taylor series truncated for instance to the second order:

$$\tilde{\varepsilon}(x+s) = \tilde{\varepsilon}(x) + \frac{\partial \tilde{\varepsilon}(x)}{\partial x} s + \frac{\partial^2 \tilde{\varepsilon}(x)}{\partial x^2} \frac{s^2}{2!} + \ldots.. \tag{19}$$

Substitution in Eq. (17) and integration with respect to variable s yields:

$$\bar{\varepsilon}(x) = \tilde{\varepsilon}(x) + c^2 \nabla^2 \tilde{\varepsilon}(x). \tag{20}$$

c is a parameter which depends on the type of weight function in Eq. (17). Its dimension is L^2 and it can be regarded as the square of an internal length. Substitution of the new expression of the nonlocal effective strain in the nonlocal damage model presented above yields a gradient damage model. Computationally, this model is still delicate to implement because it requires higher continuity in the interpolation of the displacement field. This difficulty can be solved with a simple trick devised by Peerlings et al. (1996): let us take the Laplacian of the right and left hand-sides of Eq. (20).

Because the Taylor expansion in Eq. (19) was truncated to the second order, higher derivatives can be assumed to be negligible. It follows that

$$\nabla^2\bar{\varepsilon}(x)=\nabla^2\tilde{\varepsilon}(x)+\nabla^2\left(c^2\nabla^2\tilde{\varepsilon}(x)\right)\approx\nabla^2\tilde{\varepsilon}(x). \tag{21}$$

Therefore, Eq. (20) becomes $\bar{\varepsilon}(x)-c^2\nabla^2\bar{\varepsilon}(x)=\tilde{\varepsilon}(x)$. This implicit equation which defines the nonlocal effective strain as a function of the local effective strain is easy to discretise in a finite element scheme. The implementation of the gradient damage model becomes in fact similar to the implementation of a thermomechanical (local) model in which the nonlocal effective strain replaces the nodal temperatures. This type of model has provided very good predictions of failure, for instance, of concrete structures [Peerlings et al., 1998].

Gradient damage models have also been devised differently. Frémond and co-workers (1993) started from the principle of virtual work adding the power of internal forces involving higher order terms.

A more recent approach, inspired by the mechanics of porous materials, was proposed by Pijaudier-Cabot and Burlion [1996]: assume that damage can be characterised by the variation of volume fraction of material denoted as $v(x,t)$. Failure is reached when the volume fraction is equal to zero. Starting from a reference configuration where the material is strain - free and the volume fraction is $v_R(x)$, the variation of the volume fraction of material is $\phi(x,t)=v(x,t)-v_R(x)$. This variation of volume fraction of material can be due to damage growth or straining. For constant damage, the porous material is elastic and its behaviour is modelled using the theory of elastic material with voids [Cowin and Nunziato, 1983]. The governing equations are (in the absence of body forces):

$$\sigma_{ij,j}=0, \tag{22a}$$
$$h_{i,i}+g=0, \tag{22b}$$

where h_i is the equilibrated stress vector, and g is the equilibrated body force. These variables are related to the stresses due to centres of dilation made of three couples of opposite forces without moments acting along three mutually orthogonal directions at material points. Such forces correspond to the local pressures necessary to augment the size of an existing void, in a reversible or irreversible manner and they create a local stress distribution. These forces produce void growth and a variation of the overall volume fraction of the body. The elastic (free) energy reads:

$$\rho\psi=\frac{1}{2}C_{ijkl}\varepsilon^e_{ij}\varepsilon^e_{kl}+\beta\delta_{ij}\varepsilon^e_{ij}\phi+\frac{1}{2}\alpha\phi_{,i}\phi_{,i}+\frac{1}{2}\xi\phi^2, \tag{23}$$

where C_{ijkl} are the overall stiffness coefficients of the porous material, and (α,β,ξ) are material parameters. Compared to the expression in Eq. (5), the free energy has two additional terms which correspond to the variation of the volume fraction of voids. The thermodynamic forces associated to the variables $(\varepsilon_{ij}, \phi, \phi_{,i})$ are defined as :

$$\sigma_{ij} = \rho \frac{\partial \psi}{\partial \varepsilon_{ij}}, g = -\rho \frac{\partial \psi}{\partial \phi} + g_R, h_i = \rho \frac{\partial \psi}{\partial \phi_{,i}} + h_i^R, \tag{24}$$

where g_R and h_i^R are the values of the equilibrated body forces and equilibrated stress vector in the reference state, respectively. Hence equilibrium in the reference (stress-free) state yields:

$$h_{i,i}^R + g_R = 0, \tag{25}$$

As pointed out by Cowin and Nunziato, the stiffness coefficients and the material parameters should depend on the reference volume fraction. In order to obtain an extension of this model to the case of a damaged material, let us assume that whenever damage grows, it modifies the volume fraction of the material in the reference configuration. The variation of volume fraction is rewritten as:

$$\begin{aligned} & d\phi(x,t) = d\phi^{rev}(x,t) + d\phi^{ir}(x,t), \\ & with \begin{cases} \phi^{rev}(x,t) = v(x,t) - v^R(x,t), \\ \\ \phi^{ir}(x,t) = v^R(x,t) - v^0(x), \end{cases} \end{aligned} \tag{26}$$

where ϕ^{rev} is the reversible variation of volume fraction, $v^R(x,t)$ is the volume fraction of the damaged material when it is free of loads, ϕ^{ir} is the irreversible variation of volume fraction due to the growth of damage measured when the material is free of loads and v^0 is the initial volume fraction of the material, when damage is equal to zero. For the sake of simplicity, we assume that ϕ^{rev} is very small compared to ϕ^{ir} as damage grows. The free energy of the material is now:

$$\rho\psi = \frac{1}{2}(1-d)C_{ijkl}\varepsilon_{ij}\varepsilon_{kl} + \psi(\phi^{ir}, \phi_{,i}^{ir}). \tag{27}$$

Compared to Eq. (4), $\psi(\phi^{ir}, \phi_{,i}^{ir})$ represents the variation of reversible energy due to an irreversible variation of volume fraction. In order to comply with the second principle of thermodynamics, the Clausius-Duhem inequality should be verified in all instances:

$$\frac{d\rho\psi}{dt} \leq \sigma_{ij}\dot{\varepsilon}_{ij} + h_i\dot{\phi}_{,i} + g\dot{\phi}. \tag{28}$$

The irreversible variation of volume fraction is measured when the body is free of loads. Therefore, we assume that g_R and h_i^R are expressed as follow:

$$h_i^R = \alpha\phi_{,i}^{ir},$$
$$g_R = -\xi\phi^{ir} + A. \tag{29}$$

The term A is the quantity which controls the irreversible variation of volume fraction of the material from the initial state $v^0(x)$ when loads are applied. α is a model parameter.

A simple (maybe simplistic) way to relate the variation of volume fraction to the decrease of stiffness of the material is to assume that $d = -\phi^{ir}$. The relations which follow from these assumptions are derived from Eqs. (22,24,25,29):

$$\begin{cases} \sigma_{ij,j} = 0, \\ \alpha d_{,ii} - \xi d = A, \end{cases} \qquad \sigma_{ij} = (1-d)C_{ijkl}\varepsilon_{kl}. \tag{30}$$

$\dot{A}$ is defined by a loading function and an evolution equation:

$$if\ f(\varepsilon,\chi) = 0\ and\ \frac{\partial f(\varepsilon,\chi)}{\partial\varepsilon}\dot{\varepsilon} > 0\ then \begin{cases} \dot{A} = H\dot{\varepsilon} \\ \dfrac{\partial f(\varepsilon,\chi)}{\partial\chi}\dot{\chi} = -\dfrac{\partial f(\varepsilon,\chi)}{\partial\varepsilon}\dot{\varepsilon} \end{cases} \tag{31}$$

where H is a hardening - softening modulus. In Eq. (31), ε denotes an invariant of the strain generically. Note that α in Eqs. (30) has the dimension of a length squared. It is there that an internal length has been introduced in the constitutive relations.

In the absence of body forces, the two governing equations, along with the boundary conditions, are equivalent to the following conditions: find a displacement field $\vec{u}$ and a damage field d such that for any kinematically admissible virtual displacement $\vec{u}^*$ and damage field d^*,

$$\int_\Omega \varepsilon^* : \sigma dv = \int_{\partial\Omega} \vec{u}^* . \vec{F} ds, \quad \int_\Omega \{\alpha grad(d^*).grad(d) + \xi d^* d\} dv = \int_\Omega d^* . A dv, \tag{32}$$

where $\vec{F}$ are external applied forces. For the finite element implementation, the displacement field components and damage field are discretised with different types of polynomials, same as in coupled thermomechanical problems. A similar variational principle can be obtained in an incremental fashion for the purpose of implementing the model within a Newton - Raphson procedure and of solving this non linear set of two coupled equations iteratively. It remains, however, that additional boundary conditions are required for solving Eqs. (32):

$$\begin{cases} u_i(x,t) = \tilde{u}_i(x,t) \text{ on } \partial\Omega_1 \text{ and } \sigma.\vec{n} = \vec{\tau} \text{ on } \partial\Omega_2, \\ (\partial\Omega_1 \cup \partial\Omega_2 = \partial\Omega \text{ and } \partial\Omega_1 \cap \partial\Omega_2 = \varnothing), \\ \qquad \alpha grad(d(x,t)).\vec{n} = 0 \text{ on } \partial\Omega, \end{cases} \tag{32}$$

where $\vec{n}$ is now the unit outward normal to the boundary $\partial\Omega$ of the solid Ω. The meaning of the third (natural) boundary condition remains a problem to be solved.

This gradient damage model, the one which derives from Eqs. (19-21) and the integral nonlocal model are very similar. The difference is that in the first one it is the damage variable itself on which the nonlocal treatment is applied, while in the laters, the nonlocal treatment is applied to the variable which controls damage. It follows that in the finite element implementation the damage variable is interpolated in the first model and that the nonlocal effective strain is interpolated in the second.

3.3 Strain localisation analysis

In order to exhibit the regularisation properties of the gradient damage model, we will investigate the occurrence of bifurcation in an infinite body from a homogeneous state of deformation and damage, denoted as $(\varepsilon_{ij}^0, d^0)$. Monotonic loading is assumed and body forces are omitted. Apart from the trivial solution where the strain and damage remain homogeneous, we look for velocities $(\dot{u}_1, \dot{u}_2, \dot{u}_3)$ and a damage rate distribution $\dot{d}$ of the form:

$$\dot{u}_i = U_i \exp(J\omega(n_l x_l)), \quad \dot{d} = D\exp(J\omega(n_l x_l)), \tag{34}$$

where U_i (i=1,2,3) and D are unknown constants, ω is an unknown angular frequency, $\vec{n}$ is an unknown normal vector, and J is the imaginary constant such that $J^2 = -1$. Substitution of Eqs (30,34) in the governing equations of equilibrium (Eqs. 33) yields the following homogeneous system of equations:

$$P\begin{bmatrix} \vec{U} \\ D \end{bmatrix} = \begin{bmatrix} [\omega^2(1-d^0)\vec{n}.C.\vec{n}] & [J\omega\vec{n}.(C:\varepsilon^0)] \\ [-J\omega H.\vec{n}] & \alpha\omega^2 + \xi \end{bmatrix} \begin{bmatrix} \vec{U} \\ D \end{bmatrix} = \begin{bmatrix} \vec{0} \\ 0 \end{bmatrix}. \tag{35}$$

P is a 4x4 matrix. This system admits non trivial solution if det(P)=0, which is the bifurcation condition. The bifurcation condition should yield a normal vector $\vec{n}$ for each value of the angular frequency. One can check easily that if $\alpha = 0$, the solution is independent of the angular frequency which can be fixed arbitrarily. The square root of α is an internal length of the continuum which selects the wave length of the localised solution and scales the size of the localisation band. It plays exactly the same role as the internal length in other localisation limiters.

4 CLOSURE : EXTENSION TO ANISOTROPIC GRADIENT DAMAGE

In the anisotropic damage models devised in the first section of this contribution, the evolution of damage is directional. For instance one loading function $f(\vec{n})$ is considered for each direction $\vec{n}$ in Eq. (14). Combining Eq. (14) and the definition of the nonlocal strain in Eq. (19-21) can be done: for each direction $\vec{n}$, Eq. (14) can replaced by:

$$f(\vec{n}) = \bar{\varepsilon}_N - \chi(\vec{n}), \quad \bar{\varepsilon}_N - c^2\nabla^2(\bar{\varepsilon}_N) = \vec{n}.\varepsilon^e.\vec{n}. \qquad (36)$$

For each direction in which damage growth is defined, a nonlocal variable is introduced. This extension is straightforward in the context of microplane models where directions $\vec{n}$ are defined arbitrarily and fixed in the analysis. There are as many nonlocal variables as directions considered and these variables are interpolated throughout the finite element mesh, same as for the isotropic gradient damage model. The usual set of equilibrium equations is complemented with $p \times m$ equations where p is the number of directions and m is the number of damage variable per direction [Kuhl et al., 1998].
For the orthotropic damage model proposed by Fichant et al., three directions are considered only at each material point. The directional distribution of damage is reconstructed by interpolation in between those principal directions. Hence, there is the complexity that the directions change from one finite element to another and might rotate in the course of loading too. A similar problem is encountered in nonlocal rotating crack models [de Borst, 1999]. Conceptually, this feature does not change anything in the discretisation, except that one has to keep track of the orientation of principal directions of damage. This is in fact the price to pay for the reduction of equations (and degrees of freedom) to be solved.

5 BIBLIOGRAPHY

[Bazant, 1976] Bazant, Z.P., Instability, ductility and size effect in strain softening concrete, *J. Engng. Mech. ASCE*, Vol. 102, pp. 331-344.
[Bazant and Prat, 1988] Bazant, Z.P., and Prat, P.C., Microplane model for brittle-plastic material I: theory, *J. Engng. Mech. ASCE*, Vol. 114, pp. 1672-1702.
[de Borst and Muhlhaus, 1992] de Borst, R., and Muhlhaus, H.B., Gradient dependent plasticity : formulation and algorithmic aspects, *Int. J. Num. Meth. in Engrg.*, Vol. 35, pp. 521-539.
[de Borst, 1991] de Borst, R., Simulation of strain localisation: a reappraisal of the cosserat continuum, *Eng. Comput.*, Vol. 8, pp. 317-332.
[de Borst, 1999] de Borst, R., Isotropic and anisotropic damage models for concrete fracture, in *Mechanics of Quasi-Brittle Matrials and Structures*, G. Pijaudier-Cabot et al. Eds., Hermes Sci. Pubs., pp. 39-56.
[Carol et al., 1991] Carol, I., Bazant, Z.P., and Prat, P.C., Geometrical damage tensor based on microplane model, *J. of Engrg. Mech. ASCE*, Vol. 117, pp. 2429-2448.

[Cowin and Nunziato, 1983] Cowin, S.C., and Nunziato, J.W., Linear elastic materials with voids, *J. of Elast.*, Vol. 13, pp. 125-147.

[Feenstra, 1993] Feenstra, P.H., *Computational Aspects of Biaxial Stress in Plain and Reinforced Concrete*, Dissertation, Delft University of Technology, the Netherlands.

[Fichant et al., 1997] Fichant, S., Pijaudier-Cabot, G., La Borderie, C., Continuum damage modelling: approximation of crack induced anisotropy, Mech. Res. Comm., Vol. 24, pp. 109-114.

[Frémond and Nedjar, 1993] Frémond, M., Nedjar, B., Endommagement et principe des puissances virtuelles. Comptes Rendus Acad. Sci., série II, pp. 857-864.

[Joseph et al. 1985] Joseph, D., Renardy, M., and Saut, J.C., Hyperbolicity and change of type in the flow of viscoelastic fluids, *Arch. for Rational Mech. and Analysis*, Vol. 87, pp. 213-251.

[Kuhl et al. 1998] Kuhl, E., Ramm, E., and de Borst, R., Anisotropic damage model with the microplane model, in *Computational Modelling of Concrete Structures*, de Borst et al. Eds., Balkema Pubs, pp. 103-112.

[Lemaitre, 1992] Lemaitre J., *A Course on Damage Mechanics*, Springer Verlag.

[Needleman, 1988] Needleman, A., material rate dependence and mesh sensitivity in localization problems, *Comp. Meth. Appl. Mech. Engrg.*, Vol. 67, pp. 69-86.

[Mazars, 1984] Mazars, J., *Application de la mécanique de l'endommagement au comportement non linéaire et à la rupture du béton de structure*, Thèse de Doctorat ès Sciences, Université Paris 6.

[Peerlings et al., 1996] Peerlings, R.H., de Borst, R., Brekelmans, W.A.M., de Vree, J.H.P., Gradient enhanced damage for quasi-brittle materials, *Int. J. Num. Meth. Engrg.*, Vol. 39, pp. 3391-3403.

[Peerlings et al., 1998] Peerlings, R.H., de Borst, R., Brekelmans, W.A.M, Geers, M.G.D., Gradient enhanced modelling of concrete fracture, *Mech. Coh.-Frict. Mat.*, Vol. 3, pp. 323-343.

[Pijaudier-Cabot and Bazant, 1987] Pijaudier-Cabot, G., and Bazant, Z.P., Nonlocal damage theory, *J. of Engrg. Mech. ASCE*, Vol. 113, pp. 1512 - 1533.

[Pijaudier-Cabot and Benallal, 1993] Pijaudier-Cabot, G., and Benallal, A., Strain localisation and bifurcation in a nonlocal continuum, *Int. J. Solids Struct.*, Vol. 30, pp. 1761-1775.

[Pijaudier-Cabot and Burlion, 1996] Pijaudier-Cabot, G., and Burlion, N., Damage and localisation in elastic materials with voids, *Int. J. Mech. of Cohes.-Frict. Mat.*, Vol.1, pp. 129-144.

[Rice, 1976] Rice, J.R., The localisation of plastic deformation, in *Proc. Int. Congress of Theor. and Appl. Mech.*, Delft, The Netherlands, Vol. 1, pp. 207-220.

[Sluys, 1992] Sluys, L.J., *Wave Propagation, Localisation and Dispersion in Softening Solids, Doctoral Dissertation*, Delft University of Technology, The Netherlands.

Constitutive laws, relaxation thermodynamics and Lagrange-formalism

R. Rahouadj, J.F. Ganghoffer, C. Cunat
LEMTA-CNRS, UMR 7563, ENSEM. 2, Avenue de la Forêt de Haye. BP 160.
54504 Vandoeuvre Cedex. France. E-mail: Cunat@ensem.u-nancy.fr

Abstract: An interpretation of a non-equilibrium thermodynamic approach of irreversible processes (called DNLR) in terms of the Hamiltonian principle of least action is proposed. It is shown that one possible choice for building the Lagrangian kernel is the material derivative of the internal energy density, obtained from the generalised Gibbs relation. This general framework is illustrated for the transport equations (heat and diffusion) and the rheology of solids.

Keywords: Thermodynamics of irreversible processes (TIP), Lagrange-formalism, Rheology, Heat transfer, Diffusion.

1. LEAST ACTION AND THERMODYNAMICS OF IRREVERSIBLE PROCESSES.

In this contribution we look for a variational formulation that allows us to describe the evolution of dissipative systems during non-equilibrium transformations. The study involves the framework of the thermodynamics of relaxation, and specifically the Distribution of Non-Linear Relaxations, called DNLR.

It may appear singular at the end of this century to continue to examine the problem of the evolution laws for thermodynamic systems outside equilibrium from the point of view of Hamilton's principle of least action. This action

$$S = \int_{t_1}^{t_2} \int_{\Omega_t} L(q,\dot{q})\, d\Omega_t\, dt, \qquad (1\text{-}1)$$

leads to the Euler-Lagrange equation:

$$-\frac{d}{dt}\frac{\partial L}{\partial \dot{q}} + \frac{\partial L}{\partial q} = 0, \qquad (1\text{-}2)$$

in which the integrand $L(\dot{q},q)$ represents the Lagrangian density that renders extremum the Hamiltonian action S with respect to the variable q, i.e. $\delta S = 0$ with respect to δq.

In fact, it is very often admitted that any attempt to reconcile Hamiltonian dynamics with the thermodynamics of irreversible processes - TIP - must fail. To our knowledge, one of the first conclusive attempts to reduce the principle of irreversible thermodynamics to the general principles of the mechanics is probably due to

G.A. Maugin et al. (eds.), Continuum Thermomechanics, 367-378.
© 2000 *Kluwer Academic Publishers.*

Helmholtz, [Poincaré, 1908]. This pioneer, whose works are abundantly mentioned by Poincaré, has expressed the Euler-Lagrange equations under the following form:

$$-\frac{d}{dt}\left(\frac{dL}{d\dot{q}_{\alpha}}\right)+\frac{dL}{dq_{\alpha}}+P_{\alpha}=0, \tag{1.3}$$

with $L=K-\phi$ and $\phi=\phi(q_{\alpha})$. K denotes the kinetic energy, ϕ is the potential energy related to the internal forces and P_{α} represents the external forces. The kinetic energy obeys the particular relation: $2\,K(\dot{q}_{\alpha},q_{\alpha})=\dot{q}_{\alpha}\frac{dK}{d\dot{q}_{\alpha}}$, translating a homogeneous dependence of degree –2 on time derivatives, $K=\frac{1}{2}m_{\alpha}\dot{q}_{\alpha}^{2}$. Helmholtz divided the variables q_{α} into two categories: variables with slow (q_a), or rapid (q_b) variations. He then introduced the notion of *incomplete systems*, which corresponds to the case where the work of external forces vanishes, i.e. $-\ P_{\alpha}dq_{\ \alpha}\ =\ 0$. Helmholtz showed in this case that the function L may be an odd function of $\dot{q}_b$, and consequently odd in time. Such an important result seems to open the way for the description of irreversible processes on the basis of the Euler-Lagrange equations. However, Poincaré (1908) underlined the limits of this analysis for isolated systems in which the forces P_{α} are zero, arguing that the equations of Helmholtz cannot explain the increase of entropy due to the irreversible processes. In the end, he concluded that: "...les phénomènes irréversibles et le théorème de Clausius ne sont pas explicables au moyen des équations de Lagrange".

Glansdorff et al. (1971) introduced the concept of the local potential (being for example $\bar{A}.\dot{\bar{z}}$, with $\bar{A}$ representing the non-equilibrium forces and $\bar{z}$ the dissipation variable leading to the flux $\dot{\bar{z}}$), which in a certain manner accounts for the evolution of the entropy production, in accordance with the Lyapunov stability principle.

Although not so many works are devoted to this subject - up to a point Lavenda (1993) considered these works as sporadic -, the scientific community has been debating fiercely these matters for more than a century. One can certainly consider that the formulation of the *principle of least dissipation of energy* by Onsager in 1931 constitutes the cornerstone of the modern developments of the TIP. This last principle brings a new light on the *principle of least entropy production* developed independently by Prigogine in 1955. The controversy was initially raised by Ono in 1961 and then by Gyarmati (1970); they showed that the Prigogine principle of least entropy production for stationary states can be regarded as a special case of the generalised formulation due to Onsager, called the *principle of the least dissipation of energy*.

The monograph of Gyarmati seems to be one of the few analyses that tries to sum up and summarise the extremum principles applied to the thermodynamics of irreversible processes. According to the author, the principle of the least dissipation of energy may be formulated as follows:

$$\rho\,\dot{s} + \nabla.J_s - (\psi + \phi) = \sigma(\bar{J}, \bar{X}) - [\psi(\bar{X}, \bar{X}) + \phi(\bar{J}, \bar{J})] = \max\,, \tag{1.4}$$

where J_s represents the entropy flux, σ the entropy production defined as the scalar product of the thermodynamic forces tied to irreversibility $\bar{X}$ by the associated fluxes $\bar{J}$. One further introduces phenomenological relationships between $\bar{J}$ and $\bar{X}$; the use of one Onsager's couplings allows one to write either of the alternative expressions $\bar{J} = \bar{\bar{L}}\,\bar{X}$, or $\bar{X} = \bar{\bar{R}}\,\bar{J}$, such that $\psi = \frac{1}{T}(\frac{1}{2}\bar{X}\,\bar{\bar{L}}\,\bar{X})$; $\phi = \frac{1}{T}(\frac{1}{2}\bar{J}\,\bar{\bar{R}}\,\bar{J})$. Starting from the force representation at constant fluxes, Gyarmati deduced the Fourier equation, and considered this as a sufficient basis for the fulfilment of a general formulation of thermodynamics. Before formulating the integral principle in a general form, he deduced the equation of heat conduction from a Fourier, an entropy, and an energy picture, and finally from a more general representation (with $L_\Gamma = \Gamma\rho c_v \frac{\partial T}{\partial t} - \frac{\Gamma}{2}(\nabla\Gamma)^2$, where $\Gamma = f(T)$; T denotes the absolute temperature). However, it is essential to note that the *integral principle of thermodynamics* of Gyarmati refers to a force-variation at constant fluxes combined with the balance of energy. It leads to Lagrangian densities having arguments which are not of the type $\{q, \dot{q}\}$, whereas the Hamilton principle refers to such a set of variables. Since the corresponding Euler-Lagrange equations are formally of the type:

$$-\nabla.\frac{\partial L}{\partial(\nabla q)} + \frac{\partial L}{\partial q} = 0\,, \tag{1.5}$$

they are thus truncated from their temporal dimension, leading in fact to a non-Hamiltonian action:

$$S = \int_{\Omega_t} L\, d\Omega_t\,. \tag{1.6}$$

On the other hand, it is well known that the generalised Euler-Lagrange equations for a continuum must be written in the form

$$-\frac{d}{dt}\frac{\partial L}{\partial \dot{q}} - \nabla.\frac{\partial L}{\partial \nabla q} + \frac{\partial L}{\partial q} = 0\,. \tag{1.7}$$

Since the Hamiltonian dynamics looks incompatible with irreversible thermodynamics, why then try this adventure?

The idea of this work germinated following a search pursued in a parallel manner by Rahouadj and Cunat. It follows a first study suggested by Sidoroff, to position the thermodynamic approach called DNLR developed in Nancy by Cunat (1996) with respect to the pioneering work by Biot (1954). Both approaches led apparently to constitutive equations that looked similar in form. Although, the DNLR approach relies on a generalisation of the fundamental Gibbs relationship to situations outside equilibrium, introducing internal variables, but does not need a pseudo-potential of dissipation. The Biot's approach involves both the thermodynamic and the dissipation potentials, this last one being equivalent to the *dissipativity* introduced by Rayleigh in 1873. The constitutive equations then follow from the setting up of a Lagrangian form:

$$\frac{\partial V}{\partial q}+\frac{\partial D}{\partial \dot{q}}=Q, \tag{1.8}$$

in which V plays the role of a potential energy and D denotes the dissipation function, while Q represents the thermodynamic generalised force associated with $\dot{q}$ and q, [Sidoroff, 1970]. This last usual form is often used nowadays in order to associate the Lagrangian framework with the TIP. We can see for instance the clear statement by Maugin (1990), who writes in particular:

$$L=K(v,\dot{\alpha})-\rho\,\psi(-,\alpha,\nabla\alpha,\theta), \tag{1.9}$$

where K represents the kinetic energy term, ψ is the thermodynamic potential, v the velocity field, the "-" in ψ stands for the observable variables, θ is the temperature, and α is an internal variable. The last relation is supplemented by a dissipation function D in the Euler-Lagrange equations:

$$\frac{\delta L}{\delta q}-\frac{\partial D}{\partial \dot{q}}=0, \tag{1.10}$$

where $\delta/\delta q$ denotes the Euler operator, defined as:

$$\frac{\delta}{\delta q}=\sum_{j=0}^{\infty}\left(-d_q\right)^j\frac{\partial}{\partial q_{,j}}=\frac{\partial}{\partial q}-\frac{d}{dt}\left(\frac{\partial}{\partial \dot{q}}\right)-\nabla.\left(\frac{\partial}{\partial(\nabla q)}\right)+\ldots, \tag{1.11}$$

with d_q the total derivative with respect to q.

Having concluded the comparative study of Biot's model and the DNLR, (paper under preparation; Rahouadj, Sidoroff, Cunat), a bibliographical review concerning the use of the Lagrange formalism for continua then led us to the theoretical developments by Anthony (1986). This author indicates that the Lagrange formalism fails in *traditional* TPI because of the non-self-adjointness of its fundamental field equations. Then, he writes a principle of least action in the sense of Hamilton by doubling the set of thermodynamic variables. So, for each of the original ones, a phase term is formally defined. In this way, and in the special case of heat conduction, the thermodynamic temperature T is directly replaced by a thermal excitation field $\varpi(x,t)=\sqrt{T}\,e^{i\varphi(x,t)}$, with a complex conjugate field $\varpi^*(x,t)=\sqrt{T}\,e^{-i\varphi(x,t)}$, such that the real-valued temperature becomes: $T=\varpi\,\varpi^*$.

In fact, this remark encouraged us to examine the constitutive equations arising from the DNLR approach in the light of Hamilton's principle.

The project to integrate the consideration of non-local phenomena within the DNLR approach that we take up with J-F. Ganghoffer naturally fits within this perspective.

Since the constitutive equations of the DNLR are based on a generalised thermodynamic potential, $\boldsymbol{\Psi}=\boldsymbol{\Psi}(q)$, containing the whole thermodynamic information by means of an exhaustive set of variables, it appears natural to consider the material derivative $\dot{\boldsymbol{\Psi}}=\dot{\boldsymbol{\Psi}}(q,\dot{q})$ as a good candidate for building a Hamiltonian action integral.

The aim of this work is to assess the ability of this potential to produce the suitable evolution equations involved in the irreversible processes, via a least action principle. The framework of the DNLR approach is envisaged as a focus in the present contribution.

2. THE THERMODYNAMIC POTENTIAL Ψ ACCORDING TO THE DNLR

The DNLR [Cunat, 1996] relies on the following

Postulate: *The generalised Gibbs' relation obtained by setting up internal variables complementing the thermodynamic state variables is valid for thermodynamic transformations outside equilibrium.*

In fact, one can show that the internal variables are then completely governed by the thermodynamic variables at equilibrium, whereas they can be considered as independent for non-equilibrium. In other words, it is always possible to describe transformations outside equilibrium from an *ad hoc* potential function. So, for instance, the thermodynamic potential Ψ may be selected as the specific internal energy e:

$$e = e(\underline{y}; \bar{z}), \tag{2.1}$$

where the generalised vector $\underline{y} = (s, \underline{\underline{\varepsilon}}, n_k, ...)$ represents the set of extensive variables (n_k is the number of moles of the *k-th* species). The generalised vector $\bar{z}$ denotes the whole set of internal variables which are entirely controlled by the $\underline{y}$ variables at the thermodynamic equilibrium states, i.e. $\bar{z}^{eq} = \bar{z}^{eq}(\underline{y})$. The symbols " _ "and " ‾ " used here and in the sequel allow one to distinguish the thermodynamic state variables from the pseudo-thermodynamic variables. We may easily obtain from (2.1) the rate form:

$$\dot{e}(\underline{y}, \dot{\underline{y}}; \bar{z}, \dot{\bar{z}}) = \underline{Y}(\underline{y}; \bar{z})\, \dot{\underline{y}} - \bar{A}(\underline{y}; \bar{z})\, \dot{\bar{z}} \quad ; \qquad \underline{Y} = \frac{\partial e}{\partial \underline{y}}(\underline{y}; \bar{z}); \;\; \bar{A} = -\frac{\partial e}{\partial \bar{z}}(\underline{y}; \bar{z}). \tag{2.2a,b}$$

The two terms $\underline{Y}(\underline{y}; \bar{z})$ and $\bar{A}(\underline{y}; \bar{z})$ are the intensive dual variables associated with the extensive variables $\underline{y}$ and $\bar{z}$ respectively, with $\underline{Y} = (T, \underline{\underline{\sigma}}, \mu_k, ...)$.

This postulate for non-equilibrium situations, induces a thermodynamic model similar to that of the Rational Thermodynamics, specially by incorporating internal variables (e.g. see [Coleman et al.,1967]).

The evolution equations, written in matrix form, are derived in a straightforward manner from relation (2.2b):

$$\begin{pmatrix} \dot{\underline{Y}} \\ -\dot{\bar{A}} \end{pmatrix} = \begin{pmatrix} \underline{\underline{a}}^{u} & \underline{\bar{b}} \\ {}^{t}\underline{\bar{b}} & \bar{\bar{g}} \end{pmatrix} \begin{pmatrix} \dot{\underline{y}} \\ \dot{\bar{z}} \end{pmatrix}, \tag{2.3}$$

where $\underline{\underline{a}}^u$ is the stability (coupling) matrix of Tisza (1966) for non-relaxed media, $\underline{\underline{a}}^u = \dfrac{\partial^2 e(\underline{y};\overline{z})}{\partial \underline{y}\, \partial \underline{y}}$, $\underline{\overline{b}} = \dfrac{\partial^2 e(\underline{y};\overline{z})}{\partial \overline{z}\, \partial \underline{y}}$, (rectangular coupling matrix), and $\overline{\overline{g}} = \dfrac{\partial^2 e(\underline{y};\overline{z})}{\partial \overline{z}\, \partial \overline{z}}$.

To be solvable, the system (2.3) must be complemented by an equation of evolution for the dissipative variables $\overline{z}$. As inferred above, the usual formulations often involve a convex pseudo-dissipation potential, D, which warrants a positive entropy production $\sigma = \dot{s}_i \geq 0$. However, this step is not mandatory, and it suffices to choose a suitable kinetic phenomenological model. On the other hand, in the context of the DNLR approach, the starting point can be a first order non-linear kinetic equation of the type:

$$\dot{\overline{z}} = -(\overline{\overline{L}}\,\overline{\overline{g}})(\overline{z} - \overline{z}^r) = -\overline{\overline{\tau}}^{-1}(\overline{z} - \overline{z}^r), \tag{2.4}$$

in which the identification of $\overline{z}^r$ is done from the stationary (or relaxed) condition involving the thermodynamic force $\overline{A}$, (also called the affinity by De Donder (1920)):

$$\dot{\overline{A}}^r = 0 = -{}^t\underline{\overline{b}}\,\dot{\underline{y}} - \overline{\overline{g}}\,\dot{\overline{z}}^r . \tag{2.5}$$

In equation (2.4), $\overline{\overline{L}}$ represents a phenomenological matrix of coefficients or functions whose reciprocity properties were proved by Onsager. Using this form has the advantage that it leads to a linear TIP when the coefficients of the coupling matrices are constant. The positivity of the entropy production is simply ensured by that of the bilinear form, $\overline{\overline{L}}\,\overline{\overline{g}}$, i.e. by the positivity of the relaxation time operator $\overline{\overline{\tau}}$. In fact, this condition replaces the generalised normality rule established by Ziegler (1963).

Moreover, since Meixner (1949), it is well known that the entropy production, $\sigma = \dot{s}_i = \overline{A}\,\dot{\overline{z}} \geq 0$, can be mathematically written in different bases. In particular, it is always possible to propose a modal formulation such that either the matrix $\overline{\overline{L}}$ or $\overline{\overline{g}}$ is unity. De Groot and Mazur (1962) selected $\overline{\overline{g}}$ as the unity matrix and transferred the multiplicity of dissipative modes onto the diagonalised matrix $\overline{\overline{L}}$ to account for some relaxation phenomena most prominent in the field of linear TIP. Contrary to this, the present approach selects $\overline{\overline{L}}$ as a unit-matrix in order to allow a distribution of relaxation times by mean of the matrix $\overline{\overline{g}}$. This choice is consistent with the multiple modes of the fluctuation theory, and it also permits the introduction of a shift relaxation spectrum function inherent in non-equilibrium transformations. Various kinds of non-linearities can be thus simply introduced into the coupling matrices $\underline{\underline{a}}^u$, $\underline{\overline{b}}$, $\overline{\overline{g}}$ and $\overline{\overline{L}}$, and their components may depend – in a more or less complex manner – upon the variables $\underline{y}$ and $\overline{z}$.

At this stage of description, one leaves the general framework to deal with the specific modelling of the behaviour of a given material under given thermo-mechanical loads. A number of applications have been performed successfully; they concern for instance the

modelling of the mechanical response due to multiaxial loads (e.g. the yield surface), the non-linear viscoelastic behaviour of polymers, the thermal ageing of polymers near the glass transition, [Rahouadj et al., 1999].
Moreover, it is relevant to attribute a physical sense to the internal variables, and to take care of the significant processes that actually intervene at the scale of the microstructure. For example, the viscoplastic behaviour of crystalline material has been treated in this way, using the classical theory of dislocations [Faccio-Toussaint, 1997]. The response of polymers under finite strains has also been characterised using the statistical spatial configurations of the macromolecular chains [Marceron ,1999].

3. THE LAGRANGE-FORMALISM AND THE DNLR APPROACH

The DNLR approach briefly presented above corresponds to the case of a uniform representative elementary volume; it may be improved in order to account for the gradients of intensive variables and the related boundary conditions; this is currently under development.
Let now consider the following action integral:

$$S=\int_{t_1}^{t_2}\int_{\Omega_t}\dot{e}(\underline{y},\underline{\dot{y}};\bar{z},\dot{\bar{z}})\,d\Omega_t\,dt=E(t_2)-E(t_1)\,. \tag{3.1}$$

The integrand is chosen to be the material derivative, $\dot{e}$, of the specific internal energy. In fact, this integrand represents a truncated version of a more general Lagrangian density, which will be presented in a forthcoming paper. So, without gradients of fields, one writes:

$$L=\dot{e}(\underline{y},\underline{\dot{y}};\bar{z},\dot{\bar{z}})\,, \tag{3.2}$$

and the generalised co-ordinates, $\{\underline{y},\underline{\dot{y}};\bar{z},\dot{\bar{z}}\}$, correspond to the well-known Lagrange co-ordinates $\{q_1,\dots q_n,\dot{q}_1,\dots,\dot{q}_n\}$ of the system. Note that the physical dimension of L is power, i.e. time rate of work.
The set of Euler-Lagrange equations, which ensure the extremality of the action integral, are

$$-\frac{d}{dt}\left(\frac{\partial\dot{e}}{\partial\underline{\dot{y}}}(\underline{y},\underline{\dot{y}};\bar{z},\dot{\bar{z}})\right)+\frac{\partial\dot{e}}{\partial\underline{y}}(\underline{y},\underline{\dot{y}};\bar{z},\dot{\bar{z}})=0\ ;-\frac{d}{dt}\left(\frac{\partial\dot{e}}{\partial\dot{\bar{z}}}(\underline{y},\underline{\dot{y}};\bar{z},\dot{\bar{z}})\right)+\frac{\partial\dot{e}}{\partial\bar{z}}(\underline{y},\underline{\dot{y}};\bar{z},\dot{\bar{z}})=0 \tag{3.3a,b}$$

These equations directly lead to

$$\underline{\dot{Y}}=\frac{\partial^2 e(\underline{y};\bar{z})}{\partial\underline{y}\,\partial\underline{y}}\underline{\dot{y}}+\frac{\partial^2 e(\underline{y};\bar{z})}{\partial\bar{z}\,\partial\underline{y}}\dot{\bar{z}}\,;\qquad -\dot{\bar{A}}=\frac{\partial^2 e(\underline{y};\bar{z})}{\partial\underline{y}\,\partial\bar{z}}\underline{\dot{y}}+\frac{\partial^2 e(\underline{y};\bar{z})}{\partial\bar{z}\,\partial\bar{z}}\dot{\bar{z}}\,, \tag{3.4a,b}$$

which are strictly equivalent to the evolution equations (2.3), and will be complemented by suitable boundary conditions for the generalised co-ordinates.

4. EXAMPLES OF THE GENERAL SETTING

4.1) Application to pure heat transport

Let us examine the case of a generalised extensive variable, $\underline{y}$, corresponding to the specific entropy, s, ($\underline{y} \to s$); its dual intensive variable is the temperature, $T(s;\bar{z})$, ($\underline{Y} \to T$). According to the relations (2.2a) and (3.2), the Lagrangian density relating to the pure heat transport is $L(s,\dot{s};\bar{z},\dot{\bar{z}}) = T(s;\bar{z})\dot{s} - \bar{A}(s;\bar{z})\dot{\bar{z}}$. As a consequence, the Euler-Lagrange equation (3.4a) becomes

$$\dot{T}(s;\bar{z}) = \frac{\partial T(s;\bar{z})}{\partial s}\dot{s} - \frac{\partial \bar{A}(s;\bar{z})}{\partial s}\dot{\bar{z}}, \tag{4.1}$$

which may also be written

$$\dot{T} = \frac{T}{c_v}\dot{s} + \bar{b}_{Tz}\,\dot{\bar{z}}, \tag{4.2}$$

with the phenomenological relations $\dfrac{\partial T(s;\bar{z})}{\partial s} = \dfrac{T}{c_v}$ and $\dfrac{\partial \bar{A}(\underline{y};\bar{z})}{\partial s} = \bar{b}_{Tz}$. The Euler-Lagrange equation (3.4b) governs the evolution of the non-equilibrium forces $\bar{A}$ associated with the extensive variables $\bar{z}$ and they account for the relaxation phenomena. Furthermore, considering the entropy balance,

$$\dot{s} = -\nabla(\frac{J_q}{T}) - \frac{J_q}{T^2}.\nabla T + \frac{1}{T}\bar{A}\dot{\bar{z}}, \tag{4.3}$$

where J_q denotes the heat flux through the boundary $\partial\Omega_t$ of the system, the term of exchange $\dot{s}_e = -\nabla.(\frac{J_q}{T})$ and the production $\dot{s}_i = -\frac{J_q}{T^2}.\nabla T + \frac{1}{T}\bar{A}\dot{\bar{z}}$, one recovers the so-called heat equation under the form

$$c_v\dot{T} = -\nabla.J_q + (\bar{A} + c_v\bar{b}_{Tz})\,\dot{\bar{z}}. \tag{4.4}$$

We obtain from (4.4) a generalisation of the Fourier heat equation, if the usual kinetic relation between the heat flux and the temperature field is posed from the onset:

$$J_q = -\underline{\underline{\lambda}}.\nabla T. \tag{4.5}$$

Following this condition, one has

$$c_v\,\dot{T} = \nabla(\underline{\underline{\lambda}}.\nabla T) + (\bar{A} + c_v\,\bar{b}_{Tz})\,\dot{\bar{z}}. \tag{4.6}$$

So, even when the classical assumption of a linear relation between the force and the flux is set up, the relaxation term $(\bar{A} + c_v\,\bar{b}_{Tz})\,\dot{\bar{z}}$ due to different dissipation processes, can lead to a hyperbolic like heat equation, the so-called telegraph equation. At this level, the alternative used by Maugin, $\dot{q} = -(q - \lambda\,\nabla T)/\tau$, and corresponding to the Cattaneo equation (similar to the kinetic relation of the DNLR, equation (2.4)), remains possible. In any case, we have shown that in order to obtain a hyperbolic heat equation, it does not seem necessary to have recourse to the assumptions inherent in the extended irreversible thermodynamics (EIT, [Müller, 1982]; [Jou et al., 1993],), i.e. by complementing the physical description with additive non-equilibrium quantities taking the form of fluxes.

4.2) Application to transfer of matter by diffusion:

When the current co-ordinates are identified with the number of moles n_k of each *k-th* species, ($\underline{y} \to n_k$), the dual variables correspond to the associated chemical potentials, i.e. $\underline{Y} \to \mu_k(n_k;\bar{z})$. In this case, the relation (3.4a) becomes

$$\dot{\mu}_k = \frac{\partial \mu_k(n_k;\bar{z})}{\partial n_p}\dot{n}_p + (\bar{b}_{\mu_k z})\,\dot{\bar{z}} \;; \qquad \frac{\partial \bar{A}(n_k;\bar{z})}{\partial n_k} = \bar{b}_{\mu_k z}. \tag{4.7a,b}$$

Moreover, the balance equation for the k-specie is such that:

$$\dot{n}_k = -\nabla . J_k + \dot{n}_k^{(i)}, \tag{4.8}$$

where J_k represents the flux of diffusion for the *k-th* species, and $\dot{n}_k^{(i)}$ is the internal term due to chemical reaction. Using an Onsager type relation $J_k = -L_{kl}\nabla\mu_l$, one obtains from (4.7a,b) and (4.8) the relation:

$$\dot{\mu}_k = \frac{\partial \mu_k(n_k;\bar{z})}{\partial n_p}\left[\nabla.(L_{pl}\nabla\mu_l) + \dot{n}_p^{(i)}\right] + b_{\mu_k z}\dot{\bar{z}}. \tag{4.9}$$

One recovers a generalised Fick's law, including Gyarmati's formulation when the source terms (chemical reaction) and the relaxation terms are negligible.

4.3) Classical examples of mechanical behaviour.

Using (2.4) and (2.5) and defining more specifically the relaxation time τ_j of the j-process obeying to the equipartition of the entropy production, we can find the stress response. The relaxation spectrum fundamentally depends on the applied load and its history. But, near the equilibrium, according to the fluctuation theory, each process contribution is expressed by its relative weight, p_0^j, in the overall spectrum, i.e. $p_0^j = B\sqrt{\tau^{r,j}}$, with the normalisation condition $\sum_{j=1}^{n} p_0^j = 1$, and n the number of dissipation processes. The evolution equations are obtained by combining (2.3), (2.4) and (2.5), the matrices being eventually functions of the applied strain, $\underline{\underline{\varepsilon}}$, and of the internal variables, $\underline{\bar{z}}$, [Cunat, 1996]. One obtains the response for a strain-rate controlled test, as the following tensorial equation written in rate form:

$$\dot{\underline{\underline{\sigma}}} = \sum_{j=1}^{n}\dot{\underline{\underline{\sigma}}}^j = \sum_{j=1}^{n} p_0^j \underline{\underline{a}}^u : \dot{\underline{\underline{\varepsilon}}} - \sum_{j=1}^{n}\frac{\underline{\underline{\sigma}}^j - p_0^j\underline{\underline{\sigma}}^{r,j}}{\tau^j}. \tag{4.10}$$

i) Visco-elasticity under periodic load.

We consider a mechanical load under an imposed strain $\underline{\underline{\varepsilon}}$, [Rahouadj et al., 1999]. The perturbation $\underline{y}$ and the response $\underline{Y}$ correspond to $\underline{\underline{\varepsilon}}$ and $\underline{\underline{\sigma}}$ respectively. As an example we may consider the visco-elastic behaviour in the simple case of small-applied perturbations and pure shear stress near the thermodynamic equilibrium. We present the dynamic response due to sinusoidally varying perturbations $\gamma_n = \gamma^o \exp(i\,\omega\,t)$ at the applied frequency ω. The response is obtained by integrating the above differential relationship.

The non-linearities are incorporated into the spectrum of relaxation times $\tau^j = \alpha_1(\dot{\underline{\underline{\varepsilon}}})\,\alpha_2(\underline{\underline{\varepsilon}},\underline{\underline{\varepsilon}}^r)\,\alpha_3(T,T_{ref})\,\tau^{r,j}(T_{ref})$, where the three shift functions $\{\alpha_i(.)\}_{i\in\{1,2,3\}}$ stand for the viscosity, deformation and temperature effects, respectively.

For small perturbations, we assume that the corresponding response is periodic and out of phase. For instance, for a shear test, we have $\sigma_{12} = \sigma^o \exp(i\,\omega\,t + \varphi)$, thus $\dot{\sigma}_{12} = i\,\omega\,\sigma_{12}$, where φ is the phase angle. We obtain

$$\dot{\sigma}_{12} = \sum_{j=1}^{n} \dot{\sigma}^j{}_{12} = \sum_{j=1}^{n} p_0^j\, G^u\, \dot{\varepsilon}_{12} - \sum_{j=1}^{n} \frac{\sigma^j{}_{12} - p_0^j\, G^r\, \varepsilon_{12}}{\tau^j}\,. \tag{4.11}$$

The complex modulus is evaluated as

$$G^*(\omega) = G^u + (G^r - G^u)\sum_{j=1}^{n} p_0^j \frac{1}{1 + i\,\omega\,\tau^j}\,. \tag{4.12}$$

It directly follows that its real and imaginary components are respectively

$$G'(\omega) = G^u + (G^r - G^u)\sum_{j=1}^{n} \frac{p_0^j}{1+\omega^2\,(\tau^j)^2}\,; \quad G''(\omega) = (G^r - G^u)\sum_{j=1}^{n} \frac{p_0^j\,\omega\,\tau_j}{1+\omega^2\,(\tau^j)^2}\,. \tag{4.13a,b}$$

This modelling has been compared with other well-known approaches (e.g. [Cole-Cole, 1941] and [Havriliak et al., 1966]), and it proves successful for a wide variety of polymers [Rahouadj et al., 1999].

ii) Visco-plasticity.

We now consider the example of creep, i.e. $\dot{\underline{\underline{\sigma}}} \equiv 0$. The perturbation $\underline{y}$ and the response $\underline{Y}$ correspond to $\underline{\underline{\sigma}}$ and $\underline{\underline{\varepsilon}}$ respectively. We find a coupling effect for multiaxial and anisothermal solicitations, introducing strong non-linearities, especially by the temperature dependence of the relaxation times τ^j:

$$\dot{\underline{\underline{\varepsilon}}} = \underline{\underline{\underline{\underline{a}}}}^{u-1} : \sum_{j=1}^{n}\left(\frac{\underline{\underline{\sigma}}^j - p_0^j \underline{\underline{\sigma}}^{r,j}}{a(\varepsilon)\,\tau^{r,j}}\right) - \underline{\underline{\underline{\underline{a}}}}^{u-1} : \underline{\underline{\alpha}}^{-1}\dot{s}\,. \tag{4.14}$$

$$\dot{T} = \left(\frac{T}{c_v}\right)\dot{s} + \frac{T}{c_v}\sum_{j=1}^{n}\left(\frac{s^j - s^{r,j}}{a(\varepsilon)\,\tau^{r,j}}\right). \tag{4.15}$$

where $\underline{\underline{\alpha}}$ denotes the tensor of the thermal expansion coefficients. For isothermal creep and under uniaxial applied stress σ^0, [Marceron, 1999], and neglecting the entropic coupling, the constitutive equation (4.14) becomes

$$\dot{\varepsilon} = \frac{1}{E^u} \sum_{j=1}^{n} \frac{\sigma_0^j - \sigma^{r,j}}{\tau^j}, \tag{4.16}$$

where E^u is the instantaneous Young modulus; the constant applied stress is $\sigma_0 = \sum_{j=1}^{n} \sigma_0^j$.

To describe the primary regime of creep, one can simply use a Hookean modelling of the relaxed stress $\sigma^r = E^r \varepsilon$. In addition, by considering microscopic re-organisations responsible for the softening of the relaxed stress, one can easily describe the secondary (or steady) and the tertiary creeps. Indeed, the driving force (σ^v) expressed by the difference $\sigma^v = (\sigma_0 - \sigma^r(\varepsilon))$ in relation (4.16) depends directly on the phenomenological modelling of the relaxed stress as a function of the history.

5. CONCLUSION

The goal of this work was to formulate the constitutive laws originating from the DNLR approach within a variational framework, associated to Hamilton's least action principle.
We have shown that $\dot{e}$, or more generally a potential $\dot{\Psi}$, containing the whole thermodynamic information on the continuum, is a good candidate for building a Lagrangian which optimises the evolution during a fixed time interval, $[t_1, t_2]$, even for irreversible paths. At this stage, it is not necessary to specify the complementary laws that we must introduce to close the system of equations.
The necessary main extension of the present work is the setting up of a complete variational principle, which is under way, using a point of view different from Anthony's. This formalism will be able to describe nonuniform media, provided the gradients of the thermodynamic (and pseudo-thermodynamic) variables are incorporated.

REFERENCES

[Anthony, 1986] Anthony, K.-H., *A New Approach to Thermodynamics of Irreversible Processes by Means of Lagrange-Formalism*; in *Disequilibrium and Self-Organisation*; C.W. Kilmister (ed.); by D. Reidel Publishing Company; pp. 75-82.
[Biot, 1954]. Biot, M.A.; *Theory of Stress-Strain Relations in Anisotropic Viscoelasticity and Relaxation Phenomena*; *J. of Applied Phys.*, **25**, **11**, pp. 1385-1391.
[Cole et al., 1941] Cole, K.S.; Cole, R.H.; *Dispersion and Absorption in Dielectrics;* J. Chem. Phys. **9**, pp. 341.
[Coleman et al., 1967] Coleman, B. D.; Gurtin, M. E; *Thermodynamics with Internal State Variables*; J. Chem. Phys. **47**, 2, pp. 597-613.

[Cunat, 1996] Cunat, C., *Lois Constitutives de Matériaux Complexes Stables ou Vieillissants. Apport de la Thermodynamique de la Relaxation, Rev. Gén. Therm.*, **35**, pp. 680-685.

[De Donder, 1920] De Donder, T.; *Leçon de Thermodynamique et de Chimie Physique*, Gauthiers-Villars, Paris.

[De Groot et al., 1962] De Groot, S.R., Mazur, P.; *Non-Equilibrium Thermodynamics*; North-Holland, Amsterdam (Reprint, Dover, New York, 1986).

[Faccio-Toussaint,1997] Faccio-Toussaint, E.; *Thermodynamique Nonlinéaire des Processus Irréversibles et Comportement Mécanique des Matériaux. Modélisation et Interprétation Microphysique; Thèse de Doctorat de l'INPL, LEMTA, ENSEM*, Nancy.

[Glansdorff et al., 1971] Glansdorff, P.; Prigogine I.; *Structure, Stabilité et Fluctuations*, Masson, Paris.

[Gyarmati, 1970] Gyarmati, I.; *Non-Equilibrium Thermodynamics. Field Theory and Variational Principles*; Springer-Verlag.

[Havriliak et al., 1966] Havriliak, S.; Negami, S., *A complex Plane Analysis of α-Dispersions in Some Polymer Systems,* J. Polym. Sci., Part C, **14**, Ed. R.F. Boyer, pp. 99.

[Jou et al., 1993] Jou, D.; Casas-Vazquez, J.; Lebon, G.; *Extended Irreversible Thermodynamics*, Springer Verlag, Berlin.

[Lavenda, 1993] Lavenda, B.H.; *Thermodynamics of Irreversible Processes.* Dover Publications, New York.

[Marceron, 1999] Marceron, P.; *Sur le Rôle des Potentiels Généralisés en Thermodynamique de la Relaxation. Application au Comportement Mécanique des Polymères; Thèse de Doctorat de l'INPL, LEMTA, ENSEM*, Nancy.

[Maugin, 1990] Maugin, G.A.; *Internal Variables and Dissipative Structures*; J. Non-Equilib. Thermodyn., **15**, **2**, pp.173-192.

[Meixner, 1949] Meixner, J. Z.; *Thermodynamische Theorie der Elastischen Relaxation*, Z. Naturforsch., **4a**, (1949), pp. 494-600.

[Müller, 1982]. Müller, I.; *Rational Extended Thermodynamics*, Centro Internazionale Matematico Estivo (C.I.M.E), Noto, Italy.

[Poincaré, 1908] Poincaré, H.; *Thermodynamique*, deuxième édition, Gauthier-Villars, Paris.

[Prigogine, 1955] Prigogine, I.; *Introduction to Thermodynamics of Irreversible Processes*; Springfield, ed. G. Thomaes.

[Rahouadj et al., 1999] Rahouadj, R., Cunat, C.; *A Non-Linear Visco-Elastic Model based on Fluctuating Modes*; In *Handbook of Materials Behaviour – Nonlinear Models and Properties.* Ed. J. Lemaitre. Submitted for publication.

[Rayleigh, 1873] Rayleigh, L.; *Theory of Sound*; *Proc. London math. Soc.*, **4**, pp. 357-363.

[Sidoroff, 1970] Sidoroff,. F.; *Sur Certains Modèles de Milieux Continus Dissipatifs en Déformations Finies. C.R. Acad. Sc. Paris, Série A*, t. **270**, pp. 136-139.

[Tisza, 1966] Tisza, L.; *Generalised Thermodynamics*; The M.I.T. Press.

[Ziegler, 1963] Ziegler, H.; *Some Extremum Principles in Irreversible Thermodynamics with Applications to Continuum Mechanics*; in *Progress in Solid Mechanics*; Ed. I.N. Sneddon and R. Hill; **4**; pp. 191-193, North-Holland, Amsterdam.

Second-gradient theory : application to Cahn-Hilliard fluids

P. Seppecher

Laboratoire d'Analyse Non Linéaire Appliquée
Université de Toulon et du Var
BP 132 - 83957 La Garde Cedex
seppecher@univ-tln.fr

Abstract. The second gradient theory is indispensable for describing a particular class of materials: when the energy depends on the second gradient of the displacement. We show that trying to overcome the difficulty by using extended thermodynamics is similar to trying to describe a linear elastic material without invoking the Cauchy stress tensor. We illustrate the principal features of second gradient theory with the example of Cahn-Hilliard fluids.

Keywords: Second-gradient theory, Cahn-Hilliard fluid, Extended thermodynamics

1. INTRODUCTION

Force and moment are the basic concepts in the mechanics of rigid bodies, but it is well known that these notions are not sufficient for an accurate description of internal forces in a deformable body. A great improvement in continuum mechanics has been achieved with the introduction of the Cauchy stress tensor. The aim of this paper is to show that, for some particular materials, a comparable improvement is obtained by using the second gradient theory formalized by P. Germain in [Germain, 1973 a], [Germain, 1973 b]. This theory can be formulated only in the framework of the principle of virtual power : one assumes that the power $\widetilde{\mathcal{P}}^{(i)}$ of internal forces in a domain Ω, for any possible virtual velocity field $\tilde{V}$ can be represented by two tensor fields Σ, S (of order two and three, respectively) as follows

$$\widetilde{\mathcal{P}}^{(i)} = -\int_\Omega (\Sigma : \nabla\tilde{V} + S \,\vdots\, \nabla\nabla\tilde{V})\, dv \,. \tag{1}$$

If we assume that long range external forces can be represented by a volume density $f^{(e)}$, the application of the principle of virtual power, after integration by parts, leads to the local equation:

$$\rho\frac{dV}{dt} = div(\Sigma - div(S)) + f^{(e)} \tag{2}$$

G.A. Maugin et al. (eds.), Continuum Thermomechanics, 379-388.
© 2000 *Kluwer Academic Publishers.*

where ρ denotes the mass density. In this momentum balance equation, the tensor $\sigma := \Sigma - div(S)$ plays the role of the Cauchy stress tensor. Using σ, we may write the power of internal forces as:

$$\widetilde{\mathcal{P}}^{(i)} = -\int_{\Omega} (\sigma : \nabla\tilde{V} + div(\nabla\tilde{V}^t : S))\, dv. \tag{3}$$

This power is not determined only by strain and the Cauchy stress tensor : an extra flux $\nabla\tilde{V}^t : S$ appears. We focus on this main feature of the theory in section 4 while we study boundary conditions in section 6.

The questions which naturally arise are i) what is the connection between this theory and the Cauchy stress theory? ii) is there any material which needs for such a theory? iii) is it possible to reformulate the theory without invoking the virtual power principle, and what are the connections with interstitial working or extended thermodynamics theories? iv) what are the principal mechanical features of this theory? We illustrate the answers to these different questions by the example of Cahn-Hilliard fluids.

2. SECOND-GRADIENT THEORY AND CAUCHY CONSTRUCTION OF STRESS TENSOR

To understand how one can get out of the framework of Cauchy continua, we need to reconsider the basic assumptions for Cauchy's construction of the stress tensor : the main postulate states that contact forces can be represented by a surface density of forces which depends only on the normal to the boundary of the considered domain. Noll's theorem establishes this postulate under the implicit assumption that no edge forces are present.

[Noll et al., 1990], [Dell'Isola et al., 1997] have considered the case of continuous media in which edge forces are present : it has been proved that this presence implies a dependence of contact surface forces on the curvature of the boundary of the domain. To be precisely described, the contact forces need two stress tensors (Σ, S). Moreover, edge forces imply the presence of another type of contact forces which cannot be interpreted in terms of forces or moments, and which correspond to the double forces described by P.Germain in [Germain, 1973 a]. In other words, when edge forces are present, a construction similar to Cauchy's again leads to a second-gradient theory. Moreover, as contact forces contain double forces (i.e. distributions of order one), duality (i.e. virtual power principle) cannot be avoided.

As a matter of fact, edge forces are present in some particular materials, such as Cahn-Hilliard fluids (see section 6).

3. SECOND GRADIENT MATERIALS

Let us call a material with a free energy density which depends on the second gradient of the deformation a *second gradient material.* We will see in the next section why such materials need to be described in the framework of the second gradient theory. Such materials do exist. Some have been studied for a long time : the beam model introduced by Euler, Bernouilli and Navier is a one dimensional second gradient material; the first 3D-model was introduced by Cosserat ([Cosserat et al., 1909]). However, in these examples, the way the free energy density depends on the second gradient of the deformation is special : only the gradient of the skew part of the gradient of displacement plays a role. These special cases are called *incomplete second gradient materials* in [Alibert et al., 1999].

More recently, a second gradient material has been introduced by Cahn and Hilliard for describing capillary phenomena [Cahn et al., 1959]. In this model one assumes that the free energy density ψ depends on the gradient of mass density in the simple form:

$$\psi = W(\rho) + \frac{\lambda}{2}(\nabla\rho)^2. \tag{4}$$

Note that, for sake of simplicity, we assume in this paper that the phenomena occur in isothermal conditions. Clearly, the dependence on ρ of the gradient of the displacement shows that a Cahn-Hilliard fluid is a complete second gradient material.

The simple form (4) is probably not sufficient for the study of complex fluids. However, it allows one to investigate phenomena which otherwise could not be studied in the framework of continuum mechanics : dependence of surface tension upon the radius for very small droplets [Dell'Isola et al., 1995], stability of wetting films [Seppecher, 1993], interpretation of Young's law for contact angle in terms of forces [Seppecher, 1989], removal of the moving contact line paradox [Seppecher, 1996], line tension phenomenon [Alberti et al., 1998], etc. The model can be used also for numerical studies where it removes the difficulty of tracking interfaces.

In expression (4), W is a non-convex function. This is an important point. Any energy described by a non convex function leads, in general, to ill-posed equilibrium problems : minimizing $\int_\Omega W(\rho)dx$ under some mass constraint has, in general, no solution. This phenomenon arises in every mutiphase problem, in damage or plasticity theories, etc. Adding to the energy a term depending, in a convex way, on higher gradient of deformation has two different interests : from a mathematical point of view, it leads to well-posed "regularized problems"; from a mechanical point of view, it gives a description of the transition zones which divide the phases.

Apart from their mechanical and mathematical interest, there is another reason to require that any general theory of continuum mechanics be able to describe second gradient materials. As continuum mechanics is a homogenized description of materials which are heterogeneous at the microscopic level, it is natural to expect it to be stable under homogenization or asymptotic analysis procedures. Cauchy continua do not enjoy this stability property : it is well known that the 1-D or 2-D second gradient models of beams or plates are the asymptotic limits of Cauchy models. It has recently been proved that homogenization of heterogeneous Cauchy materials can lead to a 3-D incomplete second gradient material [Pideri et al., 1997]. The fact that a complete second gradient material, and more generally any n-th gradient material, can be obtained in that way is, as far as I know, a conjecture.

In the next section we recall why second gradient materials cannot be considered as classical Cauchy continua.

4. THERMODYNAMICAL PARADOX OF SECOND GRADIENT MATERIALS

Let us write the laws of thermodynamics in a very general way. The first law states that the variation of total energy is due to the mechanical power of the external forces $\mathcal{P}^{(e)}$, and a heat supply Q_e :

$$\dot{E} + \dot{K} = \mathcal{P}^{(e)} + Q_e \tag{5}$$

where $\dot{E}$ and $\dot{K}$ denote respectively the variations of internal and kinetic energies. Taking into account the fact that the variation of kinetic energy coincides with the power of internal and external forces

$$\dot{K} = \mathcal{P}^{(i)} + \mathcal{P}^{(e)}, \tag{6}$$

we get for the variation of internal energy:

$$\dot{E} = -\mathcal{P}^{(i)} + Q_e \; . \tag{7}$$

The second law states that the variation of entropy $\dot{S}$ is larger than the entropy supply Q_s:

$$\dot{S} \geq Q_s \; . \tag{8}$$

If we accept the relatively weak assumptions that E, S, $\mathcal{P}^{(i)}$ can be represented by volume densities e, s, $p^{(i)}$; and the supplies Q_e, Q_s by fluxes J_e, J_s, then inequality (8) provides the Clausius-Duhem inequality :

$$T div(J_s) - div(J_e) - p^{(i)} + \rho \left(T \frac{d}{dt}(\frac{s}{\rho}) - \frac{d}{dt}(\frac{e}{\rho}) \right) \geq 0 \tag{9}$$

and in isothermal conditions

$$div(T J_s - J_e) - p^{(i)} - \rho \frac{d}{dt}(\frac{\psi}{\rho}) \geq 0 \tag{10}$$

where ψ denotes the volume free energy: $\psi = e - Ts$ and T the absolute temperature. The thermodynamical paradox of second gradient materials lies in the incompatibility of the Clausius-Duhem inequality (10) with the three following assumptions:

(H_1) The free energy density ψ depends on the second gradient of the displacement.

(H_2) $p^{(i)} = -\sigma : \nabla V$ (where σ denotes the Cauchy stress tensor).

(H_3) $TJ_s = J_e$, which classically results from two assumptions: J_e coincides with the heat flux ($J_e = q$) and J_s coincides with the ratio of heat flux and absolute temperature.

Indeed, under assumption (H_3), the term $div(TJ_s - J_e)$ vanishes. Due to assumption (H_1), $\frac{d}{dt}(\frac{\psi}{\rho})$ contains a term depending linearly on $\nabla\nabla V$ which cannot be balanced by $p^{(i)}$ under assumption (H_2).

This becomes clear when considering the example of Cahn-Hilliard fluids : when the dependence of ψ on the displacement is given by (4), $\frac{d}{dt}(\frac{\psi}{\rho})$ can easily be computed and, under assumptions (H_2) and (H_3), inequality (10) becomes

$$\Big(\sigma - (W(\rho) - \rho\frac{dW}{d\rho} - \frac{\lambda}{2}\nabla\rho^2)\mathrm{Id} + \lambda\nabla\rho\otimes\nabla\rho\Big) : \nabla V$$
$$+ \lambda\rho(\mathrm{Id}\otimes\nabla\rho) \vdots \nabla\nabla V \geq 0 \tag{11}$$

which cannot be assured by any constitutive equation for σ.

As, for a second gradient material, (H_1) is a basic assumption, the only possibilities for solving the paradox involve revising (H_2) or (H_3). One can revise (H_2) by using the second gradient theory, assuming that $\mathcal{P}^{(i)}$ has the form (1), i.e., in a local form:

$$p^{(i)} = -\Sigma : \nabla V - S \vdots \nabla\nabla V \tag{12}$$

or, in the equivalent form (3),

$$p^{(i)} = -\sigma : \nabla V - div(\nabla V^t : S) \; . \tag{13}$$

One can revise (H_3) in two ways: either, following [Dunn, 1986], by introducing an "interstitial working" flux J_i, writing

$$J_e = q + J_i \tag{14}$$

or, as in [Müller, 1985], by writing

$$TJ_s = J_e - J_i \; . \tag{15}$$

In these two ways, which we can call extended-thermodynamics methods, one has to search for constitutive equations for σ and J_i which assure inequality (10) while, in the second gradient theory, one has to look for constitutive equations for Σ and S. Clearly, the two extended-thermodynamics methods are equivalent. The only difference is a question of nomenclature : what is called "heat flux".

At first glance, the second gradient method seems also to be equivalent to the extended-thermodynamics methods. The term $\nabla V^t : S$ plays the role of J_i, and the difference is a question of nomenclature: what is called "power of internal forces". This is not true. Constitutive equations for Σ and S concern all admissible velocity fields, and not only the real one; the second gradient theory is stronger. To make this point clear, we show in the next section the consequences of an application of extended thermodynamics to classical Cauchy continua.

Before showing this, let us conclude this section by writing possible constitutive equations for a Cahn-Hilliard fluid. For the sake of simplicity we only recall the non dissipative constitutive equations (for the dissipative case refer to [Seppecher, 1996]). Then inequality (10) is actually an equality,

$$div(TJ_s - J_e) - p^{(i)} - \rho\frac{d}{dt}(\frac{\psi}{\rho}) = 0 \tag{16}$$

which is satisfied if we set :
either, in the second-gradient theory,

$$\begin{aligned} &TJ_s - J_e = 0, \quad p^{(i)} = -\Sigma : \nabla V - S \vdots \nabla\nabla V, \\ &\Sigma = (W(\rho) - \rho\frac{dW}{d\rho} - \frac{\lambda}{2}\nabla\rho^2)\,\mathrm{Id} - \lambda\,\nabla\rho\otimes\nabla\rho, \quad S = -\lambda\rho(\mathrm{Id}\otimes\nabla\rho)\ , \end{aligned} \tag{17}$$

or, in the extended-thermodynamics framework,

$$\begin{aligned} &TJ_s - J_e = J_i, \quad J_i = -\lambda\rho\, div(V)\nabla\rho, \quad p^{(i)} = -\sigma : \nabla V, \\ &\sigma = (W(\rho) - \rho\frac{dW}{d\rho} + \frac{\lambda}{2}\nabla\rho^2 + \lambda\rho\,\Delta\rho)\,\mathrm{Id} - \lambda\,\nabla\rho\otimes\nabla\rho\ . \end{aligned} \tag{18}$$

5. A WEAKNESS OF EXTENDED THERMODYNAMICS

In this section we show that applying the methods of extended thermodynamics proposed in (14) or in (15) to Cauchy materials is a vain attempt to describe these materials without invoking the Cauchy stress tensor. For this purpose, let us imagine, for a moment, that we are beginners in continuum mechanics; we try to describe a classical material, for instance a linear elastic one, and we make the following (frequent) error : the divergence of the Cauchy stress tensor $div(\sigma)$ corresponds to the volume density of internal forces; denoting it by $f^{(i)}$), we write that the power of internal forces is

$$\mathcal{P}^{(i)} = \int_\Omega f^{(i)} \cdot V\, dv. \tag{19}$$

Is that really an error? If we use extended-thermodynamics methods (14) or (15), we may introduce an extra flux J_i (considering it, following the different points of view, as an interstitial working or an extra entropy flux) and get the following set of equations

$$\rho\frac{d}{dt}V = f^{(i)} \tag{20}$$

$$\rho\frac{d}{dt}(\frac{e}{\rho}) = f^{(i)} \cdot V - div(J_i) - div(q) \tag{21}$$

where $f^{(i)}$ and J_i are given by adequate constitutive equations.

It is remarkable that i) this set of equations is (with the right constitutive equations) totally equivalent to the classical set of equations (we let the reader compute the right constitutive equations which give a set of equations totally equivalent to the equations of linear elasticity), ii) in such a presentation the notion of Cauchy stress tensor is not needed!

Actually, the only but important weakness of the formulation is that boundary conditions *cannot* be written. Extended-thermodynamics is not able to detect the original mechanical error (19). We claim that the same phenomenon occurs when extended-thermodynamics are used to describe a second gradient material, i.e. when σ and J_i are used instead of Σ and S for the representation of the mechanical state. The set of local equations is the same in both cases, but only the second gradient theory gives the right boundary conditions. We recall these conditions in the next section.

6. BOUNDARY CONDITIONS IN SECOND-GRADIENT THEORY

Let us recall the boundary conditions given by the second-gradient theory (details of computation can be found in [Germain, 1973 a] or [Seppecher, 1989]). Integration by parts of the virtual power of internal forces given by (3) yields

$$\widetilde{\mathcal{P}}^{(i)} = \int_\Omega div(\sigma) \cdot \tilde{V}\, dv - \int_{\partial\Omega} (\tilde{V} \cdot \sigma + \nabla\tilde{V}^t : S) \cdot n\, ds \tag{22}$$

where n denotes the external normal to the domain Ω. A second integration by parts on the boundary (assumed to be sufficiently regular) gives

$$\widetilde{\mathcal{P}}^{(i)} = \int_\Omega div(\sigma) \cdot \tilde{V}\, dv - \int_{\partial\Omega} \tilde{V} \cdot (\sigma \cdot n - div^s((\mathrm{Id} - n \otimes n) \cdot S \cdot n)))\, ds$$
$$- \int_{\partial\Omega} (n \cdot \nabla\tilde{V}^t) \cdot (S : (n \otimes n))\, ds \tag{23}$$

where div^s denotes the intrinsic surface divergence operator. As $\tilde{V}$ and $n \cdot \nabla\tilde{V}^t$ are independent fields on $\partial\Omega$, the dual quantities $\sigma \cdot n - div^s((\mathrm{Id} - n \otimes n) \cdot S \cdot n))$ and

$S:(n\otimes n)$ should be given there. If the exterior is again a second-gradient material, these quantities should be continuous across $\partial\Omega$. In all generality they are fixed by external contact interaction:

$$\sigma \cdot n - div^s((\mathrm{Id} - n \otimes n) \cdot S \cdot n)) = F^{(e)} \tag{24}$$

$$S : (n \otimes n) = G^{(e)} \tag{25}$$

The quantity $-F^{(e)}$ corresponds to the surface density of forces exerted by the material on the boundary. It is remarkable that, as $F^{(e)}$ depends on the curvature of $\partial\Omega$ (i.e. on $div^s(n)$), Cauchy's postulate is not verified. When the curvature of the boundary is very large, $F^{(e)}$ becomes singular; indeed when $\partial\Omega$ is piecewise regular, a line density of forces appears on the edges $\partial\partial\Omega$: in equation (23) one must add the term

$$\int_{\partial\partial\Omega} \tilde{V}^t \cdot S : [[n \otimes \nu]]\, d\ell. \tag{26}$$

Here, n and the vector ν, which is tangent to the boundary and normal to the edge, are discontinuous across the edge and $[[n\otimes\nu]]$ denotes their jump. On the edges the line density of forces

$$S : [[n \otimes \nu]] = \mathcal{F}^{(e)} \tag{27}$$

is exerted. If the velocity vanishes on the boundary, the boundary conditions (24) and (27) are replaced by the dual condition $\tilde{V} = 0$ on $\partial\Omega$ but the condition (25) remains valid.

The quantity $G^{(e)}$ has the dimension of a moment. Its tangent part actually corresponds to a surface density of moment but the normal part is a surface density of double forces.

For a non dissipative Cahn-Hilliard fluid, equation (17) leads to the boundary conditions :

$$\sigma \cdot n - div^s((\mathrm{Id} - n \otimes n) \cdot S \cdot n)) = F^{(e)} \tag{28}$$

$$-\lambda\rho\,(n \cdot \nabla\rho)\, n = G^{(e)} \tag{29}$$

$$-\lambda\rho\,[[(n \cdot \nabla\rho)\, \nu]] = \mathcal{F}^{(e)} \tag{30}$$

Note that a non-dissipative Cahn-Hilliard fluid can be subjected only to normal double forces $G^{(e)}$ (just like a non-viscous fluid can be subjected only to normal forces). Hence let us set $G^{(e)} = g^{(e)} n$. The equilibrium state of a Cahn-Hilliard fluid in a rigid container, in micro-gravity conditions, is then given by the following equations:

$$div(\sigma) = \rho\nabla(-\frac{dW}{d\rho} + \lambda\Delta\rho) = 0 \tag{31}$$

$$-\lambda\rho\,(n \cdot \nabla\rho) = g^{(e)} \tag{32}$$

Let us compare these boundary conditions with the natural boundary conditions which are obtained using the extended-thermodynamics method (18):

$$\sigma \cdot n = \overline{F}^{(e)} \tag{33}$$

$$J_i \cdot n = \overline{g}^{(e)} \tag{34}$$

The first condition disappears when the velocity vanishes on the boundary, and the second condition reads

$$-\lambda\rho\, div(V)(n \cdot \nabla\rho) = \overline{g}^{(e)} \tag{35}$$

This condition is close to (32), but becomes trivial in case of equilibrium. Equation (32) is stronger as it implies an equation of type (35) for all admissible velocity fields and not only the real one.

Clearly the equilibrium problem is well-posed only when the boundary condition (32) is used. The effect of the value of $g^{(e)}$ on equibrium is quite important: it tunes the value of Young's contact angle. This parameter is characteristic of the wetting properties of the wall of the container; this has been shown numerically in [Seppecher, 1989] [Seppecher, 1996] and rigourously in [Modica, 1985].

The mechanical concepts of double forces and hyperstress are not purely theoretical concepts. The Cahn-Hilliard fluid provides an example of material for which these concepts correspond to experimental evidence and are as useful as the classical concepts of surface forces and stress tensor.

References

[Alberti et al., 1998] Alberti G., Bouchitté G., Seppecher P.; Phase transition with line tension effect; *Arch. Rational Mech. Anal.* , **Vol.144**, pp. 1-46.

[Alibert et al., 1999] Alibert J.J., dell'Isola F. and Seppecher P.; Truss modular beams the deformation energy of which depends on higher displacement gradients; Submitted to *Math. and Mech. of Solids* .

[Cahn et al., 1959] J.W. Cahn and J.E. Hilliard; Free energy of a non-uniform system, *J. Chem. Phys.* , **Vol.31**, pp. 688-699.

[Cosserat et al., 1909] Cosserat E. and Cosserat F.; Sur la Théorie des Corps Déformables, *Herman* , Paris.

[Dell'Isola et al., 1995] Dell'Isola F., Gouin H. and Seppecher P.; Radius and surface tension of microscopic bubbles by second gradient theory, *C.R.Acad.Sci.* , **t. 320, Série II b**, pp. 211-216.

[Dell'Isola et al., 1997] Dell'Isola F. and Seppecher P.; Edge contact forces and quasi-balanced power, *Meccanica* , **Vol.32**, pp. 33-52.

[Dunn, 1986] Dunn J.E.; Interstitial working and non classical continuum thermodynamics, *New Perspectives in Thermodynamics* , Springer Verlag, Berlin, pp. 187-222.

[Germain, 1973 a] Germain P.; La méthode des puissances virtuelles en mécanique des milieux continus. Premiere partie: Théorie du second gradient, *Journal de Mécanique* , **Vol. 12, N. 2**, pp. 235-274.

[Germain, 1973 b] Germain P.; The method of virtual power in continuum mechanics. Part 2: Microstructure, *S.I.A.M. J. Appl. Math.*, **Vol. 25, N. 3**, pp. 556-575.

[Modica, 1985] Modica L.; Gradient theory of phase transitions with boundary contact energy, *Ann. inst. H. Poincaré, anal. Non Linéaire* , **Vol.5**, pp. 453-486.

[Müller, 1985] Müller I.; *Thermodynamics, Pitman Ed., London.*

[Noll et al., 1990] Noll W. and Virga E.G.; On edge interactions and surface tension, *Arch. Rational Mech. Anal.* , **Vol.111, N.1**, pp. 1-31.

[Pideri et al., 1997] Pideri C. and Seppecher P.; A second gradient material resulting from the homogenization of an heterogeneous linear elastic medium, *Continuum Mech.and Thermodyn.* , **Vol.9**, pp. 241-257.

[Seppecher, 1989] Seppecher P. ; Etude des conditions aux limites en théorie du second gradient: cas de la capillarité, *C. R. Acad. Sci. Paris* , **t. 309, Série II**, pp. 497-502.

[Seppecher, 1993] Seppecher P.; Equilibrium of a Cahn and Hilliard fluid on a wall: influence of the wetting properties of the fluid upon the stability of a thin liquid film, *European Journal of Mechanics B/Fluids* , **Vol. 12, 1**, pp. 69-84.

[Seppecher, 1996] Seppecher P.; A numerical study of a moving contact line in Cahn-Hilliard theory, *Int. J. Engng. Sci.* , **Vol 34., N. 9**, pp. 977-992.

Thermodynamics and duality in finite elastoplasticity

F. Sidoroff
Laboratoire de Tribologie et Dynamique des Systèmes
Ecole Centrale de Lyon - BP 163 - F 69130 ECULLY CEDEX
e-mail : Francois.Sidoroff@ec-lyon.fr
A. Dogui
Laboratoire de Mécanique des Solides
Ecole Nationale d'Ingénieurs de Monastir - Tunisie
e-mail : Abdel.Dogui@enim.rnu.tn

Abstract : An elastic-plastic constitutive model is essentially based on the decomposition of the deformation into elastic and plastic parts, with the corresponding equations relating the elastic strain and plastic strain rate to appropriate stress tensors. For large strain formulation, the identification of these dual stress tensors is an important problem which has always been an important issue. The purpose of the present paper is to present an overview of the past 30 years of research in this field with duality as a guideline. It will be emphasised that each progress has resulted from an extended interpretation of the same initial kinematic framework completed by a duality analysis to obtain the elastic law and the dual stress tensor to be used in the plastic flow rule. Some alternative models to the main trend will also be suggested.
Keywords : Thermodynamics, duality, elastoplasticity, large strain.

1. INTRODUCTION

For more than 30 years now and since the pioneering works of Green-Naghdi, Lee, and Mandel, elastic plastic modelling at large strain has been an important and sometimes controversial subject. From a fundamental point of view, the essence of elasto-plasticity consists in an elastic behaviour (i.e. a conservative relation between strain and stress), superimposed on a fluid-like behaviour (i.e. a dissipative relation between strain rate and stress), which may be viscous (linear, Maxwell model), plastic (rate-independent) or viscoplastic (the general case, with or without a threshold).

In case of small transformations (geometrical linearity, small strain and small rotation) this superposition corresponds to an additive decomposition of the infinitesimal strain tensor

G.A. Maugin et al. (eds.), Continuum Thermomechanics, 389-400.
© 2000 *Kluwer Academic Publishers.*

$$\varepsilon = \varepsilon^e + \varepsilon^p \qquad (1)$$

where ε^e is related to σ through an elastic law while the plastic strain rate $\dot{\varepsilon}^p$ is related to σ through a viscous, viscoplastic or plastic law. This simple primary model (perfect plasticity) may then be refined for instance through the introduction of hardening.

These models are usually developed and presented within a thermodynamic framework. This is natural because the distinction between elasticity and plasticity is of thermodynamic nature : elastic = conservative, plastic = dissipative. This is also very helpful for the development of new hardening models (generalised standard materials). Thermodynamics within this framework therefore is natural and helpful. It is however not essential and even if it has been extensively used by the French School under the influence of P.Germain, [Germain, 1974] similar models have been developed elsewhere without using it.

The situation however is quite different when dealing with geometrical non-linearity and large transformations. Here the kinematic splitting of the deformation into elastic and plastic parts is not so simple and even if there is now a general agreement about the fact that the additive decomposition (1) has to be replaced by a multiplicative one, there still are many possible interpretations for this decomposition. Thermodynamics – or at least some part of it connected with duality – then becomes essential for the translation of these interpretations into a complete consistent constitutive framework and in particular to obtain both the elastic law and plastic strain rate with its associated driving force which will be called here the plastic stress.

The purpose of the present paper is to present a survey of these 30 years of development of elastic-plastic models at large strain within the light of this unifying concept : thermodynamics or duality used for obtaining the elastic law as well as the plastic stress and strain rate tensors. The presentation will follow the history starting from the original isotropic models of Mandel, Lee and Green-Naghdi, following then with the developments related to kinematic hardening and plastic spin. We shall also mention some alternative formulations outside this main stream but which may be helpful or instructing.

2. SMALL PERTURBATIONS

2.1 Perfect plasticity and isotropic hardening

Before dealing with large deformations, it is important to recall the small strain formulation [Maugin, 1992]. From a methodological point of view this will show us the way to proceed. This will also help us to clarify where and how thermodynamics and duality are used.

In the basic plastic model with isotropic hardening the energy is assumed as

$$w = \frac{1}{2}\varepsilon^e A \varepsilon^e + \hat{w}(p) \qquad (2)$$

where A is the stiffness tensor, and $\hat{w}(p)$ is the stored energy resulting from the isotropic hardening variable p. Substitution in the dissipation inequality directly results in two alternative forms

$$\begin{aligned}\phi = \sigma:\dot{\varepsilon} - \dot{w} &= \left(\sigma - A\varepsilon^e\right)\dot{\varepsilon}^e + \sigma\dot{\varepsilon}^p - R\dot{p} \\ &= \left(\sigma - A\varepsilon^e\right)\dot{\varepsilon} + A\varepsilon^e\dot{\varepsilon}^p - R\dot{p} \qquad R = d\hat{w}/dp\end{aligned} \tag{3}$$

Since dissipation is associated to plasticity, the dissipation must vanish with $\dot{\varepsilon}^p$ so that both forms give the elastic law and dissipation inequality as

$$\sigma = A\varepsilon^e \qquad \phi = \sigma\dot{\varepsilon}^p - R\dot{p}$$

$$f(\sigma, R) = f_0(\sigma) - R \leq 0 \qquad \dot{\varepsilon}^p = \lambda\frac{\partial f_0}{\partial\sigma} \qquad \dot{p} = -\lambda\frac{\partial f}{\partial R} = \lambda$$

Thermodynamics therefore has been used at two different steps of the procedure
i) to obtain (3), i.e. to derive the elastic law and to identify the dual variables to be used in the plastic evolution law. This is essentially a duality concept.
ii) to postulate the evolution equations (normal dissipation). This is a constitutive assumption which is often helpful but in no way essential. In particular it is now well known that in anisotropic plasticity better results are often obtained by using a plastic potential and a non-associated flow rule. This however has nothing to do with i) which is still required to identify σ and $\dot{\varepsilon}^p$ as the dual variables to be related by the plastic flow rule – whether associated or not.

It is this first contribution of thermodynamics which will be considered here and our primary purpose in the following will be to derive the appropriate form of (3) resulting from different kinematic interpretation of finite elastic-plastic deformations. Of course, for a specific model, these equations will have to be completed by plastic flow rules but these may take different forms according to the kind of model which is considered (plastic, viscoplastic or viscous, associated or not) and will not be further considered here. Since isotropic hardening is not concerned in this discussion we shall forget about it and limit our attention to perfect plasticity

2.2 Kinematic hardening

Kinematic hardening corresponds to the translation of the yield surface and therefore requires the introduction of a tensorial variable. Within the framework of generalised standard materials it can be presented in two ways. The first one corresponds to the Prager model introducing as hardening variable the plastic deformation

$$w = \frac{1}{2}\varepsilon^e A\varepsilon^e + \frac{1}{2}\varepsilon^p B\varepsilon^p \qquad \phi = \left(\sigma - A\varepsilon^e\right)\dot{\varepsilon}^e + \left(\sigma - B\varepsilon^p\right)\dot{\varepsilon}^p$$

$$\sigma = A\varepsilon^e \qquad \phi = \sigma^p\dot{\varepsilon}^p \qquad \sigma^p = \sigma - B\varepsilon^p \tag{4}$$

with the following plastic flow rule

$$f(\sigma^p) = 0 \qquad \dot{\varepsilon}^p = \lambda \frac{\partial f}{\partial \sigma^p}$$

The second approach uses as hardening variable a tensor α with

$$w = \frac{1}{2}\varepsilon^e A \varepsilon^e + \frac{1}{2}\alpha B \alpha \qquad \phi = \left(\sigma - A\varepsilon^e\right)\dot{\varepsilon}^e + \sigma \dot{\varepsilon}^p - X\dot{\alpha}$$

$$\sigma = A\varepsilon^e \qquad \phi = \sigma\dot{\varepsilon}^p - X\dot{\alpha} \qquad X = B\alpha \tag{5}$$

Kinematic hardening then corresponds to

$$f(\sigma, X) = f_o(q) \qquad q = \sigma - X$$

where the thermodynamic force X associated to α can be interpreted as the centre of the yield surface (back-stress). The plastic evolution laws then directly follows as

$$\dot{\varepsilon}^p = \lambda \frac{\partial f}{\partial \sigma} = \lambda \frac{\partial f_o}{\partial q} \qquad \dot{\alpha} = -\lambda \frac{\partial f}{\partial X} = \dot{\varepsilon}^p$$

showing that α indeed is ε^p so that this model coincides with the Prager model with α and q identified to ε^p and σ^p. The advantage of this second formulation lies in the fact that it allows a non-associated generalised flow rule like for instance the so-called non-linear kinematic hardening model.

3. THE ISOTROPIC CASE

3.1 Mandel's approach

Mandel's approach is based on the concept of an isoclinic stress free configuration obtained from the actual configuration by local stress relaxation and restoring the initial orientation of 3 director vectors related to the microstructure [Mandel, 1971]. In the case of a single crystal [Teodosiu et al., 1976] these vectors clearly are the crystallographic directions while in the case of a polycrystal their orientation is defined from some mean value of the grain rotations.

From a kinematical point of view this corresponds to a multiplicative decomposition of the deformation gradient **F**

$$\mathbf{F} = \mathbf{A}\,\mathbf{P} \tag{6}$$

where **A** and **P** respectively are the elastic and plastic part of the deformation gradient. The elastic energy (2) is extended as

$$w = w(\Delta^e) \qquad \Delta^e = \frac{1}{2}\left(\mathbf{A}^T\mathbf{A} - \mathbf{1}\right) \tag{7}$$

where Δ^e is the elastic Green-Lagrange strain tensor. Denoting by **T** and $\tau = J\mathbf{T}$ the Cauchy and Kirchhoff stress tensors, the dissipation ϕ is obtained as

$$\phi = \tau : \mathbf{D} - \dot{w} = \left(\tau - \mathbf{A} \frac{\partial w}{\partial \Delta^e} \mathbf{A}^T \right) : \dot{\mathbf{A}} \mathbf{A}^{-1} + \mathbf{A}^T \tau \, \mathbf{A}^{-1^T} : \dot{\mathbf{P}} \mathbf{P}^{-1} \tag{8}$$

which is the appropriate extension of (3_1).

According to the general scheme described in section 2.1 this provides the elastic law and the dual plastic stress and strain rate.

$$\tau = \mathbf{A} \frac{\partial w}{\partial \Delta^e} \mathbf{A}^T \qquad \phi = \Sigma : \dot{\mathbf{A}} \mathbf{A}^{-1} \qquad \Sigma = \mathbf{A}^T \tau \, \mathbf{A}^{-1^T} \tag{9}$$

It should be emphasised that the corresponding plastic strain rate and stress tensors are non-symmetric and that the plastic flow rule must describe in addition to the usual symmetric strain rate the skew symmetric part of the plastic velocity gradient $\dot{\mathbf{P}}\mathbf{P}^{-1}$. This is the so-called plastic spin equation to be discussed later.

This however disappears in the isotropic case for in this case Δ^e commutes with $\partial w / \partial \Delta^e$ so that Σ is symmetric and it can be proved that $\dot{\mathbf{P}}\mathbf{P}^{-1}$ which must be an isotropic function of Σ is also symmetric.

3.2 Lee's approach

Lee also defines the plastic deformation [Lee, 1969] by

$$\mathbf{F} = \mathbf{F}^e \mathbf{F}^p \tag{10}$$

defining a stress free intermediate configuration which in contradiction to Mandel's isoclinic configuration is only defined up to an arbitrary rotation. The elastic energy therefore should not depend on this rotation which requires in fact an isotropic energy

$$w(\mathbf{B}^e) \qquad \mathbf{B}^e = \mathbf{F}^e \mathbf{F}^{e^T} \tag{11}$$

where $\mathbf{B}^e$ is the elastic left Cauchy-Green tensor. Starting from this isotropic elastic energy and after some computation the dissipation is obtained as

$$\phi = \left(\tau - 2\mathbf{B}^e \frac{\partial w}{\partial \mathbf{B}^e} \right) : \mathbf{D} + 2\mathbf{B}^e \frac{\partial w}{\partial \mathbf{B}^e} : \overline{\mathbf{D}}^p \qquad \overline{\mathbf{D}}^p = \mathbf{R}^e \left(\dot{\mathbf{F}}^p \mathbf{F}^{p^{-1}} \right)^S \mathbf{R}^{e^T} \tag{12}$$

which is the appropriate extension of (3_2). The elastic law and dual variables are

$$\tau = 2\mathbf{B}^e \frac{\partial w}{\partial \mathbf{B}^e} \qquad \phi = \tau : \overline{\mathbf{D}}^p \tag{13}$$

which in fact is identical to the isotropic specialisation of Mandel's formulation

$$\tau = \mathbf{R}^e \, \Sigma \, \mathbf{R}^{e^T} \qquad \overline{\mathbf{D}}^p = \mathbf{R}^e \, \dot{\mathbf{P}} \mathbf{P}^{-1} \mathbf{R}^{e^T}$$

3.3 Green-Naghdi's approach

Green-Naghdi's approach starts from an entirely different point of view [Green et al., 1965] since it is based on an additive decomposition of the Green strain tensor and an elastic energy postulated as

$$\mathbf{E} = \mathbf{E}^{e} + \mathbf{E}^{p} \qquad\qquad w(\mathbf{E}, \mathbf{E}^{p}) \tag{14}$$

The dissipation inequality then obviously is

$$\phi = \left(\mathbf{S} - \frac{\partial w}{\partial \mathbf{E}}\right) : \dot{\mathbf{E}} - \frac{\partial w}{\partial \mathbf{E}^{p}} : \dot{\mathbf{E}}^{p} \geq 0$$

where **S** is the second Piola-Kirchhoff stress tensor. This is the appropriate extension of (3_2) directly leading to

$$\mathbf{S} = \frac{\partial w}{\partial \mathbf{E}} \qquad \phi = \mathbf{S}^{p} : \dot{\mathbf{E}}^{p} \qquad \mathbf{S}^{p} = \frac{\partial w}{\partial \mathbf{E}^{p}} \tag{15}$$

In order to reconcile this with Lee's approach the plastic strain $\mathbf{E}^{p}$ is interpreted as

$$\mathbf{E}^{p} = \frac{1}{2}\left(\mathbf{C}^{p} - \mathbf{1}\right) \qquad \mathbf{C}^{p} = \mathbf{F}^{p^{T}} \mathbf{F}^{p}$$

and the energy is assumed to only depend on the elastic deformation, which in this framework can be defined by [Sidoroff, 1973]

$$w = w\left(\mathbf{g}^{e}\right) \qquad \mathbf{g}^{e} = \mathbf{R}^{p^{T}} \mathbf{C}^{e} \mathbf{R}^{p} = \mathbf{U}^{p^{-1}} \mathbf{C} \mathbf{U}^{p^{-1}} \tag{16}$$

Substituting this isotropic form of the energy in (15) shows, after some computation, that this model coincides with Lee's model as presented in section 2.2.

It should be noted that if w is not isotropic, then (19) still provides a consistent model which can describe anisotropic plasticity. This is an alternative approach to the currently developed model which will be mentioned below and which can probably be interesting for further development.

4. ROTATING FRAME AND PLASTIC SPIN

4.1 Kinematic hardening

The isotropic elastic-plastic model therefore may be presented under different formulations, but they all result in the same final model. The situation is much more complex when some kind of anisotropy – either initial anisotropy or anisotropic hardening – is involved even in a rigid plastic model without elasticity.

Kinematic hardening, as described in section 2.2 is the simplest case of anisotropic hardening. Starting from the second formulation a tensor internal variable α is introduced with, in the isotropic case,

$$w = \frac{1}{2} b\,\alpha : \alpha \qquad \phi = \sigma : \mathbf{D} - \mathbf{X} : \dot{\alpha} \qquad \mathbf{X} = b\,\alpha \tag{17}$$

with one material constant b. The evolution equation however cannot be directly extended with the time derivative $\dot{\alpha}$ and an objective rate $\overset{\circ}{\alpha}$ must be used. In order to keep duality, this objective rate must be such that

$$\mathbf{X} : \dot{\alpha} = \mathbf{X} : \overset{\circ}{\alpha} \qquad \phi = \sigma : \mathbf{D} - \mathbf{X} : \overset{\circ}{\alpha} \tag{18}$$

Using von Mises yield criterion the evolution equation will then directly follow as

$$f = \overline{q} - Y_o \qquad \overline{q} = \sqrt{\frac{3}{2}} \sqrt{(\tau - \mathbf{X})^D : (\tau - \mathbf{X})^D} \qquad \overset{\circ}{\alpha} = \mathbf{D} = \frac{3\lambda}{2Y_o}(\tau - \mathbf{X}) \tag{19}$$

Among the usual objective rates the Jaumann rate α^J is the only one to satisfy condition (18) – the convected derivatives in particular would introduce complementary terms in the dissipation and cannot be used here –. Since the Jaumann rate also naturally appears in the isotropic case when deriving rate type constitutive equations [Sidoroff, 1982], it obviously seemed to be the right choice. Unfortunately it also leads just as in the hypoelastic case to unreasonable oscillations. This difficulty is usually overcome through the introduction of other objective rate satisfying (21), the rotated derivatives defined as

$$\overset{\circ}{\alpha} = \dot{\alpha} - \Omega\,\alpha + \alpha\,\Omega \tag{20}$$

where $\mathbf{\Omega}$ is a rotation rate which behaves like the rotation rate tensor $\mathbf{W}$ in a change of frame and which may be chosen in two ways

i) directly from the kinematics, the most popular being the proper rotation or Naghdi's derivative

$$\Omega = \dot{\mathbf{R}}\mathbf{R}^T \qquad \mathbf{F} = \mathbf{RU} = \mathbf{VR}$$

resulting from the polar decomposition – but many other choices have been proposed.

ii) Through an appropriate constitutive equation for the difference $\Omega - \mathbf{W}$

$$\Omega = \mathbf{W} + \hat{\Omega}(\tau, \mathbf{X}, \mathbf{D})$$

the so-called plastic spin equation which in the rigid plastic case can be considered as the skew symmetric part of the evolution equation for $\dot{\mathbf{P}}\mathbf{P}^{-1}$ in Mandel's approach [Dafalias, 1983].

4.2 Initial anisotropy

A similar problem is encountered in case of initial anisotropy. Indeed the natural rigid plastic large strain model is

$$\phi = \tau : \mathbf{D} \qquad f(\tau) \leq 0 \qquad \mathbf{D} = \lambda \frac{\partial f}{\partial \tau} \tag{21}$$

which in the isotropic case is the usual plastic model but cannot be used in case of anisotropy due to lack of objectivity. A complete lagrangian formulation, on the other hand, relating the second Piola-Kirchhoff stress **S** to the time derivative of the Green strain tensor $\dot{\mathbf{E}}$, cannot be used either, due to the inadequacy of the total deformation for describing a fluid-like behaviour. The proper formulation must be Lagrangian for the orientation, in order to ensure objectivity, but Eulerian for the principal stresses and strain rates which are physically meaningful [Dogui et al., 1986] . This suggests replacing (21) by

$$\begin{aligned} &\phi = \overline{\tau} : \overline{\mathbf{D}} \qquad f(\overline{\tau}) \leq 0 \qquad \overline{\mathbf{D}} = \lambda \frac{\partial f}{\partial \overline{\tau}} \\ &\tau = \mathbf{Q}\,\overline{\tau}\,\mathbf{Q}^{\mathrm{T}} \qquad \mathbf{D} = \mathbf{Q}\,\overline{\mathbf{D}}\,\mathbf{Q}^{\mathrm{T}} \end{aligned} \tag{22}$$

i.e. formulating the anisotropic plastic flow rule in a rotated frame $\overline{C}(t)$ obtained from the present configuration C(t) by a rotation $\mathbf{Q}^{\mathrm{T}}(t)$. Of course (22) is identical to (21) in the isotropic case but will be objective if the rotation tensor **Q** is chosen in such a way that it behaves like **F** in a change of frame

$$\mathbf{F} \rightarrow \mathbf{F}' = \mathbf{q}\,\mathbf{F} \qquad \mathbf{Q} \rightarrow \mathbf{Q}' = \mathbf{q}\,\mathbf{Q}$$

The simplest way to ensure this is to obtain **Q** from the differential equation

$$\dot{\mathbf{Q}}\,\mathbf{Q}^{\mathrm{T}} = \mathbf{W} \tag{23}$$

resulting in the corotational or Zaremba-Jaumann rotating frame [Anglès d'Auriac, 1970]. Similarly the proper rotating frame is defined from the polar decomposition by

$$\mathbf{Q} = \mathbf{R} \tag{24}$$

More generally a rotating frame is defined by the following differential equation

$$\dot{\mathbf{Q}} = \mathbf{W}\mathbf{Q} - \mathbf{Q}\overline{\mathbf{W}} \qquad \overline{\mathbf{W}} = \mathscr{K}\left[\overline{\mathbf{D}}\right]$$

where the "plastic spin" tensor $\overline{\mathbf{W}}$ is defined in $\overline{C}(t)$ and, therefore, is invariant in a change of frame. In all known cases it is also assumed to be linear with respect to the rotated deformation rate tensor, therefore ensuring rate-independence of the rotating frame definition.

The kinematic hardening model as described in section 4.1 can also be formulated in this way replacing (22) with

$$\begin{aligned} &\dot{\overline{\alpha}} = \overline{\mathbf{D}} = \frac{3\lambda}{2}\left(\overline{\tau}^{D} - \overline{\mathbf{X}}\right) \qquad \overline{\mathbf{X}} = b\,\overline{\alpha} \\ &\Omega = \dot{\mathbf{Q}}\,\mathbf{Q}^{\mathrm{T}} \qquad \overset{\circ}{\alpha} = \mathbf{Q}\,\dot{\overline{\alpha}}\,\mathbf{Q}^{\mathrm{T}} \end{aligned} \tag{25}$$

Again this plastic spin can be treated as an ordinary constitutive equation [Dafalias, 1985], usually through the use of tensor representation theorem or directly postulated from the kinematics [Dogui, 1988 ; Dogui et al., 1984].

5. PLASTIC STRAIN AS STATE VARIABLE

5.1 General framework

A basic feature in the physics of plasticity is the fact that elastic deformation is a state variable while the plastic deformation is not. It is nevertheless found convenient in many simple mechanical models, like for instance in the Prager model interpretation of kinematic hardening, section 2.2, to use it as a hardening variable. It corresponds in fact to the introduction of a tensor state variable which in a first approximation is assumed to follow the plastic strain. Limiting ourselves to the isotropic case a corresponding large strain formulation can be obtained starting from the intermediate configuration as defined in section 3.3, the problem being to define an Eulerian plastic strain tensor. Starting, just as in section 4.2 from the idea that the principal strains are the relevant physical quantities, this can be done by using two different tensors

$$\begin{aligned} &\mathbf{G}^p = \mathbf{R}\mathbf{C}^p\mathbf{R}^T = \mathbf{V}\mathbf{B}^{e^{-1}}\mathbf{V} && \widetilde{\mathbf{R}}^p = \mathbf{R}^e\mathbf{R}^p\mathbf{R}^T \\ &\mathbf{H}^p = \mathbf{R}^e\mathbf{B}^p\mathbf{R}^{e^T} = \mathbf{V}^{e^{-1}}\mathbf{B}\mathbf{V}^{e^{-1}} && \mathbf{H}^p = \widetilde{\mathbf{R}}^p\mathbf{G}^p\widetilde{\mathbf{R}}^{p^T} \end{aligned} \tag{26}$$

with obvious notations (**C** and **B** Cauchy-Green tensors, **R** rotation, **V** left stretch tensor associated to the corresponding total, elastic or plastic deformations). These tensors may also be considered as the left and right Cauchy-Green tensors associated to $\widetilde{\mathbf{F}}^p = \mathbf{V}^{e^{-1}}\mathbf{V}$ [Sidoroff, 1984 ; Naghdi et al., 1974].

Two different elastic-plastic models can be defined by assuming the energy w as an isotropic function $w(\mathbf{B}^e, \mathbf{G}^p)$ or $w(\mathbf{B}^e, \mathbf{H}^p)$. Using duality the constitutive equation can be obtained in both cases from the dissipation

$$\phi = \tau : \mathbf{D} - \dot{w} = (\tau - \tau^e) : \mathbf{D} + \tau^p : \overline{\mathbf{D}}^p$$

Using for instance $\mathbf{H}^p$, the elastic and plastic stress are, after some computations, finally obtained as

$$\tau^e = 2\,\mathbf{V}^e \frac{\partial w}{\partial \mathbf{B}^e}\mathbf{V}^e \qquad \tau^p = 2\left(\mathbf{B}^e \frac{\partial w}{\partial \mathbf{B}^e} - \frac{\partial w}{\partial \mathbf{H}^p}\mathbf{H}^p\right) \tag{27}$$

which are non-trivial large strain extensions of (4). It should be noted that the first of these two equations can be easily obtained from Green-Naghdi's approach (section 3.3). Indeed w being an isotropic function of $(\mathbf{B}^e, \mathbf{H}^p)$ is also the same isotropic function of $(\mathbf{C}^e, \mathbf{B}^p)$ or $(\mathbf{g}^e, \mathbf{C}^p)$ where $\mathbf{g}^e$ has been defined in (16). Since $\mathbf{g}^e$ can be expressed from $\mathbf{C}$ and $\mathbf{C}^p$, the energy w can also be considered as a – different – function of $\mathbf{C}$ and $\mathbf{C}^p$ so that Green-Naghdi's initial formulation can be used leading to

$$\mathbf{S}^e = 2\frac{\partial w}{\partial \mathbf{C}} = 2\,\mathbf{U}^{p^{-1}} \frac{\partial w}{\partial \mathbf{g}^e} \mathbf{U}^{p^{-1}}$$

from which (27_1) directly follows. Such an easy treatment however does not seem to exist for τ^p and a detailed computation is required. Similar relations can be obtained using $\mathbf{G}^p$.

5.2 Applications

It is obvious from (27) that if the energy function $w(\mathbf{B}^e, \mathbf{H}^p)$ is such that $\partial w / \partial \mathbf{B}^e$ vanishes for $\mathbf{B}^e = \mathbf{1}$, whatever $\mathbf{H}^p$ may be, then $\boldsymbol{\tau} = \tau^e$ also vanishes. This is less obvious but also true when using $\mathbf{G}^p$. This means that if this condition is imposed, which will be assumed from now on, then the intermediate configuration is the local natural stress-free configuration and the essential features of plasticity are maintained.

An important special case corresponds to the case where the energy can be split in two parts

$$w = w_1\left(\mathbf{B}^e\right) + w_2\left(\mathbf{H}^p\right) \tag{28}$$

which corresponds to a natural extension of Prager's model [Dogui et al., 1985]. In this case it directly follows from (27) that

$$\boldsymbol{\tau}^p = \boldsymbol{\tau} - \mathbf{X} \qquad\qquad \mathbf{X} = 2\frac{\partial w_2}{\partial \mathbf{H}^p}\mathbf{H}^p \tag{29}$$

Thus providing an alternate formulation for kinematic hardening.

More generally a Taylor expansion around $\mathbf{B}^e = \mathbf{1}$ gives

$$w = w_0\left(\mathbf{H}^p\right) + \frac{1}{2}\varepsilon^e : A\left(\mathbf{H}^p\right) : \varepsilon^e + \ldots \qquad\qquad \varepsilon^e = \frac{1}{2}\left(\mathbf{B}^e - \mathbf{1}\right) \tag{30}$$

where as discussed above w_0 will account for kinematic hardening while the dependence of the elasticity tensor A with respect to $\mathbf{H}^p$ will account for induced elastic anisotropy. Induced plastic anisotropy can be consistently obtained as well by assuming a parametric dependence of the yield function f on the plastic strain $\mathbf{H}^p$.

5.3 Poynting and Zener

Similar ideas are used in viscoelasticity (the so-called hyperviscoelastic materials used for elastomeric materials). Two different kinds of models are considered, respectively corresponding to the Zener and Poynting model [Maugin, 1992]. The model which has been developed above directly provides the Poynting model (serial combination of a spring with a Kelvin model) which is currently used with the uncoupling assumption

(28). The Zener model (parallel combination of a spring with a Maxwell model) can be obtained in a similar way from an energy $w(\mathbf{B},\mathbf{B}^e)$. The dissipation in this case finally is obtained as

$$\phi = \left(\tau - \tau^e\right) : \mathbf{D} + \tau^p : \overline{\mathbf{D}}^p \geq 0$$

$$\tau^o = 2\left(\mathbf{B}\frac{\partial w}{\partial \mathbf{B}} + \mathbf{B}^e\frac{\partial w}{\partial \mathbf{B}^e}\right) \qquad \tau^p = 2\,\mathbf{V}^e\frac{\partial w}{\partial \mathbf{B}^e}\mathbf{V}^e \tag{31}$$

directly generalising the small strain linear formulation. The intermediate configuration however loses us interpretation as a relaxed stress-free configuration.

6. CONCLUSIONS

The more complex case of a completely anisotropic elastic-plastic model is not yet entirely clear and the extension of the rotating frame formulation presented in section 4 requires the definition of the rotating frame from the plastic deformation – which is natural in Mandel's formulation but not so easy when dealing with a kinematic rotating frame like for instance the proper rotation frame [Morando et al., 1988]. In this case again, duality allows a correct identification of the appropriate stress tensors to be used in constitutive modelling [Sidoroff et al., 1999].

Like in many other cases discussed in this volume, duality (and strictly speaking not thermodynamics) is a basic tool for the formulation of constitutive models and for identification of consistent generalised forces associated to the postulated internal variables.

REFERENCES

[Anglès d'Auriac, 1970] Anglès d'Auriac, P.; Les principes en mécanique des milieux continus; *La Houille Blanche*, vol.5,pp.427-432.

[Dafalias, 1983] Dafalias, Y.F.; Corotational rates for kinematic hardening at large plastic deformations; *Journal of Applied Mechanics*, vol.50, pp.561-565.

[Dafalias, 1985] Dafalias, Y.F.; The plastic spin; *Journal of Applied Mechanics*, vol.52, pp.865-871.

[Dogui, 1988] Dogui, A; Cinématique bidimensionnelle en grandes déformations – Application à la traction hors axes et à la torsion; *Journal de Mécanique Théorique et Appliquée*, vol.7 (1), pp.43-64.

[Dogui et al., 1984] Dogui, A.; Sidoroff, F.; Quelques remarques sur la plasticité anisotrope en grandes déformations; *Compte-Rendus de l'Académie des Sciences*, vol. 299, série II, n° 18, pp.1225-1228.

[Dogui et al., 1985] Dogui, A.; Sidoroff, F.; Kinematic hardening in large elastoplastic strain; *Engineering Fracture Mechanics*, vol.21 (4), pp.685-699.

[Dogui et al., 1986] Dogui, A.; Sidoroff, F.; Rhéologie anisotrope en grandes déformations; *Actes du 19ème Colloque GFR, Paris 1984 : Rhéologie des matériaux anisotropes*; ed. C.Huet, D.Bourgoin, S.Richemond, Editions Cepadues, Toulouse, pp.69-78.

[Germain, 1974] Germain, P.; Thermodynamique des milieux continus; *Entropie*, vol.55, pp.7-14.

[Green et al., 1965] Green, A.E.; Naghdi, P.M.; A general theory of an elastic-plastic continuum; *Archives of Rational Mechanics and Analysis*, vol.18, pp.251-281.

[Lee, 1969] Lee, E.H.; Elastic-plastic deformation at finite strains; *Journal of Applied Mechanics*, vol.36, pp.1-6.

[Mandel, 1971] Mandel, J.; Plasticité et viscoplasticité; *Cours CISM*, 97, Udine. Springer, New York.

[Maugin, 1992] Maugin, G.A.; *The Thermomechanics of Plasticity and Fracture*, Cambridge University Press, U.K., 1992.

[Morando et al., 1988] Morando, A.; Debordes, O.; Etude numérique d'un problème de collerette d'emboutissage présentant des instabilités par bandes de cisaillement; *Journal de Mécanique Théorique et Appliquée*, vol.7 (4), pp.409-441.

[Naghdi et al., 1974] Naghdi, P.M.; Trapp, J.A.; On finite elastic-plastic deformations of metal; *Journal of Applied Mechanics*, vol.41, pp.245-260.

[Sidoroff, 1984] Sidoroff, F.; Ecrouissage cinématique et anisotropie induite en grandes déformations élastoplastiques; *Journal de Mécanique Théorique et Appliquée*, vol.3 (1), pp.117-133.

[Sidoroff, 1973] Sidoroff, F.; The geometrical concept of intermediate configuration and elastic-plastic finite strain; Archives of Mechanics, vol.25, pp.299-308.

[Sidoroff, 1982] Sidoroff, F.; Incremental constitutive equation for large strain elastoplasticity; *International Journal of Engineering Science*, vol.20(1), pp.19-26.

[Sidoroff et al., 1985] Sidoroff, F.; Teodosiu, C.; Microstructure and phenomenological models for metals; Proceedings *Physical basis and modelling of finite deformation of aggregates*, CNRS, Paris.

[Sidoroff et al., 1999] Sidoroff, F.; Dogui, A.; Some issues about anisotropic elastic plastic models at finite strain; to be published.

[Teodosiu et al., 1976] Teodosiu, C.; Sidoroff, F.; A theory of finite elastoviscoplasticity of single crystals; *International Journal of Engineering Science*, vol.14 (1), pp.165-176.

Thermodynamical description of running discontinuities.

C. Stolz
Ecole Polytechnique, CNRS UMR 7649 F-91128 Palaiseau Cedex, France
stolz@lms.polytechnique.fr

Abstract : Determination of the evolution of a system is studied through the definition of functionals presented here in the case of non linear dynamics. After a short account of the necessary notions for the description of the motion and of the mechanical interactions, the paper is devoted to the Hamiltonian functional of the system. The evolution of discontinuities along moving surfaces generally generates an entropy production which is a linear function of the speed of the moving interface. The thermodynamical force accompanying this motion is an energy release rate, the expression of which is related to the gradient of Hamiltonian of the system. The canonical equations of shock waves and moving discontinuities are established for any continuum. This provides a useful way of establishing constitutive laws for moving interfaces.
Keywords : Hamiltonian, shock-waves, jump, discontinuity, energy-release rate.

1. SOME GENERAL FEATURES

In order to explain and to predict the motion and the equilibrium of bodies or structures subjected to various physical interactions, a kinematical description of the motion is first performed. Usually one looks for the motion of a material point M from a reference configuration by describing its displacement $u(M,t)$.

After the kinematical description of the body, one has to deal with the mechanical interactions. Many statements permit the description of these interactions ; we can use for example the virtual-power statement. This describes the mechanical interaction between each material point of the body with respect to a given loading distribution. For sake of simplicity and conciseness of this presentation, a thermodynamical description of interaction is adopted.

First, the local state is defined by a set of state variables such as the strain ε, a set of internal parameters α, and the absolute temperature θ. The local interaction is defined by a thermodynamical potential or the free energy $w(\varepsilon,\alpha,\theta)$ per unit of mass. The thermodynamical forces are defined by the state equations :

$$\sigma_r = \rho\frac{\partial w}{\partial \varepsilon}, \quad A = -\rho\frac{\partial w}{\partial \alpha}, \quad s = -\frac{\partial w}{\partial \theta}$$

σ_r is the reversible stress, A is the thermodynamical forces associated to the internal vari-

G.A. Maugin et al. (eds.), Continuum Thermomechanics, 401-412.
© 2000 *Kluwer Academic Publishers.*

ables, and s is the entropy. In the case of reversibility, the knowledge of the free energy is sufficient to determine the local state of equilibrium defined by the stress $\sigma = \sigma_r$. In nonlinear mechanics, the internal state is generally associated with irreversibility. Then the fundamental inequality of thermodynamics implies that the internal production of entropy must be non negative. The equations of state do not provide the full constitutive equations ; some complementary laws are necessary to describe the irreversibility. In the total dissipation, we distinguish the part due to the conduction and the part due to internal forces. The two parts are assumed to be separately non negative. The mechanical part has the form :

$$D_m = \sigma : \dot{\varepsilon} - \rho(\dot{w} + s\dot{\theta}) = (\sigma - \sigma_r) : \dot{\varepsilon} + A.\dot{\alpha} \geq 0.$$

The state of stress during the evolution has been decomposed as $\sigma = \sigma_r + \sigma_{ir}$. Let us assume that the behavior belongs to the class of the so-called *generalized standard materials*. This ensures the existence of a potential of dissipation $d(\dot{\varepsilon}, \dot{\alpha})$. The potential d is a convex function of the variables, with a minimum value at the origin. The evolution of the internal state is given by the normality rule :

$$(\sigma_{ir}, A) \in \partial d(\dot{\varepsilon}, \dot{\alpha}),$$

this means that the subdifferential $\partial d(\dot{\varepsilon}, \dot{\alpha})$ of d is the set of state (σ_{ir}, A) such that :

$$d(\dot{\varepsilon}, \dot{\alpha}) + \sigma_{ir} : (\varepsilon^* - \dot{\varepsilon}) + A.(\alpha^* - \dot{\alpha}) \leq d(\varepsilon^*, \alpha^*),$$

for all admissible fields ε^*, α^*. The existence of such a potential for the dissipation ensures the positivity of the entropy production :

$$\sigma_{ir} : \dot{\varepsilon} + A.\dot{\alpha} \geq d(\dot{\varepsilon}, \dot{\alpha}) - d(0,0) \geq 0.$$

2. EQUILIBRIUM AND QUASISTATIC EVOLUTION

As time evolves the displacement field and the internal state must satisfy the following :

- the state equations :

$$\sigma_r = \rho\frac{\partial w}{\partial \varepsilon}, \quad A = -\rho\frac{\partial w}{\partial \alpha}, \quad s = -\frac{\partial w}{\partial \theta},$$

- the conservation of the momentum :

$$div\ \sigma = 0, \qquad \sigma = \sigma_r + \sigma_{ir},$$

- the evolution equations of the state variables.
- all the boundary conditions.

For the overall system the rule of the free energy is replaced by the global free energy :

$$W(\tilde{\varepsilon}, \tilde{\alpha}, \tilde{\theta}) = \int_\Omega \rho w(\varepsilon(u), \alpha, \theta)\, d\omega$$

In a global description the equations of state possess the same form as in the local one, but the state of the system is defined by *fields* of state variables. The equations of state are relationships between fields of state variables :

$$\widetilde{\sigma}_r = \frac{\partial W}{\partial \widetilde{\varepsilon}}, \quad \widetilde{A} = -\frac{\partial W}{\partial \widetilde{\alpha}}, \quad \widetilde{s} = -\frac{\partial W}{\partial \widetilde{\theta}},$$

These relations are obtained by the following definition : $\frac{\partial W}{\partial \widetilde{q}} \bullet \widetilde{q}^* = \int_\Omega \rho \frac{\partial w}{\partial q}\, q^* d\omega$. By using the properties of the characterization of the evolution of the internal state and integration over the body, we can define the dissipative function :

$$D(\widetilde{\dot{\varepsilon}}, \widetilde{\dot{\alpha}}) = \int_\Omega d(\dot{\varepsilon}, \dot{\alpha})\, d\omega,$$

for which the evolution is given in terms of fields of state variables as :

$$(\widetilde{\sigma}_{ir}, \widetilde{A}) \in \partial D(\widetilde{\dot{\varepsilon}}, \widetilde{\dot{\alpha}}),$$

where the set $\partial D(\widetilde{\dot{\varepsilon}}, \widetilde{\dot{\alpha}})$ is defined obviously by :

$$D(\widetilde{\dot{\varepsilon}}, \widetilde{\dot{\alpha}}) - D(\widetilde{\varepsilon}^*, \widetilde{\alpha}^*) + \widetilde{\sigma}_{ir} \bullet (\widetilde{\varepsilon}^* - \widetilde{\dot{\varepsilon}}) + \widetilde{A} \bullet (\widetilde{\alpha}^* - \widetilde{\dot{\alpha}}) \leq 0, \quad \forall (\widetilde{\varepsilon}^*, \widetilde{\alpha}^*),$$

then we have for regular functions :

$$\frac{\partial D}{\partial \widetilde{\dot{\varepsilon}}} \bullet \widetilde{\varepsilon}(\delta u) = \int_\Omega \sigma_{ir} : \varepsilon(\delta u)\, d\omega, \quad \frac{\partial D}{\partial \widetilde{\dot{\alpha}}} \bullet \delta\widetilde{\alpha} = \int_\Omega A.\delta\alpha\, d\omega.$$

Let us consider that the external loading derives from a potential given in terms of traction T^d applied on the external surface of the body. Then, the global free energy can be replaced by the potential energy of the system :

$$P(\widetilde{u}, \widetilde{\alpha}, \widetilde{\theta}, T^d) = \int_\Omega w(\varepsilon(u), \alpha, \theta)\rho\, d\omega - \int_{\partial\Omega_T} T^d.u\, da.$$

By combining all the equations in terms of fields of state variables, we can state the quasistatic evolution in a global manner by the variational system :

$$\frac{\partial P}{\partial \widetilde{u}} * \delta\widetilde{u} + \frac{\partial D}{\partial \widetilde{\dot{\varepsilon}}} \bullet \widetilde{\varepsilon}(\delta u) = 0, \quad \frac{\partial P}{\partial \widetilde{\alpha}} \bullet \delta\widetilde{\alpha} + \frac{\partial D}{\partial \widetilde{\dot{\alpha}}} \bullet \delta\widetilde{\alpha} = 0.$$

These equations are defined on a set of admissible fields. The displacement is subject to boundary conditions $u = u^d$ over $\partial\Omega_u$. These equations are general ; they contain the essential structure of a problem of quasistatic evolution. The first equation of this system explains the conservation of the momentum taking into account the constitutive law, the second explains the thermodynamic forces associated with the internal parameters.

3. THE DYNAMICAL CASE

The Hamiltonian is the total energy of the system [Stolz, 1988]:

$$H(\widetilde{p}, \widetilde{u}, \widetilde{\alpha}, \widetilde{s}, T^d) = \int_\Omega \frac{1}{2}\frac{p^2}{\rho}\, d\omega + \int_\Omega \rho\, e(u, \alpha, s)\, d\omega - \int_{\partial\Omega_T} T^d.u\, da.$$

The first term is the kinetic energy $K(v) = \int_\Omega \frac{1}{2}p^2/\rho \, d\omega = \int_\Omega \frac{1}{2}\rho v^2 \, d\omega$, $p = \rho v$ is the momentum, the second is the internal energy with the density : $e = w + \theta s$, and the last is the potential energy due to prescribed loading. The equation of motion are given by

$$\begin{aligned}
\frac{\partial H}{\partial p} \bullet \delta p &= \int_\Omega \frac{p}{\rho} . \delta p \, d\omega, \\
\frac{\partial H}{\partial u} * \delta u &= \int_\Omega \rho \frac{\partial w}{\partial \varepsilon} : \varepsilon(\delta u) \, d\omega - \int_{\partial\Omega_T} T^d . \delta u \, da
\end{aligned}$$

Taking account of the decomposition of the stress $\sigma = \sigma_r + \sigma_{ir}$, of the conservation of the momentum inside the volume and of the boundary conditions

$$div\sigma = \rho\gamma \text{ over } \Omega, \; \sigma.n = T^d \text{ over } \partial\Omega_T$$

we modify the expressions :

$$\frac{\partial H}{\partial u} * \delta u = -\int_\Omega \sigma_{ir} : \varepsilon(\delta u) \, d\omega - \int_\Omega \rho\gamma . \delta u \, d\omega,$$

then with the relations :

$$\frac{d}{dt} \int_\Omega p . \delta u \; d\omega = \int_\Omega \rho\gamma . \delta u \, d\omega.$$

we obtain the conservation of momentum in the Hamiltonian form :

$$\frac{\partial H}{\partial u} * \delta u = -\frac{\partial D}{\partial \dot{\varepsilon}} \bullet \varepsilon(\delta u) - \frac{d}{dt} \int_\Omega p . \delta u \, d\omega,$$

Finally, the Hamiltonian formulation of the evolution of the system is obtained :

$$\begin{aligned}
\frac{\partial H}{\partial p} \bullet \delta p &= \frac{d}{dt} \int_\Omega u . \delta p \, d\omega, \\
\frac{\partial H}{\partial u} * \delta u &= -\frac{\partial D}{\partial \dot{\varepsilon}} \bullet \varepsilon(\delta u) + \frac{d}{dt} \int_\Omega p . \delta u \, d\omega, \\
\frac{\partial H}{\partial \alpha} \bullet \delta \alpha &= -\frac{\partial D}{\partial \dot{\alpha}} \bullet \delta \alpha, \\
\frac{\partial H}{\partial s} \bullet \delta s &= \int_\Omega \rho \theta \delta s \, d\omega.
\end{aligned}$$

A conduction law must be given and the positivity of the entropy production must be verified to determine the evolution of the system :

$$\frac{d}{dt} \int_\Omega \rho s \, d\omega + \int_{\partial\Omega} \frac{q.n}{T} da \geq 0.$$

For the real motion, the value of the Hamiltonian is the sum of the kinetic energy, of the internal energy and of the potential energy of the external (given) load. The conservation of the energy of the system can be easily rewritten as

$$\frac{dH}{dt} - \frac{\partial H}{\partial T^d} . \dot{T}^d = -\int_{\partial\Omega} q.n \, da,$$

where q is the heat flux, and we consider that no external volume heat source are prescribed.

The definition of the Hamiltonian can be extended to generalized media. The proposed description can be performed in all the cases in which we can define the behavior by two potentials : a global free energy and a dissipative function. If some particular internal constraints exist, this description must be revised.

4. MECHANICAL TRANSFORMATIONS ALONG MOVING SURFACES

We consider a moving interface along which mechanical transformation occurs [Pradeilles-Duval et al, 1995]. Two materials coexist at any time in the structure and the body is heterogeneous. The evolution of the surface along which the transformation takes place is characterized in the energy analysis. Some connections can be made with the notion of configurational forces, [Gurtin, 1995, Maugin, 1995,Truskinovski, 1987,Grinfeld, 1980, 1991].

Let Ω denote the domain, composed of two distinct volumes Ω_1 and Ω_2 which are occupied by two materials with different mechanical properties. The perfect interface between them is assumed to be regular and is denoted by Γ. Material 1 changes to materials 2 along Γ by an irreversible process. Hence Γ moves with a normal velocity ϕ positive along Γ. The state of the system is characterized by the displacement field u, from which a strain field is derived ε. The other parameters are the temperature θ, the internal parameters α and the position of the boundary Γ.

When the surface Γ propagates, with velocity $c = \phi N$, (N normal outward to Ω_2), mechanical quantities can have a jump, ($[f] = f_1 - f_2$) and all volume integrals ($F = \int_\Omega f d\omega$) have rates defined by :

$$\dot{F} = \int_\Omega \dot{f}\, d\omega - \int_\Gamma [f]\, c.N\, da.$$

The mass conservation is defined by the continuity of the mass flux :

$$m = \rho c.N = \rho\phi.$$

Then the conservation of energy and the entropy production are rewritten as

$$\int_\Omega (\dot{e} + \gamma.v)\rho\, d\omega - \int_\Gamma [e + \frac{1}{2}v^2] m\, da = -\int_{\partial\Omega} q.n\, da + P_{ext},$$

$$\int_\Omega \rho\dot{s}\, d\omega - \int_\Gamma [s]\, m\, da + \int_{\partial\Omega} \frac{q.n}{T} da \geq 0.$$

By the momentum conservation we have

$$P_{ext} = \int_\Omega \sigma : \dot{\varepsilon}\, d\omega + \int_\Gamma N.[\sigma.v]\, da + \int_\Omega \rho\gamma.v\, d\omega.$$

This is true for all volume Ω then we deduce the local equations of conservation :

$$\rho\dot{e} \;=\; \sigma : \dot{\varepsilon} - divq,$$

$$0 = m[e + \frac{1}{2}v^2] + N.[\sigma.v] - [q].N.$$

The entropy production takes the form :

$$\int_\Omega (\rho\dot{s} + \frac{divq}{\theta}) - \frac{q.\nabla\theta}{\theta^2}\, d\omega + \int_\Gamma (-[s]\,m + [\frac{q}{\theta}].N)\, da \geq 0.$$

The interface is perfect, then the displacement is continuous along the interface $[u] = 0$, and the temperature is assumed to be continuous $[\theta] = 0$.

Some dissipation can occur along the moving surface. The mechanical discontinuities must satisfy some constraints, they must be kinematically admissible and ensure the positivity of the entropy production. The jumps of the mechanical quantities must verify the Hadamard equations :

$$[v] + [\nabla u].c = 0, [\dot{\theta}] + [\nabla\theta].c = 0,$$

and the momentum equation :

$$[\sigma].N + m[v] = 0.$$

Combining the local equations of conservation, we rewrite the production of entropy in terms of volume and surface contribution :

$$D_\Omega = \sigma : \dot{\varepsilon} - \rho(\dot{w} + s\dot{\theta}) - \frac{q.\nabla\theta}{\theta^2}, \forall x \in \Omega,$$

$$D_\Gamma = \frac{1}{\theta}(\rho[w]c.N - N.\overline{\sigma}.[\nabla u].c) = \frac{G_s}{\theta}\phi, \forall x \in \Gamma,$$

where the quantity $\overline{\sigma}$ is $\frac{1}{2}(\sigma_1 + \sigma_2)$. Along the surface, the production of entropy is defined in a similar form as in a quasistatic thermomechanical coupled evolution by replacing the tension along the surface by the mean tension, and as in an isothermal analysis the results of Abeyaratne and Knowles [Abeyaratne et al.,1990].

The behavior of each phase is defined by the free energy w_i as a function of strain ε, temperature θ and a set of internal variables α, the evolution of which is governed by a pseudo potential of dissipation $d_i(\dot{\alpha})$, a convex function of $\dot{\alpha}$. As before, the state equations are

$$\sigma = \rho\frac{\partial w_i}{\partial\varepsilon}, \; s = -\frac{\partial w_i}{\partial\theta}, \; A = -\rho\frac{\partial w_i}{\partial\alpha},$$

where A, the thermodynamical forces associated with the internal parameter α satisfies : $A = \partial d_i/\partial\dot{\alpha}$. The dissipation due to conduction is given by a conduction law.

As before, the variation of the Hamiltonian H determines the power heat supply. We decompose the volume Ω into the two volumes Ω_i (n_i the normal outward to Ω_i), the position Γ of the surface is an internal variable and the Hamiltonian takes the form :

$$H(\tilde{u}, \tilde{p}, \tilde{\alpha}, \tilde{s}, \Gamma, T^d) = H_1 + H_2 - \int_{\partial\Omega_T} T^d.u\, da,$$

$$\text{where } H_i = \int_{\Omega_i} \frac{1}{2}\frac{p^2}{\rho} d\omega + \int_{\Omega_i} \rho\, e(\varepsilon(u), \alpha, s)\, d\omega,$$

The equations of motion are then

$$\frac{\partial H_i}{\partial p}.\delta p = \int_{\Omega_i} v.\delta p \, d\omega,$$

$$-\frac{\partial H_i}{\partial u} * \delta u = \int_{\Omega_i} \rho\gamma.\delta u \, d\omega + \int_{\partial\Omega_i} n_i.\sigma.\delta u \, da.$$

Each volume has a contribution to the global Hamiltonian :

$$\frac{dH_i}{dt} = \int_{\partial\Omega_i} n_i.\sigma.v \, da - \int_{\Omega_i} (-A.\dot{\alpha} + \rho\theta\dot{s}) \, d\omega + \int_\Gamma (\frac{1}{2}\frac{p^2}{\rho} + \rho \, e)c.n_i da.$$

And for the overall system we obtain the equation :

$$\frac{dH}{dt} - \frac{\partial H}{\partial T^d}.\dot{T}^d = \frac{\partial H}{\partial \alpha} \bullet \dot{\alpha} + \frac{\partial H}{\partial s} \bullet \dot{s} - \int_\Gamma [e + \frac{1}{2}v^2]m \, da - \int_\Gamma N.[\sigma.v] \, da$$

$$= -\int_\Omega divq \, d\omega - \int_\Gamma [q].N \, da.$$

Making the distinction between volume and surface terms, we find

$$\frac{\partial H}{\partial \alpha} \bullet \dot{\alpha} + \frac{\partial H}{\partial s} \bullet \dot{s} = -\int_\Omega A.\dot{\alpha} \, d\omega + \int_\Omega \rho\theta\dot{s} \, d\omega = -\int_\Omega divq \, d\omega,$$

$$\frac{\partial H}{\partial \Gamma}.\dot{\Gamma} = -\int_\Gamma G_{th}\phi \, da = -\int_\Gamma [q].N \, da.$$

The gradient of the Hamiltonian with respect to the position of the moving surface determines the intensity of the heat source due to the propagation of mechanical discontinuities. This gradient takes the form of a release rate of internal energy :

$$G_{th} = \rho[e] - N.\bar{\sigma}.[\nabla u].N.$$

The production of entropy has a surface term which is explained as

$$\int_\Gamma (-[s]m + [\frac{q}{\theta}].N \, da = \int_\Gamma \frac{G_s}{\theta}\phi \, da.$$

In the same spirit, the intensity of the source of entropy production is a release rate of free energy $G_s = \rho[w] - N.\bar{\sigma}.[\nabla u].N$, divided by the temperature. Without more hypotheses this term is not directly connected with the variation of the Hamiltonian.

In the isothermal case, the dissipation and the heat sources are related by the uniform temperature, the heat flux is zero and $G_{th} = 0$. The dissipation is a jump of entropy along the moving surface given by G_s ; defining the isothermal Hamiltonian by

$$\hat{H} = \int_\Omega \frac{p^2}{\rho} d\omega + \int_\Omega \rho(w(\varepsilon(u), \alpha) \, d\omega - \int_{\partial\Omega_T} T^d.u \, da;$$

the dissipation is expressed by the variation of the isothermal Hamiltonian :

$$\frac{d\hat{H}}{dt} - \frac{\partial \hat{H}}{\partial T^d}.\dot{T}^d = \frac{\partial \hat{H}}{\partial \alpha} \bullet \dot{\alpha} - \int_\Gamma \{[w + \frac{1}{2}v^2]m + [N.\sigma.v]\} da.$$

By considering the surface and volume terms, we see that the energy release rate is the thermodynamical force associated with the velocity of Γ, and is the derivative of the Hamiltonian with respect to the position of the surface. The expression G_s is recovered.

5. QUASISTATIC EVOLUTION

In isothermal evolution, complementary relations must be considered to describe irreversibility. An energy criterion is chosen as a generalized form of the well known Griffith theory. Then , we assume

$$\phi \geq 0, \text{if } G = G_c \text{ on } \Gamma, \phi = 0, \text{ otherwise.}$$

This is a local energy criterion. At each equilibrium state, the interface Γ can be decomposed into two subsets where the propagation is either possible or not. Let denote by Γ^+, the subset of Γ where the critical value G_c is reached. The evolution of the interface is governed by the consistency condition, during the evolution of Γ; if, at the geometrical point $x_\Gamma(t)$, the criterion is reached

$$G(x_\Gamma(t), t) = G_c,$$

then the derivative of G following the moving surface vanishes $D_\phi G = 0$. This leads to the consistency condition written for all point inside Γ^+ :

$$(\phi - \phi^*) D_\phi G \geq 0, \forall \phi^* \geq 0, \text{ over } \Gamma^+;$$

otherwise $\phi = 0$. With Hadamard relations this derivative takes on the final form

$$\begin{aligned} D_\phi G &= [\sigma] : \nabla v_1 - \dot{\sigma}_2 : [\nabla u] - \phi K \\ K &= t.div_\Gamma \sigma_2.[\nabla u].N - t.\sigma_2.\nabla([\nabla u].N) + [\sigma : \nabla \varepsilon.N] - N.\sigma.(\nabla\nabla u.N).N], \end{aligned}$$

where $div_\Gamma \sigma_2 = div \sigma_2 - N.\nabla\sigma_2.N$. In that case, the evolution is determined by the functional :

$$\begin{aligned} F(v, \phi, \dot{T}^d) &= \int_\Omega \frac{1}{2} \varepsilon(v) : \frac{\partial^2 w}{\partial \varepsilon \partial \varepsilon} : \varepsilon(v) \, d\omega - \int_{\partial \Omega_T} \dot{T}^d.v \, da \\ &+ \int_\Gamma \phi[\sigma] : \nabla v_1 + \frac{\phi^2}{2} K \, da \end{aligned}$$

Then the evolution is given by

$$\frac{\partial F}{\partial v} * (v - v^*) + \frac{\partial F}{\partial \phi}.(\phi - \phi^*) \geq 0$$

for all v* kinematically admissible field, and $\phi^*(s) \geq 0$ along Γ^+. The discussion of the stability and bifurcation along an evolution process can be now investigated as proposed by Pradeilles and Stolz [Pradeilles et al, 1995]. Consider the rate of displacement v solution of the boundary value problem for any given velocity ϕ of propagation, v satisfies :

$$div \dot{\sigma} = 0, \ \dot{\sigma} = \frac{\partial^2 w}{\partial \varepsilon \partial \varepsilon} : \varepsilon(v), \text{ inside } \Omega,$$

$$v = v^d, \text{ on } \partial\Omega_u, \dot{\sigma}.N = \dot{T}^d \text{ on } \partial\Omega_\sigma,$$

and non classical boundary conditions on Γ :

$$D_\phi(\sigma.N) = 0, \; D_\phi[u] = [v] + \phi[\nabla u].N = 0.$$

Consider the value W of F for this solution $v(\phi, \dot{T}^d)$ then $W(\phi, \dot{T}^d) = F(v(\phi, \dot{T}^d), \phi, \dot{T}^d)$, and the stability of the actual state is determined by the condition of the existence of a solution

$$\delta\phi.\frac{\partial^2 W}{\partial\phi\partial\phi}.\delta\phi \geq 0, \; \delta\phi \geq 0 \text{ on } \Gamma^+, \delta\phi \neq 0,$$

and the uniqueness and non bifurcation is characterized by

$$\delta\phi.\frac{\partial^2 W}{\partial\phi\partial\phi}.\delta\phi \geq 0, \; \delta\phi \neq 0 \text{ on } \Gamma^+.$$

6. THE CANONICAL EQUATIONS OF SHOCK WAVES.

The local equations of shock waves will be derived from a thermodynamical potential and a dissipation pseudopotential. The canonical equations rule the jump relations and the constitutive behavior in a section of the shock [Germain, 1972]. Let us consider a shock wave, we propose to study the evolution of the shock. Locally the surface of discontinuity is replaced by its tangent plane, and one considers a frame moving with a velocity ϕ which is the normal speed of the shock. Along the line of discontinuities, some relationships between the jumps of quantities must be verified :

- for the momentum :

$$[\sigma].N + m[v] = 0,$$

- for the energy :

$$-[q].N + m[\frac{1}{2}v^2 + e] + N.[\sigma.v] = 0,$$

- for the entropy production :

$$-m[s] + [\frac{q}{\theta}].N \geq 0.$$

The quantity m denotes the mass flux and accounts for the mass conservation. The shock is governed by the constants mV^d, mQ^d which are related respectively to the flux of momentum and to the flux of energy. The jump conditions are rewritten as

$$mV^d = \sigma^{\pm}.N - mv^{\pm},$$

$$-mQ^d = -q^{\pm}.N + m(\frac{1}{2}v^2 + e)^{\pm} - N.\sigma^{\pm}.v^{\pm}.$$

The main problem is to determine the state (-) if the state (+) and the constant mV^d, mQ^d are given while respecting the positivity of the entropy production in an anelastic material. The jump conditions give us only the jump of entropy, but no direct relation between this discontinuity and the jump of internal parameters. One must determine the loading path or the history of all the quantities inside the shock ; this describes the internal structure of the shock.

To solve this problem we can consider the discontinuity surface as a layer normal to the direction of the propagation of the shock ; one has to study the inner expansion of all the quantities in a continuous process in the frame translating with the shock surface, assuming that all quantities depend only on the local normal coordinate $X = x - \phi t$. For the inner expansion, X varies from $-\infty$ to $+\infty$. We are interested in a one dimensional motion.

A constitutive law and a pseudo-potential of dissipation being given to describe the internal behavior, the dissipation is known inside the shock and therefore the jump of entropy is given as

$$-m[s] + [\frac{q}{\theta}].N = \int_{-\infty}^{+\infty} (\frac{D_m}{\theta} - \frac{q.grad\theta}{\theta^2})dx \geq 0.$$

Using the other jump conditions expressed in terms of the given constant of the shock, we can rewrite the jump of entropy as

$$mQ^d[\frac{1}{\theta}] + mV^d.[\frac{v}{\theta}] + m[\frac{(w - \frac{1}{2}v^2)}{\theta}] = \int_{-\infty}^{+\infty} (\frac{D_m}{\theta} - q.\frac{grad\theta}{\theta^2})d\omega.$$

This defines the shock generating function P :

$$P = \frac{mQ^d}{\theta} + mV^d.\frac{v}{\theta} + m\frac{(w - \frac{1}{2}v^2)}{\theta},$$

in the steady state analysis, $\dot{f} = \phi f_{,x}$, and P has the following form :

$$P(u_{,x}, \alpha, \theta) = \frac{mQ^d}{\theta} + mV^d.\frac{\phi u_{,x}}{\theta} + m\frac{(w - \frac{1}{2}\phi^2 u_{,x}^2)}{\theta}.$$

The jump of P is the total dissipation :

$$P(x^+) - P(x^-) = \int_{-\infty}^{+\infty} (\frac{D_m}{\theta} - q.\frac{grad\theta}{\theta^2})\ dX$$

The function P is supposed to be a continuous function of X, and the value of P on a section dX has the following form :

$$PdX = m\frac{Q}{\theta}dX + \frac{mV^d}{\theta}.(v(X + dX) - v(X)) + \frac{1}{\theta}m(w - \frac{1}{2}v^2)dX.$$

Then P is related to the Lagrangian defined on the section dX :

$$L = (\frac{1}{2}v^2 - w)\ mdX - mV^d.[v]_X^{X+dX}$$

Defining the dissipative function D as previously :

$$D(\varepsilon(X), \alpha(X), \theta(X)) = \frac{1}{\theta}(d(\phi\varepsilon_{,x}, \phi\alpha_{,x}) + \frac{1}{2}\frac{K}{\theta}(\theta_{,x})^2),$$

we adopt a Fourier law for the thermal conduction $q = -K.grad\theta$.

The shock structure is determined by the canonical equations :

$$\frac{\partial P}{\partial \theta} = \frac{\partial D}{\partial \theta_{,x}}, \quad \frac{\partial P}{\partial u} = \frac{\partial D}{\partial u_{,x}}, \quad \frac{\partial P}{\partial \alpha} = \frac{\partial D}{\partial \alpha_{,x}},$$

and we obtain the local property

$$\frac{dP}{dX} = \frac{D_m}{\theta} - q.\frac{grad\theta}{\theta^2}.$$

If the temperature is continuous i.e. $[\theta] = 0$, the jump of P is exactly related to the definition of G_s. If a continuous process governs the internal structure of the discontinuity, the quantity G_s is a global characteristic of the moving discontinuity. When local behavior is given in term of w and D, the behavior of all the jumps are given. This is a way to build global behavior for shock waves or moving discontinuities inside a continuum. It is the analogue of homogenization for dynamical behavior.

A typical example has been given in plasticity. In this case the determination of compatible states (+) and (-) is not easily studied from a general viewpoint. For propagation of longitudinal waves, if we suppose that the loading process is monotonic, we obtain a curve which gives the relation between the jump of the quantities, as in the adiabatic Hugoniot curve in gas dynamics [Mandel, 1978]. This result is obtained by the assumption of a radial loading path during the shock. This is the structure of the shock. But in a general case, the internal structure for shock wave in plasticity must be studied as proposed by Germain and Lee [Germain et al., 1973].

7. CONCLUSIONS

We have studied the thermodynamics of running discontinuities from a global point of view and related all the quantities to a Hamiltonian principle generalized to nonlinear behavior. In quasistatic evolution we have given a boundary value problem for the evolution of a moving surface, the motion being governed by an energy criterion. Some connections with fracture mechanics can be also given in the same formalism [Stolz et al, 1996 ; Bui et al, 1987]. Finally the internal structure of discontinuities has been investigated to provide a global behavior of the moving surface in agreement with the positivity of the production of entropy.

REFERENCES

[Abeyratne et al., 1990]Abeyratne, R., Knowles, J., On the driving traction acting on a surface of strain discontinuity in a continuum; *J. Mech. Phys. Solids*, 38, 3, pp. 345-360.

[Bui et al., 1987]Bui, H.D., Ehrlacher, A., Nguyen, Q.S., Thermomechanical coupling in fracture mechanics; In: *Thermomechanical coupling in solids*, H.D. Bui and Q.S. Nguyen (Editors) Elsevier Sci. Pub. (North Holland).

[Germain, 1972]Germain, P., Shock Waves, Jump Relations, and Structure; In: *Advances in Applied Mechanics*, 12, pp. 131-194, Academic Press, Inc., New York and London.

[Germain et al, 1973]Germain, P., Lee, E.H., On shock waves in elastic plastic solids, *J. Mech. Phys. Solids*, 21, pp. 359- 382.

[Germain, 1998]Germain, P., Functional concepts in continuum mechanics,*Meccanica*, 33, 5, pp. 433-444.

[Grinfeld, 1980]Grinfeld, M., Thermodynamic equilibrium conditions of the phases of a nonlinearly elastic material, *Dokl. Akad.Nauti.SSSR,* 251, pp. 1-5.

[Grinfeld, 1991]Grinfeld, M., *Thermodynamic methods in the theory of heterogeneous systems*, ISIMM Series, Longmir, Harrow, Essex.

[Gurtin, 1995]Gurtin, M. E., The nature of configurational forces ; *Arch. Rat. Mech. Anal.*, 131, pp 67-100.

[Halphen et al, 1975]Halphen, B., Nguyen, Q.S., Sur les matériaux standard généralisés *Journal de Mécanique*, 14, 1, pp. 254-259.

[Mandel, 1978]Mandel, J., Ondes de choc longitudinales dans un milieu élastoplastique, *Mech. Res. Comm.,* vol 5(6), pp. 353-359.

[Maugin, 1995]Maugin, G. A., Material forces : concept and applications, *ASME Appl. Mech. Rev.,* 48, pp. 213-245.

[Nguyen, 1981]Nguyen, Q.S., A thermodynamic description of a running crack problem, IUTAM Symposium, Dourdan, Ed North Holland Publ, Three dimensional constitutive relations and ductile fracture, pp. 315-330.

[Pradeilles-Duval et al, 1995]Pradeilles-Duval, R.M., Stolz, C., Mechanical transformations and discontinuities along a moving surface; *J. Mech. Phys. Solids*, 43, 1, pp. 91-121.

[Stolz, 1988]Stolz, C., Sur les équations générales de la dynamique des milieux continus anélastiques; *C. R. Acad. Sci. Paris*, Série II 307, pp. 1997-2000.

[Stolz, 1989]Stolz, C., Sur la propagation d'une ligne de discontinuité et la fonction génératrice de choc pour un solide anélastique; *C. R. Acad. Sci. Paris*, Série II 308, pp. 1-3.

[Stolz, 1995]Stolz, C., Functional approach in non linear dynamics; *Arch. Mech*, 47, 3, pp. 421-435

[Stolz et al, 1996]Stolz, C., Pradeilles-Duval, R.M., Approches énergétiques des discontinuités mobiles en dynamique non linéaire; *C.R. Acad. Sci. Paris*, t 322, Série II,b, pp.525-532

[Truskinovski, 1987]Truskinovski, L.M., Dynamics of non equilibrium phases boundaries in a heat conducting nonlinearly elastic medium, *P.M.M,* 51, pp.777-784.

List of invited contributors (corresponding authors)

Professor Pierre ALART, (p.247)
Laboratoire de Mécanique et Génie Civil, Université de Montpellier 2, Bâtiment 9, Case 048, Place Eugène Bataillon, F-34095 Montpellier Cédex, France.
(alart@lmgc.univ-montp2.fr)

Professor Marcel BERVEILLER, (p.101)
Laboratoire de Physique et Mécanique des Matériaux, Université de Metz, Ile du Saulcy, F-57045 Metz Cedex 01, France.
(berveiller@lpmm.univ-metz.fr)

Professor René BILLARDON, (p.209)
Laboratoire de Mécanique et Technologie, Ecole Normale Supérieure de Cachan, 61 avenue du Président Wilson, F-94235 Cachan, France.
(Rene.Billardon@lmt.ens-cachan.fr)

Professor Pierre-Antoine BOIS, (p.29)
Université des Sciences et Techniques de Lille, USTL-UFR de Mécanique Bâtiment M3-LML, F-59655 Villeneuve d'Ascq Cedex, France.
(Pierre-Antoine.Bois@univ-lille1.fr)

Dr Louis C.J. BRUN, (p.55)
CEA, Centre d'Etudes de Vaujours, B.P.7, F-77181 Courtry, France.
(brun@bruyeres.cea.fr)

Professor H.D.BUI, (p.63)
Laboratoire de Mécanique des Solides, Ecole Polytechnique F-91128 Palaiseau Cedex, France.
(bui@lms.polytechnique.fr)

Professor George CAILLETAUD, (p.75)
Ecole Nationale Supérieure des Mines de Paris/CNRS, Centre des Matériaux, B.P.87, F-91003 Evry Cedex, France.
(cailletaud@mat.ensmp.fr)

Dr Jean-Louis CHABOCHE, (p.87)
ONERA, B.P.72, F-92322 Chatillon Cedex, France.
(jlc@onera.fr)

Professor André CHRYSOCHOOS, (p.247)
Laboratoire de Mécanique et Génie Civil, Université de Montpellier II,
Case courrier 081, Place E.Batiallon, F-34095 Montpellier Cedex 05, France.
(chryso@lmgc.univ-montp2.fr)

Professor Christian CUNAT, (p.367)
ENSEM-Nancy, B.P.160, 2 avenue de la Forêt de Haye,
F-54504 Vandoeuvre-les-Nancy Cedex, France.
(cunat@ENSEM.U-Nancy.fr)

Professor André DRAGON, (p.113)
Laboratoire de Mécanique et Physique des Matériaux, UMR 6617,
ENSMA, Teleport 2, B.P.109, F-86960 Futuroscope Cedex, France .
(dragon@lmpm.ensma.fr)

Professor Raymonde DROUOT, (co-editor)
UFR Sciences Versailles , Université de Versailles-St Quentin,
Direction Mécanique, Bâtiment Descartes, 45 avenue des Etats Unis,
F-78035 Versailles Cedex, France.
(Raymonde.Drouot@meca.uvsq.fr)

Professor Georges DUVAUT, (p.137)
Université Pierre et Marie Curie, Laboratoire de Modélisation en Mécanique des Solides et Structures, Immeuble CEA , Case 300, 8 rue du Capitaine Scott,
F-75015 Paris, France.
(duvaut@ccr.jussieu.fr)

Professor Alain EHRLACHER, (p.43)
CERMOO, Ecole Nationale des Ponts et Chaussées, 8 avenue Blaise Pascal
F-77445 Champs-sur-Marne, France.
(ehrlacher@ceram.enpc.fr)

Professor Marcelo EPSTEIN, (p.153)
Department of Mechanical Engineering, University of Calgary,
University Avenue West , Calgary, Alberta T2N 1N4, Canada.
(epstein@enme.ucalgary.ca)

Dr Samuel FOREST, (p.163)
Ecole Nationale Supérieure des Mines de Paris/CNRS, Centre des Matériaux,
B.P.87, F-91003 Evry Cedex, France.
(Samuel.forest@mat.ensmp.fr)

Dr Michel FREMOND, (p.177)
Laboratoire des Matériaux et Structures du Génie Civil, Laboratoire Central des Ponts et Chaussées, Cité Descartes, Parc Club de la Haute Maison, 2 Allée Kepler, F-77420 Champs-sur-Marne Cedex 15, France .
(fremond@lcpc.inrets.fr)

Professor Alain GERARD, (p.189)
Laboratoire de Mécanique Physique, Université de Bordeaux 1, 351 Cours de la Libération, F-33405 Talence Cedex, France.
(gerard@lmp.u-bordeaux.fr)

Professor Paul GERMAIN, (p.1)
Université Pierre et Marie Curie, Laboratoire de Modélisation en Mécanique, Case 162, 4 place Jussieu, 75252 Paris Cedex 05, France.

Professor Hélène LANCHON-DUCAUQUIS, (p.197)
Ecole Nationale Supérieure d'Electricité et de Mécanique, 2 avenue de la Forêt de Haye, F-54516 Vandoeuvre-les-Nancy Cedex, France.
(Salida.Simonigh@ensem.u-nancy.fr)

Professor Jean LEMAITRE, (p.209)
Laboratoire de Mécanique et Technologie, Ecole Normale Supérieure de Cachan, 61 avenue du Président Wilson, F-94235 Cachan Cedex, France.
(genin@lmt.ens-cachan.fr)

Professor Christian LEXCELLENT, (p.225)
Laboratoire de Mécanique Appliquée Robert Chaléat, Université de Franche-Comté, 24 Chemin de l'Epitaphe, F-25030 Besançon Cedex, France.
(Christian.Lexcellent@univ-fcomte.fr)

Professor Daniel LHUILLIER, (.237)
Laboratoire de Modélisation en Mécanique, Université Pierre et Marie Curie, UMR 7607 CNRS, Immeuble CEA, Case 162M8 rue du capitaine Scott, F-75015 Paris , France.
(lhuilier@ccr.jussieu.fr)

Professor Benjamin LORET, (p.265)
LSSS, Université de Grenoble 1, Domaine Universitaire de St.Martin d'Hères, Bâtiment E.1, B.P.53, 1025 rue de la Piscine, F-38041 Grenoble Cedex 09. France
(loret@hmg.inpg.fr)

Professor Jean-Jacques MARIGO, (p.289)
Laboratoire de Physique, Mécanique et thermodynamique des Matériaux, Université de Paris 13, Institut Galilée, Avenue J.B.Clément, F-93430 Villetaneuse, France.
(jean.jacques.marigo@lpmtm.univ-paris13.fr)

Professor Gérard A. MAUGIN, (p.153)
Head: Laboratoire de Modélisation en Mécanique, UMR 7607 CNRS, Université Pierre et Marie Curie, Case 162, 8 rue du capitaine Scott, F-75015 Paris, France .
(gam@ccr.jussieu.fr)

Professor Alain MOLINARI, (p.313)
Laboratoire de Physique et Mécanique des Matériaux - I.S.G.M.P, Université de Metz, Ile du Saulcy, F-57045 Metz Cedex, France.
(molinari@lpmm.univ-metz.fr)

Professor Patrick MULLER, (p.331)
Laboratoire de Modélisation en Mécanique des Solides et Structures, Université Pierre et Marie Curie, Immeuble CEA, case 300, 8 rue du Capitaine Scott, F-75015 Paris.
(muller@ccr.jussieu.fr)

Professor NGUYEN QUOC SON, (p.343)
Laboratoire de Mécanique des Solides, Ecole Polytechnique, F-91128 Palaiseau Cedex, France
(son@lms.polytechnique.fr)

Professor Gilles PIJAUDIER-CABOT, (p.355)
Laboratoire de Génie Civil de Nantes Saint-Nazaire, Ecole Centrale de Nantes, BP.92101, F-44321 Nantes, France.
(Gilles.Pijaudier-Cabot@ec-nantes.fr)

Professor Michel POTIER-FERRY, (p.125)
Laboratoire de physique et Mécanique des Matériaux- I.S.G.M.P. Université de Metz, Ile du Saulcy, F-57045 Metz cedex 1, France.
(potier-ferry@lpmm.univ-metz.fr)

Professor Pierre SEPPECHER, (p.379)
Laboratoire d'Analyse Non Linéaire Appliquée, Université de Toulon et du Var, Avenue de l'Université, F-83130 Lagarde, France.
(seppeche@univ-tln.fr)

Professor François SIDOROFF, (p.389)
Laboratoire de Tribologie et Dynamique des Systèmes, Ecole Centrale de Lyon, B.P.163, 36 avenue Guy Collongue, F-69131 Ecully Cedex , France.
(Francois.sidoroff@ec-lyon.fr)

Professor Claude STOLZ, (p.401)
Laboratoire de Mécanique des Solides, Ecole Polytechnique, F-91128 Palaiseau Cedex, France.
(stolz@lms.polytechnique.fr)

Professor Pierre SUQUET, (p.301)
Laboratoire de Mécanique et d'Acoustique-CNRS, 31 Chemin Joseph Aiguier, F-13402 Marseille Cedex 20, France.
(suquet@lma.cnrs-mrs.fr)

Professor Maurice TOURATIER, (p.277)
Ecole Nationale Supérieure des Arts et Métiers, LMS, 151 Boulevard de l'Hôpital, F-75013 Paris, France.
(lms@paris.ensam.fr)

Mechanics

SOLID MECHANICS AND ITS APPLICATIONS

Series Editor: G.M.L. Gladwell

Aims and Scope of the Series

The fundamental questions arising in mechanics are: *Why?*, *How?*, and *How much?* The aim of this series is to provide lucid accounts written by authoritative researchers giving vision and insight in answering these questions on the subject of mechanics as it relates to solids. The scope of the series covers the entire spectrum of solid mechanics. Thus it includes the foundation of mechanics; variational formulations; computational mechanics; statics, kinematics and dynamics of rigid and elastic bodies; vibrations of solids and structures; dynamical systems and chaos; the theories of elasticity, plasticity and viscoelasticity; composite materials; rods, beams, shells and membranes; structural control and stability; soils, rocks and geomechanics; fracture; tribology; experimental mechanics; biomechanics and machine design.

1. R.T. Haftka, Z. Gürdal and M.P. Kamat: *Elements of Structural Optimization*. 2nd rev.ed., 1990 ISBN 0-7923-0608-2
2. J.J. Kalker: *Three-Dimensional Elastic Bodies in Rolling Contact*. 1990 ISBN 0-7923-0712-7
3. P. Karasudhi: *Foundations of Solid Mechanics*. 1991 ISBN 0-7923-0772-0
4. *Not published*
5. *Not published.*
6. J.F. Doyle: *Static and Dynamic Analysis of Structures*. With an Emphasis on Mechanics and Computer Matrix Methods. 1991 ISBN 0-7923-1124-8; Pb 0-7923-1208-2
7. O.O. Ochoa and J.N. Reddy: *Finite Element Analysis of Composite Laminates*. ISBN 0-7923-1125-6
8. M.H. Aliabadi and D.P. Rooke: *Numerical Fracture Mechanics*. ISBN 0-7923-1175-2
9. J. Angeles and C.S. López-Cajún: *Optimization of Cam Mechanisms*. 1991 ISBN 0-7923-1355-0
10. D.E. Grierson, A. Franchi and P. Riva (eds.): *Progress in Structural Engineering*. 1991 ISBN 0-7923-1396-8
11. R.T. Haftka and Z. Gürdal: *Elements of Structural Optimization*. 3rd rev. and exp. ed. 1992 ISBN 0-7923-1504-9; Pb 0-7923-1505-7
12. J.R. Barber: *Elasticity*. 1992 ISBN 0-7923-1609-6; Pb 0-7923-1610-X
13. H.S. Tzou and G.L. Anderson (eds.): *Intelligent Structural Systems*. 1992 ISBN 0-7923-1920-6
14. E.E. Gdoutos: *Fracture Mechanics*. An Introduction. 1993 ISBN 0-7923-1932-X
15. J.P. Ward: *Solid Mechanics*. An Introduction. 1992 ISBN 0-7923-1949-4
16. M. Farshad: *Design and Analysis of Shell Structures*. 1992 ISBN 0-7923-1950-8
17. H.S. Tzou and T. Fukuda (eds.): *Precision Sensors, Actuators and Systems*. 1992 ISBN 0-7923-2015-8
18. J.R. Vinson: *The Behavior of Shells Composed of Isotropic and Composite Materials*. 1993 ISBN 0-7923-2113-8
19. H.S. Tzou: *Piezoelectric Shells*. Distributed Sensing and Control of Continua. 1993 ISBN 0-7923-2186-3
20. W. Schiehlen (ed.): *Advanced Multibody System Dynamics*. Simulation and Software Tools. 1993 ISBN 0-7923-2192-8
21. C.-W. Lee: *Vibration Analysis of Rotors*. 1993 ISBN 0-7923-2300-9
22. D.R. Smith: *An Introduction to Continuum Mechanics*. 1993 ISBN 0-7923-2454-4
23. G.M.L. Gladwell: *Inverse Problems in Scattering*. An Introduction. 1993 ISBN 0-7923-2478-1

Mechanics

SOLID MECHANICS AND ITS APPLICATIONS

Series Editor: G.M.L. Gladwell

24. G. Prathap: *The Finite Element Method in Structural Mechanics.* 1993 ISBN 0-7923-2492-7
25. J. Herskovits (ed.): *Advances in Structural Optimization.* 1995 ISBN 0-7923-2510-9
26. M.A. González-Palacios and J. Angeles: *Cam Synthesis.* 1993 ISBN 0-7923-2536-2
27. W.S. Hall: *The Boundary Element Method.* 1993 ISBN 0-7923-2580-X
28. J. Angeles, G. Hommel and P. Kovács (eds.): *Computational Kinematics.* 1993 ISBN 0-7923-2585-0
29. A. Curnier: *Computational Methods in Solid Mechanics.* 1994 ISBN 0-7923-2761-6
30. D.A. Hills and D. Nowell: *Mechanics of Fretting Fatigue.* 1994 ISBN 0-7923-2866-3
31. B. Tabarrok and F.P.J. Rimrott: *Variational Methods and Complementary Formulations in Dynamics.* 1994 ISBN 0-7923-2923-6
32. E.H. Dowell (ed.), E.F. Crawley, H.C. Curtiss Jr., D.A. Peters, R. H. Scanlan and F. Sisto: *A Modern Course in Aeroelasticity.* Third Revised and Enlarged Edition. 1995 ISBN 0-7923-2788-8; Pb: 0-7923-2789-6
33. A. Preumont: *Random Vibration and Spectral Analysis.* 1994 ISBN 0-7923-3036-6
34. J.N. Reddy (ed.): *Mechanics of Composite Materials.* Selected works of Nicholas J. Pagano. 1994 ISBN 0-7923-3041-2
35. A.P.S. Selvadurai (ed.): *Mechanics of Poroelastic Media.* 1996 ISBN 0-7923-3329-2
36. Z. Mróz, D. Weichert, S. Dorosz (eds.): *Inelastic Behaviour of Structures under Variable Loads.* 1995 ISBN 0-7923-3397-7
37. R. Pyrz (ed.): *IUTAM Symposium on Microstructure-Property Interactions in Composite Materials.* Proceedings of the IUTAM Symposium held in Aalborg, Denmark. 1995 ISBN 0-7923-3427-2
38. M.I. Friswell and J.E. Mottershead: *Finite Element Model Updating in Structural Dynamics.* 1995 ISBN 0-7923-3431-0
39. D.F. Parker and A.H. England (eds.): *IUTAM Symposium on Anisotropy, Inhomogeneity and Nonlinearity in Solid Mechanics.* Proceedings of the IUTAM Symposium held in Nottingham, U.K. 1995 ISBN 0-7923-3594-5
40. J.-P. Merlet and B. Ravani (eds.): *Computational Kinematics '95.* 1995 ISBN 0-7923-3673-9
41. L.P. Lebedev, I.I. Vorovich and G.M.L. Gladwell: *Functional Analysis.* Applications in Mechanics and Inverse Problems. 1996 ISBN 0-7923-3849-9
42. J. Menčik: *Mechanics of Components with Treated or Coated Surfaces.* 1996 ISBN 0-7923-3700-X
43. D. Bestle and W. Schiehlen (eds.): *IUTAM Symposium on Optimization of Mechanical Systems.* Proceedings of the IUTAM Symposium held in Stuttgart, Germany. 1996 ISBN 0-7923-3830-8
44. D.A. Hills, P.A. Kelly, D.N. Dai and A.M. Korsunsky: *Solution of Crack Problems.* The Distributed Dislocation Technique. 1996 ISBN 0-7923-3848-0
45. V.A. Squire, R.J. Hosking, A.D. Kerr and P.J. Langhorne: *Moving Loads on Ice Plates.* 1996 ISBN 0-7923-3953-3
46. A. Pineau and A. Zaoui (eds.): *IUTAM Symposium on Micromechanics of Plasticity and Damage of Multiphase Materials.* Proceedings of the IUTAM Symposium held in Sèvres, Paris, France. 1996 ISBN 0-7923-4188-0
47. A. Naess and S. Krenk (eds.): *IUTAM Symposium on Advances in Nonlinear Stochastic Mechanics.* Proceedings of the IUTAM Symposium held in Trondheim, Norway. 1996 ISBN 0-7923-4193-7
48. D. Ieşan and A. Scalia: *Thermoelastic Deformations.* 1996 ISBN 0-7923-4230-5

Mechanics

SOLID MECHANICS AND ITS APPLICATIONS

Series Editor: G.M.L. Gladwell

49. J.R. Willis (ed.): *IUTAM Symposium on Nonlinear Analysis of Fracture*. Proceedings of the IUTAM Symposium held in Cambridge, U.K. 1997 ISBN 0-7923-4378-6
50. A. Preumont: *Vibration Control of Active Structures*. An Introduction. 1997 ISBN 0-7923-4392-1
51. G.P. Cherepanov: *Methods of Fracture Mechanics: Solid Matter Physics*. 1997 ISBN 0-7923-4408-1
52. D.H. van Campen (ed.): *IUTAM Symposium on Interaction between Dynamics and Control in Advanced Mechanical Systems*. Proceedings of the IUTAM Symposium held in Eindhoven, The Netherlands. 1997 ISBN 0-7923-4429-4
53. N.A. Fleck and A.C.F. Cocks (eds.): *IUTAM Symposium on Mechanics of Granular and Porous Materials*. Proceedings of the IUTAM Symposium held in Cambridge, U.K. 1997 ISBN 0-7923-4553-3
54. J. Roorda and N.K. Srivastava (eds.): *Trends in Structural Mechanics*. Theory, Practice, Education. 1997 ISBN 0-7923-4603-3
55. Yu.A. Mitropolskii and N. Van Dao: *Applied Asymptotic Methods in Nonlinear Oscillations*. 1997 ISBN 0-7923-4605-X
56. C. Guedes Soares (ed.): *Probabilistic Methods for Structural Design*. 1997 ISBN 0-7923-4670-X
57. D. François, A. Pineau and A. Zaoui: *Mechanical Behaviour of Materials*. Volume I: Elasticity and Plasticity. 1998 ISBN 0-7923-4894-X
58. D. François, A. Pineau and A. Zaoui: *Mechanical Behaviour of Materials*. Volume II: Viscoplasticity, Damage, Fracture and Contact Mechanics. 1998 ISBN 0-7923-4895-8
59. L.T. Tenek and J. Argyris: *Finite Element Analysis for Composite Structures*. 1998 ISBN 0-7923-4899-0
60. Y.A. Bahei-El-Din and G.J. Dvorak (eds.): *IUTAM Symposium on Transformation Problems in Composite and Active Materials*. Proceedings of the IUTAM Symposium held in Cairo, Egypt. 1998 ISBN 0-7923-5122-3
61. I.G. Goryacheva: *Contact Mechanics in Tribology*. 1998 ISBN 0-7923-5257-2
62. O.T. Bruhns and E. Stein (eds.): *IUTAM Symposium on Micro- and Macrostructural Aspects of Thermoplasticity*. Proceedings of the IUTAM Symposium held in Bochum, Germany. 1999 ISBN 0-7923-5265-3
63. F.C. Moon: *IUTAM Symposium on New Applications of Nonlinear and Chaotic Dynamics in Mechanics*. Proceedings of the IUTAM Symposium held in Ithaca, NY, USA. 1998 ISBN 0-7923-5276-9
64. R. Wang: *IUTAM Symposium on Rheology of Bodies with Defects*. Proceedings of the IUTAM Symposium held in Beijing, China. 1999 ISBN 0-7923-5297-1
65. Yu.I. Dimitrienko: *Thermomechanics of Composites under High Temperatures*. 1999 ISBN 0-7923-4899-0
66. P. Argoul, M. Frémond and Q.S. Nguyen (eds.): *IUTAM Symposium on Variations of Domains and Free-Boundary Problems in Solid Mechanics*. Proceedings of the IUTAM Symposium held in Paris, France. 1999 ISBN 0-7923-5450-8
67. F.J. Fahy and W.G. Price (eds.): *IUTAM Symposium on Statistical Energy Analysis*. Proceedings of the IUTAM Symposium held in Southampton, U.K. 1999 ISBN 0-7923-5457-5
68. H.A. Mang and F.G. Rammerstorfer (eds.): *IUTAM Symposium on Discretization Methods in Structural Mechanics*. Proceedings of the IUTAM Symposium held in Vienna, Austria. 1999 ISBN 0-7923-5591-1

Mechanics

SOLID MECHANICS AND ITS APPLICATIONS

Series Editor: G.M.L. Gladwell

69. P. Pedersen and M.P. Bendsøe (eds.): *IUTAM Symposium on Synthesis in Bio Solid Mechanics*. Proceedings of the IUTAM Symposium held in Copenhagen, Denmark. 1999 ISBN 0-7923-5615-2
70. S.K. Agrawal and B.C. Fabien: *Optimization of Dynamic Systems*. 1999 ISBN 0-7923-5681-0
71. A. Carpinteri: *Nonlinear Crack Models for Nonmetallic Materials*. 1999 ISBN 0-7923-5750-7
72. F. Pfeifer (ed.): *IUTAM Symposium on Unilateral Multibody Contacts*. Proceedings of the IUTAM Symposium held in Munich, Germany. 1999 ISBN 0-7923-6030-3
73. E. Lavendelis and M. Zakrzhevsky (eds.): *IUTAM/IFToMM Symposium on Synthesis of Nonlinear Dynamical Systems*. Proceedings of the IUTAM/IFToMM Symposium held in Riga, Latvia. 2000 ISBN 0-7923-6106-7
74. J.-P. Merlet: *Parallel Robots*. 2000 ISBN 0-7923-6308-6
75. J.T. Pindera: *Techniques of Tomographic Isodyne Stress Analysis*. 2000 ISBN 0-7923-6388-4
76. G.A. Maugin, R. Drouot and F. Sidoroff (eds.): *Continuum Thermomechanics*. The Art and Science of Modelling Material Behaviour. 2000 ISBN 0-7923-6407-4
77. N. Van Dao and E.J. Kreuzer (eds.): *IUTAM Symposium on Recent Developments in Non-linear Oscillations of Mechanical Systems*. 2000 ISBN 0-7923-6470-8
78. S.D. Akbarov and A.N. Guz: *Mechanics of Curved Composites*. 2000 ISBN 0-7923-6477-5
79. M.B. Rubin: *Cosserat Theories: Shells, Rods and Points*. 2000 ISBN 0-7923-6489-9

Kluwer Academic Publishers – Dordrecht / Boston / London

MIX
Papier aus verantwortungsvollen Quellen
Paper from responsible sources
FSC® C105338

If you have any concerns about our products,
you can contact us on
ProductSafety@springernature.com

In case Publisher is established outside the EU,
the EU authorized representative is:
Springer Nature Customer Service Center GmbH
Europaplatz 3, 69115 Heidelberg, Germany

Printed by Libri Plureos GmbH
in Hamburg, Germany